Methods in Enzymology

Volume 181
RNA PROCESSING
Part B
Specific Methods

METHODS IN ENZYMOLOGY

EDITORS-IN-CHIEF

John N. Abelson   Melvin I. Simon

DIVISION OF BIOLOGY
CALIFORNIA INSTITUTE OF TECHNOLOGY
PASADENA, CALIFORNIA

FOUNDING EDITORS

Sidney P. Colowick and Nathan O. Kaplan

Methods in Enzymology

Volume 181

RNA Processing

Part B
Specific Methods

EDITED BY

James E. Dahlberg

DEPARTMENT OF PHYSIOLOGICAL CHEMISTRY
UNIVERSITY OF WISCONSIN-MADISON
MADISON, WISCONSIN

John N. Abelson

DIVISION OF BIOLOGY
CALIFORNIA INSTITUTE OF TECHNOLOGY
PASADENA, CALIFORNIA

ACADEMIC PRESS, INC.
Harcourt Brace Jovanovich, Publishers
San Diego   New York   Boston
London   Sydney   Tokyo   Toronto

This book is printed on acid-free paper.

ACADEMIC PRESS, INC.
San Diego, California 92101

United Kingdom Edition published by
ACADEMIC PRESS LIMITED
24-28 Oval Road, London NW1 7DX

LIBRARY OF CONGRESS CATALOG CARD NUMBER: 54-9110

ISBN   0-12-182082-3   (alk. paper)

PRINTED IN THE UNITED STATES OF AMERICA
90 91 92 93   9 8 7 6 5 4 3 2 1

Table of Contents

Section I. Processing of Messenger RNAs

A. Mammalian Cell Extracts and Assays

B. Yeast Extracts and Assays

F. Heterogeneous Nuclear Ribonucleoproteins

Section II. Processing of Ribosomal RNAs

Section III. Processing of Transfer RNAs

A. Genetics

B. Enzymology of Processing

Section IV. Ribozymes

Contributors to Volume 181

Article numbers are in parentheses following the names of contributors.
Affiliations listed are current.

JOHN N. ABELSON (8, 9, 38, 39), *Division of Biology, California Institute of Technology, Pasadena, California 91125*

MARKUS AEBI (5), *Institut für Molekularbiologie I, Universität Zürich, Hönggerberg, CH-8093 Zürich, Switzerland*

SIDNEY ALTMAN (46), *Department of Biology, Yale University, New Haven, Connecticut 06511*

JOHN G. ARNEZ (46), *Department of Molecular Biophysics and Biochemistry, Yale University, New Haven, Connecticut 06511*

MONTSERRAT BACH (19), *Institut für Molekularbiologie und Tumorforschung, D-3550 Marburg, Federal Republic of Germany*

MADELINE F. BAER (46), *Department of Biology, Yale University, New Haven, Connecticut 06511*

I. M. BALDI (42), *Institute of Cell Biology, National Research Council, Rome, Italy*

STANLEY F. BARNETT (24), *Department of Molecular Biophysics and Biochemistry, Yale University School of Medicine, New Haven, Connecticut 06510*

MARLENE BELFORT (43), *Wadsworth Center for Laboratories and Research, New York State Department of Health, Albany, New York 12201*

GARY BREWER (16), *Department of Microbiology and Immunology, Bowman Gray School of Medicine, Wake Forest University, Winston-Salem, North Carolina 27103*

PETER BRINGMANN (19), *Plant Science Center, Department of Biology, University of Pennsylvania, Philadelphia, Pennsylvania 19104*

CLAUDE BRUNEL (21), *Laboratoire de Biochimie, Centre Val d'Aurelle-Paul Lamarque, 34095 Montpellier Cedex, France*

GUY CATHALA (21), *Laboratoire de Biochimie, Centre Val d'Aurelle-Paul Lamarque, 34095 Montpellier Cedex, France*

THOMAS R. CECH (45), *Department of Chemistry and Biochemistry, University of Colorado, Boulder, Colorado 80309*

P. SCOTT CHANDRY (43), *Wadsworth Center for Laboratories and Research, New York State Department of Health, Albany, New York 12201*

SOO-CHEN CHENG (8), *Institute of Molecular Biology, Academia Sinica, Nankang, Taiwan 11529, Republic of China*

YANG DO CHOI (26), *Department of Agricultural Chemistry, Seoul National University, Suwon, South Korea 170*

LYNN COOLEY (37), *Department of Human Genetics, Yale University School of Medicine, New Haven, Connecticut 06511*

CHRISTOPHER DAVIES (47), *John Innes Institute, Norwich NR4 7UH, England*

MURRAY P. DEUTSCHER (34, 35), *Department of Biochemistry, University of Connecticut Health Center, Farmington, Connecticut 06032*

GIDEON DREYFUSS (26, 27), *Department of Biochemistry, Molecular Biology and Cell Biology, Northwestern University, Evanston, Illinois 60208*

MARY EDMONDS (12), *Department of Biological Sciences, University of Pittsburgh, Pittsburgh, Pennsylvania 15260*

KAREN EHRENMAN (43), *Fred Hutchinson Cancer Research Center, Seattle, Washington 98104*

DAVID R. ENGELKE (36), *Department of Biological Chemistry, University of Michigan, Ann Arbor, Michigan 48109*

ix

CLAIRE F. EVANS (36), *Department of Biological Chemistry, University of Michigan, Ann Arbor, Michigan 48109*

WITOLD FILIPOWICZ (11, 41), *Friedrich Miescher-Institut, CH-4002 Basel, Switzerland*

ANTHONY C. FORSTER (47), *Department of Biology, Yale University, New Haven, Connecticut 06520*

HENRY M. FURNEAUX (40), *Department of Molecular Biology, Memorial Sloan-Kettering Cancer Center, New York, New York 10021*

D. GANDINI-ATTARDI (42), *Institute of Cell Biology, National Research Council, Rome, Italy*

S. E. GOELZ (9), *Biogen Corporation, Cambridge, Massachusetts 02142*

GREGORY J. GOODALL (11), *Friedrich Miescher-Institut, CH-4002 Basel, Switzerland*

MICHAEL R. GREEN (2, 14), *Department of Biochemistry and Molecular Biology, Harvard University, Cambridge, Massachusetts 02138*

PHILLIP R. GREEN (39), *Corporate Research Division, The Procter & Gamble Company, Cincinnati, Ohio 45239*

CHRIS L. GREER (38), *Department of Biological Chemistry, California College of Medicine, University of California at Irvine, Irvine, California 92717*

CECILIA GUERRIER-TAKADA (46), *Department of Biology, Yale University, New Haven, Connecticut 06511*

JÖRG HAMM (22), *European Molecular Biology Laboratory, D-6900 Heidelberg, Federal Republic of Germany*

SALLIE O. HOCH (20), *The Agouron Institute, La Jolla, California 92037*

ANITA K. HOPPER (33), *Department of Biological Chemistry, The Milton S. Hershey Medical Center, Pennsylvania State University, Hershey, Pennsylvania 17033*

Z. HOSTOMSKY (9), *Agouron Pharmaceuticals, Inc., La Jolla, California 92037*

JERARD HURWITZ (10), *Graduate Program in Molecular Biology, Memorial Sloan-*

Kettering Cancer Center, New York, New York 10021

CHERYL J. HUTCHINS (47), *Institute of Medical and Veterinary Science, Adelaide, South Australia 5000, Australia*

WALTER KELLER (1), *Department of Cell Biology, Biocenter of the University of Basel, CH-4056 Basel, Switzerland*

ROGER D. KORNBERG (28), *Department of Cell Biology, Stanford University School of Medicine, Stanford, California 94305*

ANGELA KRÄMER (1, 18, 23), *Department of Cell Biology, Biocenter of the University of Basel, CH-4056 Basel, Switzerland*

JOHN A. LATHAM (45), *Gilead Sciences, Foster City, California 94404*

KEVIN A. W. LEE (2), *Department of Biochemistry and Molecular Biology, Harvard University, Cambridge, Massachusetts 02138*

P. LEGRAIN (10), *Institut Pasteur, Département de Biologie Moléculaire, Unité de Génétique Moléculaire des Levures, 75724 Paris Cedex 15, France*

WALLACE M. LESTOURGEON (24), *Department of Molecular Biology, Vanderbilt University, Nashville, Tennessee 37235*

REN-JANG LIN (8), *Department of Microbiology, University of Texas, Austin, Austin, Texas 78712*

REINHARD LÜHRMANN (19), *Institut für Molekularbiologie und Tumorforschung, D-3550 Marburg, Federal Republic of Germany*

ELSEBET LUND (4), *Department of Physiological Chemistry, University of Wisconsin-Madison, Madison, Wisconsin 53706*

BERNHARD LÜSCHER (7), *Department of Biochemistry, University of California, Berkeley, Berkeley, California 94720*

WILLIAM F. MARZLUFF (3), *Department of Chemistry, Florida State University, Tallahassee, Florida 32306*

IAIN W. MATTAJ (22), *European Molecular Biology Laboratory, D-6900 Heidelberg, Federal Republic of Germany*

E. MATTOCCIA (42), *Institute of Cell Biol-*

ogy, National Research Council, Rome, Italy

MICHAEL J. MATUNIS (27), *Department of Biochemistry, Molecular Biology and Cell Biology, Northwestern University, Evanston, Illinois 60208*

GLORIA D. MCFARLAND (8), *Division of Biology, California Institute of Technology, Pasadena, California 91125*

CLAIRE L. MOORE (6), *Department of Molecular Biology and Microbiology, Tufts University, Boston, Massachusetts 02111*

BERNARD MOSS (13), *Laboratory of Viral Diseases, National Institute of Allergy and Infectious Diseases, National Institutes of Health, Bethesda, Maryland 20892*

ANDREW NEWMAN (8), *Laboratory of Molecular Biology, Medical Research Council Center, University Medical School, Cambridge CB2 2QH, England*

MARK NICHOLS (31), *Institute of Cell and Tumor Biology, German Cancer Research Center, D-6900 Heidelberg, Federal Republic of Germany*

STEPHANIE J. NORTHINGTON (24), *Department of Molecular Biology, Vanderbilt University, Nashville, Tennessee 37235*

MUTSUHITO OHNO (17), *Department of Biophysics, Faculty of Science, Kyoto University, Kyoto 606, Japan*

BERNADETTE PACE (30), *Department of Biology, Indiana University, Bloomington, Indiana 47405*

NORMAN R. PACE (30), *Department of Biology, Indiana University, Bloomington, Indiana 47405*

PHILIP L. PAINE (4), *Department of Biological Sciences, St. Johns University, Jamaica, New York 11439*

PHILIP S. PERLMAN (44), *Department of Molecular Genetics, The Ohio State University, Columbus, Ohio 43210*

ERIC M. PHIZICKY (38), *Department of Biochemistry, University of Rochester Medical School, Rochester, New York 14642*

LESLIE PICK (40), *Department of Biology,*

Biocenter of the University of Basel, CH-4056 Basel, Switzerland

C. PIKIELNY (10), *Department of Biochemistry, Harvard University, Cambridge, Massachusetts 02138*

SERAFÍN PIÑOL-ROMA (26, 27), *Department of Biochemistry, Molecular Biology and Cell Biology, Northwestern University, Evanston, Illinois 60208*

HUGH D. ROBERTSON (15), *Department of Biochemistry, Cornell University Medical College, New York, New York 10021*

M. ROSBASH (10), *Howard Hughes Medical Institute, and Department of Biology, Brandeis University, Waltham, Massachusetts 02254*

JEFFREY ROSS (16), *McArdle Laboratory for Cancer Research, and Department of Pathology, University of Wisconsin-Madison, Madison, Wisconsin 53706*

S. W. RUBY (9), *Division of Biology, California Institute of Technology, Pasadena, California 91125*

BARBARA RUSKIN (14), *Department of Biochemistry and Molecular Biology, Harvard University, Cambridge, Massachusetts 02138*

B. C. RYMOND (10), *Department of Biological Sciences, University of Kentucky, Lexington, Kentucky 40506*

ALAN B. SACHS (28), *Whitehead Institute for Biomedical Research, Massachusetts Institute of Technology, Cambridge, Massachusetts 02142*

DAVID SCHLESSINGER (29), *Department of Molecular Microbiology, Washington University School of Medicine, St. Louis, Missouri 63110*

JERZY T. SCHÖNEICH (25), *Department of Molecular Genetics, Hoffmann-LaRoche Inc., Nutley, New Jersey 07110*

DANIEL SCHÜMPERLI (7), *Zoologisches Institut, Abteilung für Entwicklungsbiologie, Universität Bern, 3012 Bern, Switzerland*

B. SERAPHIN (10), *Institut Curie, Section de Biologie, Campus Universitaire, 91405 Orsay Cedex, France*

YOSHIRO SHIMURA (17, 32), *Department of Biophysics, Faculty of Science, Kyoto University, Kyoto 606, Japan*

HIDEAKI SHIRAISHI (32), *National Institute for Basic Biology, Myodaiji, Okazaki 444, Japan*

STEWART SHUMAN (13), *Program in Molecular Biology, Sloan-Kettering Institute, New York, New York 10021*

DOMINIQUE SOLDATI (7), *Institut für Molekularbiologie II, Universität Zürich, Hönggerberg, 8093 Zürich, Switzerland*

DIETER SÖLL (31, 37), *Department of Molecular Biophysics and Biochemistry, Yale University, New Haven, Connecticut 06511*

ANAND K. SRIVASTAVA (29), *Department of Molecular Microbiology, Washington University School of Medicine, St. Louis, Missouri 63110*

CLAUDIA STAUBER (7), *Regulatory Biology Laboratory, The Salk Institute, La Jolla, California 92037*

MAURICE S. SWANSON (27), *Department of Biochemistry, Molecular Biology and Cell Biology, Northwestern University, Evanston, Illinois 60208*

ROBERT H. SYMONS (47), *Department of Biochemistry, University of Adelaide, Adelaide, South Australia 5000, Australia*

JOHN O. THOMAS (25), *Department of Biochemistry, New York University School of Medicine, New York, New York 10016*

G. P. TOCCHINI-VALENTINI (42), *Institute of Cell Biology, National Research Council, Rome, Italy*

OSCAR VICENTE (41), *Friedrich Miescher-Institute, CH-4002 Basel, Switzerland*

AGUSTIN VIOQUE (46), *Departamento de Bioquímica, Facultad de Biología, 41080 Seville, Spain*

KARIN WIEBAUER (11), *Friedrich Miescher-Institut, CH-4002 Basel, Switzerland*

JACINTA B. WILLIAMS (37), *Merck, Sharp and Dohme Research Laboratories, West Point, Pennsylvania 19486*

IAN WILLIS (31), *Department of Biochemistry, Albert Einstein College of Medicine, Bronx, New York 10461*

QI XU (38), *The Institute of Basic Theory of TCM, China Academy of Traditional Chinese Medicine, Dongzhimen Nei, Beijing, People's Republic of China*

ARTHUR J. ZAUG (45), *Department of Chemistry and Biochemistry, University of Colorado, Boulder, Colorado 80309*

Preface

That RNA processing is one of the most active and rapidly developing areas of biology can be seen from the great increase in papers published on the subject and the high attendance at meetings and workshops in this field.

Because RNAs rarely, if ever, function as primary transcription products, they all must undergo various forms of processing. This can be as simple as the removal of one or a few nucleotides at one end or as complex as splicing (*cis* or *trans*), modification of nucleotide bases, and 3' cleavage and polyadenylation. Hence, RNA processing plays a key role in the expression of genetic information.

Numerous methods have been developed or adapted for the study of RNA processing which pertain to preparation of substrates, to preparation and purification of processing enzymes or factors, and to analysis of the resulting products. Methods have also been developed that allow investigators to study the structures of the precursors, the cofactors, and the enzymes or complexes that catalyze function at individual steps in processing. In many instances, approaches or procedures have very wide applicability to other areas of research, such as in cell structure and physiology, in transcription and translation, and in enzymology. In other instances, the methods are defined by the scope of the problem at hand, so the methods are primarily applicable to the processing of RNA. As in most fields it is impossible to predict the breadth of applications for which a procedure or method will be used. Nevertheless, we have tried to assemble the contributions according to whether they are more general (Volume 180) or more specific (Volume 181).

In an effort to keep the size of these *Methods in Enzymology* volumes manageable, we had to make many difficult and arbitrary decisions in the selection of topics to be included, and are gratified by the number of excellent chapters received. The care and dedication of the authors are evident in the quality of their papers and their willingness to share with others the methods used in their laboratories. We thank them for their contributions to these volumes.

As with any publication in a rapidly evolving field, many of the methods described will undoubtedly become supplemented by new, more powerful, and more informative ones. But most, perhaps all, of these methods will remain very useful for several years and will serve as fundamental procedures from which new techniques can be developed in this very exciting field.

JAMES E. DAHLBERG
JOHN N. ABELSON

METHODS IN ENZYMOLOGY

VOLUME 117. Enzyme Structure (Part J)
Edited by C. H. W. HIRS AND SERGE N. TIMASHEFF

VOLUME 118. Plant Molecular Biology
Edited by ARTHUR WEISSBACH AND HERBERT WEISSBACH

VOLUME 119. Interferons (Part C)
Edited by SIDNEY PESTKA

VOLUME 120. Cumulative Subject Index Volumes 81–94, 96–101

VOLUME 121. Immunochemical Techniques (Part I: Hybridoma Technology and Monoclonal Antibodies)
Edited by JOHN J. LANGONE AND HELEN VAN VUNAKIS

VOLUME 122. Vitamins and Coenzymes (Part G)
Edited by FRANK CHYTIL AND DONALD B. MCCORMICK

VOLUME 123. Vitamins and Coenzymes (Part H)
Edited by FRANK CHYTIL AND DONALD B. MCCORMICK

VOLUME 124. Hormone Action (Part J: Neuroendocrine Peptides)
Edited by P. MICHAEL CONN

VOLUME 125. Biomembranes (Part M: Transport in Bacteria, Mitochondria, and Chloroplasts: General Approaches and Transport Systems)
Edited by SIDNEY FLEISCHER AND BECCA FLEISCHER

VOLUME 126. Biomembranes (Part N: Transport in Bacteria, Mitochondria, and Chloroplasts: Protonmotive Force)
Edited by SIDNEY FLEISCHER AND BECCA FLEISCHER

VOLUME 127. Biomembranes (Part O: Protons and Water: Structure and Translocation)
Edited by LESTER PACKER

VOLUME 128. Plasma Lipoproteins (Part A: Preparation, Structure, and Molecular Biology)
Edited by JERE P. SEGREST AND JOHN J. ALBERS

VOLUME 129. Plasma Lipoproteins (Part B: Characterization, Cell Biology, and Metabolism)
Edited by JOHN J. ALBERS AND JERE P. SEGREST

Section I

Processing of Messenger RNAs

A. Mammalian Cell Extracts and Assays
Articles 1 through 7

B. Yeast Extracts and Assays
Articles 8 through 10

C. Extracts and Assays from Other Systems
Article 11

D. Purified Enzymes
Articles 12 through 17

E. Small Nuclear Ribonucleoproteins
Articles 18 through 23

F. Heterogeneous Nuclear Ribonucleoproteins
Articles 24 through 28

[1] Preparation and Fractionation of Mammalian Extracts Active in Pre-mRNA Splicing

By ANGELA KRÄMER and WALTER KELLER

With the development of cell-free systems for the *in vitro* splicing of messenger RNA precursors (pre-mRNAs) it has become feasible to analyze the individual parts of the splicing machinery. Two main strategies have been used in the past. First, a variety of methods have been developed that are suitable for the direct analysis of splicing factors in unfractionated nuclear extracts. Some of the results from these studies are summarized in the following section. Second, splicing extracts have been fractionated chromatographically and the resulting fractions tested for their activity in splicing complex assembly and in the splicing reaction. This approach has the advantage that an activity, once identified, can be analyzed in molecular detail.

By employing nuclear extracts, several of the basic components of the splicing machinery have been characterized. For example, removal of small nuclear ribonucleoprotein particles (snRNPs) from nuclear extracts with specific antibodies resulted in the inhibition of splicing *in vitro*.[1-4] A role for U1, U2, U4, and U6 snRNPs in early steps of the splicing reaction was revealed by degrading the RNA moiety of the snRNP particles in a nuclear extract with complementary oligodeoxynucleotides and RNase H.[2,5-12] The RNA species present in splicing complexes were analyzed by 3'-end-labeling or Northern blotting. Such procedures were used to demonstrate that U1 and U2 snRNPs are essential parts of the presplicing

[1] R. A. Padgett, S. M. Mount, J. A. Steitz, P. A. Sharp, *Cell* **35,** 101 (1983).
[2] A. Krämer, W. Keller, B. Appel, and R. Lührmann, *Cell* **38,** 299 (1984).
[3] A. R. Krainer, *Nucleic Acids Res.* **16,** 9415 (1988).
[4] J. A. Steitz, this series, Vol. 180, p. 468.
[5] A. R. Krainer and T. Maniatis, *Cell* **42,** 725 (1985).
[6] D. L. Black, B. Chabot, and J. A. Steitz, *Cell* **42,** 737 (1985).
[7] B. Chabot, D. L. Black, D. M. LeMaster, and J. A. Steitz, *Science* **230,** 1344 (1985).
[8] S. M. Berget and B. L. Robberson, *Cell* **46,** 691 (1986).
[9] D. L. Black and J. A. Steitz, *Cell* **46,** 697 (1986).
[10] B. Chabot and J. A. Steitz, *Mol. Cell. Biol.* **7,** 281 (1987).
[11] M. Zillmann, S. Rose, and S. Berget, *Mol. Cell. Biol.* **7,** 2877 (1987).
[12] A. Krämer, this volume [23].

complex.[13-15] This complex is converted to the spliceosome on binding of U4, U5, and U6 snRNPs.[16] Subsequently U4 snRNP is released from the spliceosome and the splicing reaction is initiated.[16-19]

Nuclear extracts have also been used to identify several protein factors that participate in the splicing reaction. Mild heat treatment of an extract results in the inhibition of 3′ cleavage and exon ligation, whereas 5′ cleavage and lariat formation occur normally.[5] This observation had several implications: first, it indicated that the two cleavage and ligation reactions are indeed separate events; second, distinct components are required for the two steps; and third, at least one of the factors required late in the reaction is heat labile.

Several proteins that interact with specific sequences in the pre-mRNA have been identified as well. Antibody inhibition and depletion studies demonstrated a requirement for the C class heterogeneous nuclear ribonucleoproteins (hnRNP) in the splicing reaction.[20,21] These proteins (as well as hnRNP A1 and D proteins) appear to interact with the 3′ region of the intron (the polypyrimidine stretch)[22]; however, their precise role in splicing is not understood. Two proteins that bind to the same region in the pre-mRNA as the hnRNP proteins were first detected in nuclear extracts because they protect the 3′-terminal intron sequences from RNase digestion.[7,23,24] Blotting of nuclear proteins to nitrocellulose and probing of the filters with radiolabeled pre-mRNAs led to the identification of a protein of M_r 100,000 or 70,000 (intron-binding protein or IBP), which binds to the polypyrimidine stretch.[25,26] This protein has been partially purified and can be detected either as a free protein or in association with U5 snRNP. The second factor that specifically interacts with the 3′ end of an intron mediates the binding of U2 snRNP to the branch site and was termed U2 snRNP auxiliary factor (U2AF).[23]

[13] P. J. Grabowski and P. A. Sharp, *Science* **233**, 1294 (1986).
[14] M. M. Konarska and P. A. Sharp, *Cell* **46**, 845 (1986).
[15] A. Bindereif and M. R. Green, *EMBO J.* **6**, 2415 (1987).
[16] M. M. Konarska and P. A. Sharp, *Cell* **49**, 763 (1987).
[17] C. W. Pikielny, B. C. Rymond, and M. Rosbash, *Nature* (*London*) **324**, 341 (1986).
[18] S. C. Cheng and J. Abelson, *Genes Dev.* **1**, 1014 (1987).
[19] A. I. Lamond, M. M. Konarska, P. J. Grabowski, and P. A. Sharp, *Proc. Natl. Acad. Sci. U.S.A.* **85**, 411 (1988).
[20] Y. D. Choi, P. J. Grabowski, P. A. Sharp, and G. Dreyfuss, *Science* **231**, 1534 (1986).
[21] H. Sierakowska, W. Szer, P. J. Furdon, and R. Kole, *Nucleic Acids Res.* **14**, 5241 (1986).
[22] M. S. Swanson and G. Dreyfuss, *EMBO J.* **7**, 3519 (1988).
[23] B. Ruskin, P. D. Zamore, and M. R. Green, *Cell* **52**, 207 (1988).
[24] K. Parker and J. A. Steitz, this series, Vol. 180, p. 454.
[25] J. Tazi, C. Alibert, J. Temsamani, I. Reveillaud, G. Cathala, C. Brunel, and P. Jeanteur, *Cell* **47**, 755 (1986).
[26] V. Gerke and J. A. Steitz, *Cell* **47**, 973 (1986).

With a filter-binding technique, Mayeda *et al.*[27] and Tatei *et al.*[28] detected several activities in nuclear extracts and chromatographic fractions which bind to the 5′ and 3′ splice sites of synthetic pre-mRNAs. However, the relevance of these components for the splicing reaction is not yet clear.

All of these studies show that unfractionated nuclear extracts can be used for analysis of the splicing pathway and for the initial identification and characterization of some of the components involved. But a detailed understanding of the processes leading to the generation of spliced RNA will emerge only when the complete reaction can be reconstituted with purified components. Attempts to fractionate extracts by several laboratories have revealed that at least four protein factors apart from the snRNPs are essential for the generation of spliced RNA *in vitro*. Krainer and Maniatis[5] resolved HeLa cell nuclear extracts into five fractions, three of which are required for the production of splicing intermediates (SF1, SF2, and SF4b).[29] Fraction SF1 contains snRNPs, whereas the other factors appear to be proteins. The remaining two fractions (SF3 and SF4a) are needed only for the generation of spliced RNA. In a similar approach, Hurwitz and collaborators[30,31] obtained three fractions: their fraction II assembles the pre-mRNA into a presplicing complex. In combination with fraction Ib, the spliceosome forms and cleavage at the 5′ splice site and lariat formation occur, whereas fraction Ia is necessary only to produce spliced RNA. The work of Abelson's group has resulted in the separation of distinct splicing activities from yeast extracts.[32,33]

We have separated nuclear extracts from HeLa cells into six fractions that have to be combined to yield splicing intermediates and reaction products.[34] Two of the active fractions contain the snRNPs whereas the remaining four appear to consist of protein factors. Figure 1 summarizes the separation of these components, the details of which are described below.

---

[27] A. Mayeda, K. Tatei, H. Kitayama, K. Takemura, and Y. Ohshima, *Nucleic Acids Res.* **14**, 3045 (1986).

[28] K. Tatei, K. Takemura, H. Tanaka, T. Masaki, and Y. Ohshima, *J. Biol. Chem.* **262**, 11667 (1987).

[29] Note that the nomenclature for the splicing activities is based on the original publication. Similar designations by different laboratories do not imply physical or functional identity.

[30] H. M. Furneaux, K. K. Perkins, G. A. Freyer, J. Arenas, and J. Hurwitz, *Proc. Natl. Acad. Sci. U.S.A.* **82**, 4351 (1985).

[31] K. K. Perkins, H. M. Furneaux, and J. Hurwitz, *Proc. Natl. Acad. Sci. U.S.A.* **83**, 887 (1986).

[32] S.-C. Cheng and J. N. Abelson, *Proc. Natl. Acad. Sci. U.S.A.* **83**, 2387 (1986).

[33] S.-C. Cheng, A. Newman, R.-J. Lin, G. McFarlane, and J. N. Abelson, this volume [8].

[34] A. Krämer, M. Frick, and W. Keller, *J. Biol. Chem.* **262**, 17630 (1987).

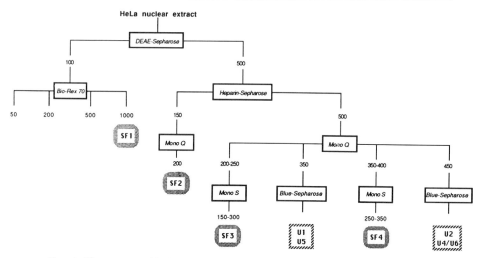

Fig. 1. Chromatographic fractionation of splicing activities from HeLa cell nuclear extracts. The different chromatographic resins are shown in boxes. Numbers designate the millimolar KCl concentration at which a putative activity is eluted from a resin. When splicing factors are desorbed with a salt gradient, the approximate KCl concentration required for elution is indicated. (See text for further explanations.)

Cell Culture

Human HeLa cells (S3 line, obtained from the Cold Spring Harbor Laboratory) are grown in three parallel 5-liter batches of suspension culture in minimum essential medium (MEM)–Joklik's medium, supplemented with 5% newborn calf serum. As culture vessels we use 10-liter round-bottomed chemical flasks (Duran glass, Schott Glaswerke, Mainz, FRG) with a neck of 10 cm length and 6 cm diameter, closed at the top with a silicon stopper wrapped in aluminum foil. The flasks contain a magnetic stirring bar (7 cm long) in the bottom, and the cell suspensions are stirred at 80 rpm on standard magnetic plates, placed in a 37° warm room. The cultures are inoculated from a 3-liter suspension grown in a 6-liter flask to a density of approximately 5×10^5 cells/ml. The initial density should be adjusted to $0.9–1.2 \times 10^5$ cells/ml. Under these conditions the cells have a doubling time of 18–20 hr. The cells must be harvested when they reach a concentration of $5–6 \times 10^5$/ml. This is crucial since extracts prepared from cells grown to higher numbers are inactive in splicing or in 3′ processing. Before harvesting, 1 liter of culture is removed and used to grow a 3-liter inoculum for the next cycle of mass culture.

The cells are collected by centrifugation for 10 min at room temperature in 1-liter plastic bottles (Nalgene) in a low-speed centrifuge (Heraeus-Christ Cryofuge 6-6) at 2800 rpm. The pelleted cells are resuspended with a pipet in a small volume of PBS [phosphate-buffered saline; 137 mM NaCl, 2.7 mM KCl, 4.3 mM Na$_2$HPO$_4$, 1.4 mM KH$_2$PO$_4$ (pH 7.3)] and combined into a separate centrifuge bottle. After all cells have been harvested, the suspension is diluted with about 400 ml of PBS and centrifuged for 10 min at 2,000 rpm. The pellet is resuspended in 500 ml of PBS and centrifuged again. The final pellet is resuspended in 60 ml of PBS, transferred into two preweighed 50-ml screw-cap Falcon tubes, and centrifuged for 10 min at 2,500 rpm in an IEC Centra 7R centrifuge. The supernatants are carefully removed, and the tubes are weighed. From 14 liters of culture grown to a density of 5 × 10^5 cells/ml (7 × 10^9 cells total), we obtain approximately 30 g of packed cells. When not used directly to prepare extracts, the pelleted cells are stored frozen at −80°.

Extract Preparation

Preparation of Nuclear Extract. Nuclear extracts active in pre-mRNA splicing are prepared by a modification of the procedure originally reported by Dignam *et al.*[35] (see also chapters by Moore[36] and Stauber *et al.*[37] in this series). Here, we describe the protocol routinely used in our laboratory. The same method is also applied for the preparation of extracts active in 3′ processing.

Thirty grams (7 × 10^9 cells, 30 ml packed cell volume) of HeLa cells in two 50-ml Falcon centrifuge tubes (see above) are resuspended by filling the tubes to 40 ml with cold PBS. When starting with frozen cells, the pellets are thawed as quickly as possible under warm tap water. The tubes are centrifuged for 10 min at 2,000 rpm. (Unless indicated otherwise, all centrifugations are carried out with an IEC Centra 7R centrifuge at 2–4°.) The supernatants are removed with a pipet, and the cell pellets are resuspended in 150 ml of buffer A [10 mM N-2-hydroxyethylpiperazine-N'-2-ethanesulfonic acid (HEPES)–KOH (pH 7.9), 1.5 mM MgCl$_2$, 10 mM KCl, 0.5 mM dithiothreitol] and transferred into a beaker. The suspension is kept on ice for 10 min with occasional swirling, distributed into four 50-ml Falcon tubes, and centrifuged for 10 min at 2,500 rpm. The supernatants are carefully removed, and the pellets are transferred with 60 ml of buffer A into an ice-cold Dounce homogenizer [60-ml cylinder, S-type plunger (tight fitting), Braun-Melsungen AG, Melsungen, FRG]. The cells

[35] J. D. Dignam, R. M. Lebovitz, and R. G. Roeder, *Nucleic Acids Res.* **11**, 1475 (1983).
[36] C. L. Moore, this volume [6].
[37] C. Stauber, D. Soldati, B. Lüscher, and D. Schümperli, this volume [7].

are broken with 15 strokes. The homogenate is transferred to two 50-ml Falcon tubes and centrifuged at 2,500 rpm. The supernatants are placed in a beaker using a pipet, without disturbing the loose pellet of nuclei. (This supernatant is set aside and can be used to prepare cytoplasmic S100; see below.) The nuclei are resuspended with 17.5 ml (2.5 ml per 10^9 cells) of buffer C [20 mM HEPES–KOH (pH 7.9), 25% glycerol, 0.42 M NaCl, 1.5 mM MgCl$_2$, 0.2 mM ethylenediaminetetraacetic acid (EDTA), 0.5 mM phenylmethylsulfonyl fluoride (from a freshly prepared 0.2 M solution in ethanol), 0.5 mM dithiothreitol] and transferred into a Dounce homogenizer (30-ml cylinder).

After 10 strokes of homogenization, the suspension is transferred into a beaker and stirred for 30 min on ice. Nuclear debris are removed by centrifugation for 30 min at 15,000 rpm in a Sorvall SS34 rotor at 2°. The supernatant is dialyzed for 5 hr on ice against two changes of 1 liter of buffer D [20 mM HEPES–KOH (pH 7.9), 20% glycerol, 0.1 M KCl, 0.2 mM EDTA, 0.5 mM dithiothreitol]. After dialysis, the extract is cleared of insoluble material by centrifugation for 30 min at 12,000 rpm in a Sorvall HB4 rotor. The supernatant (~20 ml) is stored in aliquots at −80°.

Cytoplasmic S100. The supernatant obtained after centrifugation of the nuclei from the cell homogenate is mixed with 0.11 volume of buffer B [0.3 M HEPES–KOH (pH 7.9), 1.4 M KCl, 30 mM MgCl$_2$] and centrifuged for 1 hr at 35,000 rpm in a Kontron TFT 65.38 rotor at 2°. The supernatant is dialyzed on ice against two changes of buffer D, cleared of precipitates, and stored, as described above for the nuclear extract.

Small-Scale Procedure for Preparation of Nuclear Extracts. Lee *et al.*[38] have recently described a procedure for the small-scale preparation of nuclear extracts. This method is useful to compare extracts from different lines of tissue culture cells or from cells grown under different conditions for *in vitro* transcription efficiency and for splicing or 3′ processing of pre-mRNAs. We have modified the original protocol and employed it to prepare nuclear extracts that are active in pre-mRNA splicing as well as in 3′ processing.

The entire procedure is carried out in 1.5-ml Eppendorf tubes. HeLa cells (1×10^8) from a 200-ml culture grown to a density of 5×10^5 cells/ml are collected by low-speed centrifugation, washed with PBS, and recentrifuged. The cell pellet is resuspended in 1 ml of PBS, transferred to an Eppendorf tube, and collected by centrifugation for 5 min at 3,000 rpm and 2° in a Sigma 2 MK centrifuge (M. Christ, Osterode, FRG). After the supernatant is discarded, the cells are resuspended in 0.5 ml buffer A,

[38] K. A. W. Lee, A. Bindereif, and M. R. Green, *Gene Anal. Tech.* **5**, 22 (1988).

briefly treated with a vortex mixer, and recentrifuged. The cell pellets are resuspended in 0.5 ml of buffer A and kept on ice for 15 min.

Cell lysis is carried out as follows. A 1-ml disposable syringe with a 25-gauge × 5/8 inch (0.5 × 16 mm) hypodermic needle (attached by Luer lock) is filled with buffer A and emptied again by ejection to remove the air from the syringe barrel and the needle. The cell suspension is drawn slowly into the syringe and then ejected back into the Eppendorf tube with a rapid push on the plunger. The aspiration–ejection step is repeated 4 times. The crude nuclei are collected by centrifugation for 20 sec at full speed in an Eppendorf table-top centrifuge at room temperature. The supernatant is taken off and discarded. The nuclear pellet is resuspended with 0.17 ml of buffer C and kept on ice for 30 min with occasional mixing. The suspension is centrifuged for 15 min at 13,000 rpm and 0° in the Sigma 2 MK centrifuge. The supernatant is dialyzed for 4 hr against 2 changes of buffer D. Insoluble material is removed by centrifugation (10 min, 13,000 rpm), and the extract is stored in aliquots at −80°.

Assays for Identification of Splicing Activities

Preparation of ^{32}*P-Labeled Pre-mRNA Substrate.* The standard pre-mRNA substrate is RNA 1, derived from the adenovirus 2 major late transcription unit.[39] It is transcribed from *Sca*I-cleaved DNA of the plasmid pSP62Δi1 by SP6 RNA polymerase[40,41] and consists of 102 nucleotides of exon 1, 113 nucleotides of intron, and 38 nucleotides of exon 2. Reaction mixtures contain 40 mM tris(hydroxymethyl)aminomethane hydrochloride (Tris-HCl) (pH 7.5), 6 mM MgCl$_2$, 2 mM spermidine, 10 mM dithiothreitol, 0.5 mM each of ATP and CTP, 0.1 mM GTP, 2 mM G(5′)ppp(5′)G (P-L Biochemicals), 20 μM UTP, 20 μCi of [α-^{32}P]UTP (800 Ci/mmol; Amersham Buchler), 5 units of SP6 RNA polymerase (Boehringer Mannheim), 40 units of RNAguard (Promega Biotec), and 1 μg of linearized plasmid DNA template in a total volume of 20 μl. After incubation for 1 hr at 41°, 27.5 μl of 7.5 M ammonium acetate, 153 μl of distilled water, and 1 μl of glycogen carrier (Renner, Dannstadt, FRG) are added.

The mixtures are extracted with phenol and chloroform–isoamyl alcohol, and the RNA is precipitated with ethanol. The RNA pellet is washed once with 1 ml of 70% ethanol, dried, resuspended in 8 μl of sample buffer (90% formamide, 0.2% xylene cyanol, 0.2% bromphenol blue in

[39] D. Frendewey and W. Keller, *Cell* **42**, 355 (1985).
[40] P. A. Krieg and D. A. Melton, this series, Vol. 155, p. 397.
[41] J. K. Yisraeli and D. A. Melton, this series, Vol. 180, p. 42.

1× TBE[42]). The sample is heated to 90° for 3 min and loaded onto a 6% polyacrylamide–8.3 M urea sequencing gel (prepared with 4 times crystallized acrylamide; Serva, Heidelberg, FRG). The RNA is located by brief autoradiography (30 sec). The RNA is eluted from the excised gel slice into 0.4 ml of 0.75 M ammonium acetate, 10 mM magnesium acetate, 0.1 mM EDTA, 0.1% sodium dodecyl sulfate (SDS), 1% phenol, and 25 μg/ml of *Escherichia coli* tRNA for 2–12 hr at 37°. Gel pieces are removed by centrifugation, the supernatant is extracted with phenol–chloroform, and the RNA is precipitated with ethanol and stored under ethanol at −20°.

In Vitro Splicing. The standard splicing reaction is carried out in a total volume of 50 μl. A reaction mixture (25 μl) containing 1 μl of 20 mM MgATP, 1 μl of 0.5 M creatine phosphate, 1 μl of 50 mM $MgCl_2$, 7.5 μg of *E. coli* tRNA (Boehringer Mannheim) and 10,000 Cerenkov counts per minute (cpm) of uniformly labeled pre-mRNA (5–10 fmol) is added, on ice, to the appropriate amount of nuclear extract (usually 15 μl), adjusted to 25 μl with buffer D. Incubations are performed for 60 min at 30°.

The assay for splicing activity in chromatographic fractions is performed with slight modifications of the procedure described above. The volume of individual or combined fractions is adjusted, on ice, to 25 μl with the buffer the fractions are dialyzed into (see below). Ten microliters of a 13% poly(vinyl alcohol) solution (Sigma; low mol. wt., type II), which is used instead of tRNA to obtain optimal splicing activity in the presence of the fractions, is added, followed by the addition of a mixture (15 μl) containing 1 μl of 20 mM MgATP, 1 μl of 0.5 M creatine phosphate, 1 μl of 50 mM $MgCl_2$, and radioactive RNA substrate as above. (The inclusion of RNase inhibitor in the reactions did not increase the splicing activity, nor did it alleviate the degradation of the pre-mRNA which is often observed in reconstitution assays performed with fractions.) Splicing is performed for 1–2 hr at 30°.

Reactions are terminated by the addition of 120 μl of 2% (w/v) sarcosyl, 0.1 M Tris-HCl (pH 7.5), 20 mM EDTA, and 40 μg proteinase K (Boehringer Mannheim), followed by a 10-min incubation at 30°. RNA is isolated by phenol–chloroform extraction and precipitated with ethanol in the presence of 3 M ammonium acetate (and 1 μl of glycogen carrier for reactions that contain chromatographic fractions). The reaction products are separated in denaturing 12% polyacrylamide–8.3 M urea sequencing gels (in TBE)[42] and visualized by autoradiography on Kodak X-AR5 or

---

[42] T. Maniatis, E. F. Fritsch, and J. Sambrook, "Molecular Cloning: A Laboratory Manual." Cold Spring Harbor Laboratory, Cold Spring Harbor, New York, 1982.

Fuji RX film with Du Pont intensifying screens. Exposure is typically overnight at $-80°$.

In Vitro Complex Formation. Mixtures similar to those of the splicing reactions are used, except that the final volume of the reaction is 10 or 20 μl. Uniformly labeled RNA is used at 50,000 Cerenkov cpm/reaction (50–100 fmol). Incubations are performed for 30–60 min at 30°. The reactions are terminated by addition of heparin (Sigma) to a final concentration of 200 μg/ml, followed by incubation at room temperature for 10 min. At this point the reactions can either be frozen at $-80°$ or loaded onto a native polyacrylamide gel after 1 μl of xylene cyanol and bromphenol blue in distilled water have been added to the reaction mixtures. We use 4% polyacrylamide gels (acrylamide : bisacrylamide, 80 : 1 v/v) in 25 mM Tris-OH, 25 mM boric acid, and 1 mM EDTA (pH 8). Gel dimensions are 18 × 18 × 0.1 cm. Before loading, gels are prerun for 30 min at 300 V at room temperature. Electrophoresis is carried out at 20 mA (300–450 V) until the xylene cyanol dye reaches the bottom of the gel. Reaction products are visualized by autoradiography as above.

Assays for snRNP Activity. The function of partially purified snRNPs in splicing can be tested in extracts that have been depleted of the particles by incubation with specific antibodies or after treatment of an extract with micrococcal nuclease. These assays are described in detail elsewhere in this volume.[43]

Chromatographic Procedures

General Remarks. The protocol detailed below describes our current procedure for the preparation of splicing factors. Smaller column volumes have been used and the preparation can also be scaled up. All steps are carried out in a cold room at 4–8°. For convenience all columns are connected to a FPLC system (Pharmacia/LKB) equipped with a gradient programmer, UV monitor, and recorder. Chromatographic resins are precycled according to the manufacturer's instructions and are equilibrated prior to use with the buffer in which the sample is applied.

After chromatography, fractions are immediately dialyzed against buffers containing 100 mM KCl. This is necessary because splicing *in vitro* is inhibited at salt concentrations above 50 mM KCl.[44] Fractions are centrifuged after dialysis for 20 min at 12,000 rpm in a Sorvall SS34 rotor to clear the dialysate and then frozen at $-80°$, at which temperature they stay active for more than 6 months. The fractions obtained are designated

[43] A. Krämer, this volume [18].
[44] N. Hernandez and W. Keller, *Cell* **35**, 89 (1983).

by an abbreviation of the resins used and the KCl concentration at which they elute (see Fig. 1).

Buffers. All stock solutions and buffers are made with sterile components. Please note that buffers A–D described below are different from those used for the preparation of the extracts. The designation is the same as that used by Krämer et al.[34]

Buffer A: 10% (v/v) glycerol, 20 mM HEPES–KOH (pH 7.9), 3 mM MgCl$_2$, 0.1 mM EDTA, 0.5 mM dithiothreitol, 0.5 mM phenylmethylsulfonyl fluoride, and KCl at the concentrations indicated in the text

Buffer B: equivalent to buffer A, but containing 50 mM Tris-HCl (pH 7.9)

Buffer C: equivalent to buffer A, but containing 20% (v/v) glycerol

Buffer D: equivalent to buffer B, but containing 20% (v/v) glycerol

Step 1: DEAE-Sepharose Chromatography. DEAE-Sepharose or DEAE-Sepharose Fast Flow (Pharmacia/LKB; 100 ml in a column 20 × 2.5 cm) is equilibrated with buffer A plus 100 mM KCl. The column is loaded with nuclear extract at approximately 10 mg protein/ml DEAE-Sepharose (~80–100 ml of extract, corresponding to 2.5 × 10^{10} cell equivalents) at a flow rate of 1 column volume/hr followed by a wash with 1.5 column volumes of buffer A plus 100 mM KCl. Bound material is eluted with buffer A plus 500 mM KCl (2 column volumes). The fraction size throughout this step is 10 ml. Fractions containing more than 1.0 A_{280} unit/ml are pooled and dialyzed against 3 changes (1.5 hr each) of 1 liter of buffer C plus 100 mM KCl. The pooled flow-through fractions (~20% of the protein applied to the column) are designated DS100; the combined fractions eluting from the column are designated DS500 (~50% of the applied protein).

Splicing activity in the fractions is assayed in a standard *in vitro* reaction. The DS100 fraction by itself is inactive, whereas the DS500 fraction generates a small amount of spliced RNA which varies among different preparations. When both fractions are combined in a reaction, the pre-mRNA is spliced as efficiently as in a nuclear extract.[34] The optimal ratio of the fractions for the reconstitution of splicing activity can be determined in a titration experiment. Usually, 5–10 μl of DS100 and 15–20 μl of DS500 combined in a 50-μl reaction give the best results. The activity that resides in the DS100 fraction has been designated splicing factor (SF) 1. This terminology does not take into account that a fraction could contain more than one activity. SF1 can be further purified by chromatography on Bio-Rex 70 (see below). The DS500 fraction contains the remaining splicing factors and the bulk of the snRNPs.

Step 2: Heparin-Sepharose Chromatography. Heparin-Sepharose (Pharmacia/LKB; 50 ml in a column 10 × 2.5 cm) is equilibrated with buffer A plus 150 mM KCl. The salt concentration of the dialyzed DS500 fraction is adjusted to 150 mM KCl with a 2 M KCl solution, and the fraction is loaded onto the column at 10 mg of protein/ml heparin-Sepharose at 2 column volumes/hr. Unbound material is removed with 2 column volumes of buffer A plus 150 mM KCl, and bound components are eluted with 2 column volumes of buffer A plus 500 mM KCl. The fraction size is 5 ml. Flow-through (HS150) and step fractions (HS500) containing more than 0.5 A_{280} units/ml are pooled separately and dialyzed against buffer D plus 100 mM KCl as above. About 25% of the protein applied to the column is recovered in HS150; 40–50% of the protein is found in HS500.

For the generation of spliced mRNA *in vitro*, HS150, HS500, and DS100 have to be combined.[34] No reaction products are generated on testing HS100 or HS500 individually or in combination, nor when one of these fractions is combined with DS100. Thus, a second activity (SF2) can be separated from the remaining factors. The snRNPs are almost exclusively contained in the HS500 fraction. Splicing with the combined DS100, HS150, and HS500 fractions is not very efficient; this is probably due to the dilution of factors during heparin-Sepharose chromatography.

Step 3: Mono Q Chromatography. An 8-ml Mono Q FPLC column (Pharmacia/LKB) is equilibrated with buffer B plus 100 mM KCl. Fractions are loaded at 20–30 mg of protein/ml Mono Q and 45 ml/hr. The column is washed with 3–4 column volumes of buffer B plus 100 mM KCl. Bound material is eluted with a linear gradient (120 ml) of 100–500 mM KCl in buffer B followed by a 10-ml wash of buffer B plus 500 mM KCl at 60 ml/hr. The fraction size is 5 ml during loading and washing and 2 ml during elution. Protein-containing fractions are dialyzed individually against 3 changes of buffer C plus 100 mM KCl (1.5 liters each).

Chromatography of HS500 on Mono Q results in a further separation of splicing factors. SF3 and SF4 elute at 200–250 and 350–500 mM KCl, respectively. In addition, a partial separation of the snRNPs is achieved. Fractions enriched in U1 and U5 snRNPs (Q-U1) elute at approximately 350 mM KCl whereas U2, U4, and U6 snRNPs (Q-U2) are detected in fractions eluting at 450 mM KCl.[34,43] For pre-mRNA splicing, fractions containing SF1 (DS100), SF2 (HS150), SF3, and SF4 (both Mono Q) and the Mono Q fractions enriched in snRNPs have to be combined. Compared to the nuclear extract, the kinetics of the splicing reaction in the presence of the partially purified components is slower,[34] hence splicing reactions are performed for 2 hr. For optimal splicing efficiency it is useful at this point to determine the ratio of the different fractions to be used in a

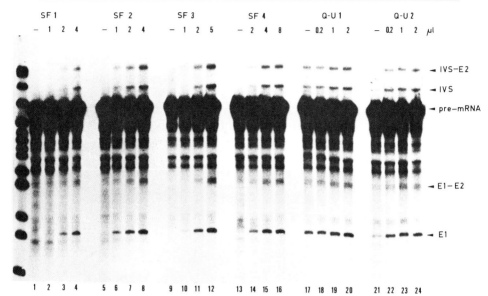

FIG. 2. Titration of splicing components. In each set of experiments the concentration of one of the factors was varied as indicated above each lane, while the concentration of the other components was kept constant. The complete 50-μl splicing reactions contained 4 μl each of SF1 (DS100) and SF2 (HS150), 5 μl of SF3, 8 μl of SF4 (both Mono Q), and 2 μl each of Q-U1 and Q-U2. Reactions were performed for 2 hr at 30°. The reaction products were analyzed in a 12% polyacrylamide–8.3 M urea gel. The splicing products generated in the reaction are indicated at right. E1, exon 1; E2, exon 2; IVS, intervening sequence. (From Ref. 34.)

splicing reaction. The titration experiment shown in Fig. 2 reveals that 2–4 μl of each of the splicing factors have to be added to a standard reaction before splicing can proceed. In the absence of the snRNP-enriched fractions, spliced RNA is produced to some degree. This can be explained by the presence of these particles in the DS100, HS150, and SF4 fractions.[34,45]

All of the fractions active in splicing contain factors that are required for cleavage of the pre-mRNA at the 5′ splice site and lariat formation, as is evident from the absence of splicing intermediates when individual fractions are omitted from the reaction (Fig. 2). It has previously been shown that the two cleavage and ligation steps of pre-mRNA splicing can be uncoupled on fractionation of splicing extracts.[5,31] Thus, distinct activities exist that participate only in the 3′ cleavage and exon ligation step. From this observation it can be concluded that one or more of the splicing

[45] A. Krämer, unpublished observations.

activities described here consists of components required for the first as well as for the second step of the reaction. Until now we have not been able to separate these activities clearly. However, results obtained recently indicate that SF3 activity might consist of two factors. On chromatography on Mono Q or Mono S columns, this activity sometimes fractionates into a factor that will allow only the formation of splicing intermediates (in combination with SF1, SF2, SF4, and the snRNPs) and into a factor that is necessary to obtain the spliced product.[34,45] The former activity shows a rather broad elution profile on both Mono Q and Mono S columns and partially overlaps with the latter activity.

Further Purification of Splicing Activities

Bio-Rex 70 Chromatography of SF1. Bio-Rex 70 (Bio-Rad; 100–200 mesh; 30 ml in a column 15 × 1.6 cm) is extensively equilibrated with buffer A plus 100 mM KCl until the outflowing buffer has reached pH 7.9. The DS100 fraction is loaded at 5–10 mg of protein/ml Bio-Rex at 2 column volumes/hr, and unbound material is removed with 2 column volumes of buffer A plus 100 mM KCl. Components are eluted by successive washing of the column with buffer A containing 200 mM, 500 mM, and 1 M KCl (1.5 column volumes each), respectively. Fractions of 3 ml are collected. Protein-containing fractions (>0.3 A_{280} units/ml) of each step are pooled and dialyzed against buffer C plus 100 mM KCl. The BR100, BR200, BR500, and BR1000 fractions contain approximately 20, 10, 55, and 5% of the protein applied to the column, respectively.

The fractions are tested for SF1 activity in combination with 5 μl of HS150 and 10 μl of HS500. Splicing activity is detected in the BR1000 fraction. Neither of the other fractions is active in splicing. In one case both BR500 and BR1000 fractions were required to complement HS150 and HS500[46]; however, we have no explanation for this observation.

Mono S Chromatography of SF3 and SF4. A 1-ml Mono S FPLC column (Pharmacia/LKB) is equilibrated with buffer A plus 100 mM KCl. Mono Q fractions containing either SF3 or SF4 activity are applied to the column at 60 ml/hr with a protein concentration of up to 10 mg/ml Mono S. Unbound material is removed with 10–15 ml of buffer A plus 100 mM KCl. Bound components are eluted with a 20-ml gradient of 100–500 mM KCl in buffer A followed by a 5-ml wash with buffer A plus 500 mM KCl. Fractions of 0.5 ml are collected. Protein-containing fractions are individually dialyzed against buffer C plus 100 mM KCl.

Splicing activity is tested in the presence of DS100, HS150, Q-U1, and

---

[46] A. Krämer, *Genes Dev.* **2**, 1155 (1988).

Q-U2. Assays for SF3 activity are supplemented with SF4 (usually derived from Mono Q fractions); assays for SF4 activity are supplemented with SF3. SF3 activity elutes from the Mono S column at 150–300 mM KCl. We sometimes observe an enrichment of splicing intermediates in early eluting fractions, whereas later eluting fractions promote the complete splicing reaction.[34] SF4 activity is found in fractions eluting at 250–350 mM KCl and appears to fractionate as a single component.

Mono Q Chromatography of SF2. SF2 can be further purified from HS150 by fractionation on a Mono Q column. The procedure is identical to that described above for HS500. SF2 activity elutes as a single component at approximately 200 mM KCl.[47]

We have previously shown that SF2 activity distributes between cytoplasmic S100 and nuclear extract.[47] On fractionation of SF2 from the S100 fraction, we discovered an activity that highly stimulated *in vitro* splicing in the presence of the partially purified factors.[48] This activity was purified to homogeneity and identified as a creatine kinase[49] by protein sequencing (in collaboration with Drs. H. Kocher and P. Graff, Sandoz AG, Basel).

The identity of this activity was further supported by the observation that commercial creatine kinase stimulates the *in vitro* splicing reaction as well. We believe that this enzyme is required in our *in vitro* system because ATPases (present in the nuclear extract as well as in several fractions) deplete the ATP pool in a splicing reaction. ATP can be regenerated in the presence of creatine phosphate and creatine kinase.

Further Purification of snRNPs. The snRNPs obtained after Mono Q chromatography can be further purified on Blue-Sepharose. The details for this procedure are described elsewhere in this volume.[43]

Purity of Splicing Factors. The most extensively purified fractions with SF1, SF2, SF3, or SF4 activity still contain numerous proteins as judged by SDS–polyacrylamide gel electrophoresis and staining with Coomassie blue or silver. As yet, no polypeptides have been identified that might correspond to a splicing activity.

Characterization of Splicing Activities

We believe that splicing factors SF1–SF4 represent proteins because their activity is resistant to treatment with micrococcal nuclease.[34,46] However, the presence of a minor RNA component associated with a

[47] A. Krämer and W. Keller, *EMBO J.* **4**, 3571 (1985).
[48] U. Utans and A. Krämer, unpublished results.
[49] G. Villarreal-Levy, T. S. Ma, S. A. Kerner, R. Roberts, and M. B. Perryman, *Biochem. Biophys. Res. Commun.* **144**, 1116 (1987).

splicing activity cannot be totally excluded. For example, it has been shown that U5 snRNP is relatively resistant to micrococcal nuclease digestion.[7] Also, the activity of U11 snRNP, which may be involved in the polyadenylation of pre-mRNAs, cannot be inhibited by nuclease treatment.[50]

Requirement of Splicing Factors at Different Steps in Splicing Reaction. As a step toward an understanding of the function of individual splicing activities, we have tested the contribution of the factors to splicing complex assembly. Two protein factors (SF1 and SF3) are essential for the formation of a presplicing complex (complex A) in conjunction with U1 and U2 snRNPs (Fig. 3).[45,46] Assembly of the spliceosome (complex B) involves additional components. First, there is a requirement for creatine kinase or elevated levels of ATP in this reaction, which is not found for the generation of complex A[48]; this supports our previous observation that higher levels of ATP are needed for the formation of complex B than for complex A.[51] Second, U4, U5, and U6 snRNPs are necessary to convert complex A to the spliceosome.[13,15,16,19,52,53] A participation of SF2 in the conversion of complex A to complex B has not yet been clearly demonstrated because the results with factor from different preparations have been variable.[48] We cannot exclude the possibility that low concentrations of SF2, which may be present as contaminant of other fractions, are sufficient to allow spliceosome formation. Thus, the assembly of the pre-mRNA into the spliceosome requires the major snRNPs U1–U6, SF1 and SF3, and creatine kinase or elevated levels of ATP, and perhaps SF2. Recent results indicate that SF4 acts in the splicing reaction after spliceosome assembly. A similar observation has been made in yeast, where at least one component (prp2) is necessary for S′ cleavage and lariat formation but not for spliceosome formation.[18]

A candidate for an activity that functions in 3′ cleavage and exon ligation is the component detected in SF3-containing fractions which is required only for the generation of spliced RNA and not for splicing intermediates (see above).

Comparison of SF1, SF2, SF3, and SF4 to Splicing Activities Identified by Other Laboratories. It is difficult to compare splicing factors fractionated by other groups with the activities described here because different fractionation schemes have been employed. To date the results suggest that the SF1 activity as described here is present in fraction Ib of

[50] G. Christofori and W. Keller, *Cell* **54,** 875 (1988).
[51] D. Frendewey, G. Lahr, and W. Keller, unpublished observation.
[52] A. Krämer, *J. Mol. Biol.* **196,** 559 (1987).
[53] D. Frendewey, A. Krämer, and W. Keller, *Cold Spring Harbor Symp. Quant. Biol.* **52,** 287 (1987).

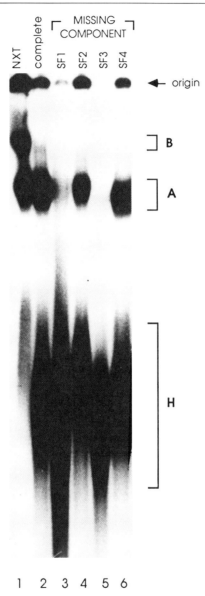

1   2   3   4   5   6

FIG. 3. Formation of splicing complexes in the presence of nuclear extract and of fractionated components. Complex reactions were performed for 30 min at 30° in the presence of the following components: nuclear extract (lane 1); complete reaction with SF1, SF2, SF3, SF4, Q-U1, and Q-U2 (lane 2); complete reaction minus SF1 (lane 3); complete reaction minus SF2 (lane 4); complete reaction minus SF3 (lane 5); complete reaction minus SF4 (lane 6). Reaction products were analyzed in a 4% polyacrylamide gel. Complexes H (unspecific), A (presplicing complex) and B (spliceosome) are indicated at right. (From Ref. 46.)

Perkins et al.[31] and in fraction SF4b[29] of Krainer and Maniatis.[5] SF3 may correspond to the SF2 factor described by Krainer and Maniatis. A more detailed comparison of the splicing factors can be found in Steitz et al.[54]

Since SF1 and SF3 act during presplicing complex formation,[46] it is possible that one or both correspond to other proteins which have been shown to interact with specific sequences in the pre-mRNA. SF3 appears to correspond to U2AF,[23] because it is required for binding of U2 snRNP to the branch point and because its elution behavior during Mono Q chromatography is similar to that of U2AF. SF1 may be the same as IBP[25,26]; however, some of its chromatographic properties cannot be reconciled with those described for IBP.[26,46]

Comparison of Splicing Factors to hnRNP Proteins. We routinely monitor all chromatographic fractions for the presence of hnRNP A and C proteins by protein blotting.[55,56] The A1 and C proteins distribute evenly between DS100 and DS500. On chromatography of DS100 on Bio-Rex 70, the A and C proteins are enriched in BR500, whereas only low concentrations are detected in BR100, BR200, and BR1000. The proteins present in DS500 bind to heparin-Sepharose and are subsequently resolved by Mono Q chromatography. A1 elutes from this column at 150–300 mM KCl, whereas the C proteins are found in fractions eluting between 350 and 500 mM KCl, with a peak at approximately 400 mM KCl.[48] These results demonstrate that hnRNP proteins are present in fractions that exhibit splicing activity. We believe that factors SF1, SF2, SF3, and SF4 are distinct from the hnRNP proteins, because fractions enriched in these proteins (e.g., BR500) cannot substitute for either of the splicing activities in an *in vitro* reaction. However, this does not rule out a function for the hnRNP proteins in splicing. This question can be resolved once a complete separation of the splicing factors from the hnRNP proteins has been achieved.

Acknowledgments

We thank Marion Frick for excellent and dedicated assistance throughout this work; Ulrike Utans for participation in the fractionation experiments; Drs. Hans Kocher and Patrick Graff (Sandoz AG, Basel) for protein sequencing; and Dr. Graeme Bilbe for comments on the manuscript. This work was supported by the Deutsche Forschungsgemeinschaft and the Schweizerischer Nationalfonds.

[54] J. A. Steitz, D. L. Black, V. Gerke, K. A. Parker, A. Krämer, D. Frendewey, and W. Keller, *in* "Structure and Function of the Major and Minor Small Nuclear Ribonucleoprotein Particles" (M. L. Birnstiel, ed.), p. 115. Springer-Verlag, Berlin and New York, 1988.
[55] H. Towbin, T. Staehelin, and J. Gordon, *Proc. Natl. Acad. Sci. U.S.A.* **76**, 4350 (1979).
[56] J. Kyhse-Anderson, *J. Biochem. Biophys. Methods* **10**, 203 (1984).

[2] Small-Scale Preparation of Extracts from Radiolabeled Cells Efficient in Pre-mRNA Splicing

By KEVIN A. W. LEE and MICHAEL R. GREEN

Gaining an understanding of processing of messenger RNA precursors (pre-mRNAs) at the biochemical level depends on *in vitro* studies. This has been made possible by the development of methods for preparing soluble extracts from cultured mammalian cells that support RNA processing.[1,2] These extracts support the basic RNA processing reactions.[3-5] It is now of interest to obtain regulated *in vitro* systems that recapitulate the events required for such phenomena as differential or tissue-specific RNA splicing. The development of such systems will be largely empirical and will employ different cell types, growth conditions, and extract preparation protocols.

Preparation of nuclear extracts by the conventional procedure[1] requires relatively large numbers of cells and imposes practical difficulties in varying the parameters outlined above. For example, many cell types do not grow in suspension, making it cumbersome to obtain sufficient cells for extract preparation. Systematic variation of cell growth conditions or extraction procedures is time-consuming and requires large amounts of starting material. In addition, experiments that exploit radioactive labeling of RNAs and proteins are invariably possible with only analytical amounts of material. Radioactive labeling of cells to high specific activity is extremely expensive and wasteful for the amounts of cells required for extract preparation by the conventional procedure.

The following protocol describes a method that enables small-scale preparation of biologically active nuclear extracts. This procedure can be performed with as few as 3×10^7 HeLa cells and employs only standard laboratory supplies. It is feasible to prepare many different nuclear ex-

[1] J. D. Dignam, R. M. Lebovitz, and R. G. Roeder, *Nucleic Acids Res.* **11**, 1475 (1983).
[2] M. R. Green, *Annu. Rev. Genet.* **20**, 671 (1986).
[3] A. R. Krainer, T. Maniatis, B. Ruskin, and M. R. Green, *Cell* **36**, 993 (1984).
[4] N. Hernandez and W. Keller, *Cell* **35**, 89 (1983).
[5] R. A. Padgett, S. F. Hardy, and P. A. Sharp, *Proc. Natl. Acad. Sci. U.S.A.* **80**, 5230 (1983).

tracts simultaneously. This is advantageous in situations demanding the economic use of scarce or expensive materials, including (1) preparation of extracts from cells that cannot be grown in large amounts in suspension, (2) preparation of biologically active extracts, radioactively labeled to high specific activity, and (3) preparation of extracts from virally infected cells using mutant viruses or different cell culture/extraction conditions.

By several criteria, nuclear extracts prepared by the small-scale procedure are comparable to extracts prepared by the conventional procedure. Miniextracts are fully active in pre-mRNA splicing and also efficiently transcribe class II promoters. These and other properties described here suggest that known characteristics of conventionally prepared extracts will apply to miniextracts. For example, it is likely that miniextracts will support other complex biological reactions, such as DNA replication. We have successfully used the procedure for entirely different cell types, indicating that it will be generally applicable to mammalian cells.

Buffers and Solutions

Buffer A: 10 mM N-2-hydroxyethylpiperazine-N'-2-ethanesulfonic acid (HEPES) (pH 8.0), 1.5 mM MgCl$_2$, 10 mM KCl, 1 mM dithiothreitol (DTT)

Buffer C: 20 mM HEPES (pH 8.0), 1.5 mM MgCl$_2$, 25% (v/v) glycerol, 420 mM NaCl, 0.2 mM ethylenediaminetetraacetic acid (EDTA) (pH 8.0), 1 mM DTT, 0.5 mM phenylmethylsulfonyl fluoride (PMSF)

Buffer D: 20 mM HEPES (pH 8.0), 20% (v/v) glycerol, 100 mM KCl, 0.2 mM EDTA, 1 mM DTT, 0.5 mM PMSF

1 M stock solution of DTT in water

100 mM solution of PMSF in ethanol

Buffers A, C, and D are essentially as previously described[1] and can be stored in the cold room for 1 week or so without the addition of DTT and PMSF. DTT and PMSF should be added fresh at the time of extract preparation. The 100 mM stock solution of PMSF should be prepared fresh and added to buffers slowly with vigorous stirring.

Materials

1-ml single-use hypodermic syringes

25-gauge 5/8 (15.9 mm) hypodermic needles

Very small magnetic stirring bar, small enough to stir efficiently 0.3 ml in a 1.5-ml Eppendorf tube

Procedures

Preparation of Nuclear Extracts by Conventional Procedures

Preparation of nuclear extracts from large amounts of HeLa cells grown in suspension is carried out essentially as previously described,[1] with some convenient modifications. Harvested cells are resuspended in a volume of buffer A equal to the packed cell volume, left to swell on ice for 15 min, and then lysed with 30 strokes of a B-type Dounce homogenizer. The crude nuclear fraction is resuspended in two-thirds of one packed cell volume (determined at the time of cell harvest) of buffer C containing 420 mM NaCl and incubated on ice for 30 min with stirring. We find it convenient and more reproducible to add buffer C according to the cell volume at harvest as opposed to the cell count determined using a hemocytometer. Following extraction, the nuclear extract is dialyzed for 2 hr against 100 volumes of buffer D containing 100 mM KCl. The short dialysis period results in less precipitation of denatured protein and is sufficient to adjust the extract to the desired ionic conditions. Precipitate that forms during dialysis can be removed by centrifugation of the extract for 10 min at 2000 rpm in a Beckman J6B centrifuge using a JS 4.2 rotor.

Preparation of Nuclear Miniextracts from Monolayer HeLa Cells

Preparation of extracts is also described elsewhere.[5a] Extracts can be prepared for in vitro assays in as little as 4 hr (including dialysis) and from as few as 3 × 10^7 cells. The procedure requires only minimal equipment and standard laboratory supplies and involves the same types of manipulations required to make an extract according to the conventional protocol[1] outlined above. Most centrifugation steps are performed in a microcentrifuge with reduced times compared to the conventional procedure (or even complete omission of steps), which greatly reduces the extract preparation time. All steps are performed on ice unless otherwise indicated.

Monolayer cells at 80% confluence are harvested using a rubber policeman or by trypsinization. Cells are washed in 30 volumes of phosphate-buffered saline (PBS) and the packed cell volume determined by pelleting for 5 min at 1200 rpm in a J6B centrifuge using a JS 4.2 rotor. Packed cells are resuspended in one packed cell volume of buffer A and allowed to swell on ice for 15 min. Cells are then lysed by rapidly pushing them through a narrow-gauge hypodermic needle as follows: a 1-ml hypodermic syringe with needle attached is filled with buffer A and the syringe

5a K. A. W. Lee, A. Bindereif, and M. R. Green, Gene. Anal. Tech. 5, 22 (1988).

plunger is used to displace the buffer as fully as possible. This removes all of the air from the syringe and prevents excess air from being "pumped" into the cell suspension during lysis. Needles of 23- to 26-gauge diameter are suitable for breaking open cells. We routinely use a 25-gauge 5/8 (15.9 mm) single-use Yale hypodermic needle. The cell suspension is drawn slowly into the syringe from a 1.5-ml Eppendorf tube (this minimizes the amount of air drawn into the syringe) and then ejected with a single rapid stroke. Five rapid strokes are usually sufficient to achieve approximately 80% cell lysis, which can be monitored using a light microscope and is sufficient to produce active extracts. Considerable amounts of air are forced into the cell suspension during the lysis procedure, but apparently without effect on the activity of extracts.

The cell homogenate is centrifuged for 20 sec in a microcentrifuge (12,000 g) or at low speed for 5 min in a Micro-Centaur Eppendorf centrifuge to produce a crude nuclear pellet and postnuclear supernatant. We routinely perform this step at room temperature for convenience. The crude nuclear pellet is resuspended in two-thirds of one packed cell volume (determined at the time of cell harvest) of buffer C containing 420 mM NaCl followed by incubation on ice with stirring for 30 min. The nuclear debris is then pelleted by spinning for 5 min in an Eppendorf microcentrifuge (12,000 g), and the supernatant (nuclear extract) is dialyzed against buffer D containing 100 mM KCl for 2 hr. The dialyzed extract can be directly used for transcription or splicing assays, or it is quick-frozen in liquid nitrogen and stored at $-70°$. Freezing of nuclear extracts does not result in significant loss of transcriptional activity. To an approximation, 10 plates (100 mm diameter) of cells at 80% confluence yield 400 μl of nuclear extract with a protein concentration of 8 mg/ml.

Preparation of Miniextracts from Radioactively Labeled Cells

Depending on the amount of labeled protein that is required for a particular experiment, between 1 and 10 100-mm plates of cells can be labeled. If one plate is sufficient for a particular purpose, then the labeled cells can be added to unlabeled cells to provide sufficient material for extract preparation. This has the advantage of minimizing the number of cells that are required during the labeling process.

1. Starve cells of endogenous unlabeled precursors by growth in phosphate-free or methionine-free medium containing 2% dialyzed fetal bovine serum [dialyzed against 100 volumes of 150 mM NaCl, 20 mM HEPES (pH 7.2)].
2. Add radioactive label at 2 mCi/ml [^{32}P]phosphate or 0.5 mCi/ml [^{35}S]methionine in 2 ml of media per 100-mm plate (the reduced

volume of media enables the concentration of label to be maximized) in a well-humidified incubator with CO_2 adjusted to less than 5% to avoid having the media turn acidic. Labeling time should be empirically determined for particular experiments.

3. Prepare nuclear extract as described.

SP6 Transcription and in Vitro Pre-mRNA Splicing

SP64-H$\beta\Delta6^3$ cut with *Bam*HI was used as the DNA template in SP6 transcription following the protocol described.[6] Transcripts were labeled at low specific radioactivity ($1-2 \times 10^6$ cpm/μg). *In vitro* pre-mRNA splicing reactions were carried out as previously described[3,7] with 30 ng of low-specific activity RNA per 1$\times$ reaction. For the splicing reactions shown in Fig. 1, 2.7% poly(vinyl alcohol) (PVA) was included. When splicing complexes were fractionated by sedimentation (Fig. 2), PVA was omitted from the splicing reaction. Sucrose-gradient sedimentation analysis was performed as described.[8] RNAs were analyzed on 5% denaturing polyacrylamide gels.

Anti-Sm Immunoprecipitation and 3'-End-Labeling of Small
   Nuclear RNAs

Immunoprecipitation by anti-Sm antiserum was done as previously described.[8] Small nuclear RNAs (snRNAs) were prepared from the immunoprecipitate by proteinase K treatment, ^{32}P-end-labeled at the 3' end by [5'-^{32}P]pCp and T4 RNA ligase,[9] and analyzed on 9% polyacrylamide gels.

Characterization of Miniextracts from HeLa Cell Nuclei

Comparison of Pre-mRNA Splicing in Miniextracts and Conventional Extracts. The pre-mRNA splicing reaction results in identical RNA processing products[7] and follows essentially the same time course in either normal nuclear extracts or miniextracts (Fig. 1). Splicing intermediates, which are the intron–exon 2 RNA species in the lariat configuration (380 RNA) and the first exon RNA species (155 RNA), are detectable after a 30-min incubation. Accurately spliced RNA (370 RNA) as well as excised intron lariat (143 RNA) accumulate during the time course of splicing to similar levels in the two extracts. The accumulation of splicing intermedi-

[6] A. Bindereif and M. R. Green, *EMBO J.* **6,** 2415 (1987).
[7] B. Ruskin, A. R. Krainer, T. Maniatis, and M. R. Green, *Cell* **38,** 317 (1984).
[8] A. Bindereif and M. R. Green, *Mol. Cell. Biol.* **6,** 2582 (1986).
[9] T. E. England, A. G. Bruce, and O. C. Uhlenbeck, this series, Vol. 65, p. 65.

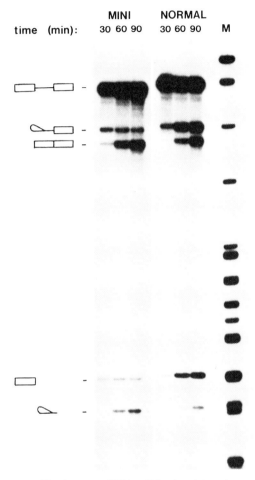

Fig. 1. Time course of *in vitro* pre-mRNA splicing in mini- and normal nuclear extracts. ³²P-Labeled SP64-HβΔ6 pre-mRNA (497 nucleotides) was synthesized and spliced *in vitro*, using nuclear extract prepared by the miniextract (MINI) or by the conventional procedure (NORMAL). At the times indicated, aliquots of the splicing reaction were taken, and the RNA was extracted, analyzed by denaturing gel electrophoresis using ³²P-labeled HpaII-digested pBR322 DNA as size markers (M), and detected by autoradiography. Schematic representations of pre-mRNA, splicing intermediates, and products are shown at left.

ates to different levels (Fig. 1) reflects a slightly different efficiency of these two particular extracts in the second step of pre-mRNA splicing (exon ligation and intron excision). Such variation occurs using conventionally prepared nuclear extracts (data not shown). In summary, we conclude that miniextracts are fully active and indistinguishable from standard nuclear extracts in supporting the pre-mRNA splicing reaction.

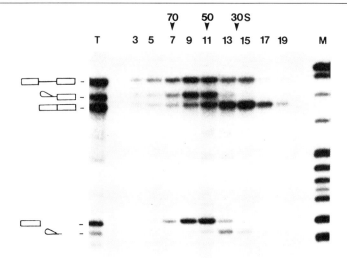

FIG. 2. Sedimentation analysis of pre-mRNA, splicing intermediate, and product ribonucleoprotein complexes. ^{32}P-Labeled SP64-H$\beta\Delta$6 pre-mRNA was synthesized and spliced *in vitro* for 60 min using a HeLa cell miniextract. Ribonucleoprotein complexes containing pre-mRNA, splicing intermediates, and products (as schematically outlined at left) were fractionated by sucrose-gradient sedimentation (sedimentation size markers are indicated above the autoradiogram). ^{32}P-Labeled RNA was prepared from equal aliquots of gradient fractions (numbered 1–20 from the bottom to the top of the gradient, as indicated above the lanes), analyzed by denaturing gel electrophoresis using ^{32}P-labeled *Hpa*II-digested pBR322 DNA as size markers (M), and detected by autoradiography.

Assay of Splicing Complex Formation in Miniextracts. Pre-mRNA splicing requires the assembly of the pre-mRNA into a functional ribonucleoprotein (RNP) complex (~60 S) before the first covalent modification of the pre-mRNA.[8,10,11] After 5′ splice site cleavage and lariat formation, the two splicing intermediates are held together in a 60 S RNP complex.[8,10,11] Accurately spliced RNA (exon 1–exon 2) and the excised intron are released from this complex in the form of smaller RNP complexes of 15–50 S.[8]

Separation of the RNP complexes in a crude splicing reaction by sucrose-gradient sedimentation, at a time point when the pre-mRNA, splicing intermediates, and products could be detected, is shown in Fig. 2. At this time point, most of the unprocessed pre-mRNA is present in 40 and 60 S RNP complexes. The two splicing intermediates (intron–exon 2 lariat

[10] D. Frendewey and W. Keller, *Cell* **42,** 355 (1985).
[11] P. J. Grabowski, S. R. Seiler, and P. A. Sharp, *Cell* **42,** 345 (1985).

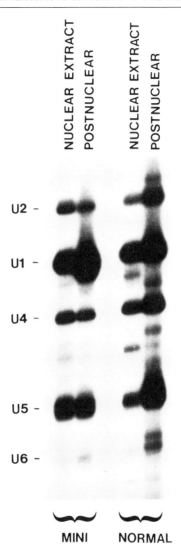

FIG. 3. Distribution of nuclear components in nuclear versus postnuclear fractions. Sm type snRNPs were immunoprecipitated from cell-equivalent amounts of nuclear and postnuclear fractions prepared by either the miniextract procedure (MINI) or the conventional procedure (NORMAL); RNA was prepared from the immunoprecipitate, ^{32}P-labeled, and analyzed by denaturing gel electrophoresis. The positions of immunoselected snRNAs are indicated at left.

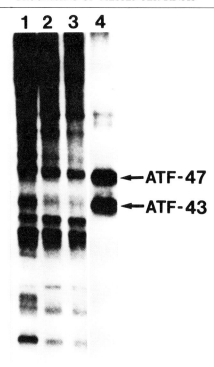

FIG. 4. Purification of ^{32}P-labeled ATF from 293 cells. ATF was purified according to the described protocol from ^{32}P-labeled 293 cells. Lane 1: ^{32}P-Labeled nuclear extract. Lane 2: ^{32}P-Labeled heat-denatured nuclear extract. Lane 3: ^{32}P-Labeled material that flows through the DNA affinity column. Lane 4: ^{32}P-Labeled ATF eluted from the DNA affinity column with 1.5 M KCl. The autoradiogram was exposed overnight, and the material shown in lane 4 was obtained from approximately 5×10^6 cells.

and exon 1) are exclusively in a 60 S RNP complex. Accurately spliced RNA, however, distributes in smaller complexes between 25 S and 40 S, and RNP complexes of the excised intron lariat species sediment heterogenously between 15 S and 50 S. We conclude that the RNP complexes formed in a miniextract and a standard nuclear extract are identical.[8,10,11]

Retention of Splicing Factors in Nuclei during Miniextract Preparations. The main difference between the miniextract and the conventional extract procedure is that cell lysis is achieved by pushing cells through a narrow-gauge hypodermic needle as opposed to using a Dounce homogenizer. It has been shown that during cell lysis by the conventional procedure,[1] "nuclear-localized" components leak out of the nuclei into the postnuclear fraction. We tested the extent to which this happens during preparation of miniextracts, by analyzing the subcellular fractionation of snRNAs that are known to be localized in the nucleus and involved in splicing.[12–16]

We determined the distribution of the Sm-type snRNPs (U1, U2, U4, U5, and U6 snRNPs) between nuclear and postnuclear supernatant fractions (Fig. 3). snRNPs were immunoprecipitated from cell-equivalent amounts of nuclear and postnuclear fractions. snRNAs were prepared from the immunoprecipitate, 3'-end-labeled, and analyzed by denaturing gel electrophoresis. It is apparent that the normal and the miniextract procedures result in very similar levels of the Sm-type snRNAs in the nuclear fractions.

Use of Miniextracts to Detect and Purify Nuclear Phosphoproteins

Factors involved in pre-mRNA processing or transcription will often be present at very low levels. As an initial screen or as an assay for such factors, it is useful to carry out procedures on an analytical scale. We have used radioactively labeled miniextracts to develop an assay for a transcription factor (ATF) (Fig. 4). ATF binds to sequences in many E1a-inducible adenovirus early promoters and cAMP-inducible cellular promoters and activates transcription. ATF can be purified from many established cell lines owing to its sequence-specific DNA-binding activity, using a DNA affinity column.[17]

1. Prepare ^{32}P-labeled nuclear extract as described, without dialysis.
2. Adjust the extract to approximately 100 mM KCl by the addition of 2 volumes of buffer D (without KCl), 2 μg/ml poly(dI)–poly(dC), and 0.01% Nonidet P-40 (NP-40).
3. Heat-denature the extract at 60° for 5 min (ATF is heat stable under these conditions). Precipitate the denatured protein (microcentrifuge, 12,000 g for 1 min).
4. Pass the supernatant over about 0.2 ml of ATF affinity matrix (prepared as described[17]) placed in a disposable column.
5. Wash the column with 15 ml of buffer D containing 100 mM KCl followed by 15 ml of buffer D containing 300 mM KCl.
6. Elute the bound ATF with 0.5 ml of a 1 : 1 mixture of buffer D and

[12] A. Kraemer, W. Keller, B. Appel, and R. Luehrmann, *Cell* **38**, 299 (1984).
[13] A. R. Krainer and T. Maniatis, *Cell* **42**, 725 (1985).
[14] D. L. Black, B. Chabot, and J. A. Steitz, *Cell* **42**, 737 (1985).
[15] D. L. Black and J. A. Steitz, *Cell* **46**, 697 (1986).
[16] S. M. Berget and B. L. Robberson, *Cell* **46**, 691 (1986).
[17] J. T. Kadonaga and R. Tjian, *Proc. Natl. Acad. Sci. U.S.A.* **83**, 5889 (1986).

3 M KCl (final KCl concentration 1 M) containing 100 μg/ml bovine serum albumin (BSA).
7. Concentrate the sample using a Centricon 10 microconcentrator.
8. Analyze by sodium dodecyl sulfate–polyacrylamide gel electrophoresis and autoradiography.

Acknowledgments

K.A.W.L. is supported by a Jane Coffin Childs Fund postdoctoral fellowship. This work was supported by a grant from the National Institutes of Health to M.R.G.

[3] Preparation of Active Nuclei

By WILLIAM F. MARZLUFF

The isolation of nuclei that retain the ability to perform a number of reactions as they occur in the cell provides a potentially useful cell-free system for studying several aspects of RNA metabolism which are not accessible for study in more purified systems. Nuclear preparations deficient in particular activities (e.g., RNA transport) may provide a system for eventual *in vitro* complementation that could result in a better understanding of these complex processes. In addition, for preparation of active extracts for both transcription and processing, it may be necessary to start with highly purified nuclei to remove general inhibitors (e.g., nucleases and proteases). In this chapter I describe methods for purifying nuclei that have been best characterized for their activity in accurate transcription. These nuclear preparations have been useful for the measurement of relative transcription rates as well as other aspects of nuclear RNA metabolism.[1-4] The methods stress speed of isolation with a minimum of handling, and the resulting nuclei are highly purified, unlike most of the preparations used for preparation of extracts for RNA processing or DNA-dependent transcription. However, as a result, some of the components necessary for RNA processing may not be retained in the nuclei [although the small nuclear ribonucleoproteins (snRNPs), for example, are all retained].

The methods consist of lysis of the cells under relatively isosmotic

[1] W. F. Marzluff, E. C. Murphy, and R. C. C. Huang, *Biochemistry* 12, 3440 (1973).
[2] D. L. Cooper and W. F. Marzluff, *J. Biol. Chem.* 253, 8375 (1978).
[3] G. F. Morris and W. F. Marzluff, *Biochemistry* 22, 645 (1983).
[4] A. Brown and W. F. Marzluff, *Biochemistry* 21, 4303 (1982).

conditions in the presence of divalent cations to stabilize the nuclei. The homogenate is then adjusted to a density above 1.17 and the nuclei recovered by a brief centrifugation through a cushion of sucrose of high density. Nuclei are the only organelles that pellet through the sucrose cushion during the short centrifugation, and many membranous organelles are either less dense than the homogenate or else collect at the interface. The sucrose concentration in the cushion is determined empirically depending on the density of the nuclei being prepared; for example, rat liver nuclei are denser than 2.3 M sucrose whereas nuclei from Chinese hamster ovary (CHO) cells are less dense than 2.0 M sucrose and are recovered by centrifugation through 1.9 M sucrose.

The methods we have developed for tissue culture cells, which generally lack large amounts of degradative activities, and sea urchin embryos,[3,5,6] from which crude preparations of nuclei were completely inactive and which contain more degradative activities, are described in this chapter. A similar procedure developed by Schibler and co-workers[7] for rat liver, another tissue with extensive degradative activities, is also discussed. In the latter two cases, the highly purified nuclei were a necessary prerequisite for subsequent preparation of DNA-dependent transcription systems. Even in the case of mouse myeloma cells, the highly purified nuclear preparations are more than 10 times as active in RNA synthesis as nuclei prepared by low-speed centrifugation after homogenization, although HeLa cell nuclei of high activity can be prepared either by low speed centrifugation[8] or by the method described here.

Solutions

Phosphate-buffered saline (PBS): 0.14 M NaCl, 3 mM KCl, 8 mM Na$_2$HPO$_4$, 1.5 mM KH$_2$PO$_4$; adjust to pH 7.2 with HCl

Swelling buffer: 25 mM KCl, 10 mM N-2-hydroxyethylpiperazine-N'-2-ethanesulfonic acid (HEPES) (pH 7.6), 2 mM Mg(CH$_3$-COO$^-$)$_2$, 1 mM dithiothreitol (DTT); the Mg^{2+} may be replaced by 1 mM spermidine and 1 mM ethylene glycol bis(β-aminoethyl ether)-N,N,N',N'-tetraacetic acid (EGTA) may be added if desired

Buffer A: 0.32 M sucrose, 10 mM HEPES (pH 7.6), 2 mM Mg-(CH$_3$COO$^-$)$_2$, 1 mM spermidine, 1 mM DTT, 0.1 mM ethylenediaminetetraacetic acid, trisodium salt, monohydrate (Na$_3$EDTA)

[5] G. F. Morris and W. F. Marzluff, *Mol. Cell. Biol.* **5**, 1143 (1985).
[6] G. F. Morris, D. H. Price, and W. F. Marzluff, *Proc. Natl. Acad. Sci. U.S.A.* **83**, 3674 (1986).
[7] K. Gorski, M. Carneiro, and U. Schibler, *Cell* **47**, 767 (1986).
[8] R. Price and S. Penman, *J. Mol. Biol.* **70**, 435 (1972).

Buffer A plus detergent includes either 0.1% Triton X-100 or 0.1% Nonidet P-40 (NP-40) added from a 10% stock solution just before use

Notes: (1) In the original procedure,[1] 3 mM CaCl$_2$ was used in place of spermidine. (2) The Mg^{2+} may be omitted and EGTA and EDTA added to a final concentration of 1 mM each, if it is desired to exclude divalent ions from the homogenate. (3) Phenylmethylsulfonyl fluoride (PMSF) may be added to a final concentration of 0.1 mM from a 0.1 M stock solution prepared in 2-propanol. (4) In all solutions DTT is added from a 1 M stock solution just before use. (5) Originally[1] 10 mM tris(hydroxymethyl)amino-methane (Tris) (pH 8.0) was used as a buffer in all solutions with identical results.

Sucrose cushion: 2.0 M sucrose, 10 mM HEPES (pH 7.6), 5 mM Mg(CH$_3$COO$^-$)$_2$, 1 mM DTT, 0.1 mM Na$_3$EDTA (pH 8.0)

Notes: (1) The optimal sucrose concentration will vary depending on the source of the nuclei (see text). (2) The Mg^{2+} may be replaced by 1 mM spermidine. (3) EGTA may be added to a concentration of 1 mM if desired.

Nuclear storage buffer: 25% glycerol, 50 mM HEPES (pH 7.6), 5 mM Mg(CH$_3$COO$^-$)$_2$, 1 mM DTT, 0.1 mM Na$_3$EDTA

Preparation of Nuclei from Cultured Mammalian Cells

The following procedure has been used both on cells growing in suspension or on cells growing on plates. The cells are harvested and then lysed by homogenization in a Dounce homogenizer. Low concentrations (0.1%) of a nonionic detergent (Triton X-100 or NP-40) may be used if desired. We have not observed major differences in RNA synthetic activity between nuclei prepared in the presence or absence of detergent.

Isolation of Nuclei Without Detergent. All operations are performed at 4°. The cells are washed with PBS and recovered by centrifugation at 500 g for 5 min. The cells are suspended in 5 volumes of swelling buffer and immediately harvested by centrifugation at 500 g for 5 min. Most tissue culture cells require swelling prior to homogenization if one is not using detergent, because of the relatively low amounts of cytoplasm. In contrast, tissues such as liver or sea urchin embryos are readily lysed by homogenization without prior swelling even in the absence of detergent.

The cells are suspended in 2 volumes of swelling buffer (2–4 × 10^8 cells/ml) and incubated on ice in a Dounce homogenizer for 5 minutes to swell the cells. The cells are then lysed by homogenization (10–20 strokes) using the tight-fitting pestle. Cell lysis is monitored by phase-contrast microscopy. The homogenate is adjusted to 1.3 M sucrose using the solution for the sucrose cushion and the homogenate adjusted to a concentration of 5×10^7 cells/ml. The homogenate is layered over a cushion of appropriate sucrose concentration (determined empirically for each cell type: for example, nuclei from mouse myeloma cells will pellet through 2.0 M sucrose but not 2.1 M sucrose and nuclei from CHO cells will not pellet through 2.0 M sucrose). In a 17-ml tube in an SW27.1 rotor we layer 12 ml of homogenate over a 5-ml cushion; in a 35-ml tube in the SW27 rotor we layer 27 ml of homogenate over an 8-ml cushion; in an SW50 rotor we layer 3 ml over a 2-ml cushion.

The homogenate is centrifuged for 45 min at 18,000 rpm (SW27 rotor) or 20,000 rpm (SW50 rotor). The tube is drained well and any interface material removed by wiping the inside of the tube with a Kimwipe. The nuclei are suspended in storage buffer at a concentration of greater than 2×10^8/ml. The nuclei are most easily suspended using 2 strokes in a 1- to 2-ml Teflon–glass homogenizer. The nuclei are stored at either $-70°$ or in liquid N_2.

Preparation of Nuclei Using Detergent. All operations are performed at 4°. The cells are harvested by centrifugation, the residual medium is drained, and the cells are immediately suspended in buffer A at a concentration of $3–5 \times 10^7$/ml. Cells may be washed once with PBS if desired. Cells are lysed by homogenization with the tight-fitting pestle in a Dounce homogenizer in buffer A containing 0.1% Triton X-100 or NP-40. Cell breakage is monitored by phase-contrast microscopy.

The homogenate is diluted with 2 volumes of the sucrose cushion, giving a final sucrose concentration of 1.3 M, and then layered over the sucrose cushion (which does not contain detergent) and centrifuged as described above. It may be necessary to use different sucrose concentrations in the cushion for nuclei for the same cells prepared with or without detergent (nuclei prepared without detergent are slightly less dense than nuclei prepared with detergent, owing to the presence of the outer nuclear membrane). The nuclei are recovered and stored as described above.

Notes: (1) Using higher concentrations of detergent (e.g., 0.5%) with cultured cells results in nuclei that are much less active in RNA synthesis. (2) It is necessary to have divalent ions present to stabilize the nuclei during the homogenization. The fragility of nuclei varies with the cell type. Mouse myeloma cell nuclei will lyse if homogenized in dilute solution with detergent and with Mg^{2+} as the only divalent ion, whereas HeLa cell nuclei remain intact. Mouse myeloma cell nuclei are stabilized by Ca^{2+} or spermine or spermidine. Although the original procedure used Ca^{2+}, spermine or spermidine (or a mixture of the two) at a final concentration of 1 mM is recommended in place of Ca^{2+}. For tissues with large amounts of Mg^{2+}- or Ca^{2+}-activated nucleases or proteases, spermidine and spermine can be used as the only polyvalent cations and endogenous metal ions can be chelated with EDTA and/or EGTA.

Preparation of Nuclei from Sea Urchin Embryos

This procedure has been adapted to the preparation of large amounts of nuclei from sea urchin embryos,[3] and a similar procedure has been developed by Schibler and co-workers for rat liver. The embryos are cultured in seawater by standard methods and harvested by centrifugation. The embryos are washed twice with 0.55 M KCl. The embryos are suspended in 5 volumes of 0.25 M sucrose–1 mM EDTA and centrifuged at 3000 g for 2 min. This removes residual KCl and helps to dissociate the embryos.

The embryos are suspended in 3–5 volumes of buffer I and homogenized with 10–20 strokes using the tight-fitting pestle (loose pestle for early stage embryos with large nuclei). The homogenate is diluted with 2 volumes of the sucrose cushion (final sucrose concentration >1.3 M), and 25 ml of the homogenate is layered over an 8-ml cushion. The homogenate is centrifuged for 45 min at 18,000 rpm in the SW27 rotor. The nuclei are suspended and stored as described above, at a final DNA concentration of 1–2 mg/ml (10^9 nuclei/ml since sea urchin nuclei contain less DNA than mammalian nuclei).

Note: This procedure was necessary to obtain nuclei with high activity in RNA transcription. The omission of detergent, removal of all divalent metal ions, and inclusion of protease inhibitors are all essential parts of the procedure. The resulting nuclei contain very high molecular weight DNA, lack proteases as judged by the failure to degrade histones, and lack nonspecific RNases and DNases as judged by the stability of both endogenous and exogenous nucleic acids when the nuclei are incubated at 25° for 1 hr. It is very likely that this type of procedure will be generally applicable to many tissues, as the similar procedure described below for rat liver nuclei demonstrates.

Preparation of Nuclei from Rat Liver

This preparation is from Ref. 7. Minced tissue (10–15 g) is adjusted to 30 ml final volume with homogenization buffer [2 M sucrose, 10% (w/v) glycerol, 10 mM HEPES (pH 7.6), 25 mM KCl, 0.15 mM spermine, 0.5 mM spermidine, 1 mM EDTA]. The tissue is homogenized with a motor-driven Teflon–glass homogenizer. Cell breakage is monitored by phase-contrast microscopy. The homogenate is then diluted to 85 ml with homogenization buffer, and 37 ml is layered over a 10-ml cushion of homogenization buffer. The nuclei are recovered by centrifugation at 24,000 rpm for 30 min. The nuclear pellet is resuspended in 50 ml of homogenization buffer, using a Teflon–glass homogenizer, and then centrifuged again through a 10-ml pad of homogenization buffer.

Nuclei prepared in this manner from rat brain, liver, and spleen have been used as a starting material for preparation of active DNA-dependent transcription systems[8] and will probably also be a useful starting material for preparation of extracts containing processing activities.

Properties of Isolated Nuclei. The nuclei isolated using this method from cultured mammalian cells (most of our experience has been with mouse myeloma cells, but similar results have been obtained with CHO cells and mouse L cells) are very active in transcription. Elsewhere we have given detailed procedures for measuring relative transcription rates and characterizing the RNA products.[9] There is efficient initiation of tRNA and 5 S RNA synthesis by RNA polymerase III[10,11] and also initiation of U1 and U2 snRNA synthesis by RNA polymerase II.[12] These nuclei are deficient in a number of RNA processing reactions including tRNA processing,[10] polyadenylation,[2] and methylation.[4] All of these reactions can be restored by the addition of an extract of crude nuclei,[2,4,10] suggesting that some necessary factors were extracted during preparation of the nuclei in dilute hypotonic solutions. These factors are not the snRNPs, which are retained in large amounts in the nuclei. These experiments made it possible to demonstrate the close coupling between transcription and polyadenylation.

These nuclei do carry out one reaction, the synthesis of snRNAs,[12] that thus far has not been successfully reproduced in a DNA-dependent system from mammalian cells. Only a portion of the transcripts synthesized in isolated nuclei form the correct 3′ end, and these transcripts are quantitatively released from the nuclei while the longer transcripts are quantitatively retained in the nuclei. To further characterize the 3′-end-formation of snRNAs it may be possible to complement the purified nuclei with extracts from crude nuclear preparations, providing an assay for characterization of the requirements for 3′-end-formation. In the future, systems based on nuclear preparations may be used to reconstitute processes such as RNA transport, for which there are currently no good model systems.

[9] W. F. Marzluff and R. C. C. Huang, *in* "In Vitro Transcription and Translation—A Practical Approach" (B. D. Hames and S. J. Higgins, eds.), p. 89. IRL Press, Oxford and Washington, D.C., 1984.
[10] W. F. Marzluff, E. C. Murphy, and R. C. C. Huang, *Biochemistry* **13**, 3689 (1974).
[11] M. M. Smith, R. A. Reeve, and R. C. C. Huang, *Cell* **15**, 615 (1978).
[12] S. M. Lobo and W. F. Marzluff, *Mol. Cell. Biol.* **7**, 4290 (1987).

The nuclei prepared from concentrated homogenates of sea urchin embryos retain at least some of the RNA processing activities. It is likely that the relatively high tissue concentrations in the homogenates and during the centrifugation result in less loss of material from the nuclei. These nuclei are much more active in snRNA synthesis than mammalian cell nuclei.[6] Extracts prepared from these nuclei by standard procedures[13] are very active in at least one RNA processing reaction, formation of the 3' end of histone mRNA.[14] It is likely that nuclear preparations of this type will provide excellent starting material for preparing extracts for the study of tissue-specific RNA processing events.

[13] C. Stauber, D. Soldati, B. Lüscher, and D. Schumperli, this volume [7].
[14] C. S. Santiago and W. F. Marzluff, unpublished results.

[4] Nonaqueous Isolation of Transcriptionally Active Nuclei from *Xenopus* Oocytes

By ELSEBET LUND and PHILIP L. PAINE

Introduction

Studies of RNA synthesis and processing *in vitro* have relied largely on nuclei isolated into aqueous media.[1,2] Such purified nuclei, uncomplicated by cytoplasm, simplify the biochemical fractionation and analyses of nuclear components. However, because the nuclear envelope is permeable to small solutes and proteins,[3,4] nuclei isolated into aqueous media inevitably lose contents by leakage to the environment, even if the envelope is not disrupted. Moreover, exposure of nuclei to aqueous buffers can alter intermolecular associations, precipitating some molecules and solubilizing others.[5] Although the extent of each of these problems is difficult to assess, the net loss of nuclear constituents is rapid and massive,[6,7] and it presumably accounts for the inability of highly purified

[1] W. F. Marzluff, this volume [3].
[2] E. H. Birkenmeier, D. D. Brown, and E. Jordan, *Cell* **15**, 1077 (1981).
[3] P. L. Paine and S. B. Horowitz, in "Cell Biology: A Comprehensive Treatise" (D. M. Prescott and L. Goldstein, eds.), Vol. 4, p. 299. Academic Press, New York, 1980.
[4] I. Lang and R. Peters, in "Information and Energy Transduction in Biological Membranes" (H. J. Helmreich, ed.), p. 377. Alan R. Liss, New York, 1984.
[5] J. Nelson, R. Clarke, G. R. Dickson, H. W. van den Berg, and R. F. Murphy, *J. Steroid Biochem.* **25**, 619 (1986).
[6] R. W. Merriam, *J. Biophys. Biochem. Cytol.* **6**, 353 (1959).
[7] P. L. Paine, C. F. Austerberry, L. J. Desjarlais, and S. B. Horowitz, *J. Cell Biol.* **97**, 1240 (1983).

nuclei to carry out many *in vivo* nuclear processes.[1,8] In contrast to aqueously isolated nuclei, *Xenopus* oocyte nuclei isolated under paraffin oil retain their *in vivo* composition of water, small solutes, and proteins (owing to the negligible solubility of these materials in oil) and also retain their *in vivo* structure and many *in vivo* functions.[8,9] We describe herein methods for using the oil-isolated nucleus (OIN) to study RNA transcription, processing, and transport.

Materials

Adult female *Xenopus laevis* frogs are maintained at room temperature (18–20°) on a 12 hr day–12 hr night cycle in dechlorinated water, fed beef liver twice weekly, and kept unovulated for the 6 months prior to use. Folliculated stage V–VI oocytes[10] are manually isolated from ovaries in extracellular medium using microdissecting scissors [Biomedical Research Instruments, Inc. (BRI), Rockville, MD], and individual oocytes are handled by the remaining tabs of follicular tissue with watchmaker's forceps (No. 5, A. Dumont; BRI, Rockville, MD). Nuclear isolation and subsequent procedures under oil are carried out in polystyrene weigh boats (B2045-5, American Scientific Products) containing light paraffin oil (0–12, Fisher Scientific, Chicago, IL) which is previously saturated and washed with intracellular medium. Cross-linked agarose beads are available commercially (Sigma, St. Louis, MO; Pharmacia, Piscataway, NJ) as is small diameter polyethylene tubing (Fisher Scientific, Chicago, IL).

Media

Extracellular: 82.5 mM NaCl, 2.5 mM KCl, 1.0 mM CaCl$_2$, 1.0 mM MgCl$_2$, 1.0 mM Na$_2$HPO$_4$, 3.3 mM NaOH, 5.0 mM N-2-hydroxyethylpiperazine-N'-2-ethanesulfonic acid (HEPES), and 1.0 mM sodium pyruvate (pH 7.6)[11,12]
Intracellular: 10.0 mM NaCl, 125.0 mM KCl, 1.0 mM KH$_2$PO$_4$, 2.0 mM NaHCO$_3$ (pH 7.2)[7]
TE buffer: 10.0 mM tris(hydroxymethyl)aminomethane hydrochloride (Tris-HCl), 0.1 mM ethylenediaminetetraacetic acid (EDTA) (pH 7.6)

[8] E. Lund and J. E. Dahlberg, *EMBO J.* **8**, 287 (1989).
[9] P. L. Paine, D. S. Miller, M. E. Johnson, Y.-T. Lau, L. J. M. Tluczek, and S. B. Horowitz, manuscript in preparation; to our knowledge, this work includes the first demonstration that the oocyte nucleus isolated in oil may be a useful transcriptional system—experiments carried out by M. E. Johnson in the laboratory of S. B. Horowitz, Michigan Cancer Foundation.
[10] J. N. Dumont, *J. Morphol.* **136**, 153 (1972).
[11] R. A. Wallace, D. W. Jared, J. N. Dumont, and M. W. Sega, *J. Exp. Zool.* **184**, 321 (1973).
[12] J. J. Eppig and M. L. Steckmann, *In Vitro* **12**, 173 (1976).

Proteinase K buffer: 50.0 mM Tris-HCl, 10.0 mM EDTA (pH 7.6) containing 1% sodium dodecyl sulfate (SDS) and 0.8 mg/ml proteinase K.[13]

Transcription buffer: 75.0–100.0 mM potassium acetate, 5.0 mM MgSO$_4$, 0.5 mM spermidine, 0.15 mM spermine, 2.0 mM ethylene glycol bis(β-aminoethyl ether)-N,N,N',N'-tetraacetic acid (EGTA), 0.2 mM EDTA, 4.0 mM ATP, 1.5% polyvinylpyrrolidone (molecular weight 360,000), 1.0 mM dithiothreitol (DTT), 1.0 mM phenylmethylsulfonyl fluoride (PMSF), 20.0 mM creatine phosphate, 80 μg/ml creatine phosphokinase, 8–10% glycerol, 40.0 mM HEPES (pH 7.8)[8]

Note. For the preparation of OIN homogenates, spermidine, spermine, and polyvinylpyrrolidone can be omitted from the transcription buffer, and additional protease inhibitors, e.g., 0.7 μg/ml of leupeptin and pepstatin, can be included.

Methods

Isolation of Oocyte Nuclei under Oil

The puncture method for isolating the nucleus under oil (Fig. 1a–c) is derived from the long-utilized technique for manual isolation of *Xenopus* oocyte nuclei (germinal vesicles) into aqueous solutions.[9,14] First, individual oocytes are dissected from the ovary under extracellular medium. Then the cells are carefully blotted on filter paper (Whatman No. 1) to remove extracellular fluid and immersed under paraffin oil. Utilizing a dissecting microscope, the operator steadies the oocyte with forceps (behind the vegetal hemisphere) and makes a puncture 30–45° from the animal pole with a 22-gauge hypodermic needle. The nucleus sometimes emerges through the puncture hole spontaneously, but it is usually extruded by slowly and gently squeezing the cell about the equator with forceps. Care is taken to control the angle and depth of the puncture to (1) avoid damaging the nucleus (~250–350 μm beneath the animal pole) and (2) assure that the puncture hole is small enough to strip cytoplasm from the nucleus as it passes through, but not so small that the nucleus is ruptured. Because the oocyte is blotted prior to puncture, the nucleus passes from the cell into the oil without being exposed to an aqueous phase. After the nucleus is extruded, the cell is gently shaken to free the nucleus from the pedestal of cytoplasm which emerges after it.

[13] J. E. Mertz and J. B. Gurdon, *Proc. Natl. Acad. Sci. U.S.A.* **74**, 1502 (1977).
[14] C. M. Feldherr and P. A. Richmond, *in* "Methods in Cell Biology" (G. Stein, J. Stein, and L. J. Kleinsmith, eds.), Vol. 17, p. 75. Academic Press, New York, 1978.

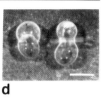

a   b   c   d

FIG. 1. Isolating the *Xenopus* oocyte nucleus under oil. (a) The oocyte is blocked with forceps behind the vegetal hemisphere and punctured near the animal pole. (b) The nucleus extrudes and (c) is shaken free from the pedestal of cytoplasm. (d) OIN–agarose bead heteropairs, with and without a polyethylene "collar." Bars, 500 μm.

With some experience, intact nuclei can be isolated from over 90% of the cells attempted, at a rate of 2–3 per minute. Mechanical damage to the nucleus during isolation inevitably results in a marked change termed "nuclear collapse,"[15] wherein the nuclear cortex collapses within a few minutes toward the center, leaving a peripheral, optically clear substance surrounding the condensed, opaque materials in the central 10% of the volume.

An alternate "tear-isolation" method can be employed when some cytoplasmic contamination of the nucleus is tolerable and the chance of damaging the nucleus must be minimized (e.g., when microinjection or other treatment protocols have been previously invested in the cell). In the tear method, the oocyte is grasped on both sides of the animal pole with two pairs of forceps, and a tear is made about 60–90° in arc length. The nucleus, covered by a layer of cytoplasm, protrudes through the tear and can be separated from the remainder of the cell using the tips of the forceps or a similar microtool. Cytoplasm is not stripped from the nucleus by passage through a size-restricted hole as in the puncture method. However, if the nucleus is not immediately separated, but is left connected to the cell through the cytoplasmic isthmus, much of the cytoplasm migrates from the nuclear surface back to the cell within a few minutes.

Notes. (1) While nuclei can be isolated from folliculated or defolliculated oocytes, remaining follicular tissue permits handling of individual cells with forceps. (2) Glass surfaces are avoided, because the isolated nucleus adheres to glass and can seldom be freed without damage. Thus, nuclear isolation and subsequent procedures are carried out in plastic weigh boats. Once isolated into oil, the nucleus can be moved by waving the puncture needle or any other fine-tipped tool near it; actually touching the nucleus with the tool risks physical damage. (3) While the success rate

[15] S. B. Horowitz, L. J. M. Tluczek, and P. L. Paine, *J. Cell Biol.* **103**, 46a (1986).

of puncture isolation is a function of operator skill, we have also noted it to be variable from frog to frog, presumably owing to nutritional, hormonal, and/or other physiological differences among animals.

Characteristics of Oil-Isolated Nuclei

As viewed under the dissecting microscope, the OIN appears optically transparent, with cortical, refractile nucleoli giving it an appearance somewhat reminiscent of a translucent golf ball. If left intact, the OIN can remain unchanged in appearance for 24 hr or more after isolation. Both light and electron microscopic examinations of the OIN reveal typical *in vivo* morphology, and careful measurements of composition demonstrate that its *in vivo* concentrations of water, small solutes, and proteins are completely retained.[9] The only compositional change detected over 24 hr is that of ATP, whose concentration decreases from an initial value of approximately 4.5 to about 1 mM[9] because of continuing metabolic consumption. Permeability studies (using OIN–cytoplasm and OIN–bead heteropairs; see following section) showed that the intact OIN retains a size-selective surface permeability barrier: the nuclear entry of 20-kDa but not 3-kDa dextran is rate-limited by the nuclear envelope.[9] This is consistent with previous *in vivo* measurements of passive diffusion through the nuclear pores.[16] The OIN also retains the ability to accumulate specifically nuclear proteins that contain a nuclear localization sequence.[17]

The OIN is transcriptionally very active and can continue to synthesize RNAs for up to 20 hr after isolation.[9] All three RNA polymerases (I, II, and III) remain active, as evidenced by transcription of endogenous ribosomal RNA genes by RNA polymerases I and III[8,9] and by expression of exogenous RNA polymerase II and III genes preinjected into the oocyte.[8] Significantly, the RNA polymerase II activity in the OIN includes highly efficient and accurate initiation and termination of small nuclear RNA (snRNA) transcription, which is very inefficient (and frequently nonexistent) in aqueously isolated nuclei.[1,8] Furthermore, unfractionated OIN homogenates and extracts retain the capacity for faithful snRNA synthesis and have provided the first DNA-dependent *in vitro* transcription system for vertebrate snRNAs.[8,18] The OIN also is capable of carrying out coupled transcription and processing, including correct 5′ and 3′ end trimming and splicing, of *Xenopus* tRNA[Tyr] with kinetics similar to those observed *in vivo* in the intact oocyte.[18]

[16] P. L. Paine, L. C. Moore, and S. B. Horowitz, *Nature* (*London*) **254,** 109 (1975).
[17] P. L. Paine and C. M. Feldherr, unpublished results.
[18] E. Lund, unpublished results.

Experimental Manipulation of Oil-Isolated Nuclei

In this section, we describe the use of the OIN and micromethods that are applicable both to studies of RNA transcription and processing, which cannot be assayed in conventional preparations of aqueously isolated nuclei, and to measurements of RNA export from the nucleus.

Labeling and Analysis of RNA Synthesized by Oil-Isolated Nuclei. To label individual (or pooled) OINs, a 5- to 8-nl microdroplet of intracellular medium or TE buffer containing 0.25–0.50 μCi of [³H]- (30–45 Ci/mmol) or [α-³²P]- (400–800 Ci/mol) GTP or UTP is fused to an OIN under oil. This method of precursor addition increases total nuclear GTP or UTP pools by about 130 and 50%, respectively.[9,19] Incubation for *in vitro* RNA synthesis by the OIN is at room temperature (18–22°) for 30 min or longer, as required.[8,9]

To measure the incorporation of label, single or pooled nuclei are homogenized in ice-cold 10% trichloroacetic acid (TCA) (in the presence of carrier RNA when only one or a few nuclei are assayed), and the precipitate is collected, washed, and counted on GF/C glassfibre filters (Whatman) according to standard procedures. Alternatively, incorporation by a single OIN is measured by moving the nucleus into a 100-μl volume of ice-cold 10% TCA. The white nuclear precipitate remains intact throughout subsequent manipulations. To remove unincorporated radiolabel, the nucleus is transferred with a fine copper wire loop through several 500-μl volumes of 10% TCA. The washed precipitate is then solubilized in 0.1 M NaOH, neutralized with 0.1 M HCl, and counted by standard liquid scintillation methods. This method is useful if wash solutions are to be counted in order to monitor elution kinetics or to determine the precise amount of labeled precursor added to each OIN.

To analyze the nature of the RNA products, transcription is terminated by transfer of the OINs (or OIN homogenate, see below) into 100 μl of proteinase K buffer (for up to 20 OINs) containing 30 μg of carrier RNA. After digestion for 1–2 hours at 37°, total RNAs are isolated by phenol–chloroform–isoamyl alcohol (24 : 24 : 1) extraction and ethanol precipitation and analyzed by polyacrylamide gel electrophoresis.

Note. Isolation and incubation of the OIN at higher temperatures (>25–27°) inactivates snRNA synthesis.[8]

Transcription of Exogenous DNA by Oil-Isolated Nuclei Homogenates. To monitor transcription in homogenates of nuclei (which were preinjected with template DNA within the intact oocytes), groups of 10 OINs are transferred under oil into 5-μl droplets of transcription buffer

¹⁹ H. R. Woodland and R. Q. W. Pestle, *Biochem. J.* **127,** 597 (1972).

supplemented with 0.5 mM each of ATP, CTP, and UTP, 20 μM GTP, and 5 μCi of [α-^{32}P]GTP. Homogenization is accomplished by repeatedly pipetting through a fine tip (pulled from a glass capillary), and incubation is for 60–120 min at 18–22° under oil.[8] Alternatively, pools of 10 OINs (isolated from uninjected oocytes) are homogenized as above in 5 μl of transcription buffer containing circular template DNAs at 0.5–50.0 ng DNA per OIN equivalent (dependent on the type of gene assayed). After preincubation of the homogenate for approximately 30 min at room temperature (18–22°), unlabeled ribonucleotide triphosphates and [α-^{32}P]GTP are added to the same final concentrations as above by fusion with a 0.5- to 1.0-μl droplet of transcription buffer, and incubation is continued under oil for an additional 60–90 min.[8]

Transcription in Oil-Isolated Nuclei Extracts. Crude extracts active for RNA polymerase II and III transcription of snRNA, tRNA, and 5 S RNA genes can be prepared by homogenizing OINs in transcription buffer followed by a brief centrifugation to remove nucleoli membranes and any contaminating cytoplasmic yolk proteins.[18] For this preparation, 150–200 OINs are collected under oil and then transferred (still in oil) to a 0.5-ml Eppendorf tube (on ice) using a siliconized 20-μl capillary pipet. Excess paraffin oil is removed, and the OINs are homogenized in transcription buffer (1 μl per 2 OINs) by pipetting with a yellow Eppendorf tip. After centrifugation for 10–15 sec at 10,000–12,000 g in a microcentrifuge, the clarified supernatant is used directly for transcription (see above) or aliquoted for storage at $-70°$.

Transport Studies with Heteropairs. Because the oocyte nucleus is 90% water and interacts weakly with paraffin oil, the OIN and a droplet of oocyte cytoplasm are readily recombined under oil, with such "heteropairs" making contact along an interface devoid of paraffin oil.[9] Similarly, the OIN can be paired with a cross-linked agarose bead (Fig. 1d) which has been preequilibrated in intracellular medium, transferred to the plastic weigh boat containing the nucleus under oil, blotted with a wick of filter paper to remove external fluid, and then abutted to the OIN.[9,20] To limit the heteropair contact region to a defined area, an annulus or "collar" shaved from the end of polyethylene tubing (i.d. 0.011 inches, o.d. 0.024 inches) is placed between nucleus and cytoplasm (or bead). In the case of OIN–cytoplasm heteropairs, the collar is fitted with a diaphragm of fine platinum grid (200 mesh, Electron Microscopy Sciences) which counteracts the tendency of the cytoplasm to move slowly through the collar and spread over the surface of the nucleus.

[20] P. L. Paine, *in* "Molecular Mechanisms in the Regulation of Cell Behavior" (C. Waymouth, ed.), p. 169. Alan R. Liss, New York, 1987.

Because materials diffuse readily between the nucleus and a paired bead or cytoplasm, heteropairs have been employed to measure equilibrium and transport properties of the nucleus.[9] Likewise, because the OIN retains *in vivo* structure and surface permeability, the transport of RNAs from the nucleus to the cytoplasm (or bead) may be measured to investigate the *in vivo* process of RNA exit from the nucleus. To measure RNA transport, the heteropair can be manually separated, either under oil or by freezing and separation by cryomicrodissection,[7,21,22] and each compartment assayed for RNA content as a function of time.

Summary and Prospectus

The OIN constitutes a unique and valuable experimental system for the study of RNA metabolism. The isolation of the oocyte nucleus under oil, its titration with aqueous microdroplets, and the construction of OIN–cytoplasm and OIN–bead heteropairs provide the means to (1) study isolated nuclei that retain their *in vivo* contents and are unaltered with respect to *in vivo* structure and (2) work with nuclei from individual cells as well as pooled samples. The combined methods thus closely approximate the capabilities of working with intact living cells and additionally facilitate experimental control of physiological variables and analyses of the separate nuclear and cytoplasmic compartments. Appropriate supplementation of the internal milieu of the OIN with metabolites and energy sources via manipulation of the composition of fused microdroplets or conjoined beads should further optimize the system for analyses of RNA synthesis, processing, and transport.

[21] P. L. Paine, *J. Cell Biol.* **19**, 188s (1984).
[22] S. B. Horowitz and T. W. Pearson, *Mol. Cell. Biol.* **1**, 769 (1981).

[5] Transient Transfection of HeLa Cells for Analysis of Splicing Reaction

By Markus Aebi

Introduction

Originally, the sequence requirements for the specificity in splicing reactions was studied by the analysis of naturally occurring mutations that affect splicing. Most of such mutations have been found in α- or β-

thalessemia genes.[1] The finding that higher eukaryotic cell lines can be either transiently or stably transformed using plasmid DNA made it possible to alter the sequence of a gene, to introduce it back in cells, and to analyze the effect of the mutations introduced on the efficiency and accuracy of the splicing reaction.[2] The development of *in vitro* splicing systems[3–5] has made it possible to study the splicing pathway in greater detail and has led to the formulation of a two-step reaction mechanism.

The *in vitro* splicing systems allow analysis of the reaction intermediates and products, and, therefore, the effects of various mutations in messenger RNA precursors (pre-mRNAs) on the different steps of the splicing reaction can be tested. *In vitro* splicing systems offer several advantages over *in vivo* methods: the *in vitro* splicing reaction intermediates and products can be analyzed directly, since radioactively labeled substrate is used. In addition, analysis of the products is facilitated by the large amounts of substrate used for *in vitro* splicing reactions. This is in contrast to analysis *in vivo,* where the products of the splicing reaction of a specific gene are only a small fraction of the total mRNA produced in the cell, requiring the use of sensitive but indirect methods of analysis, for example, S1 mapping or primer extension. *In vivo* the splicing reaction is only one step in the production of mRNA. Transcription, mRNA transport, translation, or mRNA stability may influence the appearance of a specific splicing product and may mask the effects of a mutation in the splicing reaction,[6] whereas *in vitro* the focus is specifically on the splicing reaction. In addition, it is known that the intermediates of the splicing reaction as well as the free lariat intron are only rarely detected *in vivo.*[7,8] Nevertheless, it seems important to analyze the effect of substrate alterations not only *in vitro* but also *in vivo,* since differences in the splicing pathway may lead to the identification of principles in the splicing pathway that are dependent on the *in vivo* situation and are not detected *in vitro*. In addition, some minor splicing events, for example, the use of

[1] R. Treisman, S. H. Orkin, and T. Maniatis, *in* "Globin Gene Expression and Hematopoietic Differentiation" (G. Stamatoyannopoulos and A. W. Nienhuis, eds.), p. 99. Alan R. Liss, New York, 1983.

[2] B. Wieringa, F. Meyer, J. Reiser, and C. Weissmann, *Nature (London)* **301,** 38 (1983).

[3] N. Hernandez and W. Keller, *Cell* **35,** 89 (1983).

[4] R. A. Padgett, S. F. Hardy, and P. A. Sharp, *Proc. Natl. Acad. Sci. U.S.A.* **80,** 5230 (1984).

[5] A. R. Krainer, T. Maniatis, B. Ruskin, and M. R. Green, *Cell* **36,** 993 (1984).

[6] C. W. Pikielny and M. Rosbash, *Cell* **41,** 119 (1985).

[7] S. Zeitlin and A. Efstratiadis, *Cell* **39,** 589 (1985).

[8] M. Aebi, H. Hornig, R. A. Padgett, J. Reiser, and C. Weissmann, *Cell* **47,** 555 (1986).

cryptic splice sites, can be studied *in vivo,* since these splicing events are difficult to detect *in vitro.*[8]

In this chapter, I describe the methods I have used to analyze the effect of point mutations in the rabbit β-globin large intron *in vivo,* namely, the transient transfection of HeLa cells and the isolation of the RNA from these cells. The different methods for the analysis of the spliced RNA are described in great detail elsewhere in this volume.

Materials

Plasmid DNA. For the study of mutations in the large intron of the rabbit β-globin gene, the construct shown in Fig. 1 is used. This vector, designed by Wieringa *et al.,*[2] contains the entire transcription unit of the rabbit as well as the mouse β-globin genes. The transcription of the two genes is enhanced by the SV40 enhancer. Mutations are introduced in the large intron of the rabbit β-globin gene, and the resulting constructs are used for the transient transformation of HeLa cells. The transcripts from the mouse β-globin genes are used as an internal standard to measure the effect of mutations in the rabbit β-globin gene on the expression of the test gene.

Cells. HeLa cells are grown in Dulbecco's minimum essential medium (DMEM) containing 10% fetal calf serum in 9-cm tissue culture dishes. A confluent plate is divided 1 : 3 at 24 hr prior to transfection in order to get a 50–70% confluent plate for the transfection procedure. Four hours before transfection, the cells are fed with fresh medium.

Methods

Transient Transformation of HeLa Cells. The method described is based on the one developed by Wigler *et al.*[9] Five micrograms of plasmid DNA and 20 μg carrier DNA (calf thymus DNA) are coprecipitated and dissolved in 440 μl of distilled water. Sixty microliters of 2 M CaCl$_2$ is added, and the mixture is kept on ice for 5 min. The DNA solution is added dropwise with intermittent mixing to 500 μl of 2× HBS [280 mM NaCl, 50 mM N-2-hydroxyethylpiperazine-N'-2-ethanesulfonic acid (HEPES), 1.5 mM Na$_2$HPO$_4$ (pH 7.1)]. The solution is kept on ice for 10 min and then added to a 9-cm tissue culture dish containing a 50–70% confluent monolayer of HeLa cells in 10 ml of DMEM. The cultures are

[9] M. Wigler, A. Pellicer, S. Silverstein, R. Axel, G. Urlaub, and L. Chasin, *Proc. Natl. Acad. Sci. U.S.A.* **76,** 1373 (1979).

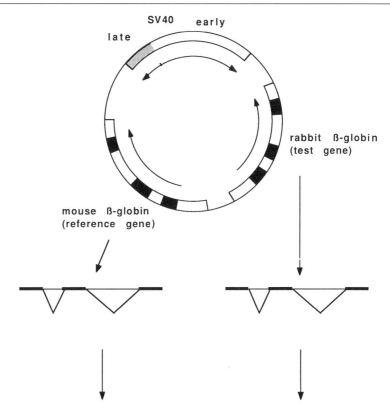

S1 mapping using mouse and rabbit ß-globin-specific probes

primer extension analysis

cDNA cloning

FIG. 1. General map of the vector used to study the effect of mutations in the rabbit β-globin large intron *in vivo*. The scheme shows the two globin transcripts derived from the test (rabbit β-globin) and the reference gene (mouse β-globin). They can be analyzed by the different methods described in the text.

checked microscopically for the presence of a fine calcium phosphate precipitate, which should stick to the surface of the cells. After incubating for 4 hr at 37°, the medium is removed from the cells; the cells are washed with 5 ml of 1× TBS buffer [140 mM NaCl, 5 mM KCl, 0.5 mM Na$_2$HPO$_4$, 25 mM tris(hydroxymethyl)aminomethane hydrochloride (Tris-HCl) (pH 7.5)], shocked with 5 ml of 1× TBS–20% dimethyl sulfoxide (DMSO) for 3 min at room temperature, washed twice with 5 ml of 1× TBS buffer, and fed with 10 ml of DMEM containing 10% fetal calf serum.

Twenty-four hours posttransfection, the medium in the dishes is replaced, and the RNA can be isolated 48 hr after transfection.

RNA Isolation. The LiCl procedure described below is based on the method by Auffray and Rougeon.[10] The cells are washed once with 10 mM Tris-HCl (pH 7.5), 140 mM NaCl and then lysed by the addition of 5 ml of lysis buffer [6 M urea, 3 M LiCl, 50 mM sodium acetate (pH 5), 200 μg/ml heparin, 0.2% sodium dodecyl sulfate (SDS, added to the buffer just prior to the lysis)]. After leaving the solution on the cells for 10 min, the lysate is scraped from the plate using a rubber policeman. In order to shear the DNA, the lysate is blended on ice for 1 min in a Sorvall omnimixer (level 5, small beaker). The solution is then transferred to a 15-ml Corex tube and left overnight at 4°. After centrifugation (16,000 g, 15 min, 4°) the pellet is washed in 5 ml of 8 M urea–4 M LiCl by strong vortexing. After centrifugation as above, the pellet is dissolved in 3 ml of 200 mM sodium acetate (pH 5), 0.2% SDS, 1 mM ethylenediaminetetraacetic acid (EDTA), extracted once with an equal volume of phenol, and extracted once with phenol–chloroform (1 : 1), and the RNA is precipitated with ethanol. One confluent 9-cm dish of HeLa cells usually yields 200 μg of RNA. RNA isolated by this procedure can be used without further purification for S1 mapping or primer extension. Twenty micrograms of total RNA is routinely used for S1 mapping or primer extension analysis, with ^{32}P-end-labeled probes or primers.

Analysis of Splicing Products. In order to analyze the effect of mutations on the splicing pathway of the rabbit β-globin gene, several indirect methods can be used.[11] I would like to mention that four main methods can be applied in the analysis of splicing products *in vivo*. Northern blot analysis can be used to test for the presence or absence of sequences in the final mRNA; however, this method does not allow for the determination of splice sites used. S1 analysis[12] and primer extension[13] allow for the identification of normal and cryptic splice sites; however, these methods are usually not accurate enough to determine the exact cleavage sites. Therefore, the isolation and analysis of cDNA sequences corresponding to specific splicing products is necessary. Several methods exist for the isolation of specific cDNA sequences,[14] but the use of specific primers for the synthesis and isolation of corresponding cDNA clones[15] seems to be the method of choice.

[10] C. Auffray and F. Rougeon, *Eur. J. Biochem.* **107,** 303 (1980).
[11] B. Wieringa, E. Hofer, and C. Weissmann, *Cell* **37,** 915 (1984).
[12] A. J. Berk, this series, Vol. 180, p. 334.
[13] W. R. Boorstein and E. A. Craig, this series, Vol. 180, p. 347.
[14] See this series, Vol. 154.
[15] A. Schmid, R. Cattaneo, and M. A. Billeter, *Nucleic Acids Res.* **15,** 3987 (1987).

In vivo, splicing intermediates and the lariat intron product can rarely been detected. In some cases, for example, mutations in the 5' splice regions of the rabbit β-globin large intron[8] and the yeast actin intron,[16] lariat intermediates can be detected *in vivo.* In a primer extension analysis, the branched substrate causes a stop of the reverse transcriptase at the branch point.

Conclusion

The development of *in vitro* splicing systems has led to the formulation of a splicing pathway and has allowed for detailed analysis of the functions of different regions on the pre-mRNA. This is mainly due to the detection of intermediates from the *in vitro* splicing reaction. Moreover, *in vivo,* only the final splicing products can be analyzed. The level of these products is influenced not only by the splicing reaction but also by several other factors as, for example, mRNA stability, mRNA transport from the nucleus to the cytoplasm, or transcription efficiency. Therefore, the *in vitro* splicing reaction may be superior to *in vivo* analysis with regard to definition of the basic mechanism of the splicing reaction. However, some aspects of the splicing reaction, such as differential splicing, the ordered intron excision of a multiintron pre-mRNA, or the effect of transcription on the splicing reaction, can so far be studied only *in vivo.*

The method presented above, namely, the transient transformation of HeLa cells using the described vector, is a basic application of the *in vivo* analysis of the splicing reaction. Several other approaches have been used.[17–20] The selection of the cell type or the vector may depend greatly on the problem to be analyzed. It has been reported that, in the case of the alternative splicing of the SV40 early region, the ratio of the two splicing products may be influenced by the cell type used.[18] In addition, one has to make the choice between stable and transient transformation or even the use of transgenic animals[21] to study the regulation of tissue-specific, differential splicing.

Acknowledgments

I would like to thank Usha Vijayraghavan for critical reading of the manuscript. This work was supported by the Swiss National Foundation.

[16] A. J. Newman, R.-J. Lin, S.-C. Cheng, and J. Abelson, *Cell* **42,** 335 (1985).
[17] Y. Zhuang and A. M. Weiner, *Cell* **46,** 827 (1986).
[18] X.-Y. Fu and J. L. Manley, *Mol. Cell. Biol.* **7,** 738 (1987).
[19] D. Solnick, *Cell* **43,** 667 (1985).
[20] L. P. Eperon, I. R. Graham, A. D. Griffiths, and I. C. Eperon, *Cell* **54,** 393 (1988).
[21] E. B. Crenshaw III, A. F. Russo, L. W. Swanson, and M. G. Rosenfeld, *Cell* **49,** 389 (1987).

[6] Preparation of Mammalian Extracts Active in Polyadenylation

By CLAIRE L. MOORE

Introduction

The formation of most eukaryotic mRNA 3′ ends, referred to as polyadenylation, requires cleavage of a precursor followed by the sequential addition of adenylate residues.[1] Kinetic labeling studies showed that precursor RNA from the adenovirus major late transcriptional unit extended beyond the final mRNA 3′ end,[2] and they provided the initial evidence for this type of processing mechanism. Experiments in isolated nuclei demonstrated that RNA polymerase II transcription of cellular genes also continued many nucleotides beyond the poly(A) addition site.[1] Thus, polyadenylation was not coupled to transcription termination, and the extra 3′ sequence had to be removed. Because polyadenylation *in vivo* is so rapid,[3] it was not possible to determine the steps of the processing pathway in intact cells. However, this processing reaction could be demonstrated in isolated nuclei[4] and on adenovirus transcriptional complexes.[5]

It was not until the development of *in vitro* systems using cell extracts that further analysis of the pathway was possible. The original systems used the whole cell extracts developed for *in vitro* transcription.[6] In these extracts, poly(A) was synthesized onto the ends of exogenous precursor RNA.[7,8] A low level of cleavage at the adenovirus type 2 L3 poly(A) site could be detected only if the RNA was transcribed in the extracts from a viral promoter.[9] The system was improved by using HeLa nuclear extract and highly radioactive precursor synthesized with bacteriophage RNA

[1] J. L. Manley, *Biochem. Biophys. Acta* **950**, 1 (1988).
[2] J. E. Darnell, *Prog. Nucleic Acids Res.* **22**, 327 (1979).
[3] J. R. Nevins and J. E. Darnell, *Cell* **15**, 1477 (1978).
[4] J. L. Manley, P. A. Sharp, and M. L. Gefter, *J. Mol. Biol.* **159**, 581 (1982).
[5] S. Chen-Kiang, D. J. Wolgemuth, M.-T. Hsu, and J. E. Darnell, *Cell* **28**, 575 (1982).
[6] J. L. Manley, A. Fire, A. Cano, P. A. Sharp, and M. L. Gefter, *Proc. Natl. Acad. Sci. U.S.A.* **77**, 3855 (1980).
[7] J. L. Manley, *Cell* **3**, 595 (1983).
[8] J. L. Manley, H. Yu, and L. C. Ryner, *Mol. Cell. Biol.* **5**, 373 (1985).
[9] C. L. Moore and P. A. Sharp, *Cell* **36**, 581 (1984).

polymerases.[10] When incubated with ATP and nuclear extract, this precursor RNA was accurately and efficiently cleaved and polyadenylated. The sensitivity of the *in vitro* system allowed detection of a product containing only sequence downstream of the poly(A) site.[10-14] This indicated that the poly(A) site was formed by endonucleolytic cleavage rather than exonucleolytic degradation from the 3' end. Like the *in vivo* reaction, polyadenylation in this cell-free system requires the hexanucleotide sequence AAUAAA as well as downstream elements,[1,8,12,14-25] and is thus an authentic model for RNA processing at the poly(A) site. The following sections describe the preparation of nuclear extract active for polyadenylation, the synthesis of RNA precursor, reaction conditions, assays for different steps of the reaction, and fractionation of extracts into active components.

Preparation of Extracts

Nuclear extract is prepared from HeLa cells by a modification of the method of Dignam *et al.*,[26] developed for *in vitro* transcription. It is very similar to the protocol used to prepare mammalian splicing extracts.[28] Typically, 8–10 liters of HeLa cells are grown in suspension in Joklik's or Dulbecco's minimal essential media (MEM) plus 5% horse serum to a density of 3–4 × 10^5 cells/ml. The time to reach this density can be estimated by using an approximate doubling time for HeLa cells of 18–24 hr. For the most active extracts, it is important that the cells are not

[10] C. L. Moore and P. A. Sharp, *Cell* **41**, 845 (1985).
[11] C. L. Moore, H. Skolnik-David, and P. A. Sharp, *EMBO J.* **5**, 1929 (1986).
[12] D. Zarkower, P. Stephenson, M. D. Sheets, and M. P. Wickens, *Mol. Cell. Biol.* **6**, 2317 (1986).
[13] M. D. Sheets, P. Stephenson, and M. P. Wickens, *Mol. Cell. Biol.* **7**, 1518 (1987).
[14] F. Zhang and C. N. Cole, *Mol. Cell. Biol.* **7**, 3277 (1987).
[15] R. P. Hart, M. A. McDevitt, and J. R. Nevins, *Cell* **43**, 677 (1985).
[16] A. O. Sperry and S. M. Berget, *Mol. Cell. Biol.* **6**, 4734 (1986).
[17] T. Humphrey, G. Christofori, V. Kucijanic, and W. Keller, *EMBO J.* **6**, 4159 (1987).
[18] L. C. Ryner and J. L. Manley, *Mol. Cell. Biol.* **7**, 495 (1987).
[19] H. Skolnik-David, C. L. Moore, and P. A. Sharp, *Genes Dev.* **1**, 672 (1987).
[20] T. L. Green and R. P. Hart, *Mol. Cell. Biol.* **8**, 1839 (1988).
[21] J. McLauchlan, C. L. Moore, and J. Clements, *Nucleic Acids Res.* **16**, 5323 (1988).
[22] J. Stefano and D. E. Adams, *Mol. Cell. Biol.* **8**, 2052 (1988).
[23] Y. Takagaki, L. C. Ryner, and J. L. Manley, *Cell* **52**, 731 (1988).
[24] J. Wilusz and T. Shenk, *Cell* **53**, 221 (1988).
[25] D. Zarkower and M. P. Wickens, *J. Biol. Chem.* **263**, 5780 (1988).
[26] J. D. Dignam, R. M. Lebovitz, and R. G. Roeder, *Nucleic Acids Res.* **11**, 1475 (1983).
[27] G. Christofori and W. Keller, *Mol. Cell. Biol.* **9**, 193 (1989).
[28] A. Krämer and W. Keller, this volume [1].

overgrown. Cells are harvested by centrifugation in 1-liter bottles for 10 min at 1500 rpm. They are then transferred to 50-ml tubes and washed once in cold phosphate-buffered saline [137 mM NaCl, 2.7 mM KCl, 4.3 mM Na$_2$HPO$_4 \cdot$ 7H$_2$O, and 1.4 mM KH$_2$PO$_4$ (pH 7.3)]. Cells are resuspended in 5 packed cell volumes (PCV) of hypotonic buffer A [10 mM N-2-hydroxyethylpiperazine-N'-2-ethanesulfonic acid (HEPES) (pH 7.6), 1.5 mM MgCl$_2$, 10 mM KCl, and 0.5 mM dithiothreitol]. The buffers are chilled to 4° before using. The dithiothreitol (DTT) should be added just before use. The cells are swelled in buffer A for 10 min at 0°, pelleted, resuspended in 2 PCV of buffer A, and homogenized on ice in 10 strokes with a tight-fitting, B-type pestle. The apparatus is baked at 177° overnight to inactivate ribonucleases and then chilled before use. Lysis of the cells can be checked by examination under the light microscope. HeLa nuclei are much smaller than the unlysed cells, and by comparing aliquots taken before and after douncing, it is easy to tell if the cells have broken open.

The nuclei are pelleted by centrifugation at 2000 rpm in a swinging-bucket rotor for 10 min at 4°. Most of the cloudy supernatant is carefully removed (the nuclei pellet will not be tightly packed). The pellet is then subjected to centrifugation for 20 min at 16,000 rpm in a Sorvall SS34 rotor. The nuclei are resuspended by homogenizing for 10 strokes in buffer C [20 mM HEPES (pH 7.6), 25% glycerol, 420 mM NaCl, 1.5 mM MgCl$_2$, 0.2 mM ethylenediaminetetraacetic acid (EDTA), and 0.5 mM DTT] at a ratio of 3 ml buffer C per 10^9 cells. The pellet at this stage is a slimy mass, and homogenizing will be difficult. The resuspended nuclei are stirred gently for 30 min on ice. Alternatively, the nuclei can be transferred to a screw-cap tube and rotated or gently rocked for 30 min at 4°. Using the same tube, the nuclear debris is pelleted by centrifugation at 16,000 rpm for 30 min at 4°. The supernatant is then dialyzed at 4° against 1 liter of buffer D [20 mM HEPES (pH 7.6), 20% glycerol, 100 mM KCl, 0.2 mM EDTA, 1.5 mM MgCl$_2$, 0.5 mM DTT] for 4–6 hr with 1 change of dialysis buffer over this time. The magnesium can be omitted from buffers A and D without affecting extract activity[13,16,18,29] and added later for the polyadenylation reaction. Active extracts from HeLa cells have also been made by extracting nuclei with 250 mM NaCl.[29] Finally, the KCl in buffer D can be replaced with 42 mM (NH$_4$)$_2$SO$_4$[18] or 230 mM NaCl.

Each liter of HeLa cells at 3–4 × 10^5 cells/ml should yield approximately 1 ml of nuclear extract containing about 20 mg/ml of protein. The extract is divided into 50- to 200-μl aliquots in prechilled microcentrifuge tubes, quick-frozen in liquid nitrogen, and stored at −70°. To avoid

---

[29] G. M. Gilmartin, M. A. McDevitt, and J. R. Nevins, *Genes Dev.* **2**, 578 (1988).

changes in temperature when the freezer is opened, it is preferable to keep the extract in a chest-type freezer rather than an upright one. The extract should retain activity for over 1 year. Aliquots will lose some activity if they are thawed and refrozen. Modifications which might be useful to obtain better extracts include addition of protease inhibitors such as 1 mM phenylmethylsulfonyl fluoride (PMSF), 0.6 μM leupeptin, and 2 μM pepstatin A to the buffers, dialysis of extracts on ice instead of at 4°, and inclusion of 18% Ficoll during nuclei isolation to minimize leakage.

Extracts active for polyadenylation have also been made from lymphoid cells grown in tissue culture or harvested from the spleens of inoculated mice.[30] For these cells, extraction of the nuclei with 300 mM NaCl and one-third as much buffer C per cell gave better activity. Very recently, cleavage and polyadenylation activity has been detected in yeast whole cell extract.[31] This extract was prepared using a protocol that gives extracts active for *in vitro* splicing.[32]

Currently, the processing of precursor in the yeast extract is not as efficient as it is in extracts obtained from mammalian cells, and different in that significant amounts of cleaved, unpolyadenylated product accumulates. Also, precursor containing mammalian polyadenylation signals is not processed in yeast extracts.

Preparation of RNA Substrates

For *in vitro* synthesis of RNA substrate, it is convenient to have the polyadenylation region cloned into one of the commercially available vectors with bidirectional phage promoters (Pharmacia-LKB, Piscataway, NJ; Promega Biotec, Madison, WI; Stratagene, La Jolla, CA). With such a construct, substrate can be synthesized in one direction and complementary RNA probe for hybridization experiments from the other promoter. The following parameters should be considered in cloning and transcribing the DNA to be used as template. First, to give runoff RNAs of a defined length, there should be a convenient restriction site downstream of the poly(A) site. Short precursors of 100–300 bases are more manageable than longer ones when it is necessary to analyze the sequence of cleavage products. However, the segment spanning the poly(A) site must contain the AATAAA sequence upstream of this site and sufficient sequence downstream of the poly(A) site to include any downstream signal elements. Several studies suggest that specific sequence upstream

[30] A. Virtanen and P. A. Sharp, *EMBO J.* **7**, 1421 (1988).
[31] J. S. Butler and T. Platt, *Science* **242**, 1270 (1988).
[32] S.-C. Cheng, A. Newman, R.-J. Lin, G. McFarlane, and J. N. Abelson, this volume [8].
[33] N. F. Lue and R. D. Kornberg, *Proc. Natl. Acad. Sci. U.S.A.* **84**, 8839 (1987).

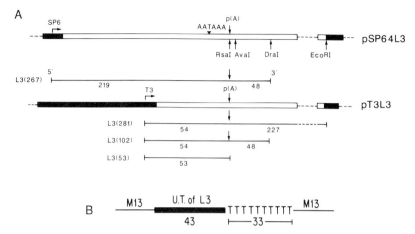

FIG. 1. (A) DNA templates and substrate RNAs for *in vitro* polyadenylation reactions. Recombinant DNAs pSP64L3 and pT3L3 containing the adenovirus type 2 L3 polyadenylation site were digested with the indicated restriction enzymes. The various runoff RNAs (indicated under each DNA) were transcribed *in vitro* with SP6 or T3 RNA polymerase. Shaded region, plasmid sequence; open region, adenovirus type 2 sequence. (B) M13 cDNA recombinant used for hybridization analysis. This single-stranded DNA is complementary to 43 nucleotides of the 3′ untranslated region (U.T.) of L3 and a tract of 33 adenosine residues.

of the AAUAAA signal is not critical.[8,34,35] However, it is likely that spacer sequence is needed between the 5′ cap of the precursor RNA and the AAUAAA. In functional precursor containing the adenovirus type 2 L3 poly(A) site, this distance is 28 nucleotides, and only 14 nucleotides upstream of the AAUAAA are of viral origin (Fig. 1A). If downstream signal elements have not been identified for a particular site by previous deletion experiments *in vivo,* a substrate which extends 100 nucleotides beyond the poly(A) site will probably contain the essential sequences. It should be kept in mind that cleavage at the poly(A) site can be less efficient on substrates which extend more than 50–100 nucleotides beyond the poly(A) site.[16]

To synthesize the substrate, the DNA template is incubated with polymerase and nucleotide triphosphates.[36] Since capped substrates are generally more stable in nuclear extracts, the dinucleotide primer G(5′)ppp(5′)G can be included in the transcription reaction. Most,[10,18,37]

[34] B. M. Bhat and W. S. Wold, *Mol. Cell Biol.* **5,** 3183 (1985).
[35] M. M. Kessler, R. C. Beckendorf, M. A. Westhafer, and J. L. Nordstrom, *Nucleic Acids Res.* **14,** 4939 (1986).
[36] J. K. Yisraeli and D. A. Melton, this series, Vol. 18, p. 42; P. A. Krieg and D. A. Melton, this series, Vol. 155, p. 397.
[37] L. Conway and M. Wickens, *EMBO J.* **6,** 4177 (1987).

but not all,[29] investigators have reported that the cap is not essential for polyadenylation activity. The RNA can be body-labeled by including ^{32}P-labeled nucleotide triphosphates in the transcription reaction. A 20-μl transcription reaction containing 1 μg of linearized plasmid DNA, 12 μM unlabeled UTP, and 50 μCi of [α-^{32}P]UTP, in addition to the other unlabeled nucleotide triphosphates, will give RNA with a specific activity useful for most experiments. After the transcription reaction, DNA is removed by adding RNase-free DNase I (Pharmacia-LKB or Promega Biotec) to 100 units/ml for 10 min at 37°. The transcripts are purified from unincorporated label by 2–3 ethanol precipitations in the presence of 2 M ammonium acetate. Alternatively, nonradioactive substrate can be labeled at the 5' end with T4 polynucleotide kinase and [γ-^{32}P]ATP[37,38] or at the 3' end with T4 RNA ligase and [5'-^{32}P]pCp.[37–39] These types of substrates have been useful in studying the effect of chemical modifications of particular bases on polyadenylation activity and specific complex formation[37] and could be used to footprint complexes on RNA. The precursor RNA can also be tagged with biotin for streptavidin affinity chromatography. This approach has been used to identify the components of spliceosomes[41] but has not yet been applied to polyadenylation complexes.

Assays for Polyadenylation-Specific Events

Polyadenylation Reaction Conditions

A typical 25-μl polyadenylation reaction contains 32–60% (v/v) nuclear extract and 0.1–5 nM of substrate RNA. The amount of cleavage and polyadenylation increases linearly with substrate concentration in this range.[14,30] For substrates of 300 nucleotides or less, approximately 0.5–5 ng or 10,000–50,000 Cerenkov counts per minute is adequate for most experiments using an efficient poly(A) site. The amount of extract necessary for the best activity can vary from batch to batch and should be titrated. Dialysis buffer D or its components are added as necessary to maintain the reaction at 40–60 mM KCl, 0.5–1 mM MgCl$_2$, 0.1 mM EDTA, 8.8–10% glycerol, and 0.2 mM DTT. The optimal cation concentrations may vary depending on the poly(A) site and the source of the extract.[30] The endogenous ATP in the extract, measured using a firefly

[38] J. F. Milligan and O. C. Uhlenbeck, this series, Vol. 180, p. 51.
[39] P. J. Romaniuk and O. C. Uhlenbeck, this series, Vol. 100, p. 52.
[40] T. Humphrey and N. J. Proudfoot, *Trends Genet.* **4**, 243 (1988).
[41] P. J. Grabowski and P. A. Sharp, *Science* **233**, 1294 (1986).

luciferase assay (Sigma, St. Louis, MO),[42] is less than 1 μM. Therefore, ATP must be added to 1 mM. The inclusion of 20 mM creatine phosphate (Sigma) greatly stimulates the reaction and probably helps to regenerate the ATP pool through endogenous creatine phosphokinase present in the extract. For some poly(A) sites, the addition of 1–4% poly(vinyl alcohol) or polyethylene glycol can significantly enhance the efficiency of processing.[12,16,20,21,30]

When incubated under the above conditions at 30°, RNA should be cleaved at the poly(A) site and a poly(A) tract of 200–400 residues added. The efficiency of cleavage will vary depending on the poly(A) site used. For the adenovirus L3 and the SV40 late sites, it can be as high as 80% of the input RNA.[10,12] For the SV40 early site, only 5–30% of the input RNA will be cleaved, and much of the uncleaved precursor is polyadenylated at its 3' end.[16,18,29,43] The behavior of the SV40 early site *in vitro* is also unusual in that accurate cleavage requires magnesium or spermidine and will occur in the presence of ATP or AMP(CH$_2$)P$_2$, but not 3'-dATP.[16,18,43]

Detection of Cleavage and Polyadenylation

Direct Gel Analysis of Polyadenylated RNAs. After incubation with nuclear extract as described above, ^{32}P-labeled RNAs can be examined directly by electrophoresis on denaturing urea–polyacrylamide gels. To purify the RNA from a 25-μl reaction mixture, the volume is increased to 200 μl by the addition of digestion buffer [50 mM tris(hydroxymethyl)-aminomethane hydrochloride (Tris-HCl) (pH 8), 10 mM EDTA, 10 mM NaCl, 0.2% sodium dodecyl sulfate (SDS)], proteinase K is added to 200 μg/ml, and the sample incubated for 10 min at 30–37°. The sample is then extracted once with phenol–chloroform–isoamyl alcohol (50 : 48 : 2) and once with chloroform–isoamyl alcohol (48 : 2), recovered by ethanol precipitation, resuspended in sample buffer [80% formamide, v/v, 0.1% xylene cyanol, 0.1% bromphenol blue, 1 mM EDTA (pH 8)], and electrophoresed on a polyacrylamide gel containing 8.3 M urea. The best percentage of acrylamide depends on the size of the precursor and products. Higher percentage gels have the advantage of compressing the heterogeneous length polyadenylated product into a tighter band and thus making it more easily detectable.

A typical time course for the *in vitro* polyadenylation of RNA containing the adenovirus L3 poly(A) site is shown in Fig. 2. Within 15 min, polyadenylated RNA appears. Even though this RNA has been cleaved at

[42] G. A. Kimmich, J. Randles, and J. S. Grand, *Anal. Biochem.* **69,** 187 (1975).
[43] C. Hashimoto and J. A. Steitz, *Cell* **45,** 581 (1986).

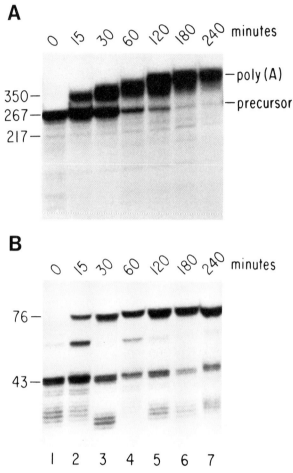

FIG. 2. Time course of the *in vitro* polyadenylation reaction. Precursor RNA was incubated at 30° in a 25-μl reaction containing 11 μl HeLa nuclear extract, 1 mM ATP, 20 mM creatine phosphate, 44 mM KCl, 0.7 mM MgCl₂, 8.8% glycerol, 8.8 mM HEPES (pH 7.6), 0.1 mM EDTA, 0.2 mM DTT, and 10⁴ cpm of L3(267) precursor (Fig. 1A). The reactions were stopped at the indicated times, and aliquots were analyzed (A) directly by electrophoresis on a 15% polyacrylamide–8.3 M urea gel or (B) with a hybridization–nuclease protection assay using the M13 cDNA clone shown in Fig. 1B. In B, the 43-nucleotide band represents precursor, and the 76-nucleotide band is derived from correctly processed RNA. The band at 60 nucleotides is an artifact of hybridization. (Reproduced, with permission, from Moore and Sharp.[10])

the poly(A) site, it migrates more slowly than the precursor RNA because of the poly(A) tail of 200–400 adenosines. After 2–3 hr, the maximum level of processed RNA is attained. Generally, for the sites that have been examined *in vitro,* cleaved, unpolyadenylated RNA does not accumulate when conditions are optimal for polyadenylation. Depending on the length of the precursor and cleaved product, the polyadenylated RNA may migrate near the precursor band. In these cases, the RNA can be analyzed by a hybridization–nuclease protection protocol, as described later, or separated into poly(A)$^+$ and poly(A)$^-$ RNAs by chromatography on poly(U)-Sephadex[9,44] or oligo(dT)-cellulose.[12,45,46]

Selection of Polyadenylated RNAs by Chromatography. Selection of poly(A) RNAs can be conveniently accomplished without columns. The protocol described below is for poly(U)-Sephadex but could be modified for oligo(dT)-cellulose. Poly(U)-Sephadex has the advantage of retaining smaller poly(A) tracts than oligo(dT)-cellulose.

Purified RNA from a polyadenylation reaction is resuspended in 100 μl of 10 mM Tris (pH 7.5), 1 mM EDTA, 0.2% SDS and heated at 65° for 5 min. The solution is adjusted to 200 mM NaCl and mixed with 100 μl of binding buffer [200 mM NaCl, 10 mM Tris (pH 7.5), 1 mM EDTA, 0.2% SDS] and 100 μl of poly(U)-Sephadex beads [Bethesda Research Laboratories (BRL), Gaithersburg, MD; Pharmacia-LKB, Piscataway, NJ] that have been washed with 20 volumes of elution buffer and 20 volumes of binding buffer to remove any free poly(U). The tubes containing this mixture are rotated for 10 min at room temperature. The beads are centrifuged, and the supernatant containing the poly(A)$^-$ RNA is precipitated with ethanol. After rinsing the beads once with 1 ml of binding buffer and twice with 1 ml each of a wash solution [10% formamide, 10 mM Tris (pH 7.5), 1 mM EDTA, 0.2% SDS], poly(A)$^+$ RNA is released in 200 μl of elution buffer [90% formamide, 10 mM Tris (pH 7.5), 1 mM EDTA, 0.2% SDS] and then precipitated with ethanol.

Analysis of Cleavage by Hybridization–Nuclease Protection Assays. The accuracy of the cleavage reaction can be verified by hybridization–nuclease protection analysis. While probes derived from genomic DNA can be used, the most sensitive assay employs a single-stranded probe derived from a cDNA cloned into either an M13 phage vector or a transcriptional expression vector to give, respectively, single-stranded DNA or RNA complementary to correctly cleaved and polyadenylated mRNA.

[44] C. L. Moore, J. Chen, and J. Whoriskey, *EMBO J.* **7,** 3159 (1988).
[45] H. Aviv and P. Leder, *Proc. Natl. Acad. Sci. U.S.A.* **69,** 1408 (1972).
[46] T. Maniatis, E. R. Fritsch, and J. Sambrook, "Molecular Cloning: A Laboratory Manual." Cold Spring Harbor Laboratory, Cold Spring Harbor, New York, 1982.

Such a probe for the L3 site (diagrammed in Fig. 1B) contains 43 nucleotides of sequence complementary to the 3′ untranslated region and a stretch of 33 thymidines. If the cleavage site is known but a cDNA does not exist, it can be constructed using a protocol described by Virtanen and Sharp.[30] Deoxyadenosine residues are added to an appropriate oligonucleotide either by the terminal transferase tailing reaction[46,47] or with the DNA synthesizer. These poly(dA)-tailed single-stranded oligonucleotides can be annealed to M13mp18 double-stranded DNA that is restricted at the unique *Kpn*I site and 3′-end-tailed with poly(dT). The resulting hybrids are made double stranded by filling in the single-stranded region using deoxynucleotides and Klenow DNA polymerase.[46] The vector DNA is then digested with *Hinc*II restriction enzyme and the blunt-ended fragments ligated with T4 DNA ligase to the *Hinc*II site. This DNA is transfected into *Escherichia coli* strain JM109, and clones with the desired structure are identified by restriction enzyme digestion and DNA sequencing. Such synthetic constructs have been used to quantitate the *in vitro* processing of RNAs containing the immunoglobulin poly(A) sites.[30]

For analysis using a single-stranded cDNA probe, RNA is mixed with 0.5 μg of DNA in 15 μl of 750 m*M* NaCl, 50 m*M* HEPES (pH 7.6), and 1 m*M* EDTA, heated at 85° to be denatured, and hybridized for 3 hr at 50°. It may be necessary to determine empirically the optimal hybridization temperature. The sample is then diluted with 200 μl of 200 m*M* NaCl, 10 m*M* HEPES (pH 7.6), and 1 m*M* EDTA and treated with ribonuclease T2 (Calbiochem, San Diego, CA), which will digest unhybridized RNA at the 5′ end as well as any overhanging poly(A). Suitable ribonuclease concentrations and digestion temperatures (from 25 to 37°) should be determined for each different precursor. For the L3 precursor, this requires digestion with 50 units/ml of ribonuclease T2 for 30 min at room temperature.[10] For immunoglobulin poly(A) sites, better results were obtained with 5 units/ml ribonuclease T2 and 5 μg/ml ribonuclease A (Pharmacia).[30] To ensure that all the ribonuclease is inactivated, the sample is then treated with proteinase K at 200 μg/ml and 1% SDS for 15–60 min at 30–37° and extracted once with PCIA before ethanol precipitation with 20 μg of carrier tRNA. Protected fragments are resolved on a polyacrylamide–8.3 *M* urea gel. The protocol is similar when riboprobes are used, except that 0.5 ng of probe is sufficient and hybridization is in 80% formamide, 40 m*M* piperazine-*N,N*′-bis(2-ethanesulfonic acid) (PIPES) (pH 6.4), 0.4 *M* NaCl, and 1 m*M* EDTA for at least 8 hr.[48]

[47] T. Nelson and D. Brutlag, this series, Vol. 68, p. 41.
[48] F. M. Ausubel, R. Brent, R. E. Kingston, D. D. Moore, J. A. Smith, J. G. Seidman, and K. Struhl, "Current Protocols in Molecular Biology." Greene and Wiley (Interscience), New York, 1987.

This method provides very sensitive quantitation of accurate cleavage and polyadenylation, since RNAs that are correctly polyadenylated will yield a single band instead of the broad smear observed when RNAs from an *in vitro* reaction are examined directly. In addition, correctly processed RNAs produce the largest protected fragment, and will be easily distinguished from RNA resulting from unprocessed, incorrectly processed, or degraded precursor. Such an analysis for the kinetics of cleavage and polyadenylation of L3 precursor is depicted in Fig. 2B. Precursor RNA yields a protected band of 43 nucleotides, and processed RNA gives a fragment of 76 nucleotides.

When single-stranded cDNA probes are not available, nuclease protection assays can be performed with single-stranded genomic probes using the hybridization conditions described above or with double-stranded genomic probes using S1 nuclease analysis.[49] To obtain a clean signal when the processing efficiency is low, it may be necessary to select the poly(A)$^+$ RNA first.

Direct Analysis of Cleavage Products. When radioactively labeled RNA is incubated with nuclear extract under the appropriate conditions, the cleaved, unpolyadenylated products of the *in vitro* reaction can be visualized directly. A set of small RNAs with 5'-terminal phosphates can be detected at short incubation times (Fig. 3, lane 5). These contain only downstream sequence and have a 3' end identical to that of the original precursor. These RNAs differ in length at their 5' termini, with the longest RNAs beginning just after the poly(A) site.[11,13] These downstream cleavage products are rapidly degraded, probably by a 5' exonuclease, but can be stabilized for longer times by the addition of EDTA to the reaction.[10-12] The cleaved-off RNAs are sometimes more prominent if the precursor is labeled only at the 3' end by the T4 RNA ligase-catalyzed addition of [^{32}P]pCp.[50]

To detect the upstream cleavage product, the poly(A) addition step can be blocked by including 2 mM EDTA in the reaction or replacing ATP with 0.5–1 mM of an analog such as α,β-methyleneadenosine 5'-triphosphate [AMP(CH$_2$)P$_2$, Sigma], cordycepin triphosphate (3'-dATP, Pharmacia-LKB or Sigma), or 2',3'-dideoxyadenosine triphosphate (ddATP).[10-13,17] In the presence of the nonhydrolyzable AMP(CH$_2$)P$_2$, the product terminates at the cleavage site and contains a 3'-hydroxyl (Fig. 3, lane 2). The exact nucleotide at this 3' end has been determined for product derived from the adenovirus L3 and the SV40 late polyadenylation precursor.[11,13] For both, it is an adenosine, which then becomes the

---

[49] A. Berk, this series, Vol. 180, p. 334.
[50] D. Zarkower and M. Wickens, *EMBO J.* **6**, 4185 (1987).

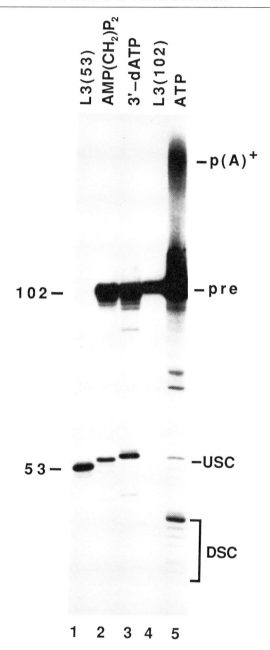

first A of the poly(A) tract. In the presence of the chain terminators (3'-dATP or ddATP), the product is extended by the addition of only one residue of adenosine (Fig. 3, lane 3).

The identity of cleavage products eluted from preparative gels can be verified by ribonuclease T1 or A digestion followed by RNA fingerprint-ing.[10,12] This analysis is best accomplished by separating the oligonu-cleotides with two-dimensional chromatography,[51] in which the first di-mension is high-voltage electrophoresis on a cellulose-acetate strip (Schleicher & Schuell, Keene, NH) in 5% pyridinium acetate (pH 3.5), 2 mM EDTA at 5000 V and the second dimension is homochromatography on Cel 300 polyethyleneimine (Brinkman Instruments, Inc.)[12] or DEAE-cellulose (Macherey-Nagel, Polygram Cell 300 DEAE/Hr-2/15)[10] thin-layer plates using a homomix of 7 M urea and 30% hydrolyzed yeast tRNA. Alternatively, diagnostic oligonucleotides can sometimes be more quickly resolved on a 20% polyacrylamide–8.3 M urea gel.[10,11] This ap-proach is especially valuable if the products are recovered in low yield. If accurate cleavage has occurred, either analysis should reveal the follow-ing: one cleavage product will contain only oligonucleotides upstream of the poly(A) site and the other only oligomers downstream of this site, and the oligonucleotide that spans the poly(A) site in the precursor will not appear in either product and will be replaced by a new oligomer.

Analysis of Poly(A) Addition without Cleavage. The poly(A) addition step alone can be studied using RNA substrate which terminates at or near the poly(A) site.[11,12,52] In the presence of ATP and nuclear extract, these RNAs are rapidly polyadenylated at their 3' ends through a mecha-nism that relies on the AAUAAA sequence and the presence of a 3' end a short distance away from the AAUAAA (Fig. 4, lanes 7–12). Thus, unlike cleavage, polyadenylation does not require the downstream signal se-quences, and it does not need to be preceded by a cleavage event. There

---

[51] M. Silberlkang, A. M. Gillum, and U. L. RajBhandary, this series, Vol. 59, p. 58.
[52] C. L. Moore, H. Skolnik-David, and P. A. Sharp, *Mol. Cell. Biol.* **8,** 226 (1988).

FIG. 3. Products of cleavage at the L3 poly(A) site. ^{32}P-Labeled RNA was electro-phoresed on a 15% polyacrylamide–8.3 M urea gel. Lane 1, L3(53) RNA, which terminates one nucleotide upstream of the poly(A) site; lane 2, L3(102) RNA incubated with nuclear extract and 4 mM AMP(CH$_2$)P$_2$ for 3 hr; lane 3, L3(102) RNA incubated with extract and 1 mM 3'-dATP for 3 hr; lane 4, untreated L3(102) RNA; lane 5, L3(102) RNA incubated with extract and 1 mM ATP for 30 min. The RNAs are labeled as follows: USC, upstream cleavage product; DSC, downstream cleavage product; pre, precursor; p(A)$^+$, polyadenyl-ated RNA. (Reproduced, with permission, from Moore *et al.*[11])

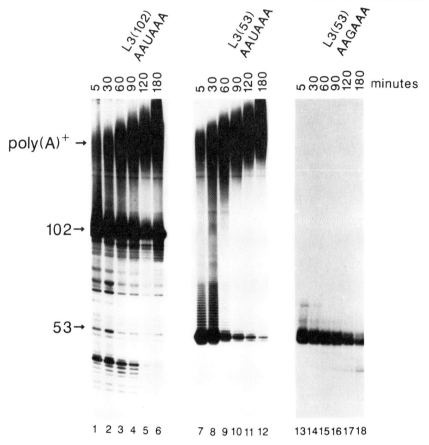

FIG. 4. Activity of "precleaved" RNA substrates. ^{32}P-Labeled substrate RNA was incubated with HeLa nuclear extract under conditions optimal for polyadenylation. The reactions were stopped at the indicated times by freezing, and purified RNAs were analyzed by electrophoresis on a 15% polyacrylamide–8.3 M urea gel. The unreacted substrate RNAs, 53 and 102 nucleotides long, and the polyadenylated RNA product, poly(A)[+], are indicated by arrows. Reactions contained the following substrates: L3(102), lanes 1–6; L3(53), lanes 7–12; L3(53) in which AAUAAA has been mutated to AAGAAA, lanes 13–18. (Reproduced, with permission, from Skolnik-David et al.[19])

is surprisingly little nonspecific polyadenylation, and substrates containing a mutation of AAUAAA to AAGAAA are not utilized (Fig. 4, lanes 13–18). There is recent evidence that polymerization of only the first 10 or so adenosines is AAUAAA-dependent.[53] After that, the poly(A) synthesis

[53] M. D. Sheets and M. Wickens, Genes Dev. 3, 1401 (1989).

shifts into a mode which probably requires only poly(A) polymerase, ATP, and a poly(A) primer.

The types of assays described above have been used to show that the *in vitro* reaction depends on the same signal sequences as the *in vivo* one[1,8,12,14–25] and to identify fractions from a crude extract which contain the cleavage and polyadenylation activities.[40]

Detection of Polyadenylation-Specific Complexes

The earliest step in the polyadenylation reaction occurs before processing can be detected and involves the formation of polyadenylation-specific complexes that contain precursor RNA, proteins, and, possibly, small nuclear RNAs (snRNAs). These complexes can be detected by their sedimentation in glycerol[52] or sucrose gradients[22] and by their electrophoretic mobility on nondenaturing polyacrylamide gels.[14,17,19,21,50] Efficient assembly of these complexes requires the AAUAAA and the downstream signal sequences.

Sedimentation Analysis of Polyadenylation-Specific Complexes. To separate complexes by sedimentation, a 120-μl polyadenylation reaction mixture is layered directly onto a 5-ml gradient containing 10–30% glycerol in 20 mM HEPES (pH 7.5), 1 mM MgCl$_2$, 25 mM KCl, and 0.1 mM EDTA. Gradients are centrifuged in an SW50.1 rotor at 49,000 rpm at 5° for 2–3 hr. These gradients are fractionated into 250-μl aliquots and the ^{32}P Cerenkov counts per minute determined to give the gradient profile. The assembly of complexes on L3 precursor is depicted in Fig. 5. A 35 S complex forms in the absence of ATP.[55] This complex is nonspecific and does not require the polyadenylation signal sequences on the RNA for its formation. After incubation in the presence of ATP, a specific complex appears which sediments at 50 S. Both precursor and polyadenylated product are found in the 50 S region of the gradient. Treatment with heparin causes this complex to sediment at 20–25 S.

Analysis of Specific Complexes by Electrophoresis. Specific and non-specific complexes can also be resolved on native gels.[56] So far, the studies of polyadenylation complexes have all used 4% polyacrylamide–0.05% bisacrylamide gels containing 25–90 mM Tris–borate and 1.2–2.5 mM EDTA (pH 8.3). It is not known how these complexes would migrate in a Tris–glycine buffer system.[56] Before loading the gel, it is necessary to

[54] M. A. McDevitt, G. M. Gilmartin, W. H. Reeves, and J. R. Nevins, *Genes Dev.* **2**, 588 (1988).

[55] Sedimentation values were calculated by comparison with 40 S and 60 S ribosomal subunit markers centrifuged under identical conditions.

[56] M. M. Konarska, this series, Vol. 180, p. 442.

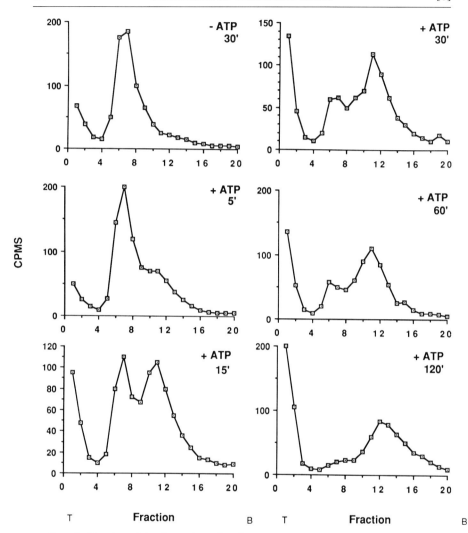

FIG. 5. Kinetics of the formation of sedimentation complexes. ^{32}P-Labeled precursor RNA L3(267) was incubated under optimal polyadenylation conditions for the times indicated at the top right of each graph. The reaction represented by the top left graph lacked ATP. The reactions were centrifuged through glycerol gradients, and the fractions were analyzed by Cerenkov counting. The top (T) of the gradient is at left. Counts per minute are in thousands. (Reproduced, with permission, from Moore *et al.*[52])

treat the samples with heparin at 5 mg/ml for 10 min on ice to remove loosely bound proteins and to prevent the complexes from aggregating at the top of the gel. Aggregation is also lessened if precipitate is cleared from the nuclear extract before incubation with RNA substrate. This is easily done by centrifugation in a microcentrifuge for 20 sec at 4°. Because complexes formed on precursor containing different poly(A) sites can migrate slightly differently, it is necessary to empirically determine the best conditions of electrophoresis. Complexes containing the L3 precursor are well resolved by electrophoresis at 4° and 18 V/cm for 4–5 hr (until the xylene cyanol marker dye reaches 15–16 cm). For complex stability, it is important that the gel remains cold, and, thus, it should not heat up during electrophoresis. After electrophoresis, the gel is dried down onto Whatman 3MM chromatography paper, and the complexes are detected by autoradiography.

Incubation of L3 precursor with nuclear extract and ATP results in the rapid formation of a nonspecific complex A and an AAUAAA-dependent complex B (Fig. 6A).[19] The specific complex B migrates between splicing complexes A and B,[17,57] and it can sometimes be resolved as two complexes.[17] As the polyadenylation reaction proceeds, a third complex, C, appears (Fig. 6B, lane 3), which contains primarily polyadenylated RNA (Fig. 6B, lane 8).[19] The faster migration of complex C compared to complex B suggests that the former has probably lost components found in complex B. Complexes similar to A, B, and C have been observed on precursor containing the poly(A) sites for the SV40 late, HSV-2 immediate early gene 5, or HSV-1 thymidine kinase genes.[14,21,50] With the SV40 late site, a postcleavage complex containing cleaved, unpolyadenylated RNA can be detected that migrates only slightly faster than complex B.[50]

Additional Methods for Detecting Polyadenylation-Specific Complexes. It has also been possible to demonstrate the interaction of extract components with the AAUAAA sequence in other ways. When precursor is incubated with extract, this sequence is not accessible for hybridization to complementary oligonucleotides,[22,58] and it is present in ribonuclease-resistant fragments which can be immunoprecipitated with antibodies against small nuclear ribonucleoprotein (snRNP) components.[17,22,43] It has also been shown that DNA oligonucleotides base-paired to AAUAAA[50,58] or chemical modifications of nucleotides in the AAUAAA signal[37] prevent processing as well as formation of the specific complex detected by electrophoresis.

[57] M. M. Konarska and P. A. Sharp, *Cell* **46,** 845 (1986).
[58] D. Zarkower and M. Wickens, *EMBO J.* **6,** 177 (1987).

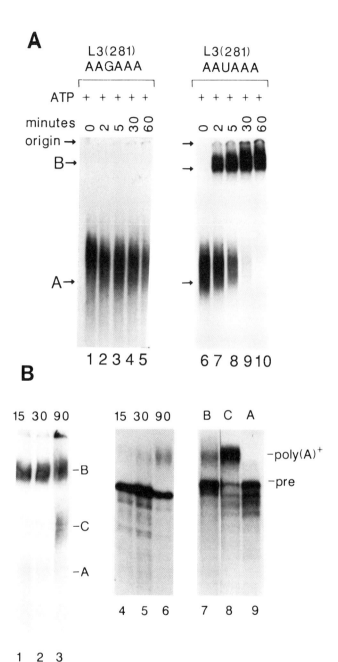

FIG. 6. (A) Time course of complex formation analyzed by gel electrophoresis. [α-³²P]UTP-labeled RNA substrate L3(281) was incubated for different times with HeLa nuclear extract in the presence of ATP. After addition of heparin to 5 mg/ml for 10 min on ice, aliquots of the reaction mixtures were loaded on a nondenaturing 4% polyacrylamide gel

Identification of Factors Responsible for Polyadenylation

Little is known about the nature of the extract components that mediate the polyadenylation reaction. Three techniques that have been employed to identify such factors include the effect of specific antibodies on the reaction, ultraviolet (UV) cross-linking of proteins to RNA, and fractionation of extracts by chromatography.

Effect of Specific Antibodies on in Vitro Polyadenylation Reaction

Polyadenylation is inhibited in extracts preincubated with antibodies against poly(A) polymerase[78] or certain snRNPs.[9,10,16] The antibodies that have such an effect include a monoclonal (designated Y12) against the Sm proteins of the U class of snRNPs,[59] polyclonal sera against the trimethyl cap structure (m_3G) unique to the U class of snRNAs,[60,61] and autoimmune sera specific for Sm, U1 RNP, or the La nuclear antigen from patients with lupus erythematosus.[62,63] When precursor is incubated with extract and then treated with ribonuclease T1, the anti-Sm and anti-m_3G sera will immunoprecipitate a ribonuclease-protected fragment that contains the AAUAAA sequence.[43] Polyadenylation is also inhibited in extracts predigested with micrococcal nuclease to destroy endogenous RNA.[18,43] These results suggest a role for a snRNP in the polyadenylation process.[79] However, no particular snRNP has been identified, and alternative explanations have been proposed. For example, it is also possible that a non-snRNP-associated protein which shares similar epitopes is

[59] E. A. Lerner, M. R. Lerner, C. A. Janeway, and J. A. Steitz, *Proc. Natl. Acad. Sci. U.S.A.* **78**, 2737 (1981).

[60] R. Luhrmann, B. Appel, P. Bringmann, J. Rinke, R. Reuter, S. Rothe, and R. Bald, *Nucleic Acids Res.* **10**, 7103 (1982).

[61] B. Chabot, D. L. Black, D. M. LeMaster, and J. A. Steitz, *Science* **230**, 1344 (1985).

[62] M. R. Lerner and J. A. Steitz, *Proc. Natl. Acad. Sci. U.S.A.* **76**, 5495 (1979).

[63] M. R. Lerner, J. A. Boyle, J. Hardin, and J. A. Steitz, *Science* **211**, 400 (1981).

---

and the complexes separated by electrophoresis. Lanes 1–5, L3(281)RNA containing the AAGAAA mutation; lanes 6–10, L3(281) RNA with the AAUAAA sequence. A, Nonspecific complex; B, specific complex. (B) Analysis of RNA species in gel complexes. L3(267) precursor was incubated with extract and ATP at 30° for the indicated times. Complexes were resolved as described in A (lanes 1–3). Purified RNA from each time point was analyzed by electrophoresis on a 15% polyacrylamide–8.3 M urea gel (lanes 4–6). The RNAs in the different complexes of the 90-min reaction, shown in lane 3 and indicated as A, B, and C, were transferred from the gel to DEAE-membrane by electroblotting and eluted as previously described [C. W. Pikielny and M. Rosbash, *Cell* **45**, 869 (1986)]. This RNA was then analyzed by electrophoresis on a 15% polyacrylamide–8.3 M urea gel. Lane 7, RNA eluted from the region containing complex B; lane 8, RNA eluted from complex C; lane 9, RNA from complex A. (Reproduced, with permission, from Skolnik-David *et al.*[19])

being recognized by these antibodies (but this is not indicated in the results with the anti-m$_3$G serum).[44] In one case, polyadenylation was restored to nuclease-treated extracts by the addition of E. coli RNA, suggesting that the release of RNA-binding proteins caused the inhibition.[18]

The immunoprecipitation of specific ribonucleoproteins is described in this series.[64] For the experiments that demonstrate inhibition of polyadenylation, different amounts of the desired antisera are incubated with extract for 30 min on ice. The other components of the polyadenylation reaction mixture are added, and after further incubation for 2–3 hr at 30°, the RNA products are analyzed. Controls for nonspecific inhibition include nonimmune sera, sera with unrelated specificities but of the same origin, or sera inactivated by heat denaturation. For micrococcal nuclease digestions, the extract is brought to 0.4–1 mM CaCl$_2$, and treated with 1–3 units of micrococcal nuclease (Boehringer Mannheim, Cooper Biomedical, or Pharmacia) per microliter of extract at 30° for 30 min. The nuclease treatment is stopped by adding ethylene glycol bis(β-aminoethyl ether)-N, N, N', N'-tetraacetic acid (EGTA) to twice the CaCl$_2$ concentration. The result of this micrococcal nuclease treatment should be compared to the effect of treatment in the presence of calcium and EGTA and to that of adding purified E. coli RNA to 40 μg/ml after the nuclease treatment.[18]

Ultraviolet Cross-Linking of Proteins to Precursor RNA

Specific proteins of 64–68 and 155 kDa, as well as the C heterogeneous nuclear ribonucleoprotein (hnRNP), can be covalently cross-linked to polyadenylation precursors by UV light.[24,44,65] The detailed protocol for this type of experiment is outlined elsewhere in this series.[66] Briefly, radioactive precursor is incubated with extract, exposed to UV light, and digested with ribonuclease. Covalently cross-linked proteins are resolved on discontinuous SDS-containing polyacrylamide gels and detected by label transfer. The proteins cross-linked to L3 precursor are shown in Fig. 7, lane 2. The proteins at 68 and 155-kDa do not bind to substrate in which the AAUAAA sequence has been mutated to AAGAAA (Fig. 7, lane 3), and excess amounts of such a substrate cannot compete with authentic precursor for this binding.[44] These proteins are found in polyadenylation-specific complexes and dissociate from the RNA once it is polyadenylated.[44] The 68-kDa protein elutes from DEAE-Sepharose at a salt concen-

[64] J. A. Steitz, this series, Vol. 180, p. 468.
[65] J. Wilusz and T. Shenk, *J. Cell. Biochem.* in press (1988).
[66] S. Pinol-Roma, S. A. Adam, Y. D. Chou, and G. Dreyfuss, this series, Vol. 180, p. 410.

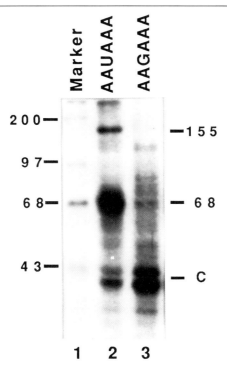

FIG. 7. Proteins cross-linked to polyadenylation precursor by UV light. ^{32}P-Labeled L3(102) RNA was incubated with nuclear extract for 5 min at 30° and exposed to UV irradiation for 10 min at 4°. The UV light source (Ultraviolet Products Model UVG-54) was supported 4.5 cm above the sample, which gave an intensity of 1.8 mW/m². The RNA was then digested for 30 min at 37° with ribonuclease T1 (2.5 units/ml) and the proteins resolved on a 10% polyacrylamide gel containing SDS (P. J. Blackshear, this series, Vol. 104, p. 237). Lane 1, ^{14}C-Labeled protein markers (BRL); lane 2, proteins cross-linked to RNA containing AAUAAA; lane 3, proteins cross-linked to RNA containing AAGAAA. (Reproduced, with permission, from Moore et al.[44])

tration that also elutes the cleavage activity.[23,24] These findings suggest a role for the 155- and 68-kDa proteins in the polyadenylation reaction, possibly in initiating assembly of the processing complex.

Fractionation of Nuclear Extract

Using the *in vitro* assays for polyadenylation, several laboratories have obtained active fractions from crude nuclear extracts by chromatography on ion-exchange or sizing columns (see review[40]). Several fractions

[67] G. Christofori and W. Keller, *Cell* **54,** 875 (1988).

are necessary to reconstitute a complete reaction that is both accurate and efficient. One of the fractions contains a nonspecific, manganese-dependent poly(A) polymerase sized at either 50K[54,67,27,76] or 300K.[23] Additional factors are needed for accurate cleavage.[67,68,77] One of these, called the specificity factor (~290K), is required for both specific cleavage and AAUAAA-dependent polyadenylation[23,67,68] and copurifies over several columns with a snRNP designated as U11.[67] This snRNP is resistant to micrococcal nuclease and oligonucleotide-directed RNase H degradation.[67] Two other factors, cleavage factor I (~130K) and II (~110K) are sufficient to reconstitute cleavage when mixed with the specificity factor.[68] A fourth factor stimulates the cleavage reaction.[68] Using a sizing column, the last four factors can be separated from the poly(A) polymerase as a complex of at least 360K.[23] Cleavage of precursors other than SV40 late pre-mRNA requires addition of the poly(A) polymerase fraction.[23,29,67] There is also a nonspecific requirement for RNA mass; E. coli total or rRNA (Pharmacia) at 40–60 μg/ml or heparin at 80 μg/ml can satisfy this requirement. The following sections discuss the assays useful for the different activities involved and the schemes utilized to achieve the fractionation.

Assays for Identifying Active Fractions

Detection of Cleavage or Coupled Cleavage and Polyadenylation. Fractions or combinations of fractions are incubated with ATP and radioactive precursor RNA using the optimal conditions for polyadenylation described previously. RNA from the reactions can be examined directly by gel electrophoresis to detect a cleaved RNA of the appropriate size. If polyadenylation is observed by a lengthening of the RNA, a hybridization–nuclease protection assay with a cDNA probe will verify that accurate cleavage has occurred. This assay will also detect lower levels of activity, especially when only short tracts of adenosines are polymerized. Alternatively, poly(A) polymerization can be prevented either by eliminating magnesium from the reaction mixture or by including AMP(CH_2)P_2 or 3'-dATP.

Assay for AAUAAA-Dependent Poly(A) Addition. To determine if a fraction contains an AAUAAA-dependent poly(A) polymerase activity, truncated precursors ending at or near the polyadenylation site can be used as substrates in a reaction containing ATP and magnesium. To verify

---

[68] Y. Takagaki, L. Ryner, and J. Manley, *Genes Dev.* **3,** 1711 (1989).

the specificity, the fraction should also be tested with a similar substrate that has an inactivating point mutation of AAUAAA.[69]

Assay for Nonspecific Poly(A) Addition. As long as the RNA has a free 3'-hydroxyl, preparations of purified poly(A) polymerase are fairly nonspecific for the type of RNA substrate that will serve as a primer.[70,71] They require a divalent cation but are most active in the presence of manganese. It is important to test for this type of activity, since fractionation studies[23,54] indicate that a fraction containing nonspecific, manganese-dependent poly(A) polymerase becomes AAUAAA-dependent for polyadenylation when mixed with magnesium and the fraction containing cleavage activity. To assay for nonspecific poly(A) polymerase, fractions can be tested for their ability, in the presence of 1 mM Mn^{2+} and ATP, to lengthen a substrate containing a mutation in the AAUAAA sequence.[23] For a more sensitive assay, the incorporation of radioactive ATP onto the ends of nonspecific unlabeled substrate can be measured.[54,72,73] The standard assay includes 0.5 mM [α-^{32}P]- or [^{3}H]ATP, 50 mM HEPES (pH 7.9), 0.5 mM MnCl$_2$, and 150 μg/ml RNA primer. The mixture is incubated at 37° for 30 min. The reaction is halted by the addition of 5% trichloroacetic acid and carrier DNA. Acid-insoluble material is collected by filtration and counted in a liquid scintillation counter.

Chromatography of Factors Involved in mRNA 3'-End-Processing

One possible fractionation (Fig. 8) utilizes ammonium sulfate precipitation and a Superose 6 sizing column for the initial steps.[23,68] Ion-exchange chromatography on DEAE matrices has also been used as a first step.[23,29,54,67] Protocols for the initial chromatography steps are given as illustrations on how to handle the crude extract and individual fractions.[74]

Treatment of Extract in Preparation for Chromatography. Because of the insoluble matter found in crude nuclear extracts after dialysis and freezing, it is necessary to first clear the extract to prevent clogging of the columns. This can be accomplished by centrifuging the extract for 10 min

[69] L. Conway and M. Wickens, this series, Vol. 180, p. 369.

[70] M. Edmonds, *in* "The Enzymes" (P. Boyer, ed.), Vol. 15, p. 217. Academic Press, San Diego, California, 1982.

[71] S. T. Jacob and K. M. Rose, *in* "Enzymes of Nucleic Acid Synthesis and Modification," Vol. 2, p. 135. CRC Press, Boca Raton, Florida, 1983.

[72] J. R. Nevins and W. K. Joklik, *J. Biol. Chem.* **252,** 6939 (1977).

[73] S. Shuman and B. Moss, *J. Biol. Chem.* **263,** 8405 (1988).

[74] The detailed protocols which follow are reproduced, with permission, from Y. Takagaki, L. C. Ryner, and J. L. Manley, *Cell* **52,** 731 (1988).

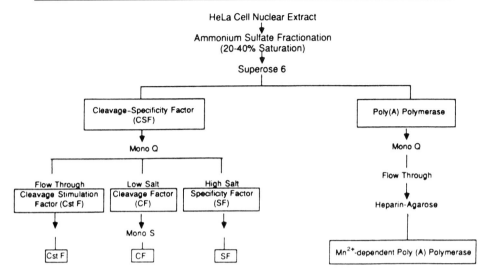

FIG. 8. Fractionation scheme for the purification of factors required for cleavage and polyadenylation, using ammonium sulfate precipitation and a Superose 6 sizing column for the initial steps.[23,77] The cleavage-specificity factor (CSF) can be separated into three activities by Mono Q and Mono S chromatography. These are designated specificity factor (SF), cleavage factor (CF), and cleavage stimulation factor (CstF). The Mn^{2+}-dependent nonspecific poly(A) polymerase (PAP) was further purified by sequential chromatography on Mono Q and heparin-agarose. The activities of combinations of the different factors are summarized as follows: PAP + SF, specific, Mg^{2+}-dependent polyadenylation; CF + SF, accurate cleavage; CstF + CF + SF, efficient cleavage; and PAP + CstF + CF + SF, reconstitution of coupled cleavage and polyadenylation.

at 4° in a microcentrifuge. Alternatively, a preliminary fractionation can be achieved by precipitation with different concentrations of $(NH_4)_2SO_4$.[23] The cleavage and polyadenylation activities remain soluble when $(NH_4)_2SO_4$ is added to extracts at 20% of saturation but are precipitated when the salt concentration is raised to 40%. To perform this step, 0.114 g/ml of solid $(NH_4)_2SO_4$ is slowly added to the extract, and the pH is brought to neutral with 1 μl/g of 1 M NaOH. After stirring for 30 min, the precipitate is removed by centrifugation at 16,500 rpm for 20 min. Then, 0.123 g/ml of $(NH_4)_2SO_4$ is added to the supernatant, and the mixture is processed at above. After centrifugation, the precipitate is resuspended in a minimal volume of a modified buffer D [20 mM Tris-HCl (pH 8.5), 20% glycerol, 50 mM $(NH_4)_2SO_4$, 0.2 mM EDTA, 0.5 mM DTT, and 0.5 mM PMSF], quick-frozen on dry ice, and stored at −70° until use.

Size Fractionation. For gel filtration, the cleared extract is passed through a cellulose acetate membrane filter (UNIFLO, pore size 0.45 μm;

Schleicher & Shuell), and 0.25 ml of the filtrate (protein concentration ~20 mg/ml) is loaded on a 130-cm FPLC Superose 6 column (Pharmacia) equilibrated with buffer D containing 0.2 M $(NH_4)_2SO_4$. Proteins are eluted at a flow rate of 0.25 ml/min, and 0.5-ml fractions are collected in 1.5-ml Eppendorf tubes. After pooling adjacent fractions, 0.36 g of solid $(NH_4)_2SO_4$ (60% saturation) is added to each tube, and the tubes are gently shaken for 1 hr. Proteins are recovered by centrifugation for 10 min in a microcentrifuge, resuspended in 0.1 ml of buffer D containing 50 mM $(NH_4)_2SO_4$, and dialyzed against the same buffer for 3 hr. A mixture of molecular weight markers should be loaded and eluted under the same conditions as above. Marker peaks can be localized by electrophoresis on an SDS–polyacrylamide gel followed by staining with Coomassie Brilliant blue.[75] The cleavage activity elutes from the column at approximately fraction 25, and well before the largest size marker. By sedimentation on glycerol gradients, it has a sedimentation value of 10 S.[23] Nonspecific poly(A) polymerase active in the presence of manganese, but not magnesium, elutes around fraction 32, corresponding to a molecular weight of 300,000.[23]

DEAE-Sepharose Chromatography. For fractionation on DEAE-Sepharose, the cleared nuclear extract is equilibrated with buffer D containing 0.1 M $(NH_4)_2SO_4$ by passage over a G-25 column (Pharmacia). The protein-containing fractions are pooled and filtered to remove insoluble material. Just prior to loading, the sample is diluted with buffer D containing no salt to a final concentration of 50 mM $(NH_4)_2SO_4$. Ten milligrams of protein is loaded onto a 2-ml DEAE-Sepharose column (Pharmacia) equilibrated with buffer D containing 50 mM $(NH_4)_2SO_4$ at a flow rate of 3 ml/hr and eluted with steps of buffer D containing 0.1, 0.15, 0.2, and 0.5 M $(NH_4)_2SO_4$. Fractions from each step are pooled, concentrated approximately 20-fold by $(NH_4)_2SO_4$ precipitation as described above, and dialyzed against buffer D containing 50 mM $(NH_4)_2SO_4$. Poly(A) polymerase activity eluted from this column in the flow-through fraction, whereas cleavage activity eluted primarily at 150 and 200 mM $(NH_4)_2SO_4$.

The fraction from the sizing column that contains the cleavage activity can be separated into three factors by FPLC ion-exchange chromatography (Fig. 8). Using the anion-exchange resin Mono Q, two fractions, eluted at 0.1 and 0.2 M $(NH_4)_2SO_4$, were necessary for accurate cleavage of SV40 late pre-mRNA. The high-salt fraction was needed to activate specific poly(A) addition when mixed with the poly(A) polymerase. An

[75] P. J. Blackshear, this series, Vol. 104, p. 237.
[76] L. C. Ryner, Y. Takagaki, and J. L. Manley, *Mol. Cell. Biol.* **9**, 1759 (1989).

additional factor from the flow-through, as well as the 0.1 and 0.2 M fractions, was needed to reconstitute efficient cleavage activity.[77]

Fractionation of nuclear extracts on DEAE-Sephacel (Pharmacia) using KCl instead of $(NH_4)_2SO_4$ as the buffer salt has produced similar results.[29,54] Nonspecific poly(A) polymerase activity was found in the flow-through. A factor, which eluted at 250 mM KCl, gave cleavage activity if combined with the poly(A) polymerase fraction. The activity in the 250 mM fraction is more sensitive to heat inactivation than the poly(A) polymerase activity. In addition, it is sensitive to micrococcal nuclease digestion and to depletion of the fraction with an Sm monoclonal antibody bound to protein A-Sepharose. The factor that reconstitutes specific polyadenylation when mixed with the 250 mM fraction coelutes with the nonspecific poly(A) polymerase after further fractionation on an FPLC Mono S column followed by a poly(A)-Sepharose column. By glycerol gradient sedimentation, this factor has a native molecular weight of 50,000,[54] similar to that of previously purified poly(A) polymerases.[54,70,71]

[77] A. Krämer, M. Frick, and W. Keller, *J. Biol. Chem.* **262**, 17630 (1987).
[78] M. P. Terns and S. T. Jacob, *Mol. Cell. Biol.* **9**, 1435 (1989).
[79] K. L. Mowry and J. A. Steitz, *Trends Biochem. Sci.* **13**, 447 (1988).

[7] Histone-Specific RNA 3' Processing in Nuclear Extracts from Mammalian Cells

By Claudia Stauber, Dominique Soldati, Bernhard Lüscher, and Daniel Schümperli

Introduction

The mature 3' ends of histone mRNAs are formed by endonucleolytic cleavage of longer precursor transcripts.[1,2] This process occurs in the nucleus and can be regarded as the equivalent of the polyadenylation reaction involved in 3'-end-generation of all other mRNAs. Initially, the sequence requirements for this process were studied by microinjection of sea urchin histone genes into *Xenopus laevis* oocytes. Two conserved sequence elements of the messenger RNA precursor (pre-mRNA) were

[1] M. L. Birnstiel, M. Busslinger, and K. Strub, *Cell* **41**, 349 (1985).
[2] M. L. Birnstiel and F. J. Schaufele, *in* "Structure and Function of Major and Minor Small Ribonucleoprotein Particles" (M. L. Birnstiel, ed.), p. 155. Springer-Verlag, Berlin, 1988.

found to be required.[3–5] The first involves the last 23–29 nucleotides of the mature mRNA and contains a dyad symmetry element capable of forming a hairpin loop structure. The second is a purine-rich sequence located in the spacer, a few nucleotides downstream from the cleavage site.

A sea urchin H3 gene that failed to be properly processed in the *Xenopus* oocyte system proved particularly useful, because it allowed the identification of a processing component from sea urchins by a complementation assay. This component turned out to be U7 small nuclear RNA (snRNA),[6,7] one of the minor U RNA species. Like most members of the U RNA family, U7 snRNA is present in ribonucleoprotein particles (snRNPs) that can be immunoprecipitated with autoimmune sera from human systemic lupus erythematosus patients (Sm antisera). Sequence comparisons and mutational analyses indicate that U7 snRNA exerts at least part of its function by base-pairing with the spacer motif of histone pre-mRNA, both in sea urchins[8] and in vertebrates.[9–11]

The first *in vitro* system for studying histone RNA 3' processing made use of nuclear extracts from *Drosophila melanogaster* K_c cells.[12] More recently, nuclear extracts from several different mammalian cell lines, including human HeLa,[13,14] mouse K21 and 21-Tb mastocytoma,[15] and C127 fibroblast cells,[16] challenged with synthetic pre-mRNA substrates, have been used in different laboratories. The main part of this chapter deals with the preparation and different applications of these mammalian processing extracts, while some key findings of these *in vitro* studies are discussed in the next paragraph.

The mature RNA species produced *in vitro* extends from the RNA cap to the precise 3' nucleotide also observed *in vivo*.[13] In addition, several 3' fragments which are colinear with the input RNA at their 3' end but differ

[3] C. Birchmeier, R. Grosschedl, and M. L. Birnstiel, *Cell* **28**, 739 (1982).
[4] C. Birchmeier, W. Folk, and M. L. Birnstiel, *Cell* **35**, 433 (1983).
[5] O. Georgiev and M. L. Birnstiel, *EMBO J.* **4**, 481 (1985).
[6] K. Strub, G. Galli, M. Busslinger, and M. L. Birnstiel, *EMBO J.* **3**, 2801 (1984).
[7] K. Strub and M. L. Birnstiel, *EMBO J.* **5**, 1675 (1986).
[8] F. Schaufele, G. M. Gilmartin, W. Bannwarth, and M. L. Birnstiel, *Nature (London)* **323**, 777 (1986).
[9] D. Soldati and D. Schümperli, *Mol. Cell. Biol.* **8**, 1518 (1988).
[10] K. L. Mowry and J. A. Steitz, *Science* **238**, 1682 (1987).
[11] M. Cotten, O. Gick, A. Vasserot, G. Schaffner, and M. L. Birnstiel, *EMBO J.* **7**, 801 (1988).
[12] D. H. Price and C. S. Parker, *Cell* **38**, 423 (1984).
[13] O. Gick, A. Krämer, W. Keller, and M. L. Birnstiel, *EMBO J.* **5**, 1319 (1986).
[14] K. L. Mowry and J. A. Steitz, *Mol. Cell. Biol.* **7**, 1663 (1987).
[15] B. Lüscher and D. Schümperli, *EMBO J.* **6**, 1721 (1987).
[16] C. Stauber and D. Schümperli, *Nucleic Acids Res.* **16**, 9399 (1988).

in their 5' ends can be observed, if the reaction is carried out in the presence of ethylenediaminetetraacetic acid (EDTA). The longest of these 3' fragments is the exact complement of the mature 5' fragment; the shorter ones start at and around the conserved spacer motif. The exact relationship between these 3' fragments is not entirely clear, so that several reaction mechanisms are possible, involving either one or several endonucleolytic cuts and possibly 3'- or 5'-exonuclease nibbling to produce all the RNA species observed *in vitro*. Moreover, the *in vitro* system is less dependent on the presence of the hairpin loop motif than *in vivo*,[11] possibly reflecting differences in the stabilities of the mature RNA product. The processing reaction is absolutely dependent on the presence of snRNPs[10,13,14,17] and an additional heat-labile component.[17] Changes in this heat-labile activity are responsible for a down-regulation of histone RNA 3' processing and hence for much of the regulation of histone gene expression seen in proliferation-arrested cells.[15,16] It is presently not clear whether U7 is the only snRNP involved in the processing reaction. RNA-binding studies revealed that the hairpin loop motif of the pre-mRNA interacts very rapidly with an Sm-precipitable but micrococcal nuclease-resistant component.[10,14] At later reaction times, a more extensive interaction, which is sensitive to micrococcal nuclease, includes the downstream spacer motif. An additional interaction, whose functional role is less evident, involves the capped 5' end of the pre-mRNA.

Preparation of Nuclear Extracts

Nuclear extracts are prepared essentially as described for transcriptionally competent extracts by Dignam *et al.*[18,19] Similar, if not identical, methods are also used to obtain splicing[20] and polyadenylation[21] extracts. The following protocol is equally suitable for human HeLa cells[13] or for the mouse mastocytoma cell line K21 and temperature-sensitive derivative 21-Tb.[15] HeLa cells are grown in suspension in Joklik's modified essential medium with 5% newborn calf serum at 37°, K21 or 21-Tb cells in P815 medium[22] with 10% horse serum at 37 or at 33°, respectively. To obtain the most active extracts, it is important to harvest cells in

[17] O. Gick, A. Krämer, A. Vasserot, and M. L. Birnstiel, *Proc. Natl. Acad. Sci. U.S.A.* **84,** 8937 (1987).
[18] J. D. Dignam, R. M. Lebovitz, and R. G. Roeder, *Nucleic Acids Res.* **11,** 1475 (1983).
[19] J. D. Dignam, P. L. Martin, B. S. Shastry, and R. G. Roeder, this series, Vol. 101, p. 582.
[20] A. Krämer and W. Keller, this volume [1].
[21] C. L. Moore, this volume [6].
[22] J. C. Schaer and R. Schindler, *Biochim. Biophys. Acta* **147,** 154 (1967).

the exponential phase, i.e., at a density of 4–6 × 10^5 cells/ml, and, during growth, cell densities should always be kept in the range of 1–6 ×10^5 cells/ml.

For a large-scale preparation, 5–20 liters of cell suspension is spun down in 1-liter bottles in a Sorvall H600A rotor at 2000 rpm for 20 min. The cell pellet is resuspended in 25 ml of cold phosphate-buffered saline [(PBS) 137 mM NaCl, 2.7 mM KCl, 4.3 mM $Na_2HPO_4 \cdot 7H_2O$, 1.4 mM KH_2PO_4 (pH 7.4)] per 10^9 cells and centrifuged at 4° and 1300 rpm (H600A rotor) for 5 min. All the following steps are performed on ice, and the solutions and the homogenizer are precooled. Additionally, all solutions, the homogenizer, and all tubes have to be RNase-free.

The washed cells are resuspended in 5 packed cell volumes (PCV) of buffer A [10 mM N-2-hydroxyethylpiperazine-N'-2-ethanesulfonic acid (HEPES)–KOH (pH 7.9), 1.5 mM $MgCl_2$, 10 mM KCl, 0.5 mM dithiothreitol (DTT), added just before use] and left to swell on ice for 10 min, after which they are centrifuged in a Sorvall HB-4 rotor at 1300 rpm for 5 min. The pellet should be 1.5–2 times larger than the initial one after washing with PBS. The swollen cells are taken up in 2 PCV (cell volumes before swelling) of buffer A and transferred to a Dounce homogenizer (S-type pestle). The cells are lysed by 5–10 strokes, but the nuclei should remain intact. This is checked under a fluorescence microscope by staining a small sample with acridine orange (~50 μg/ml, final concentration). Nuclei are centrifuged at 2500 rpm (HB-4 rotor) for 10 min. The cloudy supernatant is carefully removed by pipetting.

The nuclear pellet is resuspended in exactly 1.3 nuclear pellet volumes of buffer C [20 mM HEPES–KOH (pH 7.9), 25% (v/v) glycerol, 420 mM NaCl, 1.5 mM $MgCl_2$, 0.2 mM EDTA, 0.5 mM DTT, freshly added, 0.5 mM phenylmethylsulfonyl fluoride (PMSF), freshly added from a 0.25 mM stock solution in 2-propanol], transferred to the Dounce homogenizer, and treated by 10–15 strokes. Homogenization is considered sufficient when about 10% of the nuclei appear disrupted (checked by fluorescence microscopy as above). This nuclear fraction is then slowly stirred on ice for 40 min, transferred to a 15-ml Corex tube, and centrifuged in a Sorvall SS34 rotor at 15,000 rpm for 30 min. The supernatant is placed in dialysis membrane tubes (Spectrapor, MW cutoff ~3500) and dialyzed against 50 volumes of buffer D [20 mM HEPES–KOH (pH 7.9) 20% (v/v) glycerol, 100 mM KCl, 0.2 mM EDTA, 0.5 mM DTT, freshly added] at 4° for 5 hr. After centrifugation at 10,000 rpm (HB-4 rotor) for 15 min at 0–4°, the nuclear extract is quickly frozen in aliquots in liquid nitrogen. The extracts remain active in histone RNA processing for at least 2 years, if they are kept frozen at −70°. Thawing and refreezing slightly reduces the activity. A preparation starting from 10^{10} K21 or 21-Tb cells should yield about 30 ml of nuclear extract with a protein concentration of approximately 5 mg/ml.

Nuclear extracts from the mouse fibroblast cell line C127[16] and from smaller numbers of K21 or 21-Tb cells ($1-2 \times 10^8$ cells) can be prepared using the same protocol as above, except for a few modifications. C127 fibroblasts[23] are grown as monolayer cultures in Dulbecco's modified essential medium with 10% fetal calf serum. Twenty petri dishes (10 cm in diameter) of almost confluent C127 cells ($8-10 \times 10^7$ cells) are harvested by trypsinization and washed with 10 ml of PBS. For swelling, 3 ml of buffer A is added to the pellet, and the cell suspension is left on ice for 15 min. After centrifugation, the swollen cells are transferred with 3 ml of buffer A to a Dounce homogenizer (S-type pestle) and lysed by 100–300 strokes in the case of C127 cells or 40–70 strokes for K21/21-Tb cells. The nuclei are spun down, taken up in two pellet volumes of buffer C, and lysed in the homogenizer by 20–50 strokes for C127 cells or 40 strokes for K21/21-Tb cells. The broken nuclei are transferred to Eppendorf tubes and stirred on a rotating wheel for 1 hr. After centrifugation, the extract is dialyzed in Eppendorf tubes covered with dialysis membranes against 100 volumes of buffer D for 5 hr.

Nuclear extracts prepared from cells under various growth conditions have helped to reveal proliferation-dependent changes in the efficiency of histone RNA 3' processing.[15,16] In particular, nuclear extracts of G_1 phase-arrested 21-Tb cells were prepared after incubating the cells at 39° for 44 hr.[15] C127 fibroblasts have been similarly arrested by incubating confluent plates for 24 hr in medium containing only 0.5% fetal calf serum.[16] Such arrested extracts exhibit hardly detectable processing activity, owing to a specific lack of a heat-labile processing component. However, a normal content of snRNP component(s) can be demonstrated by a complementation assay (see below).

Standard Assay for Histone RNA 3' Processing

In Vitro Synthesis of Capped Histone Pre-mRNAs

RNA substrates for *in vitro* processing are best prepared by runoff transcription of specific DNA templates with bacterial or phage RNA polymerases. For this purpose, a restriction fragment containing the 3'-terminal region of a histone gene and including the conserved palindrome and spacer motifs is cloned into a polylinker sequence downstream of a

[23] D. R. Lowy, E. Rands, and E. M. Scolnick, *J. Virol.* **26**, 291 (1978).

strong promoter. Previous studies have used transcripts generated with *Escherichia coli*[24] or SP6 RNA polymerase,[5,9–11,13–17] but the T3 or T7 systems should be similarly useful. In earlier studies, using the *Xenopus laevis* oocyte microinjection assay, *in vitro* transcripts of sea urchin,[5,24] chicken,[25] *Xenopus*,[26] and mouse[27] origin were all processed efficiently, with the only exception, so far, of one sea urchin H3 gene. Considering the relatively high degree of sequence conservation of the RNA processing signal, it can therefore be assumed that the mammalian *in vitro* systems will display a similarly broad substrate specificity.

It may be preferable to use relatively short histone mRNA precursors (100–300 nucleotides), because they are easier to analyze and less subject to degradation than longer ones. In our laboratory, we routinely use run-off transcripts of *Bam*HI-cut plasmid SP65-H4-119/70 derived from mouse histone H4 sequences.[13] This RNA contains 119 nucleotides (25 from the plasmid and 94 from the H4 insert) preceding and 70 nucleotides (63 from the H4 insert and 7 from the plasmid) following the cleavage site.

To synthesize the substrate, the template DNA is cleaved with the appropriate restriction enzyme, purified by phenol and chloroform extractions and ethanol precipitation, and then incubated with RNA polymerase and nucleoside triphosphates. To ensure capping of the RNA, and thus to increase its stability during incubation in the nuclear extract, a cap analog of the structure m^7GpppN_1, where N_1 is the first nucleotide of the *in vitro* transcript, is included.

A standard reaction of 20 μl using plasmid SP65-H4-119/70 contains the following: 1 μg of template DNA, 40–100 μCi of $[\alpha^{32}P]GTP$ diluted with cold GTP to 50 μM (final concentration), the other unlabeled nucleoside triphosphates at 500 μM, m^7GpppG at 500 μM, 30 units of RNasin, and 20 units of SP6 RNA polymerase. The buffer for the reaction is 40 mM tris(hydroxymethyl)aminomethane hydrochloride (Tris-HCl) (pH 7.5), 6 mM $MgCl_2$, 10 mM DTT, 2 mM spermidine, and 100 μg/ml bovine serum albumin. The reaction is started by the simultaneous addition of template DNA and RNA polymerase and is allowed to proceed for 1 hr at 40°. Alternatively, nonradioactive pre-mRNA transcripts can also be 3'-end-labeled with $[5'-^{32}P]pCp$ and T4 RNA ligase.[28]

All substrate RNAs are isolated on preparative 5–8% polyacrylamide–8.3 M urea gels in TBE buffer [100 mM Tris–borate (pH 8.3), 1 mM

[24] C. Birchmeier, D. Schümperli, G. Sconzo, and M. L. Birnstiel, *Proc. Natl. Acad. Sci. U.S.A.* **81,** 1057 (1984).
[25] P. A. Krieg and D. A. Melton, *Nature (London)* **308,** 203 (1984).
[26] O. Georgiev, J. Mous, and M. L. Birnstiel, *Nucleic Acids Res.* **12,** 8539 (1984).
[27] O. Georgiev and D. Schümperli, unpublished result.
[28] T. E. England, A. G. Bruce, and O. C. Uhlenbeck, this series, Vol. 65, p. 65.

EDTA]. This is preferable to a purification by ethanol precipitations alone, because *in vitro* transcriptions can frequently produce prematurely terminated RNA species. After short autoradiography, the band containing the substrate RNA is excised, crushed, and eluted for 2 hr at room temperature in 0.3 M NaCl, 0.1% sodium dodecyl sulfate (SDS), 0.1 mM EDTA, 10 mM Tris-HCl (pH 7.5), and 20 μg/ml proteinase K. The supernatant is further extracted with phenol, phenol–chloroform, and chloroform, and the RNA is then precipitated with ethanol.

In Vitro Processing Reaction

Typically, processing reactions are started by adding, on ice, 15 μl of a radioactive mix [20,000–30,000 cpm of substrate RNA ($\sim$30 fmol), 40 mM EDTA, and 0.4 μg/μl yeast RNA carrier] to the appropriate amount of nuclear extract, adjusted to 15 μl with buffer D. The mixture is incubated at 30°.

The reaction shows no pronounced time lag and proceeds linearly with time for the first 2 hr, after which it begins to level off.[13] The appearance of the products is dependent on the amount of nuclear extract, the optimal being 15–30% of extract in the reaction. It is also dependent on the type of histone gene used as substrate.[14] Efficiency of product formation correlates with the efficiency of expression of the gene *in vivo*. The reaction does not seem to require an energy source[13] and the optimal salt concentration is rather broad, peaking around 60 mM KCl, with 50% activity at 20 mM and 80% activity at 100 mM.[14]

The reaction is stopped by adding an equal volume of 2$\times$ proteinase K stop solution [1 mg/ml proteinase K, 25 mM EDTA, 300 mM NaCl, 2% (v/v) SDS, 20 mM Tris-HCl (pH 7.5)] and incubating for another 20 min at 30°. After extraction with phenol, phenol–chloroform and chloroform, substrate and product RNAs are precipitated with ethanol.

Detection of Processed RNAs

RNAs can be directly analyzed on 8% polyacrylamide–8.3 M urea gels in TBE buffer (Fig. 1). If the reaction has been performed in the presence of 20 mM EDTA, both the capped 5' fragment (mature mRNA) and multiple species corresponding to the uncapped 3'-terminal fragments can be detected.[13] These multiple 3' fragments differ in their 5' ends, only the longest species being the exact complement of the mature 5' fragment. It is therefore possible that the reaction involves multiple endonucleolytic cuts and/or exonucleolytic nibbling of the primary cleavage products (see Introduction).

In the presence of Mg^{2+}, the cleaved-off spacer transcripts are not detected,[13] presumably because they are subject to degradation by a $5' \rightarrow 3'$-exonuclease. Moreover, in the presence of Mg^{2+}, the $5'$ fragment obtained is slightly smaller, suggesting the presence of a Mg^{2+}-dependent $3' \rightarrow 5'$-exonuclease activity for which the terminal palindrome represents a resection barrier.[2,29]

Identification of Trans-Acting Components by Inactivation and Complementation

The histone-specific RNA processing reaction can be specifically inhibited by a number of procedures (Fig. 1). Inhibition of the reaction after pretreatment of the nuclear extract with micrococcal nuclease (lane 2)[10,17] or after immunoprecipitation with antisera directed against Sm determinant(s) (lane 3)[13] or against the specific $m^{2,2,7}G(m_3G)$ cap structure (lane 4)[30] are indicative of the involvement of snRNPs of the U family. A specific requirement for one member of this family, the U7 snRNP, can be demonstrated by inhibition after a pretreatment of the extract with RNase H and a DNA oligonucleotide complementary to the $5'$ end of U7 snRNA (lane 5),[9–11] but it is not yet clear if U7 is the only snRNP involved. At least, none of the major U snRNPs seem to be required, because a highly enriched U7 snRNP preparation that was essentially free of the major snRNPs but still contained other minor snRNPs was shown to be active in histone RNA 3' processing by itself.[13]

A separate component(s) of the processing reaction can be inactivated by a mild heat treatment (lane 6).[17] This heat-labile component is resistant to micrococcal nuclease treatment and cannot be immunoprecipitated with Sm antisera. Moreover, a heat-inactivated nuclear extract can complement extracts inactivated for the snRNP component(s) by one of the methods mentioned above (lanes 7–10). As already mentioned, the heat-labile activity is subject to a specific down-regulation in proliferation-arrested cells and thus plays a major role in the proliferation-dependent regulation of histone gene expression.[15,16]

A similar type of complementation was demonstrated after separation of a HeLa cell nuclear extract on a BioGel A-1.5m sizing column (see Fractionation of Nuclear Extract).[10] It is not known, however, whether the components separated by this method are identical with the heat-labile and snRNP components mentioned above.

[29] D. Soldati and D. Schümperli, unpublished result.
[30] V. S. Meier, R. Lührmann, and D. Schümperli, unpublished result.

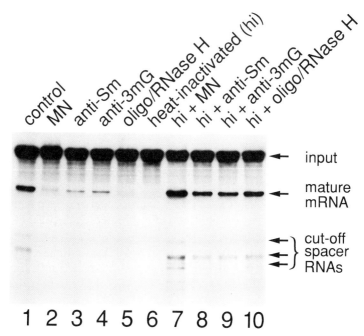

input

mature
mRNA

cut-off
} spacer
RNAs

1 2 3 4 5 6 7 8 9 10

FIG. 1. Inactivation by various procedures and mutual complementation of inactivated extracts. Precursor RNA (input) was incubated for 90 min at 30° in a 30-μl processing reaction containing 5 μl of native K21 nuclear extract (lane 1), 5 μl of inactivated extracts as indicated (lanes 2–6), or a mixture of 5 μl each of heat-inactivated extract and extract inactivated by one of the other procedures as indicated (lanes 7–10). The reactions were stopped, and RNAs were isolated and analyzed by electrophoresis on a 8% polyacrylamide–8.3 M urea gel. Mature RNA, 5' fragment extending from the cap site to the processing site; cut-off spacer RNAs, 3' fragments extending from the processing site or positions in the spacer to the end of the runoff transcript. MN, Inactivation by micrococcal nuclease digestion; anti-Sm and anti-m_3G, inactivation by immunoprecipitation with anti-Sm or anti-m_3G antibodies, respectively; oligo/RNase H, inactivation by pretreatment with a synthetic DNA oligonucleotide complementary to the first 16 nucleotides of mouse U7 snRNA (Fig. 2c) in the presence of 2.7 mM MgCl$_2$.

Micrococcal Nuclease Inactivation

For micrococcal nuclease (MN) inactivation, the nuclear extract (in buffer D) is adjusted to 1 mM CaCl$_2$ and incubated for 15 min at 30° in the presence of 3 units of MN [Pharmacia; dissolved in 5 mM Tris-HCl (pH 7.5), 25% (v/v) glycerol] per microliter of nuclear extract.[17] The digestion is stopped by chelating Ca^{2+} with ethylene glycol bis(β-aminoethyl etner)-N,N,N',N'-tetraacetic acid (EGTA) (pH 8.0) at a final concentration of 2 mM.

If histone RNA processing is assayed as above in such MN-treated extracts, there is little if any detectable activity left (Fig. 1, lane 2; Fig. 2a, lane 4)[17] and a specific shortening of U7 snRNA can be demonstrated directly (Fig. 2b, lane 4).[9,10,31] However, complementation is possible with fractions enriched in snRNPs or with heat-inactivated extracts (Fig. 1, lane 7).[17] Complementation assays are performed by mixing together 5 μl of each of the pretreated extracts, adjusting the volume to 15 μl with buffer D, and then adding 15 μl of radioactive mix to assay histone RNA processing as described previously.

Depletion of snRNPs in Nuclear Extracts by Immunoprecipitation with Anti-Sm or Anti-m₃G Antibodies

Anti-Sm antibodies precipitate all known U-type snRNPs except for U3.[32,33] Moreover, specific antibodies have been generated that are directed against the unique m_3G cap structure present on all U RNAs analyzed, except for U6.[34-36] Extracts depleted by immunoprecipitation with one or the other of these antibodies are lacking the essential U7 snRNP (and possibly other essential snRNPs as well) and are therefore not able to mature histone pre-mRNAs (Fig. 1, lanes 3 and 4). However, complementation is possible, in analogy to MN-treated extracts (Fig. 1, lanes 8 and 9).

The depletion is carried out by incubating the extract with the antibodies coupled to protein A-Sepharose beads (Pharmacia).[37,38] Before coupling, the beads are swollen for 2 hr at room temperature in NET-2 [150 mM NaCl, 50 mM Tris-HCl (pH 7.5), 0.05% (v/v) Nonidet P-40]. Then the beads are washed twice in IPP [500 mM NaCl, 10 mM Tris-HCl (pH 8.0), 0.1% (v/v) Nonidet P-40] and can be stored as a 1 : 4 suspension in IPP at 4°. Antiserum in IPP is allowed to react with an equal volume of packed beads for 2 hr at 4° on a slowly rotating wheel. The beads are then washed 3 times with ice-cold IPP and 3 times with ice-cold buffer D.

---

[31] G. M. Gilmartin, F. Schaufele, G. Schaffner, and M. L. Birnstiel, *Mol. Cell. Biol.* **8,** 1076 (1988).
[32] M. R. Lerner and J. A. Steitz, *Proc. Natl. Acad. Sci. U.S.A.* **76,** 5495 (1979).
[33] I. Petterson, M. Hinterberger, T. Mimori, E. Gottlieb, and J. A. Steitz, *J. Biol. Chem.* **259,** 5907 (1984).
[34] P. Bringmann, R. Reuter, J. Rinke, B. Appel, and R. Lührmann, *J. Biol. Chem.* **258,** 2745 (1983).
[35] R. Lührmann, *in* "Structure and Function of Major and Minor Small Ribonucleoprotein Particles" (M. L. Birnstiel, ed.), p. 71. Springer-Verlag, Berlin, 1988.
[36] A. R. Krainer, *Nucleic Acids Res.* **16,** 9415 (1988).
[37] E. M. de Robertis, S. Lienhard, and R. E. Parisot, *Nature (London)* **295,** 572 (1982).
[38] A. Krol, J. P. Ebel, J. Rinke, and R. Lührmann, *Nucleic Acids Res.* **11,** 8583 (1983).

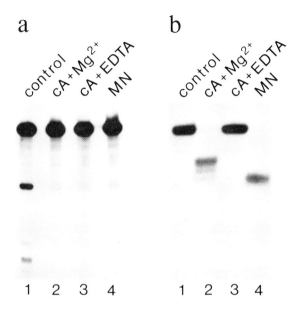

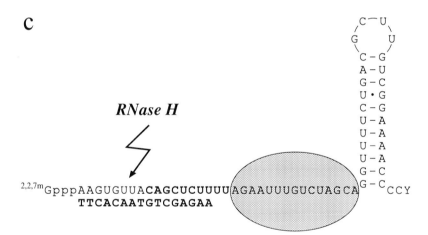

FIG. 2. (a) Inhibition of histone RNA 3' processing by oligonucleotide cA complementary to the first 16 nucleotides of mouse U7 snRNA or by micrococcal nuclease. Nuclear extract from K21 cells was pretreated in the presence of the following: lane 1, 10 mM EDTA; lanes 2 and 3, 50 ng of oligonucleotide cA plus 2.7 mM MgCl$_2$ (lane 2) or 10 mM EDTA (lane 3); lane 4, micrococcal nuclease. (b) Specific cleavage of U7 snRNA by RNase H in the presence of complementary oligonucleotide cA or by micrococcal nuclease. RNA from extracts preincubated as in (a) were separated by gel electrophoresis and transferred to a GeneScreen (New England Nuclear) filter membrane by electroblotting. The filter was hybridized with a mixture of ^{32}P-labeled oligonucleotides complementary to the entire length of U7 snRNA. Each lane contained approximately 2 × 10^7 cell equivalents of RNA. (c)

Nuclear extract is mixed 1:1 with packed beads, and the mixture is incubated for 1 hr at 4° on the wheel. The beads are removed by centrifugation, and the supernatant can be used directly in processing/complementation assays.

Specific Inactivation of U7 snRNPs by Oligonucleotide-Targeted RNase H Treatment

Single snRNAs present in the nuclear extract can be destroyed specifically by forming a DNA–RNA hybrid with a synthetic deoxyoligonucleotide and then digesting that hybrid with RNase H.[39,40] This method requires that a vital part of the U snRNA under investigation be available for base-pairing with a complementary deoxyoligonucleotide within the context of the native snRNP. This is, of course, the case when the function of that snRNP is based on base-pairing contacts with the substrate RNA. For U7, such an essential interaction is occurring between the 5′ end of U7 RNA and the conserved spacer element of the histone pre-mRNA.[8–11]

The oligomers used for this purpose are designed to mimic the pre-mRNA by being exactly complementary to the 5′ end of U7 RNA[9] (Fig. 2c). The synthesized oligonucleotides are purified on 12% polyacrylamide–8 M urea gels in TBE. After excision and elution from the gel, they are passed over a Sep-Pak column (Circartridges, Millipore) to remove urea, salts, and acrylamide. They are then lyophilized and redissolved in TE [10 mM Tris-HCl (pH 7.5), 1 mM EDTA].

The extracts should contain enough endogenous RNase H (see below) so that it is sufficient to incubate them with an excess of oligonucleotide (1–10 ng/μl of nuclear extract) for 15 min in the presence of 2.7 mM MgCl$_2$. Subsequently, the Mg^{2+} is chelated by the addition of 20 mM EDTA to inhibit the RNase H. The temperature of incubation depends on the length of the oligonucleotide. It is optimal to use a temperature just slightly below the melting point of the DNA–RNA hybrid. The RNase H is still active at quite low temperatures (~10°).

[39] A. Krämer, W. Keller, B. Appel, and R. Lührmann, *Cell* **38**, 299 (1984).
[40] A. Krämer, this volume [23].

Schematic representation of the oligo/RNase H method. Mouse U7 snRNA is shown with the m₃G cap structure at the 5′ end and with its hairpin loop structure at the 3′ end. The region base-pairing with the spacer element of histone pre-mRNA is shown in bold type, and the presumptive interaction with proteins carrying Sm antigenic determinants is indicated by the shaded region. The complementary oligonucleotide cA used in the experiments of Figs. 1 and 2 is shown below the U7 sequence. (Figure 2a,b is reprinted from Ref. 9 with permission of the American Society for Microbiology.)

Before being used in a complementation assay, the inactivated extract should be first treated with DNase I (0.5 units/μl of nuclear extract; RNase-free) for 15 min at room temperature to remove the oligonucleotides. This is necessary because an oligonucleotide that is perfectly complementary to U7 RNA inhibits the processing reaction even in the absence of RNase H cleavage (Fig. 2a, lane 3).

The same principle can be applied by using oligonucleotides complementary to the spacer element of the histone pre-mRNA. By mimicking U7 snRNP recognition, these oligomers will inhibit processing, with or without RNase cleavage. In fact, incubation of the radioactive pre-mRNA with a complementary oligonucleotide provides a convenient test to determine if there is enough RNase H in the nuclear extract.

The precise effect of digestion by RNase H or by MN on the U7 RNA can be visualized by Northern blot analysis (Fig. 2b). Total RNA from a sample of treated nuclear extract is isolated, separated on a 15% poly-acrylamide–8.3 M urea gel in TBE, transferred onto a nitrocellulose membrane by electroblotting, and hybridized with a probe recognizing U7 RNA (oligonucleotides complementary to the remainder of U7 RNA, 5'-end-labeled with [γ-^{32}P]ATP and T4 polynucleotide kinase can be used).

U7 RNA is shortened at the 5' end owing to truncation of the fragment corresponding to the DNA–RNA hybrid formed (Fig. 2b). In the case of MN treatment, the 5' end is similarly shortened, while the rest of the RNA is presumably protected from nucleolytic attack by proteins.

Heat Inactivation of Nuclear Extract

The heat-labile cofactor(s) of histone RNA 3' processing is readily inactivated by incubation at 48–50°, whereas the snRNP component withstands heating up to 65°.[17] The heat-labile component has been partly purified by column chromatography (see Fractionation of Nuclear Extract). Although a slight reduction of the processing activity can already be seen after a 15-min preincubation of the nuclear extract at 42°, the standard inactivation is carried out at 50° for 15 min.

Detection of Processing-Specific Complexes by Ribonuclease Protection of Pre-mRNA

RNase T1 digestion of processing reactions in combination with immunoprecipitation by anti-Sm or anti-m$_3$G antibodies has been used as a means to detect RNase-resistant complexes involving the radioactively labeled pre-mRNA substrate.[14] Determination of the nature of the resistant fragments by standard RNA fingerprinting techniques[41] directly reveals which parts of the substrate are involved in the formation of these complexes. This experimental approach, described in detail by Mowry and Steitz,[14] has revealed several potentially important interactions (see Introduction).

RNA substrates of very high specific activity are necessary for these protection experiments. They can be obtained by using the labeled GTP in undiluted form and possibly by adding additional labeled nucleotides. Processing reactions are then performed with 3×10^6 to 8×10^8 cpm of transcript per reaction. After appropriate times of incubation at 30°, 6.5 μl of RNase T1 (Calbiochem, San Diego, CA; 50 units/μl) is added to 20 μl of processing reaction and incubated for 30 min on ice. The entire reaction is added to the prebound antibodies on protein A-Sepharose beads and incubated for 30 min on ice with gentle mixing. The samples are pelleted, washed 4 times with NET-2, extracted with phenol–chloroform–isoamyl alcohol (50 : 49 : 1) and precipitated with ethanol.

The protected RNA fragments are resolved on a 15% polyacrylamide–8 M urea gel in TBE. After elution from the gel, the fragments are analyzed by complete digestion with RNase T1. The resulting T1 oligonucleotides are compared with the oligonucleotides generated from the full-length pre-mRNA and predicted from the sequence by electrophoresis on a 20% polyacrylamide gel or by two-dimensional fingerprinting. An unambiguous identification will sometimes require secondary digestion of reisolated oligonucleotides with RNase A.

Fractionation of Nuclear Extract

The following chromatographic fractionation procedure has been used to identify factors from HeLa cell extracts involved in polyadenylation[21] or splicing.[20] The same method also yields partly purified histone RNA 3' processing components.[17] The detailed protocol is described by Krämer *et al.*[42,43] In brief, HeLa cell extract (80 ml with a protein concentration of 4 mg/ml, corresponding to 3.1×10^{10} cells) is first passed over a DEAE-Sepharose column. The bound material is eluted with 500 mM KCl, dialyzed, and then loaded onto a heparin-Sepharose column. On both of these columns, total histone RNA processing activity and also the major

[41] G. G. Brownlee, *in* "Laboratory Techniques in Biochemistry and Molecular Biology" (T. S. Work and E. Work, eds.), Vol. 3, p. 1. North-Holland Publ., Amsterdam, 1972.
[42] A. Krämer, M. Frick, and W. Keller, *J. Biol. Chem.* **262,** 17630 (1987).
[43] A. Krämer, this volume [18].

snRNPs U1–U6 are enriched in the bound fraction. Only very minor amounts of snRNPs are detected in the flow-throughs. The bound fraction is eluted from the heparin-Sepharose column with 500 mM KCl, dialyzed, and applied onto a Mono Q column. This column is eluted with a linear salt gradient ranging from 100 to 500 mM KCl. During this step, the heat-labile component elutes slightly ahead of the snRNP activity, but the two peaks still overlap.[17] The intermediate fraction containing most of the total activity contains a number of minor snRNPs, whereas the more abundant snRNPs elute from the column at higher salt concentrations between 350 and 450 mM KCl.[43]

When the Mono Q fraction containing both heat-labile and snRNP activity is further chromatographed on a Superose 12 gel filtration column, both components and hence total processing activity are found in fractions corresponding to structures with molecular weights of approximately 300,000.[17] Additionally, the heat-labile factor is also present in a fraction corresponding to about 40,000,[17] as can be shown by complementation with a heat-inactivated extract. A similar bimodal distribution is obtained with glycerol gradients, the large form of heat-labile activity sedimenting around 14 S and the small form at less than 4 S.[17]

The analysis of snRNAs in the Superose 12 fractions revealed the presence of a minor snRNA with a length of 66/67 nucleotides, i.e., with the expected size of U7 RNA.[44] This RNA was detected in fractions corresponding to molecular weights of approximately 115,000. If this particle represents the U7 snRNP, it is obviously not competent in processing, which may be due either to incomplete assembly or to a lack of additional factor(s).

In a different approach, Mowry and Steitz[10,14] have identified several components of the nuclear extracts that interact with histone pre-mRNA (see also above) and have used biochemical separation procedures to obtain individual, mutually complementing fractions. On fractionation of crude HeLa cell extract over a BioGel A-1.5m gel filtration column (extract from 2.5 × 10⁹ cells applied to a 2.2 × 100 cm column),[45] they collected three pools of fractions. Pool A contained structures of large molecular size (>10⁶ daltons), including most snRNPs. Pool B contained some U1 snRNPs and proteins ranging between 100 and 500 kDa. Pool C contained proteins smaller than 100 kDa.

Fractions A and B, both of which were inactive by themselves, were able to complement each other, and fraction A (containing Sm-type snRNPs) additionally complemented an extract inactivated by MN treat-

[44] A. Krämer, *Proc. Natl. Acad. Sci. U.S.A.* **84,** 8408 (1987).
[45] V. Gerke and J. A. Steitz, *Cell* **47,** 973 (1986).

ment. The addition of fraction C had no obvious effect on the processing reaction. Fraction B mainly appeared to contain factor(s) capable of binding independently to the hairpin loop of histone pre-mRNA, whereas fraction A was required in addition for the extended interaction with the downstream element. As mentioned above, the hairpin-binding factor has been shown to be Sm reactive but insensitive to MN digestion, whereas the interaction with the downstream element is both sensitive to MN and Sm reactive.[14]

Acknowledgments

Work in the authors' laboratory was supported by Swiss National Science Foundation Grants 3.216.85 and 3.036.87 and by the State of Zürich.

[8] Preparation and Fractionation of Yeast Splicing Extract

By Soo-Chen Cheng, Andrew Newman, Ren-Jang Lin, Gloria D. McFarland, and John N. Abelson

Introduction

The advent of *in vitro* synthesis of messenger RNA precursor (pre-mRNA) using SP6 RNA polymerase has facilitated *in vitro* studies of pre-mRNA splicing. With the development of cell-free systems for pre-mRNA splicing in both mammals and yeast, the splicing reactions have been characterized and the spliced products analyzed.[1-7] A two-step model has been formulated for pre-mRNA splicing. The first step involves cleavage at the 5' splice site and the formation of a lariat structure.[8,9] In the second step, the 3' splice site is cleaved, and the two exons are joined.

[1] R. Kole and S. M. Weissman, *Nucleic Acids Res.* **10**, 5429 (1982).
[2] C. J. Goldenberg and S. D. Hauser, *Nucleic Acids Res.* **11**, 13337 (1983).
[3] R. A. Padgett, S. F. Hardy, and P. A. Sharp, *Proc. Natl. Acad. Sci. U.S.A.* **80**, 5230 (1983).
[4] N. Hernandez and W. Keller, *Cell* **35**, 89 (1983).
[5] S. F. Hardy, P. J. Grabowski, R. A. Padgett, and P. A. Sharp, *Nature (London)* **308**, 375 (1984).
[6] A. R. Krainer, T. Maniatis, B. Ruskin, and M. R. Green, *Cell* **36**, 993 (1983).
[7] R.-J. Lin, A. J. Newman, S.-C. Cheng, and J. Abelson, *J. Biol. Chem.* **260**, 14780 (1985).
[8] B. Ruskin, A. R. Krainer, T. Maniatis, and M. R. Green, *Cell* **38**, 317 (1984).
[9] R. A. Padgett, M. M. Konarska, P. J. Grabowski, S. F. Hardy, and P. A. Sharp, *Science* **225**, 898 (1984).

Both nuclear and whole cell extracts have been described for mammalian systems.[3,6] The nuclear extracts splice with higher efficiency than the whole cell extracts. Nevertheless, the basic features of the splicing reactions are similar. Although splicing is believed to take place in nuclei, attempts to make nuclear extracts for splicing in yeast have not been successful. Presumably, essential components of the splicing machinery leak out of nuclei during preparation. A whole cell extract, however, can efficiently splice synthetic pre-mRNAs.[7]

To understand the mechanism of the splicing reaction, it is necessary to identify components of the splicing machinery. Biochemical fractionation of the yeast whole cell extract has separated splicing activity into three fractions, termed Fractions I, II, and III.[10] Each fraction alone does not have any splicing activity. Combination of Fractions I and II gives the first step of the splicing reaction. Addition of Fraction III completes the reaction. Procedures for preparation of the whole cell extract and fractionation of the extracts are described in this chapter.

Materials and Methods

Medium and Solutions

YPD: 1% yeast extract, 2% peptone, 2% dextrose

Buffer A: 10 mM N-2-hydroxyethylpiperazine-N'-2-ethanesulfonic acid (HEPES)–K$^+$ (pH 7.9 at 4°), 1.5 mM MgCl$_2$, 10 mM KCl, 0.5 mM dithiothreitol (DTT)

Buffer D: 20 mM HEPES–K$^+$ (pH 7.9), 0.2 mM ethylenediaminetetraacetic acid (EDTA), 0.5 mM DTT, 50 mM KCl, 20% (v/v) glycerol

Buffer SB3: 50 mM tris(hydroxymethyl)aminomethane hydrochloride (Tris-HCl) (pH 8.0 at 25°), 10 mM MgCl$_2$, 1 M sorbitol, 3 mM DTT

Buffer SB30: 50 mM Tris-HCl (pH 8.0), 10 mM MgCl$_2$, 1 M sorbitol, 30 mM DTT

Buffer I: 20 mM HEPES (pH 7.9), 0.5 mM EDTA, 0.5 mM DTT, 10% (v/v) glycerol

Preparation of Whole Cell Extracts

The following procedure is for preparation of extracts from a 1-liter yeast culture. Yeast strain EJ101 (α, $trp1$, $pro1$-126, $prb1$-112, $pep4$-3,

[10] S.-C. Cheng and J. Abelson, *Proc. Natl. Acad. Sci. U.S.A.* **83,** 2387 (1986).

prc1-126) is routinely used for preparation of extracts because of its deficiency in proteases. Active extracts have been made from other strains of *Saccharomyces cerevisiae*. Yeast cells are grown at 30° in YPD medium to an A_{600} value of approximately 3 and harvested by centrifugation at 3000 rpm for 5 min (Sorvall GS3 rotor). Cells grown to higher density (up to an A_{600} of ~8) can also make active extracts when the scale of preparation is normalized to the cell density. The cell pellets are resuspended in 15 ml of SB30 buffer and incubated at 25° without shaking for 15 min. The cells are collected by centrifugation at 3000 rpm for 5 min and resuspended in 15 ml of SB3 buffer. Ninety microliters of 20 mg/ml zymolyase 100,000 (Seikagak Kogyo Co., Ltd.) is added, and the cell suspension is gently shaken at 30° for 40 min.

The following operations are carried out at 0–4°. The spheroplasts are collected by centrifugation at 3000 rpm for 5 min and washed once by gentle resuspension in 15 ml of SB3 buffer and then collected as before. The spheroplast pellets are gently resuspended in 8 ml of buffer A and lysed by homogenization in a Dounce homogenizer with 5 strokes using a tight-fitting pestle. The tightness of the pestle is important for making an active extract; loose-fitting pestles have consistently given inactive extracts. Both glass (Kontes and Wheaton) and steel (Wheaton) Dounce homogenizers have been used to give active extracts. Next, 2 M KCl is added to the lysate to give a final concentration of 0.2 M, and the mixture is stirred gently on ice for 30 min. Cell debris is removed by centrifugation at 17,000 rpm for 30 min (Sorvall SS34 rotor).

The supernatant is centrifuged at 37,000 rpm in Beckman Ti60 rotor (100,000 g) for 60 min. The clear phase of the supernatant is carefully collected with pipets. The suspensions at the top and near the bottom are also active for splicing but usually cause aggregation of RNA during the splicing reaction; they are, therefore, discarded at this step. The clear supernatant is dialyzed for 3 hr against 1 liter of buffer D, and the dialyzate is centrifuged at 17,000 rpm for 20 min (Sorvall SS34 rotor) to remove insoluble material. Prolonged dialysis usually leads to diminished activity. The dialyzate is divided into small aliquots, quick-frozen in liquid nitrogen, and stored at −70°.

Extracts made by this procedure have a protein concentration of 20–30 mg/ml and are stable for at least 2 years. NaCl can replace KCl in all buffers to give active extracts. Potassium phosphate buffer at pH 7.4 can also be used in buffers A and D, replacing HEPES. Addition of phenylmethylsulfonyl fluoride (PMSF) in buffers A and D does not increase the level of the splicing activity for extracts made from strain EJ101, probably because of the protease deficiency in that strain. Extracts made from other strains seem to have better splicing activity when PMSF (0.5 mM) is added to the buffers.

Assay of in Vitro Splicing

Actin precursor RNA is synthesized *in vitro* as runoff transcripts using SP6 RNA polymerase and labeled with [α-^{32}P]UTP at 20 Ci/mmol. The full-length transcript is purified from 5% polyacrylamide (29 : 1)–8 M urea gels. Standard splicing reactions are carried out in a solution containing 60 mM potassium phosphate (pH 7.0), 3 mM MgCl$_2$, 2 mM ATP, 1 mM spermidine, 3% (w/v) polyethylene glycol (PEG) 8000, 0.4 nM actin transcript, and 40% whole cell extract at 23° for 20 min. The reactions are stopped by addition of one-third volume of stop solution containing 1 mg/ml proteinase K, 50 mM EDTA, and 1% sodium dodecyl sulfate (SDS), and then incubated at 37° for 20 min. Two hundred microliters of a mixture containing 50 mM sodium acetate (pH 5), 1 mM EDTA, 0.1% SDS, 25 mg/ml *Escherichia coli* RNA is then added, and proteins are removed by extraction with an equal volume of phenol–chloroform (50 : 50) equilibrated with 50 mM sodium acetate (pH 5), 1 mM EDTA. Sodium acetate is added to 0.3 M, and the nucleic acids are precipitated with 2.5 volumes of ethanol, rinsed with 70% ethanol, and dried *in vacuo*. For each 10-μl splicing reaction, nucleic acids are resuspended in 3 μl of distilled water, and 7 μl of formamide containing 0.025% each of bromphenol blue and xylene cyanol is added. The RNA is then analyzed on thin 8% polyacrylamide (29 : 1)–8 M urea gels.

Partial Fractionation of Whole Cell Extract

The following (NH$_4$)$_2$SO$_4$ fractionation procedure is for a 1-liter yeast culture, and the percentage of (NH$_4$)$_2$SO$_4$ represents percent saturation at 25°. When solid (NH$_4$)$_2$SO$_4$ is added for precipitation, it is always done over a period of 20–30 min. All (NH$_4$)$_2$SO$_4$ precipitations are performed at 0° with stirring for 30 min followed by centrifugation at 4° in a Sorvall SS34 rotor at 17,000 rpm for 20 min. The whole cell extract is prepared through the 100,000 g centrifugation step. The clear supernatant is the starting material for (NH$_4$)$_2$SO$_4$ fractionation.

Preparation of Fractions I, II, and III. Solid (NH$_4$)$_2$SO$_4$ is added to the 100,000 g supernatant to yield a 40% saturated solution. After stirring and centrifugation, the soluble fraction (40W) is concentrated by precipitation with 70% saturated (NH$_4$)$_2$SO$_4$, and the pellet is dissolved in 4 ml of buffer D. The insoluble fraction (40P1) is extracted twice with 10–15 ml of 40% saturated (NH$_4$)$_2$SO$_4$, prepared by dissolving 24.2 g of solid (NH$_4$)$_2$SO$_4$ in 100 ml of 20 mM HEPES (pH 7.9), 0.2 mM EDTA, and 0.5 mM DTT, and the final pellet (40P3 or Fraction I) is dissolved in 0.15 ml of buffer D. All fractions are dialyzed for 3–4 hr against 2 changes of 1 liter each of buffer D, followed by centrifugation in an Eppendorf microcentrifuge for 10 min to remove small amounts of insoluble material, and stored in aliquots at −70°. The 40P3 and 40W fractions should have protein concentrations of approximately 10–20 and 20–30 mg/ml, respectively.

To prepare Fractions II and III, 40W is diluted by a factor of 8 with buffer I and then applied to a 1-ml heparin-agarose column preequilibrated with 50 mM NaCl in buffer I. After collection of the flow-through fraction (Fraction II), the column is washed with 3–5 column volumes of 50 mM NaCl in buffer I. The column is then eluted successively with 0.25 and 0.5 M NaCl in buffer I. The 0.25 M eluent is discarded and the 0.5 M eluent is Fraction III. Fraction II and Fraction III are then stored at −70° in aliquots.

Preparation of 35P. The 100,000 g supernatant is mixed well with one-eighth volume of 2 M potassium phosphate (pH 8.0). Solid $(NH_4)_2SO_4$ is then added to yield a 35% saturated solution. The insoluble fraction is collected by centrifugation and dissolved in 0.15 ml of buffer D. After dialyzing against 2 changes of 1 liter each of buffer D for 3–4 hr, the dialyzates are centrifuged to remove insoluble material and stored in aliquots at −70°.

Features of in Vitro Splicing Reaction

Pre-mRNA splicing reactions in the yeast whole cell extracts display rapid kinetics,[7] unlike splicing in the mammalian system, in which there is a lag of 30–45 min.[3,6] The spliced products can be seen within 5 min of the reaction (Fig. 1). ATP is required for the splicing reaction. UTP, GTP, and CTP can substitute poorly for ATP, perhaps by inefficient phosphate transfer to endogenous ADP. Nonhydrolyzable ATP analogs [β,γ-methylene-ATP (AMPPCP) and α,β-methylene-ATP (AMPCPP)], either by themselves or in combination, cannot substitute for ATP, implying a requirement for ATP hydrolysis.[7] Since the splicing reaction is usually complete within 30 min, 2 mM ATP is sufficient to support the reaction in the absence of an ATP regeneration system.

Splicing is abolished by addition of 10 mM EDTA, indicating a divalent cation requirement, and displays a broad optimum for Mg^{2+}. Since there is a small amount of Mg^{2+} in the whole cell extract, it is usually not necessary to supply Mg^{2+} in the reactions using the whole cell extract. However, Mg^{2+} is absolutely required for reactions using the ammonium sulfate fractions. Therefore, 3 mM Mg^{2+} is routinely included in the splicing reaction. Spermidine seems to stimulate the splicing reaction in some

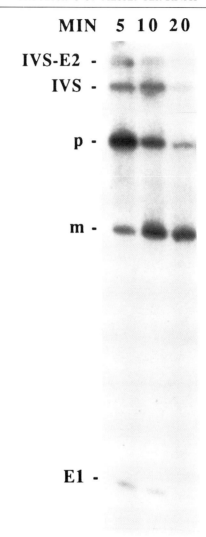

FIG. 1. Splicing of synthetic actin pre-mRNA in yeast whole cell extract.

extracts but not in others, and therefore it is included in all reactions for consistency. The optimum pH for the splicing reaction is 7.0. The optimum salt concentration is 60 mM potassium phosphate. KCl at the same ionic strength gives a lower splicing efficiency. The addition of PEG 8000 to 3% (w/v) enhanced the yield of spliced products, presumably by an excluded volume effect. The optimum temperature for splicing is 30°. At this temperature, pre-mRNA and the spliced products are degraded rapidly by nonspecific nucleases present in the extract. The reactions are, therefore, routinely performed at 23°. The placental ribonuclease inhibitor RNasin (from Promega, Madison, WI) at 1 unit/μl does not inhibit splicing *in vitro*,[7] nor does it inhibit nonspecific nucleolytic activities. Vanadyl–ribonucleoside complex (from Bethesda Research Laboratories, Gaithersburg, MD) completely inhibits splicing at a concentration of 10 mM.[7]

The pre-mRNAs carrying m[7]GpppG caps are spliced in the whole cell extract with an efficiency equal to those without caps.[7] The possibility that the transcripts might be capped in the whole cell extract during the splicing reaction cannot be ruled out. The addition of the cap analog dinucleotide m[7]GpppG to the splicing reaction does not affect the splicing activity even at concentrations as high as 250 μM.

Fractionation of Splicing Extracts

When the splicing extract is precipitated with 40% $(NH_4)_2SO_4$, the pellet fraction (40P1) is active for splicing but usually accumulates more splicing intermediates, free exon 1 and intron–exon 2, in the reaction. The supernatant fraction (40W), on the other hand, is inactive for splicing. Extraction of 40P1 twice with 40% $(NH_4)_2SO_4$, giving rise to the 40P3 fraction, removes all the splicing activity. Addition of the 40W fraction to 40P3 restores full splicing activity. Therefore, by this ammonium sulfate fractionation procedure, the splicing extract can be separated into 40P3 and 40W fractions that alone have no splicing activity but give full splicing activity in combination. The 40P3 fraction contains only about 10% of the total proteins in the whole cell extract. Since it is prepared at approximately the same protein concentration as the whole cell extract, it is a highly concentrated fraction for some specific components.

The 40W fraction can be further fractionated by chromatography on heparin-agarose. When the 40P3 fraction (Fraction I) is added to the flow-through fraction (Fraction II) from a heparin-agarose column, the first step of the splicing reaction takes place, giving rise to the splicing intermediates. Addition of the eluent from 0.5 M NaCl (Fraction III) completes the reaction (Fig. 2). It is difficult to separate Fractions II and III completely. A small amount of Fraction III activity is always associated with Fraction II, and vice versa.

Fractions I and II contain RNA species that are required in the splicing reaction, since they are sensitive to micrococcal nuclease treatment.[10] Fraction III, however, does not have essential RNA by the same criterion. Five small nuclear snRNAs have been shown to be components of

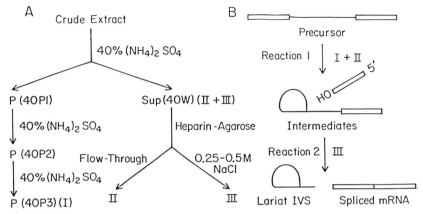

FIG. 2. (A) Scheme of fractionation of the mRNA splicing extract. (B) The mRNA splicing pathway and the involvement of Fractions I, II, and III in the pathway.

the spliceosome.[11-15] These are U1, U2, U4, U5, and U6. Analysis of snRNA constituents in the fractions reveals that the 40P3 fraction (Fraction I) is enriched for U5, U4, and U1 but contains no U2. The 40W fraction, on the other hand, contains U2 but only negligible amounts of U5, U4, and U1. RNA content in the 40W fraction reflects that in Fraction II, since 40W is a combination of Fractions II and III, of which only Fraction II has essential RNA species. U6 is distributed in both the 40P3 and 40W fractions. Analysis of ribonucleoprotein (RNP) particles by electrophoresis on native polyacrylamide gels reveals that this distribution of U6 is due to two forms of RNP, one of which is associated with U4 and is found in the 40P3 fraction.[12] The other form, not associated with U4, is found in the 40W fraction.

The 35P fraction is the fraction after precipitation with 35% (NH$_4$)$_2$SO$_4$. For preparation of 35P, the extract is kept at neutral pH in 0.25 M potassium phosphate during ammonium sulfate precipitation. The 35P fraction has full splicing activity, as does 40P1, but does not accumulate splicing intermediates in the splicing reaction, presumably owing to a higher level of Fraction III activity. Like 40P3, 35P is a highly concentrated fraction and contains only small amounts of ribosomes. It also contains less nonspecific nucleolytic activity than the whole cell extract and, therefore, is frequently used to replace whole cell extracts for various studies of the splicing reaction.

[11] P. J. Growboski and P. A. Sharp, *Science* **233**, 1294 (1986).
[12] S.-C. Cheng and J. Abelson, *Genes Dev.* **1**, 1014 (1987).
[13] C. W. Pikielny, B. C. Rymond, and M. Rosbash, *Nature* (*London*) **324**, 341 (1986).
[14] A. Binderelf and M. R. Green, *EMBO J.* **6**, 2415 (1987).
[15] S. Ruby and J. Abelson, unpublished results.

[9] Affinity Chromatography with Biotinylated RNAs

By S. W. RUBY, S. E. GOELZ, Z. HOSTOMSKY, and J. N. ABELSON

Introduction

The biotin–avidin association is an extremely powerful tool for affinity chromatography (reviewed in Ref. 1). Biotin can be conjugated to a variety of molecules including carbohydrates,[2] phospholipids,[3] proteins,[4–6] and nucleic acids.[7–9] Such biotinylated molecules often retain their biological activities. They have been used for diverse studies, from the activation of biotinylated T cells by avidin-modified T cells[10] to the detection of single-copy genes in the human genome.[11,12]

The tight binding of biotin by avidin (K_d 10^{-15} M) has made these applications possible, but conversely it has made the elution of the bound biotin inefficient unless extremely harsh conditions are used to denature or degrade avidin (reviewed in Ref. 13). An analog of biotin, 2-iminobiotin, binds avidin at pH 9 and dissociates at pH 4.[14,15] It does not, however, dissociate from avidin efficiently at physiological pH.[15] Recently, biotinylating reagents have been developed in which biotin is conjugated to a molecule via a linker arm containing a disulfide bond.[9,16] Biotinylation can thus be reversed by reducing the disulfide bond to sepa-

---

[1] M. Wilchek and E. A. Bayer, *Anal. Biochem.* **171**, 1 (1988).

[2] D. J. O'Shannessy, M. J. Dobersen, and R. H. Quarles, *Immunol. Lett.* **8**, 273 (1984).

[3] E. Bayer, B. Rivnay, and E. Skutelsky, *Biochem. Biophys. Acta* **550**, 464 (1979).

[4] E. Bayer, M. G. Zalis, and M. Wilchek, *Anal. Biochem.* **149**, 529 (1985).

[5] K. Hofmann, F. M. Finn, and Y. Kiso, *J. Am. Chem. Soc.* **100**, 3585 (1978).

[6] J. F. Rothenberg and M. Wilchek, *Biochem. Biophys. Res. Commun.* **138**, 872 (1986).

[7] A. C. Forster, J. L. McInnes, D. C. Skingle, and R. H. Symons, *Nucleic Acids Res.* **13**, 745 (1985).

[8] P. R. Langer, A. A. Waldrop, and D. C. Ward, *Proc. Natl. Acad. Sci. U.S.A.* **78**, 6633 (1981).

[9] M. L. Shimkus, P. Guaglianone, and T. M. Herman, *DNA* **5**, 247 (1986).

[10] E. Roffman and M. Wilchek, *Cell. Immunol.* **103**, 140 (1986).

[11] J. J. Leary, D. J. Brigati, and D. C. Ward, *Proc. Natl. Acad. Sci. U.S.A.* **80**, 4045 (1983).

[12] U. Landegren, R. Kaiser, J. Sanders, and L. Hood, *Science* **241**, 1077 (1988).

[13] N. M. Green, *Adv. Protein Chem.* **29**, 85 (1975).

[14] N. M. Green, *Biochem. J.* **101**, 774 (1966).

[15] K. Hofmann, S. W. Wood, C. C. Brinton, J. A. Montibeller, and F. M. Finn, *Proc. Natl. Acad. Sci. U.S.A.* **77**, 4666 (1980).

[16] T. M. Herman, E. Lefever, and M. Shimkus, *Anal. Biochem.* **156**, 48 (1986).

rate biotin from the molecule. This feature, coupled with another recent development for *in vitro* synthesis of large quantities of a specific RNA,[17] has made it possible to use small, reversibly biotinylated RNAs for affinity chromatography.

This chapter describes the use of such RNAs for analyzing small nuclear ribonucleoproteins (snRNPs) that bind to precursor messenger RNA (pre-mRNA) while it is being spliced in a yeast *in vitro* system. These snRNPs assemble on the pre-mRNA to form the spliceosome, the large ribonucleoprotein particle in which splicing occurs. We present our results from testing several materials for affinity chromatography as well as modifications of our original method.[18] Our results on the binding and elution of the small, reversibly biotinylated RNAs suggest that reversibly biotinylated ligands, in general, can have a broad application.

Principles of Method

A yeast pre-mRNA is synthesized *in vitro* and is bound to a solid support via a small complementary "anchor" RNA hybridized to either its 5' or 3' end (Fig. 1).[19] The anchor RNA is synthesized in two steps (Fig. 2) and has multiple biotin molecules, each conjugated to position 5 of a uridine residue via a linker containing a disulfide bond. It is hybridized to the pre-mRNA before the RNA hybrid is immobilized on the support. The biotinylated anchor RNA attaches to the biotinylated solid support via succinylavidin molecules. Avidin is a tetramer capable of binding four biotin moieties. To purify components involved in splicing, we add an active yeast splicing extract to the immobilized pre-mRNA. During the ensuing incubation period, the spliceosome forms on the immobilized pre-mRNA, and some of the pre-mRNA is spliced. After incubation, the support is washed quickly and extensively to remove extraneous extract components. The pre-mRNA, RNA intermediates, the mRNA product, and any components bound to them are eluted by adding dithiothreitol (DTT) to reduce the disulfide bonds linking biotin to the anchor RNA.

The method has the following features. The reagents are easily prepared or commercially available. Specific sites of the pre-mRNA are bound to the solid support. Components bound to an RNA can be rapidly isolated. The bound material may be eluted under mild conditions. The method may be used on a preparative scale.

[17] J. F. Milligan, D. R. Groebe, G. W. Witherell, and O. C. Uhlenbeck, *Nucleic Acids Res.* **15**, 8783 (1987).
[18] S. W. Ruby and J. Abelson, *Science* **242**, 1028 (1988).
[19] RNase H activity in the extract cleaves the RNA in an RNA–DNA hybrid and precludes the use of oligodeoxynucleotides to immobilize the pre-mRNA.

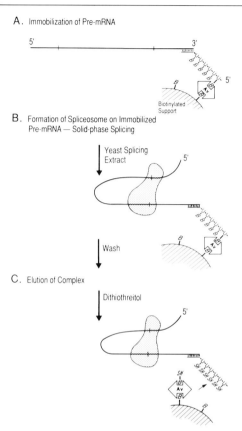

A. Immobilization of Pre-mRNA

B. Formation of Spliceosome on Immobilized
Pre-mRNA — Solid-phase Splicing

C. Elution of Complex

FIG. 1. Affinity chromatography with reversibly biotinylated RNAs: solid-phase splicing. (A) The pre-mRNA is synthesized *in vitro*. It is hybridized in solution to a small biotinylated anchor RNA (Fig. 2) complementary to either the 3' end (as shown here) or the 5' end. The RNA hybrid is immobilized on a biotinylated support by binding to succinylavidin. (B) A cell extract is added to the immobilized RNA, and during the ensuing incubation some of the RNA is spliced. The immobilized RNA is washed. (C) The RNA and any components bound to it are eluted by reducing the disulfide bonds linking the anchor RNA to the biotin residues.

These features suggest additional applications for purifying and characterizing RNA-binding factors and reconstituting ribonucleoprotein particles. For example, an immobilized RNA could be used selectively to deplete an extract of a factor that binds to RNA or to purify such a factor, as has been previously done with other biotinylated immobilized ligands.[20-22] The binding of such factors could be studied by footprinting

[20] M. S. Kasher, D. Pintel, and D. C. Ward, *Mol. Cell. Biol.* **6**, 3117 (1986).
[21] P. J. Grabowski and P. A. Sharp, *Science* **233**, 1294 (1986).
[22] A. Bindereif and M. R. Green, *EMBO J.* **6**, 2415 (1987).

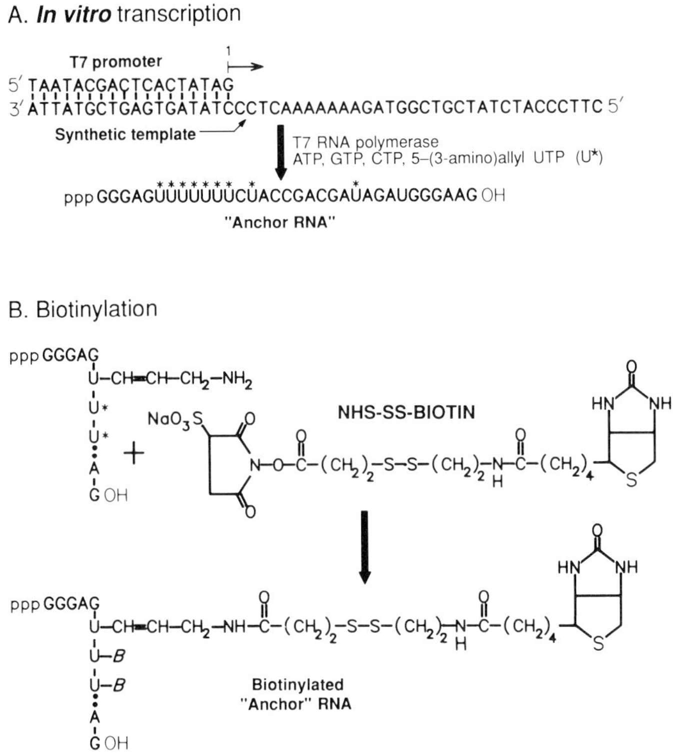

A. *In vitro* transcription

B. Biotinylation

FIG. 2. Synthesis of small, biotinylated anchor RNAs. (A) Small RNAs of 32–35 nucleotides (nt) in length are transcribed by phage T7 RNA polymerase from two synthetic oligodeoxynucleotides encoding the phage T7 promoter and the RNA as shown here for synthesizing anchor RNA 3'-21. Only the promoter must be encoded by double-stranded DNA. Allylamine-UTP is substituted for UTP in the synthesis; the substituted residues in the RNA are indicated by asterisks. (B) The RNA is biotinylated by a chemical reaction in which NHS-SS-biotin reacts preferentially with allylamines.

techniques,[23] perhaps with the anchor RNA acting as a primer for extension by reverse transcriptase. An immobilized small nuclear RNA (snRNA) could be reconstituted to a snRNP, eluted, and used in subsequent assays. We have already shown that two factors can bind sequentially to an immobilized RNA.[18] Such sequential bindings could be done repeatedly to analyze the order of addition of factors during reconstitution. Of course, some of these approaches can also be applied to immobilized ligands other than RNA.

[23] D. McPheeters, in "Methods in Nucleic Acids" (D. Chao, F. Warr, and J. Karam, eds.), in press. CRC Press, Boca Raton, Florida, 1989.

Materials

Reagents

NTP, nucleoside triphosphates (Pharmacia, Piscataway, NJ; Sigma, St. Louis, MO): dissolve in water, adjust to pH 7.8 with 10 N NaOH and a final concentration of 100 mM, and store at $-20°$; for transcription, make a mixture [NTP($-$UTP) mix] of 20 mM each ATP, GTP, and CTP and a separate 40 mM solution of UTP

AA-UTP, 10 mM allylamine-UTP [5-(3-amino)allyluridine 5'-triphosphate, Enzo Biochem Co., New York, NY]: store at $-20°$

NHS-SS-biotin, sulfosuccinimyl-2-(biotinamido)ethyl 1,3'-dithiopropionate (Pierce Chemicals, Rockford, IL): store desiccated at 4°

[α^{32}P]UTP, 3,000 Ci/mmol in Tricine [N-tris(hydroxymethyl)-methylglycine, Amersham, Arlington Heights, IL]

[5'^{32}P]pCp, cytidine bisphosphate; 7000 Ci/mmol in Tricine (New England Nuclear, Wilmington, DE)

pCp, cytidine bisphosphate (Pharmacia): dissolve at 10 mM in 1$\times$ TE (see below)

SP6 RNA polymerase (Promega, Madison, WI; U.S. Biochemicals, Cleveland, OH)

T7 RNA polymerase (US Biochemicals)

T4 RNA ligase (New England Biolabs, Beverly, MA)

RNasin, human placental ribonuclease inhibitor (Promega)

Restriction endonucleases [New England Biolabs; Bethesda Research Laboratories (BRL), Gaithersburg, MD]

Affinity Chromatography Materials

Avidin, egg-white avidin (Sigma): resuspend at 10 mg/ml in 50 mM potassium phosphate buffer (pH 7) and 1% sodium azide; store at $-20°$

Avidin-DN (Vector Laboratories, Burlingame, CA): 1 mg/ml in sodium phosphate buffer (pH 7) and 0.2% sodium azide; store at 4°

S. suc. avidin, succinylavidin (Sigma): resuspend and store like avidin; freeze–thaw a sample only twice

Streptavidin (BRL), resuspend and store like avidin; freeze–thaw a sample only twice

Suc. avidin, succinylavidin (EY Laboratories, San Mateo, CA): resuspend and store like avidin; freeze–thaw a sample only twice

EY avidin-agarose, avidin coupled to Sepharose 4B via an 11-atom hydrophilic arm (EY Laboratories): about 1 mg (15 nmol) avidin/ml of gel

EY biotin-acryl, biotin coupled to Sephacryl S-1000 via a 12-atom (8-carbon) hydrophilic arm (EY Laboratories): about 4 μmol biotin/ml of gel

EY biotin-agarose, biotin coupled to Sepharose 4B via a 10-atom (7-carbon) hydrophobic arm (EY Laboratories): binds 5 mg (75 nmol) avidin/ml of gel

Pierce biotin-agarose, biotin coupled to 6% cross-linked agarose via a 10-atom (7-carbon) hydrophobic arm (Pierce Chemicals): binds about 2 mg (30 nmol) avidin/ml of gel

Pierce biotin-cellulose, biotin coupled without a linker arm to microgranular cellulose (Pierce Chemicals): binds 1.4 mg (21 nmol) avidin/ml of gel

Sigma biotin-agarose, biotin coupled to 4% cross-linked agarose via a 12-atom (8-carbon) hydrophilic arm (Sigma): 2–6 μmol biotin/ml of gel, binds 22 mg (33 nmol) avidin/ml of gel

Streptavidin-agarose, streptavidin coupled to 4% cross-linked agarose via an 11-atom hydrophilic arm (BRL): 1.7 mg (28 nmol) streptavidin/ml of gel

Buffers and Solutions

10× TE, 100 mM tris(hydroxymethyl)amino methane hydrochloride (Tris-HCl) (pH 7.6) and 1 mM ethylenediaminetetraacetic acid (EDTA)

PEG 8000, polyethylene glycol, 8000 MW: store 30% (w/v) stock solution in water at −20° after autoclaving

10× RNase, pancreatic RNase A (Boehringer Mannheim, Indianapolis, IN): 10 mg/ml in 1× TE heated at 65° for 10 min to inactivate any DNase activity and stored at −20°

Phenol (in TE), liquid phenol (Mallinkcrodt): equilibrated with 1× TE

Chloroform–isoamyl alcohol, 25 : 1 v/v

10× SP6 transcription buffer, 0.4 M Tris-HCl (pH 7.6 at 23°), 120 mM DTT, 160 mM MgCl$_2$, and 40 mM spermidine; store at −20°

10× T7 transcription buffer, 0.5 M Tris-HCl (pH 8.1 at 23°), 0.25 M MgCl$_2$, and 20 mM spermidine; store at −20°

Gel sample loading buffer, 90% deionized formamide, 2 mM EDTA (pH 7.9), 0.05% xylene cyanol, and 0.05% bromphenol blue; store at −20°

Gel sample elution buffer, 0.5 M NaCl, 10 mM Tris-HCl (pH 7.4), 1 mM EDTA, and 1% phenol

Phenol–chloroform, 50% v/v each of crystallized phenol (Mallinckrodt) and chloroform equilibrated with 0.1 M sodium acetate (pH 6) as described[24]

---

[24] R.-J. Lin, A. J. Newman, S.-C. Cheng, and J. Abelson, *J. Biol. Chem.* **260,** 14780 (1985).

Glycogen, oyster glycogen (Sigma): incubated first at 10 mg/ml in 1×
TE with 10 μg/ml RNase A for 60 min at 37° then in 1× proteinase
K for 30 min at 55°; it is extracted 3 times with phenol, extracted
once with chloroform–isoamyl alcohol, precipitated by the addition of salt and ethanol, dried, and resuspended at 10 mg/ml in
distilled water; store at −20°

10× RNA annealing buffer, 0.7 M Tris-HCl (pH 7.6), 80 mM NaCl,
and 15 mM EDTA; store at −20°

Binding buffer, 75 mM potassium phosphate buffer (pH 7 at 23°),
1 mM EDTA, 15% glycerol, and 0.05% Nonidet P-40 (NP-40);
store at 4° for 1 week or at −20° for longer periods

Splicing buffer, 60 mM potassium phosphate buffer (pH 7.2 at 23°),
1 mM spermidine, 3 mM MgCl$_2$, 3% PEG 8000, 0.5 mM DTT, and
with or without 2 mM ATP; make up fresh

Splicing extract, splicing buffer supplemented with either 40% (v/v)
whole cell extract[24] or a fraction (35P) of whole cell extract that is
active for splicing[25,26]

Washing buffer, 20 mM N-2-hydroxyethylpiperazine-N'-2-ethane-
sulfonic acid (HEPES) (pH 7.6 at 23°), 150 mM NaCl, 3 mM
MgCl$_2$, 0.5 mM DTT, 15% glycerol, and 0.05% NP-40; store at 4°
for 1 week or at −20° for longer periods

Elution buffer A, 40 mM Tris-HCl (pH 9 at 23°), 150 mM NaCl, 3
mM MgCl$_2$, and 150 mM DTT; make up fresh each week and store
at 4°

Elution buffer B, 40 mM HEPES (pH 7.6 at 23°), 150 mM NaCl, 3
mM MgCl$_2$, 15% glycerol, and 2–50 mM DTT; make up fresh each
week and store at 4°

10× proteinase K, 2.5 mg/ml proteinase K (Boehringer) in 250 mM
EDTA (pH 8.0) and 2.5% sodium dodecyl sulfate (SDS); store
at −20°

Phenol (in sodium acetate), redistilled phenol equilibrated with distilled water, washed once with 3 M sodium acetate (pH 8), and
brought to pH 6 by washing with 0.3 M sodium acetate (pH 8)

Carrier tRNA, *Escherichia coli* tRNA (Boehringer) at 10 mg/ml in
1× TE

10× T4 RNA ligase buffer, 0.5 M HEPES (pH 8.3 at 23°), 100 mM
MgCl$_2$, and 70 mM DTT; store at −20°

Denaturing polyacrylamide gels, 29 : 1 (w/w) polyacrylamide : bisacrylamide in 90 mM Tris–borate (pH 7.9), 100 mM EDTA (TBE),
and 8 M urea; 40 V/cm is applied to 0.5-mm thick gels in TBE

[25] S.-C. Cheng, A. Newman, R.-J. Lin, G. McFarland, and J. N. Abelson, this volume [8].
[26] U. Vijayraghavan, R. Parker, J. Tamm, Y. Iimura, J. Rossi, J. Abelson, and C. Guthrie,
EMBO J. **5,** 1683 (1986).

Methods

Synthesis of Pre-mRNA

Transcripts of a gene are synthesized by a phage RNA polymerase transcribing a linear template *in vitro*. The template we use is the yeast actin gene cloned into a plasmid vector downstream of the bacterial phage SP6 promoter.[26] Plasmid DNA is prepared by banding in cesium chloride gradients[27] or by the alkaline–SDS method.[28] Using the latter method, we remove excess RNA by treating the DNA in TE with 1× RNase solution for 30–60 min at 37°, extracting twice with phenol (in TE) and once with chloroform–isoamyl alcohol, and precipitating the DNA with salt and ethanol. After resuspension in 1× TE, the plasmid DNA is linearized with the appropriate restriction enzyme according to conditions recommended by the supplier. The cut DNA is extracted once each with phenol (in TE) and chloroform–isoamyl alcohol, precipitated with salt and ethanol, washed once with 70% ethanol, dried and resuspended at about 1 mg/ml in TE.

To obtain good yields of the 500- or 600-nucleotide (nt) actin premRNA, we modified previously reported conditions[24,29] by increasing the concentrations of nucleotides and $MgCl_2$ and by adding PEG 8000. The addition of PEG increases transcription,[30] possibly by stabilizing the polymerase.[31] A typical reaction contains the following components added at 23° in the indicated order: 5 μl of water, 4 μl of 10× SP6 transcription buffer, 8 μl of 20 mM NTP(−UTP) mix, 1 μl of 40 mM UTP, 2 μl containing 32 units of RNasin, 4 μl containing 4 μg of template DNA, 4 μl of PEG, 4 μl of [^{32}P]UTP, and 3 μl containing 60–80 units of SP6 polymerase. The reaction is incubated at 37° for 2–4 hr.

Full-length transcripts must be used to maximize the efficiency of the affinity chromatographic method. They are isolated by gel electrophoresis as follows. The nucleic acids are concentrated from the transcription reaction by precipitation with salt and ethanol, dried, and then resuspended in 40 μl of sample loading buffer. The sample is heated at 100° for 2 min and immediately thereafter is loaded onto a denaturing 4% polyacrylamide gel. Full-length transcripts are visualized by a 5-min autora-

[27] T. Maniatis, E. F. Fritsch, and J. Sambrook, "Molecular Cloning: A Laboratory Manual." Cold Spring Harbor Laboratory, Cold Spring Harbor, New York, 1982.
[28] H. C. Birnboim and J. Doly, *Nucleic Acid Res.* **7**, 1513 (1979).
[29] D. A. Melton, M. R. Krieg, M. R. Rebagliati, T. Maniatis, K. Zinn, and M. R. Green, *Nucleic Acids Res.* **12**, 7035 (1984).
[30] S. W. Ruby, unpublished results (1988).
[31] B. Harrison and S. B. Zimmerman, *Nucleic Acids Res.* **14**, 1863 (1986).

diographic exposure. The desired band is excised, and the RNA is eluted in about 1 ml of gel elution buffer in an Eppendorf tube by shaking at 37° for 2 hr. The eluent is separated from the gel slice by centrifugation and extracted once with phenol–chloroform. Glycogen is added as a carrier (20 μg per 500 μl of eluted RNA). The RNA is then precipitated and washed twice with ethanol, resuspended in water, and stored at $-20°$.

Synthesis of Biotinylated Anchor RNA

The small biotinylated anchor RNA is synthesized in two steps to obtain large amounts of RNA and to maintain the disulfide bond in the linker between the uridine and biotin moieties (Fig. 2). The anchor RNA is first synthesized *in vitro* with phage T7 polymerase from two synthetic oligodeoxynucleotides.[17] Only the T7 promoter need be double stranded. Allylamine-UTP is substituted for UTP in the synthesis. In the second step, the RNA is biotinylated by a chemical reaction of allylamine in the uridine residues with NHS-SS-biotin.[8,9]

Oligodeoxynucleotides are synthesized by the phosphoramidite method on an Applied Biosystems 380B DNA synthesizer and purified by electrophoresis in a denaturing 15% polyacrylamide gel. The oligodeoxynucleotides are eluted from the excised, crushed gel band in 1–2 ml of gel elution buffer. The suspension is shaken in a siliconized glass vial for 1–4 hr at 37° and is then filtered through a QuickSep column (from Isolab). The solution is extracted once each with phenol (in TE) and chloroform–isoamyl alcohol. The oligodeoxynucleotides are precipitated with ethanol, washed once with 70% ethanol, dried, resuspended in 1× TE, and stored at $-20°$. The transcription templates are prepared by annealing the top and bottom strand oligodeoxynucleotides together in solution [22.5 μM each oligodeoxynucleotide, 100 mM NaCl, 10 mM Tris-HCl (pH 7.6), and 1 mM EDTA]. The solution is heated at 100° for 2 min, cooled gradually to 23° over 40 min, and then diluted with 1× TE to 5 μM duplex DNA and stored at $-20°$.

A typical transcription reaction in 200 μl total volume contains the reagents added at 23° in the indicated order: 81 μl of water, 20 μl of 10× T7 transcription buffer, 5 μl of 0.5 M DTT, 10 μl of template oligodeoxynucleotides, 40 μl of 20 mM NTP($-$UTP) mix, 19 μl of 10 mM allylamine-UTP, 5 μl containing 200 units of RNasin, 8 μl of [^{32}P]UTP (optional), and 12 μl containing at least 1000 units of T7 polymerase. The reaction is incubated at 37° for 3–4 hr.

Full-length RNA must be purified from the reaction as there may be several shorter products whose relative amounts depend on the template sequence.[17] The RNA can be purified by HPLC[9,17] or by gel electrophoresis. To purify by gel electrophoresis, the nucleic acids are precipitated completely from the reaction with salt and ethanol. After being washed with ethanol and dried, the pellet is resuspended in 50–60 μl of gel sample loading buffer. The sample is boiled for 2 min immediately before being loaded onto a denaturing 15% polyacrylamide gel. Full-length RNA can be detected by autoradiography as described above, by UV shadowing, or by staining with ethidium bromide. The RNA is eluted from the crushed, excised gel piece by the same method used for eluting the oligodeoxynucleotides. Glycogen is added as a carrier as described above. The eluted RNA is twice extracted with phenol–chloroform, precipitated with ethanol, washed twice with 70% ethanol, and resuspended in 225 μl of distilled water.

To biotinylate the RNA, we first add 1 M NaHCO$_3$ (pH 8) to the RNA to a final concentration of 0.1 M. Dry NHS-SS-biotin is quickly weighed and immediately added to the RNA to a final concentration of about 2 mM. After 30 min at 23°, the RNA is precipitated by the addition of one-tenth volumes each of 1 M glycine (pH 7), 3 M sodium acetate (pH 5), and dimethylformamide and three volumes of ethanol.[32] The RNA is washed 3 times with 70% ethanol, resuspended in water, and stored at $-20°$. The final yield of biotinylated RNA is 5–20 nmol.

Annealing RNAs

The pre-mRNA and anchor RNA are typically annealed in 20–300 μl of a solution of 1× RNA annealing buffer, 125–500 nM anchor RNA, and 25 nM pre-mRNA. The solution is heated in an Eppendorf tube for 2 min at 100° and then at 65° for 1 min. It is then gradually cooled from 37 to 15° for 1 hr by placing the tube in 500 ml of 37° water in a 4° room. The hybrid RNA is stored at $-20°$ and thawed at 23°. The ratio of anchor RNA to pre-mRNA that is most efficient for annealing needs to be determined empirically. If the efficiency of annealing as measured by the binding of the pre-mRNA–anchor RNA hybrid on streptavidin-agarose (see below) is not satisfactory, it may be that other factors interfere with annealing (see below).

---

[32] The biotinylating esters initially used were not sulfonated and were nearly insoluble in water. They were dissolved first in dimethylformamide before being added to an aqueous solution. We noted that excess, unreacted sulfonated ester is more readily removed when dimethylformamide is present during ethanol precipitation. It may be necessary to repeat this ethanol precipitation if a large, granular pellet appears, an indication that excess ester has not been removed. The number of biotinylated residues per RNA molecule can be determined as in D. B. McCormick and J. A. Roth, *Anal. Biochem.* **34**, 226 (1970).

Binding and Eluting RNA

We routinely use two binding protocols. (1) To assay the binding of a newly synthesized, biotinylated anchor RNA or the annealing of the pre-mRNA–anchor RNA hybrid, we bind it directly to streptavidin-agarose in buffer without extract. (2) To immobilize the pre-mRNA–anchor RNA hybrid for subsequent incubation in splicing extract, we use "bridge" binding in which the hybrid is bound to a biotinylated support with suc-cinylavidin (Fig. 1). In addition, we tested the binding and elution of the biotinylated anchor RNA and the pre-mRNA–anchor RNA hybrid in either the absence or presence of splicing extract (see below). Several variations of the bridge binding method were assayed by varying the order of addition of RNA, avidin, and extract.

To directly bind biotinylated RNA to streptavidin-agarose or another avidin-agarose support, the RNA [1–10 μl containing 10^4 to 10^5 Cerenkov counts per minute (cpm)] is added directly to 25 μl of agarose and kept on ice for at least 30 min. Four hundred microliters of binding buffer is added, and the total number of RNA Cerenkov cpm for each sample is determined.[33] The agarose is then sedimented for 7 sec in a microcentri-fuge, washed twice more with binding buffer, and suspended in 200 μl of either binding buffer or elution buffer A or B. The total number of bound cpm is counted.

To bind the pre-mRNA–anchor RNA hybrid by the bridge method preparatory for solid-phase splicing, succinylavidin is first bound to the support and then the RNA bound as follows. Five milligrams of succi-nylavidin is added to 1.5 ml of settled support and kept for 1 hr to several days at 0°. The support is washed twice with at least 10 volumes of binding buffer. RNA is then added to the settled support (from 0.2 to 0.6 × 10^4 Cerenkov cpm of hybrid per 20–30 μl of support) and incubated for 30 min at 0°. To bind an additional 10–20% of the hybrid to the support, we add a second portion of succinylavidin (1 μg/μl of support) and incu-bate the sample for an additional 20 min. The RNA–support is counted and then washed twice with binding buffer. It is resuspended in 10 vol-umes of binding buffer, recounted, and then distributed into Eppendorf tubes. The RNA–support is sedimented, washed once with 10 volumes of

---

[33] The samples and buffers are kept at 0° to reduce any RNase activity. We prefer not to use RNasin to inhibit RNase activity during binding of the RNA because this inhibitor requires at least 1 mM DTT to be active. A sample in an Eppendorf tube can be kept at a low temperature while being counted if the tube is placed in an uncapped scintillation vial in a liquid scintillation counter immediately before being counted and quickly removed there-after.

splicing buffer with or without ATP, and quickly recounted. This wash in splicing buffer is essential for the pre-mRNA to be subsequently spliced. The splicing buffer is removed, and the samples are ready for incubation in splicing extract (see below).

To elute either the biotinylated anchor RNA or the pre-mRNA–anchor RNA hybrid, the support with the bound RNA is incubated in 10 volumes of elution buffer A or B for the desired time (see below). The support is then sedimented for 7 sec in a microcentrifuge, and the supernatant is retained. The support is washed again with another 10 volumes of elution buffer for 5 min at 0°. The two eluents are combined and counted to determine the total number of eluted cpm.

The following methods were also used to test the binding and elution of RNAs under various conditions. For testing the binding and elution of the RNAs by the bridge binding method in the absence of splicing extract, the RNAs are bound as for solid-phase splicing except that the addition of the second aliquot of avidin is omitted. The samples are counted in the first wash of binding buffer to determine the total number of cpm, washed twice again with 10 volumes each of binding buffer, and then recounted to determine the total number of bound cpm. The RNA is then eluted in buffer A for 10–20 min.

To test the effects of splicing extract on RNAs already bound to a solid support, the RNA is bound to the support by one of the methods described above. The support is washed twice with binding buffer, 10 volumes of splicing buffer without ATP is added, and the samples are then counted. The slicing buffer is removed, splicing buffer supplemented with 2 mM ATP and whole cell extract is added, and the samples are incubated for 30 min at 23°. The samples are then quickly washed 3 times with washing buffer at 0°, after which they are counted and eluted for 10–20 min in elution buffer A.

When the RNAs were incubated in splicing extract before being bound to the solid support, four different binding methods were tested. (1) For binding RNAs directly to avidin- or streptavidin-agarose, 20–30 μl of agarose is added to a sample at the end of the incubation of the RNA in splicing extract. The sample is kept at 0° for 40–60 min, after which 10–20 volumes of washing buffer is added. The sample is counted and washed twice more with washing buffer. The RNA is eluted for 10–20 min in buffer A. (2) For bridge binding, any type of avidin is added to the RNA in a 10- to 20-fold molar excess of avidin to anchor RNA and a total volume of less than 5 μl. After 30 min at 0°, the RNA and avidin are added to 40 μl of splicing extract and incubated for 30 min at 23° to promote splicing. The sample is then placed on ice. Forty microliters of a biotinylated support, previously washed in binding buffer and resuspended in splicing buffer without DTT to 50% v/v agarose, is then added and incubated with the sample at 0° for 40–60 min. The samples are then processed as in (1). (3) When any type of avidin is added after the RNA is incubated in splicing extract, a 10- to 20-fold molar excess of the avidin is added to a 40 μl splicing reaction as the reaction is placed on ice. After 10 min, 40 μl of biotinylated support [prepared as in (2)] is added, and the sample is incubated for an additional 40–60 min. The sample is then processed as in (1). (4) When any type of avidin is first bound to the RNA, a 10- to 20-fold molar excess of the avidin is added to the RNA. After incubation for 20 min at 0°, the RNA is added to splicing extract, and the sample is incubated at 23° for 30 min. Forty microliters of washed biotinylated support [as in (2)] is added, and the sample is incubated for at least 40 min at 0°. The sample is then processed as in (1).

Solid-Phase Splicing

Before splicing extract is added to the RNA–support, it is incubated at 23° for 10–20 min to deplete any endogenous ATP by ATPases in the extract. Formation of some of the intermediate splicing complexes[34] and the binding of U2 snRNP[30] require low levels of ATP that are often present in yeast whole cell extracts. The amount of endogenous ATP varies batchwise, therefore, the length of time necessary for ATP depletion must be determined for each batch.

Exogenous ATP (2 nmol/μl of splicing extract) is added to the support for the appropriate reactions. Splicing extract is then added; usually 2 μl of extract is added per microliter of RNA–support containing 0.25–1 fmol of pre-mRNA. The reactions are incubated at 23° for the indicated times, during which they are gently mixed by hand at 5-min intervals. Reactions are stopped by the addition of a 10-fold excess of washing buffer at 0° and by sedimentation for 7 sec in a microcentrifuge. The RNA–support is washed twice again with washing buffer at 0°. To analyze the immobilized RNA and any factors bound to it, the RNA is eluted in elution buffer A for 10–20 min as described above.

Detecting Ribonucleoproteins Bound to Immobilized Pre-mRNA

The RNAs bound to immobilized pre-mRNA from a 40- to 60-μl solid-phase splicing reaction can be readily analyzed by ^{32}P labeling or RNA blot hybridization (Northern) analysis. The RNAs are isolated from the eluents as follows. Twenty micrograms of carrier, either glycogen (for ^{32}P labeling) or *E. coli* tRNA (for Northern analysis), and 40 μl each of 10$\times$

---

[34] S.-C. Cheng and J. Abelson, *Genes Dev.* **1**, 1014 (1987).

proteinase K solution and 10% SDS are added to 400 μl of eluent. The samples are incubated for 30 min at 37°, after which they are extracted once with phenol (in sodium acetate), twice with phenol–chloroform, and once with chloroform–isoamyl alcohol. The RNAs are precipitated from the aqueous phase by addition of sodium acetate and ethanol. The RNA pellet is washed twice with ethanol and dried. A second ethanol precipitation may be required if the RNA pellet does not look clean.

The RNAs are radiolabeled by ligating [^{32}P]pCp to their 3'-hydroxyl ends with T4 RNA ligase as follows. The RNA pellet is resuspended in 15 μl of 1× RNA ligase buffer, 12 μM ATP, 10% dimethyl sulfoxide, 5% PEG, 15 units RNasin, 1.5 units T4 RNA ligase, and 75 μCi [^{32}P]pCp and incubated for 2–4 hr at 20°. These reaction conditions are modified from previous ones[35] so that 2- to 4-fold more pCp is incorporated into total yeast RNA in one-quarter of the time originally required.[30] Presumably, PEG is stabilizing the ligase.[31] The reaction is stopped by the addition of 400 μl of phenol–chloroform followed by 200 μl of 1× TE supplemented with 0.2 M sodium acetate (pH 7), 2 μg carrier tRNA, and 100 μM cold pCp. The RNA is precipitated from the extracted aqueous phase by the addition of an equal volume of 1 M Tris-HCl (pH 8.0 at 23°) and 2.5 volumes of ethanol. The precipitation in the presence of excess cold pCp and 0.5 M Tris-HCl prevents precipitation of most of the unincorporated [^{32}P]pCp. The RNA is washed twice with 70% ethanol, dried, and resuspended in gel sample loading buffer supplemented with 100 μM cold pCp. The samples are analyzed by electrophoresis in denaturing 4% polyacrylamide gels.

The conditions for RNA blot hybridization analysis are as described previously.[18,33]

Results and Discussion

Pre-mRNA–Anchor RNA Hybrids

In designing the anchor RNAs we considered six parameters to be important: the length and base composition of the anchor RNA, including the number of biotinylated residues; the type of biotinylating agent; the location of the complementary sequence on the pre-mRNA; any other sequences on the pre-mRNA that could hybridize with the anchor RNA; and any sequence or structure of the anchor RNA that would bind any splicing factors.

We chose biotinylated uridine residues for two reasons. First, the

---

[35] T. E. England, A. G. Bruce, and O. C. Uhlenbeck, this series, Vol. 65, p. 65.

RNA could be biotinylated after synthesis. T7 polymerase does not incorporate biotinylated UTP efficiently,[8] but it incorporates allylamine-UTP equally or more efficiently than UTP.[30] The DTT concentration and the reaction time and temperature for transcription would be sufficient to reduce the disulfide bonds in biotin-SS-UTP (see below). Whereas fully biotinylated RNA remains at the interphase during phenol and chloroform extractions, RNA with allylamines remains in the aqueous phase. Biotinylation of allylamine-UTP is efficient, rapid, and reliable.[8,9] Second, biotinylation of UTP occurs preferentially at the allylamine at position 5 under the reaction conditions used here. Other bases are not detectably biotinylated[30]; therefore, the number and position of biotinylated residues could be controlled by the sequence composition of the anchor RNA. Furthermore, biotinylation at position 5 does not disrupt base pairing and has a small effect on the melting temperature (T_m) and reassociation kinetics of a nucleic acid duplex.[8] The biotin moieties are accessible to avidin when incorporated into polynucleotides.[11] An alternative approach would be to biotinylate the RNA with photobiotin,[7] which reacts with primary amines; however, such biotinylation disrupts Watson–Crick base pairs, a disadvantage for the short RNAs we use.

Although theoretically one biotinylated residue is sufficient to bind a conjugated ligand to avidin, we designed anchor RNAs containing at least 10 biotinylated residues and with at least 5 of these located in an overhanging leader or tail to further facilitate bindings. We have not determined the minimum number of biotinylated residues sufficient to retain RNAs bound to avidin and a biotinylated support in the presence of 0.5 mM DTT. Other events in addition to reduction of the disulfide bonds affect retention of the RNAs when they are incubated in splicing extract (see below).

We designed anchor RNAs with 19–24 nt complementary to the pre-mRNA because we reasoned, by analogy to DNA–DNA and DNA–RNA duplexes,[36] that this size would give the best specificity and stability for a small anchor RNA. It also allows both the template oligodeoxynucleotide, encoding the anchor RNA and polymerase promoter,[37] and the anchor RNA[30] to be synthesized with good yields. The factors determining the formation and stability of RNA–RNA duplexes are not so well understood as those for DNA–DNA duplexes. In general, the melting temperature (T_m) of an RNA–RNA duplex is slightly higher than that of a DNA–DNA duplex of equivalent base composition.[38,39] However, intra-

[36] K. Itakura, J. J. Rossi, and R. B. Wallace, *Annu. Rev. Biochem.* **53**, 323 (1984).
[37] S. Horvath, California Institute of Technology, personal communication (1988).
[38] D. K. Bodkin and D. L. Knudson, *Virology* **143**, 55 (1985).
[39] D. M. Gray, J.-J. Liu, R. L. Ratliff, and F. S. Allen, *Biopolymers* **20**, 1337 (1981).

molecular interactions including base pairing may affect the formation of an intermolecular RNA–RNA duplex. Although some intramolecular structures and intermolecular duplexes can be predicted by computer[40] for both pre-mRNAs and anchor RNAs, we have found that it is best to determine empirically whether a particular sequence can be used for an anchor RNA (see below). It is also necessary to determine experimentally if any factors bind to the anchor RNA alone.

To anneal the pre-mRNA and anchor RNAs together, we adapted techniques previously used to anneal oligodeoxynucleotides to RNAs for subsequent primer extension with reverse transcriptase.[27,41] A low salt concentration in the annealing buffer is used to maintain specificity and perhaps promote annealing of the intermolecular duplexes relative to shorter intramolecular duplexes. Although RNAs can be annealed to oligodeoxynucleotides by heating them together at 100° and then immediately putting them in an ice or dry ice–ethanol bath,[23] we found that the pre-mRNA–anchor RNA hybrids we have used (Fig. 3) do not form efficiently in this "quick-cool" method. They are formed more efficiently by the gradual cooling method we describe here. By using the equation of Suggs et al.[42] and adjusting the values for the ionic strength of the annealing buffer,[27] we estimate that the T_m values of the anchor RNAs are in the range of 26–38°. Therefore, the optimal temperatures for annealing the RNAs are probably within the temperature gradient used.

The three anchor RNAs we have tested differ in the efficiency with which they anneal to pre-mRNA. Anchor RNAs 5′-24 and 3′-19, whose sequences are homologous to the polylinker region of the plasmid vector, do not hybridize as efficiently as anchor RNA 3′-21 under the conditions we routinely use. Whereas 90% of the pre-mRNA–anchor RNA hybrid forms with 3′-21 annealing to pre-mRNA in a ratio of 20 to 1, this ratio only gives 40% hybridization for anchor 5′-24. The difference could be due to strong intramolecular associations within both the pre-mRNA and anchor 5′-24 or to effects of the position and number of biotinylated residues on the stability and hybridization kinetics of the RNAs.

Binding and Eluting RNAs

We tested several supports with covalently attached avidin or streptavidin as well as biotinylated supports combined with soluble avidin or

[40] D. H. Turner, N. Sugimoto, and S. M. Freier, *Annu. Rev. Biophys. Biophys. Chem.* **17**, 167 (1988).

[41] M. Ares, M. Mangin, and A. M. Weiner, *Mol. Cell. Biol.* **5**, 1560 (1985).

[42] S. V. Suggs, T. Hirose, T. Miyake, E. H. Kawashima, M. J. Johnson, K. Itakura, and R. B. Wallace, *in* "Developmental Biology Using Purified Genes" (D. D. Brown and C. F. Fox, eds.), p. 683. Academic Press, New York, 1981.

A. In vitro synthesized actin pre-mRNA

B. Pre-mRNA / anchor RNA hybrid structures

```
                5'pppAAUACACGGAAUUCGAGCUCGCCCGGGATC---exon1
                    I I I I I I I I I  I I I  I I  I I I I I I I I  I I I I
(5'-24)  HOCGUUUUUUAUGUGCCUUAAGCUCGAGCGGGppp5'
                    ****** * *    **          *
```

```
                exon2---CUUCCCAUCUAUCGUCGGUAG OH
                        I I I I I I I I I I  I I I  I I I I I I  I I I
(3'-21)         HOGAAGGGUAGAUAGCAGCCAUCUUUUUUUGAGGGppp5'
                          *      *          *  *******
```

```
                exon2--UCGACCUGCAGCCCAAGCU OH
                        I I I I I I I  I I I I I I I I I I I I
(3'-19)         HOAGCUGGACGUCGGGUUCGAUUUUUUUUGAGGGppp5'
                        *      *      **   ********
```

FIG. 3. Anchor RNAs hybridized to actin pre-mRNA synthesized *in vitro*. Three anchor RNAs that we have used are shown hybridized to actin pre-mRNA. (A) Each anchor RNA is designated by its position (either at the 5' or 3' end of the pre-mRNA) and the number of nucleotides complementary to the pre-mRNA. The regions of actin pre-mRNA complementary to the anchor RNAs are indicated. Anchor 5'-24 is separated from the 5' splice site by 72 nt. Anchor 3'-21 is complementary to the last 21 nt of an actin pre-mRNA synthesized from the template linearized by cutting within the exon 2 sequence with the restriction endonuclease *Acc*I. Anchor 3'-19 is complementary to the last 19 nt of an actin pre-mRNA synthesized from the DNA template linearized with *Hin*dIII restriction endonuclease in the polylinker region immediately downstream of the cloned actin gene in plasmid pSP65. Anchors 3'-21 and 3'-19 are 79 and 182 nt downstream from the 3' splice site. (B) Structures of the RNA–RNA hybrids, with the biotinylated nucleotides indicated by asterisks.

streptavidin (Tables I and II).[43] We sought a support giving both efficient binding and elution of the anchor RNA and allowing the immobilized pre-mRNA to be spliced. We found that the best binding and elution occur with the bridge binding method (Fig. 1). In this method, the binding and elution of anchor RNA depend on several factors: the length and hydrophobicity of the linker arm attaching biotin to the support; the density of

[43] Although we refer to the chromatographic materials by the names of the suppliers, there are some important chemical and physical differences among them as described in Materials.

TABLE I

BINDING AND ELUTING RNA[a]

Binding method[b]	RNA[c]	Avidin type[d]		Solid support[e]	RNA bound[f] (%)	RNA eluted[g] (%)
Bridge	**Hybrid**[h]	Suc. avidin	+	**EY biotin-acryl**	**70–80**	**>80**
			+	EY biotin-agarose	40–80[i]	>75
			+	Pierce biotin-agarose	>90	<5
			+	**Pierce biotin-cellulose**	**30–50**	**>85**
			+	**Sigma biotin-agarose**	**70–90**	**>70**
		Streptavidin	+	EY biotin-acryl	70–90	35–40
	Anchor	Avidin	+	EY biotin-acryl	65–70	40–50
			+	Sigma biotin-agarose	90–95	40–50
		S. suc. avidin	+	EY biotin-acryl	5–10	75–80
		Suc. avidin	+	Sigma biotin-agarose	5–10	65–80
			+	EY biotin-acryl	70–85	>75
			+	EY biotin-agarose	40–60	>75
			+	Pierce biotin-agarose	85–95	<10
			+	Pierce biotin-cellulose	20–30	>85
			+	Sigma biotin-agarose	80–95	>75
		Avidin DN	+	Pierce biotin-agarose	>90	<10
			+	Pierce biotin-cellulose	30–70	>80
		Streptavidin	+	EY biotin-acryl	80–95	10–15
			+	Sigma biotin-agarose	90–95	20–30
		(No avidin)		EY biotin-acryl	<1	ND[j]
				EY biotin-agarose	40–60	>75
				Pierce biotin-agarose	90–95	<10
				Pierce biotin-cellulose	<4	ND
				Sigma biotin-agarose	<2	ND
	Pre-mRNA[k]	Avidin	+	EY biotin-acryl	60–65	ND
			+	Sigma biotin-agarose	60–65	ND
		Suc. avidin,	+	EY biotin-acryl	<4	ND
		S. suc.	+	EY biotin-agarose	<5	ND
		avidin, or	+	Sigma biotin- Sepharose	<2	ND
		streptavidin		Streptavidin-agarose	<2	ND
Direct	Hybrid			Streptavidin-agarose	80–95	<20
	Anchor			Streptavidin-agarose	85–100	<10
	Anchor			EY avidin-agarose	60–80	20–60

[a] Three ^{32}P-labeled RNAs (biotinylated anchor RNA, pre-mRNA–biotinylated anchor RNA hybrid, and pre-mRNA) were tested for binding to and elution from various solid supports supplemented in some cases with different types of avidin as described in Methods.

[b] Avidin binds both the biotinylated support and the biotinylated anchor RNA in the bridge binding method. Biotinylated anchor RNA binds to a support containing covalently attached avidin in the direct binding method.

[c] Biotinylated anchor RNA 3'-21 was ^{32}P labeled to a specific activity of 1.9 × 10³ Cerenkov cpm/pmol. Actin pre-mRNA was at a specific activity of 2.6 × 10⁶ Cerenkov

biotins on the support; the type of avidin (Table I); and the order of addition of RNA, avidin, support, and extract (Table II).

The best combinations for the bridge binding method are either Sigma biotin-agarose or EY biotin-acryl combined with EY succinylavidin. These two biotinylated supports have the same 12-atom hydrophilic linker arm that minimizes steric hindrance for avidin binding. EY succinylavidin consistently gives the best binding and elution as well as one of the lowest levels of nonspecific binding. The extent of succinylation is a critical factor in the binding capacity of succinylavidin because one of the nine lysines in an avidin monomer that may be succinylated is involved in binding biotin.[13] EY Laboratories and Sigma use different methods to succinylate avidin.[44] Other forms of avidin are less desirable for use in this method. For example, streptavidin, a bacterial analog of avidin, may have a deeper binding site than avidin so that the disulfide bond is probably not exposed to solvent and reduced.[16] Unmodified avidin gives the highest nonspecific binding of RNA under all the conditions tested (Tables I and II), probably because of its high isoelectric point of 10.[13] We note that the lowest amounts of nonspecific binding of RNAs to agarose, acrylic, or cellulose supports are achieved with washing buffers containing at least 75 and 100 mM NaCl in the absence and presence of splicing extract, respectively (data not shown).

The pre-mRNA–anchor RNA hybrid binds and elutes in the same manner as the anchor RNA alone with one notable exception. If the hybrid RNA either with or without bound avidin is incubated in extract

---

[44] G. Stanton and A. Chu, EY Laboratories, Inc., personal communication (1988).

---

cpm/pmol. A 20-fold molar excess of either ^{32}P-labeled or nonlabeled biotinylated anchor RNA was hybridized to the actin pre-mRNA to obtain the hybrid. Less than 2% of the cpm was contributed by the labeled anchor RNA in the pre-mRNA–anchor RNA hybrid.

[d] The types of avidin tested are described in Materials.

[e] The types of solid supports tested are described in Materials.

[f] A total of at least 10[4] Cerenkov cpm of RNA was added to 20 μl of support. The number of cpm bound was determined after the support was washed 3 times with binding buffer. The values were determined from duplicates in at least two separate experiments.

[g] Percentage of the bound RNA that is eluted in elution buffer A in 15 min at 0°.

[h] Conditions tested for solid-phase splicing reactions are indicated in bold type.

[i] The variation was due to the use of two different batches of biotin-agarose. Such batchwise variation was not observed with the other supports. Only one batch of biotin-acryl was tested.

[j] Not determined.

[k] To monitor the levels of nonspecific binding of RNA without biotin, pre-mRNA only was incubated with the supports in the presence or absence of the indicated avidin. All supports tested retained less than 5% of the pre-mRNA in the absence of avidin.

TABLE II

EFFECTS OF SPLICING EXTRACT ON BINDING AND ELUTING RNA[a]

Order of addition (1, 2, or 3) of avidin, support, and extract to RNA[b]	RNA	Avidin type[c]	Solid support[d]	RNA bound[e] (%)	RNA eluted[f] (%)
(A) 1. Avidin plus support	**Hybrid**	**Suc. avidin**	**+ Biotin-acryl**	**40–65**	**80–90**
2. Extract		**Suc. avidin**	**+ Biotin-agarose**	**45–70**	**70–90**
	Anchor	Avidin	+ Biotin-acryl	50–55	65–75
		Avidin	+ Biotin-agarose	75–85	55–60
		Suc. avidin	+ Biotin-acryl	75–90	>75
		Suc. avidin	+ Biotin-agarose	80–95	>75
		Streptavidin	+ Biotin-acryl	60–65	15–20
		Streptavidin	+ Biotin-agarose	85–90	15–20
(B) 1. Extract	Hybrid	Suc. avidin	+ Biotin-acryl	10–20	ND[g]
2. Avidin plus support	Anchor	Suc. avidin	+ Biotin-acryl	55–65	>85
(C) 1. Avidin	Hybrid	Suc. avidin	+ Biotin-acryl	10–20	>80
2. Extract		Streptavidin	+ Biotin acryl	25–30	35–40
3. Support		(No avidin)	Biotin-acryl	<1	ND
	Anchor	Suc. avidin	+ Biotin-acryl	80–90	>75
(D) 1. Extract plus avidin	Pre-mRNA	Avidin	+ Biotin-acryl	10–20	ND
plus support		Avidin	+ Biotin-agarose	10–15	ND
		Suc. avidin, or strep-tavidin	+ Biotin-acryl	<1	ND
			+ Biotin-agarose	<1	ND
			Streptavidin-agarose	10–25	ND

[a] Three ^{32}P-labeled RNAs (as in Table I) were tested for binding to and elution from various solid supports before, during, or after incubation for 30 min at 23° in splicing extract supplemented with 40% whole cell extract as described in Methods.

[b] In (A), the RNA was first bound to the support and then incubated in extract. Those conditions similar to the ones used for solid-phase splicing are indicated in bold type. In (B), the RNA was incubated in extract and then bound to the support. In (C), avidin was first bound to the RNA, then incubated in extract, and finally bound to a support. In (D), the levels of nonspecific binding of RNA without biotin were monitored by the incubation of pre-mRNA in extract together with the support.

[c] The types of avidin tested are described in Materials.

[d] The types of solid supports tested are described in Materials. Biotin-acryl is from EY Laboratories, biotin-agarose is from Sigma, and streptavidin-agarose is from BRL.

[e] A total of at least 10^4 Cerenkov cpm of RNA was added to 20 μl of support. The number of cpm bound was determined after washing away the extract as described in Methods. The values were determined from duplicates in at least two separate experiments.

[f] Percentage of the bound RNA that is eluted in elution buffer A in 15 min at 0°.

[g] Not determined.

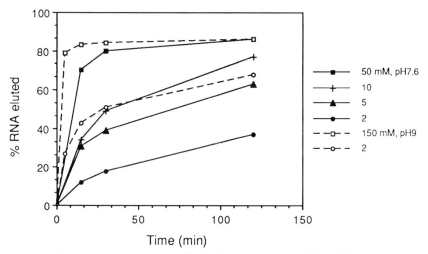

FIG. 4. Eluting biotinylated anchor RNA. Biotinylated anchor RNA 3'-21 was bound to biotin-acryl with succinylavidin as described in Methods. The amounts of RNA eluted with various concentrations of DTT at pH 7.6 (solid lines) or with 2 or 150 mM DTT at pH 9 (dashed lines) were measured as a function of time of incubation in the elution buffers.

before being added to a solid support, most of the RNA hybrid will not bind to the support (Table II). This is not due to cleavage of the disulfide bonds of the anchor RNA, as the anchor RNA alone in the presence of extract still binds to the solid support. The loss of binding of the hybrid may be due to an activity in the extract that unwinds the hybrid. This activity may be sterically hindered from unwinding a hybrid already attached to the support. Perhaps a longer region of complementarity in the hybrid would confer stability; however, such an RNA hybrid could lead to inhibition of splicing.[45–48] Some loss of the pre-mRNA–anchor RNA hybrid is also probably due to nuclease activity in the extract.

There is also a small decrease in anchor RNA binding when it is incubated with extract before or simultaneously with an avidin and a support. Some of this loss may be due to extract factors that bind biotin, because a similar reduction in binding occurs for RNAs conjugated to biotin without a disulfide bond in the linker arm.[22]

Elution by reduction of the disulfide bonds on the biotinylated anchor RNA depends on the pH of the buffer, the DTT concentration, and the

[45] S. H. Munroe, *EMBO J.* **7,** 2523 (1988).
[46] B. L. Bass and H. A. Weintraub, *Cell* **48,** 607 (1987).
[47] M. R. Regbagliati and D. A. Melton, *Cell* **48,** 599 (1987).
[48] R. W. Wagner and K. Nishikura, *in* "Antisense RNA and DNA" (D. A. Melton, ed.), p. 103. Cold Spring Harbor Laboratory, Cold Spring Harbor, New York, 1988.

time of incubation in DTT (Fig. 4) in addition to the type of avidin binding the anchor RNA. The pH dependence is probably due to the pK_a of the thiol group.[16] Although the splicing extract at pH 7.0 contains 0.5 mM DTT, this concentration is not high enough to elute most of the anchor RNA at that pH during short periods of incubation in the extract. By increasing the pH and DTT concentration of the elution buffer slightly, one can effectively elute the RNA during longer incubation times. We have yet to assay the activity of any complexes eluted with 2 mM DTT; however, splicing does proceed in 50 mM DTT.[30]

Solid-Phase Splicing

We measured the efficiency of splicing of the pre-mRNA–anchor RNA hybrid in the presence or absence of a solid support. In the absence of the solid support, hybrids with any of the three anchor RNAs splice as efficiently as pre-mRNA alone. Either biotinylated or nonbiotinylated anchor RNA 3'-19 has no effect on splicing efficiency even when added in a 200 molar excess to the pre-mRNA.[30] Succinylavidin also has no effect on splicing when added to the extract.

When substrate RNA is immobilized on a biotin-cellulose, biotin-agarose, or biotin-acryl support, it can be spliced, but not so efficiently as in the absence of a solid support.[30] When pre-mRNA without an anchor is incubated in splicing extract in the presence of either agarose or Sephacryl beads, the splicing efficiency is reduced compared to reactions without the supports. This suggests that the reduction in efficiency is due, in part, to restricted diffusion of the splicing factors and RNA. Splicing occurs on the spliceosome, a large ribonucleoprotein particle that assembles on the pre-mRNA. The spliceosome and some of its component snRNPs would be excluded owing to their sizes from the agarose matrix but not from the Sephacryl matrix.[49–51] Our results suggest that the spliceosome forms more efficiently in the presence of Sephacryl beads rather than agarose beads (see below). Although cellulose does not have an internal matrix volume and thus may circumvent part of the diffusion problem, it is not suitable for large-scale preparations because it has a slow flow rate.

snRNP Binding to Immobilized RNAs

By ^{32}P labeling the RNAs extracted from chromatographed samples, we can monitor the total ribonucleoprotein population that binds to the

[49] R. Reed and T. Maniatis, *Cell* **53**, 949 (1988).
[50] M. W. Clark, S. Goelz, and J. Abelson, *EMBO J.* **7**, 3819 (1988).
[51] C. W. Pikielny, B. C. Rymond, and M. Rosbash, *Nature (London)* **324**, 341 (1986).

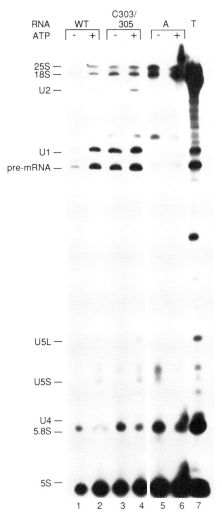

FIG. 5. Enrichment of specific RNAs with immobilized pre-mRNA. Two immobilized actin pre-mRNAs were used in splicing reactions: wild-type (lanes 1 and 2) and mutant C303/305 (lanes 3 and 4). The mutant pre-mRNA undergoes the first splicing reaction (5' splice site cleavage) and accumulates spliceosomes.[26] Biotinylated anchor RNA without pre-mRNA was used as a control (lanes 5 and 6). The reactions contained splicing buffer supplemented with a fraction (35P) of whole cell extract active for splicing. They were incubated with (+) or without (−) ATP for 30 min at 23°. The RNAs eluted from each reaction (lanes 1 to 6) and from total splicing extract (lane 7) were analyzed by 3'-end-labeling with [^{32}P]pCp and T4 RNA ligase. After electrophoresis in a denaturing polyacrylamide gel, the RNAs were visualized by autoradiography as shown here. The small nuclear RNAs U1, U2, U4, and U5 and the ribosomal RNAs are indicated.

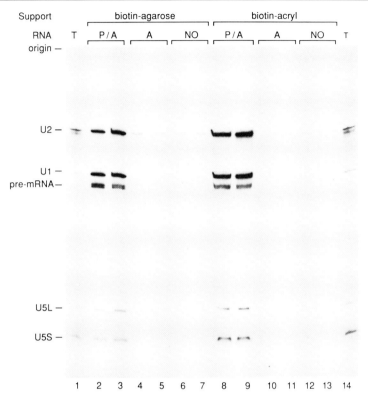

FIG. 6. snRNP binding to pre-mRNA immobilized on either biotin-agarose or biotin-acryl. Two supports, Sigma biotin-agarose (lanes 2–7) and EY biotin-acryl (lanes 8–13), were tested for their effects on snRNP binding to immobilized pre-mRNA. Either actin mutant C303/305 hybridized to anchor RNA 5'-24 (lanes 2, 3, 8, and 9), anchor RNA 5'-24 only (lanes 4, 5, 10, and 11), or no RNA (lanes 6, 7, 12, and 13) was bound to the supports. Duplicates of each were incubated in splicing buffer supplemented with whole cell extract and ATP for 20 min at 23°. The RNAs eluted from each reaction were analyzed by RNA blot hybridization with U1, U2, and U5 snRNA-specific probes. RNAs extracted from whole cell extract (lanes 1 and 14) were used as markers.

immobilized pre-mRNA in a splicing reaction. We find that pre-mRNAs which form spliceosomes bind spliceosomal snRNPs, U1 in the absence of ATP and U1, U2, U5, and U4 in the presence of ATP (Fig. 5). (U6 snRNA does not label efficiently with T4 RNA ligase, but we have detected it by RNA blot hybridization.[18]) Usually, these snRNAs can be easily discerned in the chromatographed samples, whereas their detection in splicing complexes isolated by other techniques requires an additional purification step (see, for example, Ref. 51). These snRNPs are not retained by the anchor RNA alone. Furthermore, mutant pre-mRNAs that do not form spliceosomes bind no or decreased amounts of these snRNPs in this method.[18] There is a large reduction in the amount of ribosomes, which are the major contaminants of spliceosomes isolated by other methods.[49–51] We estimate, by comparing the levels of U1 and U2 snRNAs relative to 18 and 25 S rRNAs in the chromatographed samples (Fig. 5, lanes 1–4) versus those in splicing extract (lane 7), that these snRNAs are enriched at least 100-fold relative to the ribosomal RNAs. We have also observed a large decrease in the amounts of tRNA in the chromatographed material.[30]

Enrichment for specific snRNPs can also be monitored by RNA blot hybridization (Fig. 6). The levels of snRNP binding can be readily detected from 2- to 5000-fold above background levels.[30] The background levels of nonspecific binding of the snRNPs to either anchor RNA or the support is usually somewhat lower with biotin-acryl or biotin-cellulose than with biotin-agarose. This difference is particularly noticeable for U5 snRNP (Fig. 6), which shows the highest level of nonspecific binding of the five spliceosomal snRNPs.[30]

Summary

Small, reversibly biotinylated RNAs as described here are versatile ligands for affinity chromatography of RNA-binding components. These RNAs can be attached to a solid support by binding to avidin and used as ligands, or they may be hybridized to another RNA which acts as the ligand. The incorporation of a disulfide bond in the linker arm connecting biotin to the RNA makes it possible to dissociate the RNA from avidin under mild conditions. Our results regarding the binding and elution of the biotinylated RNA may be applied to other, reversibly biotinylated molecules.

Acknowledgments

We thank the following: O. Uhlenbeck for advice on synthesis of small RNAs; S. Horvath for synthesizing oligodeoxynucleotides; G. Stanton and A. Chu for synthesizing biotin-acryl and succinylavidin-agarose; E. Rabbani and M. Brumfield of Enzo Biochem, and T. Herman for allylamine-UTP; D. Ward and T. Herman for initial advice on using biotinylated RNA; M. Ares, P. Fabrizio, C. Guthrie, B. Patterson, P. Siliciano, and U. Vijayraghavan for cloned genes; and T. Amatruda, R. Miake-Lye, and U. Landegren for comments on the manuscript. Supported by National Institutes of Health Grant GM32637 (J.A.) and American Cancer Society Fellowship PF2647 (S.R.).

[10] Measurement and Analysis of Yeast Pre-mRNA Sequence Contribution to Splicing Efficiency

By B. C. Rymond, C. Pikielny, B. Seraphin, P. Legrain, and M. Rosbash

Introduction

For over 10 years, our laboratory has been interested in RNA processing in the yeast *Saccharomyces cerevisiae*. At the outset we were motivated to explain the puzzling specificity of the *rna* mutants for ribosomal protein synthesis. These mutants were isolated and initially characterized by Lee Hartwell and colleagues, who screened a large number of temperature-sensitive mutant strains of *S. cerevisiae* for effects on RNA metabolism.[1] Jon Warner and colleagues then showed that incubation of these mutants at the nonpermissive temperature had a major impact on rRNA accumulation, owing to an effect on rRNA processing rather than rRNA synthesis.[2] Subsequent work demonstrated that the effect on rRNA processing was likely due to a more primary effect on ribosomal protein synthesis, mediated by a disappearance of ribosomal protein mRNA on incubation of the mutant strains at the nonpermissive temperature.[3]

Our contribution to this topic began as we sought to explain the apparently selective action of this set of temperature-sensitive genes on ribosomal protein mRNA levels. By cloning several yeast ribosomal protein genes, we discovered that they contained introns at an unusually high frequency.[4] Only one yeast gene, the actin gene, had been previously shown to contain an intron.[5,6] We also observed that incubation of several of these *rna* strains at the nonpermissive temperature had a dramatic effect on the steady-state distribution of ribosomal protein mRNA and pre-mRNA; pre-mRNA levels increased rapidly and dramatically whereas mature, processed mRNA levels decreased with kinetics consistent with mRNA decay.[4] Similar observations were subsequently made in

---

[1] L. H. Hartwell, C. S. McLaughlin, and J. R. Warner, *Mol. Gen. Genet.* **109,** 42 (1970).

[2] R. W. Shulman and J. R. Warner, *Mol. Gen. Genet.* **161,** 221 (1978).

[3] J. R. Warner and C. Gorenstein, *Cell* **11,** 201 (1977).

[4] M. Rosbash, P. K. W. Harris, J. L. Woolford, and J. L. Teem, *Cell* **24,** 679 (1981).

[5] R. Ng and J. Abelson, *Proc. Natl. Acad. Sci. U.S.A.* **77,** 3912 (1980).

[6] D. Gallwitz and I. Sures, *Proc. Natl. Acad. Sci. U.S.A.* **77,** 2546 (1980).

Warner's laboratory.[7] We proposed that these observations were suffi-cient to explain the apparent specificity of the *rna* mutants for ribosomal protein genes; the *RNA* genes are required for pre-mRNA splicing, and most ribosomal protein genes (in contrast to most other yeast genes) contain an intron.[4] The fact that the *rna2* mutation affects actin pre-mRNA splicing as well as ribosomal protein pre-mRNAs was consistent with this view.[8]

To generalize these findings and to provide a tool with which pre-mRNA splicing could be easily studied *in vivo*, we constructed the fusion gene HZ18[9] (Fig. 1). HZ18 contains the *rp51* intron, with some flanking exon sequences, inserted into the galactose-inducible *CYC1–lacZ* fusion gene of pLGSD5.[10] In brief, HZ18 provides the following features: (1) The plasmid is easily grown in *Escherichia coli* under ampicillin selection. (2) Owing to the presence of a yeast *URA3* gene, the plasmid can be selected for and maintained in *ura3⁻* yeast hosts. (3) The 2μ replication origin maintains the plasmid at 5–10 copies per yeast cell. (4) The GAL up-stream activator sequence puts transcription of the fusion gene under galactose control, i.e., transcription requires the presence of galactose and is repressed by glucose. (5) The presence of the *rp51* intron places the fusion gene under "splicing control," that is, gene expression (β-galacto-sidase activity) requires proper removal of the *rp51* intron. Splicing re-moves the intron (with its many translational stop codons) and leaves the initiating methionine in frame and contiguous with the β-galactosidase coding region.

To demonstrate that β-galactosidase expression is splicing dependent in the HZ18 construct, the plasmid was transformed into *rna2* and *RNA2* strains and induced with galactose at both permissive and nonpermissive temperatures (Fig. 2). The results showed that galactose induction of HZ18 gave rise to no β-galactosidase activity in an *rna2* strain at the nonpermissive temperature but did produce substantial levels of activity in an *rna2* background at the permissive temperature as well as in a wild-type strain at both permissive and nonpermissive temperatures. β-Galac-tosidase activity was induced in the pLGSD5 control (the intronless par-ent gene) in both genetic backgrounds at both temperatures. These results strongly suggested that the HZ18 gene could be used to investigate the cis- and trans-acting requirements of yeast splicing.

[7] H. M. Fried, N. J. Pearson, C. H. Kim, and J. R. Warner, *J. Biol. Chem.* **256,** 10176 (1981).
[8] J. L. Teem, J. R. Rodriguez, L. Tung, and M. Rosbash, *Mol. Gen. Genet.* **192,** 101 (1983).
[9] J. L. Teem and M. Rosbash, *Proc. Natl. Acad. Sci. U.S.A.* **80,** 4403 (1983).
[10] L. Guarente, R. Yocum, and P. Gifford, *Proc. Natl. Acad. Sci. U.S.A.* **79,** 7410 (1982).

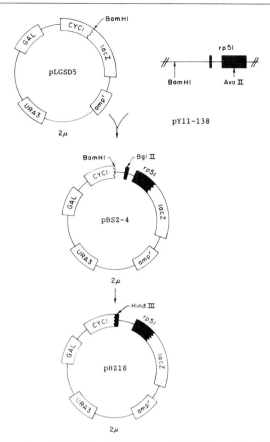

FIG. 1. Construction of pHZ18. pLGSD5 plasmid DNA was linearized with *Bam*HI at the junction where the *CYC1* start codon has been fused to the *lacZ* gene. The plasmid pY11-138 containing the *rp51* gene [J. L. Woolford, Jr., L. M. Hereford, and M. Rosbash, *Cell* **18**, 1247 (1979)] was cut with *Bam*HI and *Ava*II, and a 1.2-kb fragment containing exon I, the intron, and 63 codons of exon II was isolated. *rp51* exon sequences are indicated by the two black segments; the *rp51* intron lies in between. The *Bam*HI end of the fragment was ligated to *Bam*HI-digested pLGSD5 DNA. Klenow fragment was then used to fill in the *Ava*II end of the fragment as well as the remaining *Bam*HI end of the vector. The filled-in *Ava*II site was then ligated to the filled-in *Bam*HI site, joining the *rp51* exon II sequence in frame to the *lacZ* coding sequence and circularizing the plasmid DNA to produce pBSZ-4. pBSZ-4 DNA was cut with *Bgl*II and *Bam*HI, and the vector fragment was isolated. The *Bam*HI end and the *Bgl*II end of the fragment were filled in with Klenow fragment. *Hind*III linkers were added, and the plasmid DNA was circularized by ligation. Inclusion of a single *Hind*III linker between the filled-in *Bam*HI and *Bgl*II ends of the vector correctly joins the reading frame of the *rp51* exon to the reading frame initiated at the *CYC1* start codon, resulting in pHZ18. amp^r, Ampicillin resistance; GAL, galactose-inducible upstream activator sequence; URA3, selectable marker for yeast uracil auxotrophs; 2μ, yeast origin of DNA replication.

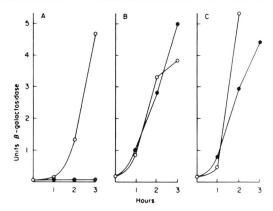

FIG. 2. β-Galactosidase activity in galactose-induced cells subsequent to a temperature shift. Cells were grown at 23° in selective medium to an OD$_{610}$ of 0.3 and were divided into equal aliquots. An equal volume of medium at 23° was added to one aliquot, and to the other was added an equal volume of medium at 45°. Immediately after the temperature shift, samples of each culture were taken. Galactose was added to a final concentration of 2% 3 min after the temperature shift, and time points were taken from each culture at 1-hr intervals. β-Galactosidase was then assayed. (A) *rna2* cells transformed with pHZ18; (B) *RNA2* cells transformed with pHZ18; (C) *rna2* cells transformed with pLGSD5. ○, Cells maintained at 23°; ●, cells maintained at 34°.

We next sought to characterize the various HZ18-derived RNA species present in wild-type cells. The most convenient and often the most informative method of analysis that we routinely employ is primer extension.[10a] By using single-stranded DNA oligonucleotides (primers) and reverse transcriptase, the HZ18-derived RNA species present in yeast (pre-mRNA, splicing intermediates, and products) can be visualized (Fig. 3). These transcripts had been initially identified by a variety of other procedures, including Northern blotting from agarose and polyacrylamide gels.[11] However, the large sizes and 3' end heterogeneity of the HZ18 pre-mRNA, lariat intermediate, and mRNA species limited the utility of these blotting procedures. Each RNA gives rise to a unique primer extension product or set of products, and the relative amount of each product (intensity of radioactive bands) reflects the relative steady-state level of its RNA template (see below). Most importantly, the two-step pathway for metazoan nuclear pre-mRNA splicing, worked out with the development

[10a] W. R. Boorstein and E. A. Craig, this series, Vol. 180, p. 347.
[11] J. R. Rodriguez, C. W. Pikielny, and M. Rosbash, *Cell* **39**, 603 (1984).

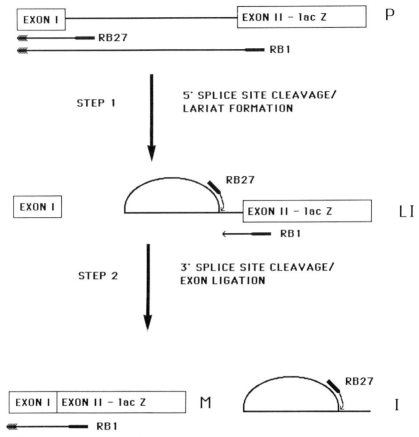

Fig. 3. Characterization of the two-step splicing pathway by primer extension analysis. The first step in splicing consists of cleavage of the pre-mRNA at the 5' splice site and the formation of a lariat structure in which the 5' terminal nucleotide of the intron (almost always a G residue) is attached by a 2'–5' phosphodiester bond to a residue (almost always A) near the 3' end of the intron. The second step of splicing consists of 3' splice site cleavage and exon ligation with release of the excised lariat intron. The presence and relative abundance of these molecules can be determined by primer extension analysis with an exon II primer (RB1) and an intron primer (RB27). The lengths of the various cDNA products (indicated by the distance between the RB1 and RB27 binding sites and the corresponding arrowheads) are functions of the primer binding sites. The relative abundance of individual RNA species (e.g., pre-mRNA and mRNA) can be estimated by the relative intensities of the cDNA signal (in this case, derived from RB1). The multiple arrowheads on RB1 pre-mRNA and mRNA species reflect the multiple transcriptional start sites of the HZ18 construct. The branch point within the lariat-containing molecules inhibits the extension by reverse transcriptase from both exon II- (RB1) and intron- (RB27) complementary oligonucleotides. P, pre-mRNA, LI, lariat intermediate; M, mRNA; I, intron. Exon I, intron, and exon II are not drawn to scale.

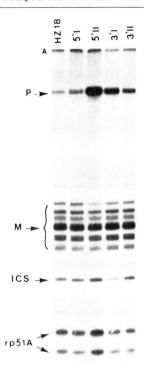

FIG. 4. Primer extension analysis of wild-type and mutant transcripts using an exon II primer. Yeast transformed with the wild-type plasmid HZ18 or with plasmids carrying point mutations in the 5' splice junction or the TACTAAC box (5'I, 5'II, 3'I, or 3'II) were grown in selective medium to an OD_{610} of 0.4–0.6, harvested, and RNA prepared. The oligonucleotide RB1 primer is complementary to exon II sequences in the fusion gene as indicated in Fig. 3 and to exon II sequences in *rp51A* as described in the text. After synthesis, the cDNA products were analyzed by electrophoresis on a 6% acrylamide sequencing gel. The arrows denote cDNA molecules corresponding to (from top to bottom) unspliced fusion transcripts (P, pre-mRNA), spliced fusion transcripts (M, mRNA), stop of the reverse transcriptase at the branch point of the lariat molecule (ICS), and spliced mRNAs from *rp51A* (*rp51A*). The plasmids are indicated above each lane.

of *in vitro* splicing procedures[12,13] and equally applicable to yeast splicing *in vitro*,[14] agrees with the pattern of RNAs we see *in vivo*.[11,15]

[12] R. A. Padgett, M. Konarska, P. J. Grabowski, S. F. Hardy, and P. A. Sharp, *Science* **225**, 898 (1984).

[13] B. Ruskin, A. R. Krainer, T. Maniatis, and M. R. Green, *Cell* **38**, 317 (1984); A. Krämer and W. Keller, this volume [1].

[14] R.-J. Lin, A. J. Newman, S.-C. Cheng, and J. Abelson, *J. Biol. Chem.* **260**, 14780 (1985); S.-C. Cheng, A. Newman, R.-J. Lin, G. McFarlane, and J. N. Abelson, this volume [8].

[15] C. W. Pikielny, J. L. Teem, and M. Rosbash, *Cell* **34**, 395 (1983).

The HZ18-derived primer extension products (exon II primer, RB1; see Fig. 3) are shown in Fig. 4, lane 1. Based on the size of the RNAs, we can assign each of the prominent bands to a particular product or intermediate.[15] P is HZ18 pre-mRNA. M is spliced mRNAs (the heterogeneity is due to the multiple start sites of the *CYC1* promoter; the pre-mRNA bands are similarly heterogeneous if they are electrophoresed for a longer time). ICS is the strong reverse transcriptase stop at the branch point sequence (UACUAAC) arising from termination at the $2'-5'$ phosphodiester bond between the last A of this sequence and the first nucleotide of the intron; this band reflects the concentration of the lariat intermediate RNA. *rp51A* is the two *rp51A* mRNAs (two principal transcriptional start sites) transcribed from the chromosome; they have homology with the RB1 primer and serve as internal controls for the efficacy of RNA extraction, primer extension, etc. Thus, a single primer, RB1, gives rise to cDNAs that reflect the steady-state level of pre-mRNA, lariat intermediate (a product of the first step of splicing), and mRNA (a product of the second step of splicing). A primer complementary to intron sequences, RB27 (see Fig. 3), can be used to determine the relative abundance of pre-mRNA and lariat molecules (lariat intermediate and excised intron).

We used this assay to analyze the effects of point mutations within the splicing consensus sequences, the 5' splice junction, and the TACTAAC box.[15,16] By oligonucleotide-directed mutagenesis we introduced mutations into the consensus sequences of HZ18 and assayed by primer extension the steady-state RNA levels.[17] Shown in Fig. 4 are four mutations, two in the 5' splice junction (5'I and 5'II) and two in the TACTAAC box (3'I and 3'II). The surprising result was that all four point mutations have no effect or only a minor effect on the steady-state levels of mRNA or lariat intermediate (reflected by the intensities of the M and ICS bands, respectively). Accordingly, β-galactosidase activity derived from the mutant genes is almost unchanged as compared to the parent gene (Table I). In contrast, these mutants have a detectable and, in some cases, pronounced effect on pre-mRNA levels (Fig. 4).

Our interpretation of these observations[18] is illustrated in Fig. 5. Pre-mRNA has two fates in the nucleus. It can be spliced into mRNA and transported to the cytoplasm (illustrated rather simply by the single, forward rate constant K_{sp}) or it can be degraded within the nucleus (illustrated by the rate constant K_p). Very recent experiments[18a] make it clear that pre-mRNA has a third fate, namely, transport to the cytoplasm; for a

[16] C. J. Langford and D. Gallwitz, *Cell* **33**, 519 (1983).
[17] A. Jacquier, J. R. Rodriguez, and M. Rosbash, *Cell* **43**, 423 (1985).
[18] C. W. Pikielny and M. Rosbash, *Cell* **41**, 119 (1985).
[18a] P. Legrain and M. Rosbash, *Cell* **57**, 573 (1989).

TABLE I
ESTIMATION OF SPLICING EFFICIENCY OF
DIFFERENT MUTANTS

Plasmid	β-Galactosidase activity[a]	M/P[b]
HZ18	1	1
5'I	1	1
5'II	1–0.75	0.1
3'I	1	0.2
3'II	1	0.3

[a] The β-galactosidase activity was assayed as described by Teem and Rosbash.[9] The numbers represent the relative activity normalized to the wild-type plasmid HZ18. Some variation occurred from experiment to experiment, so these values should be taken as averages. For 5'II the activity was most often equal to wild-type but was, in some cases, found to be slightly lower as indicated.

[b] The ratio M/P represents the ratio of mRNA to pre-mRNA levels, normalized to the ratio observed for the wild-type plasmid HZ18. Quantitation of the two RNA species was done by densitometric analyses of different exposures of several primer extension experiments. Here again, some minor variations did occur for different cultures of the same mutant, so these values vary ±20%.

variety of reasons, this possibility is not considered in the discussion presented below. We assume that under normal circumstances most of the pre-mRNA is converted to mRNA, i.e., the forward rate constant K_{sp} is much greater than the turnover rate constant K_p. Furthermore, we assume that mutant substrates are spliced less efficiently than their wild-type counterparts; in other words, the forward rate constant K_{sp} is reduced for mutant substrates. A major consequence of a reduction in K_{sp} is an increase in the steady-state concentration of pre-mRNA. In general, the rate of mRNA formation, or the flux through the splicing pathway, is the product of pre-mRNA concentration and K_{sp}. With weak mutants (like those shown in Fig. 4) this product is essentially identical to that of the wild type because the product of an increased pre-mRNA concentration and a comparably decreased K_{sp} is almost equal to the wild-type product. These arguments, both intuitively and mathematically, depend on the fact that for both wild-type and weak mutant substrates, the frac-

A

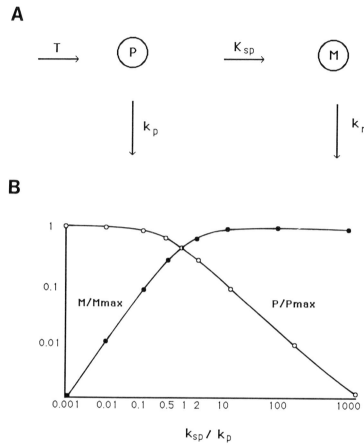

B

$$k_{sp} / k_p$$

FIG. 5. General splicing scheme. (A) Steps involved in a hypothetical splicing pathway involving a single cellular compartment. The two populations, mature mRNA (M) and pre-mRNA (P), are at steady state, that is, their concentrations do not vary as a function of time. Conversion of pre-mRNA to mRNA occurs by splicing (K_{sp}). Pre-mRNA and mRNA have specific turnover rate constants, K_p and K_m, respectively. Transcription (T) is constant. All reactions obey first-order kinetics, that is, the rate of a given reaction is proportional to the concentration of starting material, which means that none of the reactions are reversible and none of the putative enzymatic components are saturated. (B) Solving M and P in the above scheme yields

$$M = (T/K_m)(K_{sp}/K_p)/(1 + K_{sp}/K_p); \qquad M/M_{max} = (K_{sp}/K_p)/(1 + K_{sp}/K_p)$$
$$P = (T/K_p)/(1 + K_{sp}/K_p); \qquad P/P_{max} = 1/(1 + K_{sp}/K_p)$$

The maximum for M, M_{max}, is reached when $K_{sp} \gg K_p$ (splicing is not limiting), $M_{max} = T/K_m$. The maximum for P, P_{max}, is reached when $K_{sp} \ll K_p$ (splicing is limiting), $P_{max} = T/K_p$. The plot shows M/M_{max} and P/P_{max} as a function of K_{sp}/K_p, but K_p is kept constant (because P_{max} is a function of K_p), and only K_{sp} is varied. Note that $M/P = K_{sp}/K_m$ and does not depend on K_p.

tion of pre-mRNA that is degraded in the nucleus is small compared to the fraction processed into mRNA, i.e., $K_{sp} \gg K_p$. Very severe mutants, illustrated, for example, by deletions of the 5′ splice junction and the TACTAAC box, manifest maximum levels of pre-mRNA and very reduced or no detectable mRNA.[18]

We interpret these observations to indicate that with these severe mutants the forward rate constant, K_{sp}, has fallen below the turnover rate constant, K_p, such that much or most of the pre-mRNA is degraded. These arguments are further illustrated in Fig. 5B, where M/M_{max} (M_{max} is the amount of mRNA from a wild-type substrate) and P/P_{max} (P_{max} is the steady-state level of pre-mRNA from a totally splicing-deficient substrate) are plotted as a function of the K_{sp} to K_p ratio. This simple equation and the plot demonstrate that from substrates with relatively high splicing efficiencies (where K_{sp}/K_p is a large number) mRNA levels are essentially invariant, but the pre-mRNA levels are an inverse function of the splicing efficiency (i.e., Fig. 4). In contrast, with severe mutants (where K_{sp}/K_p is a small number) the pre-mRNA levels are high and essentially invariant, but mRNA levels (or β-galactosidase activity) are low and a function of the splicing efficiency. Based on these considerations we believe that the ratios of mRNA level to pre-mRNA level (M/P values) are the best indicators of relative splicing efficiencies.

This concept is further illustrated by a primer extension assay, related to that described above (Fig. 6). As in Fig. 4, this assay uses the exon II primer RB1. In addition, the extensions are done in the absence of dGTP and in the presence of dideoxy-GTP so that DNA synthesis arrests at the position of the first C downstream of the primer.[18] The precise position of the primer and the dideoxynucleotide (G) were chosen so that there is a clear distinction between the extension products from the mRNA and pre-mRNA templates. For example, the HZ18 (or *rp51A*) extension products are 40 nucleotides in length from the pre-mRNA template and 36 nucleotides from the mRNA templates (Fig. 6A). *Saccharomyces cerevisiae* has a second *rp51* gene, called *rp51B*. There is excellent homology in the coding region between *rp51A* and *rp51B* such that this primer also hybridizes to *rp51B* RNA. The *rp51B* pre-mRNA and *rp51B* mRNA extension products are both 41 nucleotides in length. This 41-nucleotide band serves as an internal control (Fig. 6a).

The results of such a modified primer extension assay on HZ18 and three mutant derivatives are shown in Fig. 6B. Some of these mutant derivatives, Δ3Act · I and Δ3Act, are deletion derivatives of the wild-type intron (HZ18) while the third, 5′.2, is a point mutant in the 5′ junction (GTATGT → GTATAT). The results show that the M/P ratio is different for each of these mutants. This abortive primer extension procedure is

RP51B   A

41 bases

ddG--------------------

P: ...GAUUAUUGCUAUUUUUAUAG/GGUAGAGUUA...

41 bases

ddG--------------------

M: ...GAAAAAAGCAGAUAAAAAUG/GGUAGAGUUA...

RP51A

40 bases

ddG--------------------

P: ...GCUUUUCGUCAUUUUAAUAG/GGUAGAGUUA...

36 bases

ddG----------------

M: ...AGACUAGCAAUAACAAAAUG/GGUAGAGUUA...

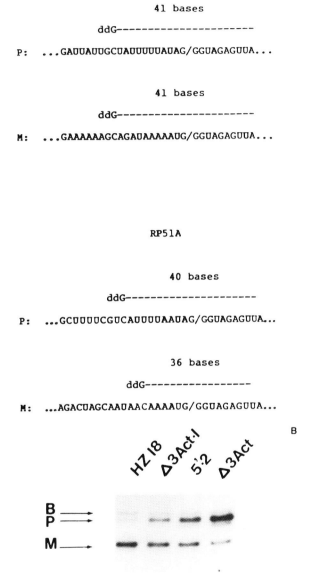

FIG. 6. "Short" reverse transcription. (A) The sequences of the pre-mRNA (P) and mature mRNA (M) of *rp51A* and *rp51B* are shown in the vicinity of the 3′ splice site up to the RB1 primer binding site. The slash (/) indicates the 3′ splice site. In the case of the pre-mRNAs (P) intron sequences lie to the left of the slash; in the case of the mRNAs (M), exon I sequences lie to the left of the slash. The cDNAs obtained in the presence of ddGTP instead

very useful for cases in which there are multiple transcripts that are due, for example, to multiple initiation sites or for cases in which strong primer extension stops occur between the primer and the 5' end of the transcripts. These circumstances make it rather difficult to generate quantitative data, e.g., scan gels, so by compacting all the pre-mRNA-derived transcripts in a single band and all the mRNA-derived transcripts in another, reliable M/P values are more easily obtained.

It is, of course, of considerable interest to understand why mutations in the consensus sequences adversely affect splicing efficiency. However, methods are only beginning to become available to analyze the early *in vivo* splicing events (the association of pre-mRNA with trans-acting factors).[18a] *In vitro* systems had established that splicing, both in metazoa and in yeast, occurs in large ribonucleoprotein (RNP) particles,[19-21] i.e., spliceosomes, and it seemed to us quite likely that mutations in the consensus sequences might affect spliceosome assembly. Because the sucrose or glycerol gradient assays currently available[19-21] did not seem to have the resolution required to examine in detail the individual assembly steps, we turned to gel electrophoresis for our analyses.[22,23] We modified the protocols originally developed by Dahlberg and colleagues[24] for the electrophoretic analysis of polyribosomes, ribosomes, and ribosome subunits. A similar procedure was independently devised by Konarska and Sharp.[25] On addition of a small, splicing-competent substrate and ATP to a yeast whole cell extract,[14] the time-dependent appearance of three splicing complexes could be visualized (Fig. 7). Based on a large number of criteria, we ordered these three complexes into an assembly pathway. This three-step pathway (complex III → complex I → complex II) results in the formation of a mature spliceosome (complex II or a subtle derivative thereof) within which the cleavage and ligation reactions take place. An important step in our procedures involves the addition of carrier RNA

[19] E. Brody and J. Abelson, *Science* **228**, 963 (1985).
[20] P. J. Grabowski, S. R. Seiler, and P. A. Sharp, *Cell* **42**, 345 (1985).
[21] D. Frendewey and W. Keller, *Cell* **42**, 355 (1985).
[22] C. W. Pikielny and M. Rosbash, *Cell* **45**, 869 (1986).
[23] C. W. Pikielny, B. C. Rymond, and M. Rosbash, *Nature (London)* **324**, 341 (1986).
[24] A. E. Dahlberg, C. W. Dingman, and A. C. Peacock, *J. Mol. Biol.* **41**, 139 (1969).
[25] M. M. Konarska and P. A. Sharp, *Cell* **49**, 763 (1987); M. M. Konarska, this series, Vol. 180, p. 442.

of dGTP are shown above the sequences. In each case transcripts from the chromosomal *rp51A* gene contribute a constant and small amount (<10% of the HZ18 value) to the P and M cDNA signals. (B) cDNAs obtained with the method described above: M, derived from HZ18 mRNA; P, derived from HZ18 pre-mRNA; B, transcripts originating from the *rp51B* gene. M/P values obtained from this experiment are as follows: HZ18, 7; Δ3Act-1, 4; 5'-2, 1; Δ3Act, 0.5.

FIG. 7. Native gel analysis of splicing complexes. Uniformly labeled *rp51A*-derived pre-mRNA [B. C. Rymond, D. D. Torrey, and M. Rosbash, *Genes Dev.* **1**, 238 (1987)] was prepared *in vitro* (J. K. Yisraeli and D. A. Melton, this series, Vol. 180, p. 42) and incubated under standard splicing conditions for 1, 5, 10, or 20 min. Complex formation was assayed on a 3% polyacrylamide–0.5% agarose composite gel. I, II, and III refer to the previously described forms of the yeast spliceosome;[23] U indicates the position of the unassembled pre-mRNA.

subsequent to spliceosome assembly but prior to electrophoretic analysis.[23] Presumably the carrier RNA eliminates nonspecific and adventitious binding of proteins to the spliceosome complexes, thereby increasing the electrophoretic mobility of the complexes and improving resolution (although we note that the addition of carrier may also remove important, splicing-specific factors from the spliceosome).

Our analyses of the snRNAs present in the three splicing complexes provided supporting evidence that the assembly pathway was correct as described (Fig. 8).[23] A surprising finding was that snRNA14 (U4), which associated with spliceosomes during the complex III → complex I transition, disappeared from the spliceosomes during the subsequent complex I → complex II transition. With the realization that each of the yeast snRNAs was a cognate of the better studied metazoan snRNAs[26–30] and with the addition of U6 snRNA to the assembly pathway[31] (yeast U6 RNA, like metazoan U6 RNA, is not capped, and therefore it escaped our initial analyses which were dependent on the presence of a trimethyl cap), it became clear that the yeast and metazoan pathways were very similar. Indeed, recent studies with HeLa cell extracts have shown that, as in yeast, metazoan spliceosomes progress through a three-step assembly pathway with the loss of U4 snRNA between the second and third steps.[32] Of course, this description is an oversimplification as protein factors which presumably associate (and disassociate) during spliceosome assembly are only beginning to be assigned to the individual assembly steps. For example, it would appear that one stage of spliceosome assembly can be further subdivided into two steps to yield a four-step reaction.[31]

It should be kept in mind that the snRNP assembly pathway is based on a gel analysis, which might introduce artifacts; for example, U4 snRNP may not leave the spliceosome but may become more loosely attached so that it dissociates under electrophoresis conditions. Likewise, very recent experiments show that U1 snRNP associates in an ATP-independent manner with yeast pre-mRNA, in part by base pairing with the 5' splice site.[31a-c] Formation of this U1 snRNP–pre-mRNA complex is required for

[26] M. Ares, Jr., *Cell* **47**, 49 (1986).
[27] L. Kretzner, B. C. Rymond, and M. Rosbash, *Cell* **50**, 593 (1987).
[28] P. G. Siliciano, M. H. Jones, and C. Guthrie, *Science* **237**, 1484 (1987).
[29] P. G. Siliciano, D. A. Brow, H. Roiha, and C. Guthrie, *Cell* **50**, 587 (1987).
[30] B. Patterson and C. Guthrie, *Cell* **49**, 613 (1987); D. Brow and C. Guthrie, *Nature* (*London*) **334**, 213 (1988).
[31] S.-C. Cheng and J. Abelson, *Genes Dev.* **1**, 1014 (1987).
[31a] B. Seraphin, L. Kretzner, and M. Rosbash, *EMBO J.* **7**, 2533 (1988).
[31b] S. Ruby and J. Abelson, *Science* **242**, 1028 (1988).
[31c] B. Seraphin and M. Rosbash, *Cell* **59**, 349 (1989).
[32] A. I. Lamond, M. M. Konarska, P. J. Grabowski, and P. A. Sharp, *Proc. Natl. Acad. Sci. U.S.A.* **85**, 411 (1988).

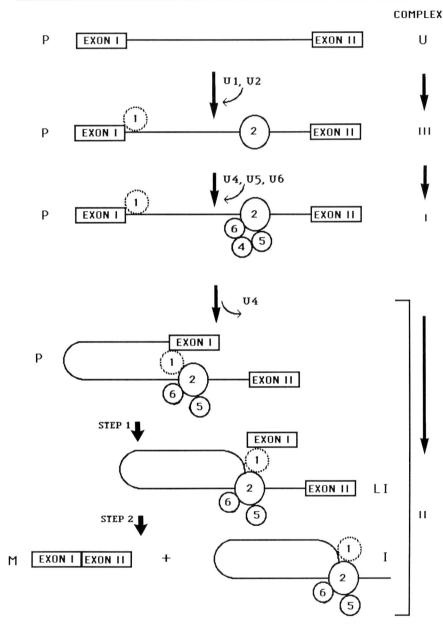

FIG. 8. Schematic of yeast splicing complex assembly. Pre-mRNA (P) added to the yeast splicing extract progresses from a nonassembled state (U) to three splicing complexes (termed I, II, and III based on the relative electrophoretic mobilities) in the order complex III → complex I → complex II. Also associated with complex II are the intermediates (exon

the subsequent ATP-dependent binding of U2 snRNP and for splicing.[31c] Thus, band III formation (Fig. 8) consists of at least two snRNP assembly steps: the ATP-independent association of U1 snRNP and the subsequent ATP-dependent association of U2 snRNP. The absence of U1 snRNP from all previous descriptions of the yeast spliceosome assembly pathway is almost certainly a gel artifact.

Preliminary *in vitro* analyses of splicing substrates bearing mutations in the consensus sequences suggested that most of these mutations do affect spliceosome assembly.[23,33–35] Reminiscent of the *in vivo* results, some consensus sequence mutants have dramatic effects on spliceosome assembly (10-fold effects or greater); others seem to show 2- or 3-fold effects, which are more difficult to quantitate and reproduce with certainty.

In order to analyze and more easily quantitate the contribution of individual nucleotides to spliceosome assembly, we changed our strategy from an analysis of site-directed point mutations to a modification interference approach,[36] analogous to procedures used to examine DNA–protein interactions.[37] The outline of this assay is diagrammed in Fig. 9. Pre-mRNA substrates, 3'-end-labeled with [^{32}P]pCp, are subjected to modification reagents [diethyl pyrocarbonate (DEPC) for purines and hydrazine for pyrimidines] under conditions such that there is less than one lesion per molecule. The modified pre-mRNA is then added (in excess) to an *in vitro* assembly reaction, and the splicing complexes are resolved on native polyacrylamide gels (see Methods). The substrates are purified

[33] A. J. Newman, R.-J. Lin, S.-C. Cheng, and J. Abelson, *Cell* **42**, 335 (1985).
[34] V. Vijayraghavan, R. Parker, J. Tamm, Y. Iimura, J. Rossi, J. Abelson, and C. Guthrie, *EMBO J.* **5**, 1683 (1986).
[35] B. C. Rymond, D. D. Torrey, and M. Rosbash, *Genes Dev.* **1**, 238 (1987).
[36] B. C. Rymond and M. Rosbash, *Genes Dev.* **2**, 428 (1988); L. Conway and M. Wickens, this series, Vol. 180, p. 369.
[37] U. Siebenlist and W. Gilbert, *Proc. Natl. Acad. Sci. U.S.A.* **77**, 122 (1980).

I; lariat intermediate, LI) and products (intron, I; mRNA, M) of the splicing reaction. The small nuclear RNAs (snRNAs) associated with each of these complexes have been determined and are indicated in this diagram using the metazoan, U-series nomenclature: U1 (yeast snR19), U2 (yeast snR20; also referred to as LSR1), U4 (yeast snR14), U5 (yeast snR7), and U6 (yeast snR6). The subtle and/or problematic association of U1 snRNA with these complexes is discussed in the text. U1 is shown associated with the 5' splice site [B. Seraphin, L. Kretzner, and M. Rosbash, *EMBO J.* **7**, 2533 (1988).] U2 is shown associated with the branch point [R. Parker, P. G. Siliciano, and C. Guthrie, *Cell* **49**, 229 (1987)]; U4, U5, and U6 are shown associated with U2 [H. Bindereif and M. R. Green, *EMBO J.* **6**, 2415 (1987)]. The unusually large size of yeast U1 and U2 snRNAs is indicated by the relatively large size of these two circles.

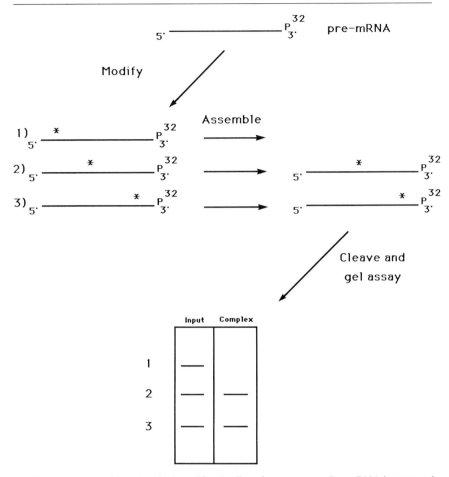

FIG. 9. Outline of the chemical modification/interference assay. Pre-mRNA is prepared *in vitro*, 3'-end-labeled, and randomly modified with base-specific chemical modifying agents. The modified RNAs are then used as substrates for spliceosome assembly. Pre-mRNA lesions (*) that are inhibitory (1) are underrepresented or absent in the RNA cleavage pattern of the splicing complex-derived pre-mRNA (Complex), relative to the input pre-mRNA (Input). Lesions that are not inhibitory (2, 3) are not underrepresented.

from the individual spliceosome complexes, subjected to chemically induced cleavage at the sites of modification, and analyzed on sequencing gels. The expectation is that the assembly process should be sensitive to chemical modification at nucleotide positions required for the formation of spliceosomes. If a particular modification is extremely deleterious to assembly it should be absent from the complex-derived RNA. If the modified residue has a relatively minor influence on the rate or extent of complex formation, then this modification should be only modestly reduced from control values. Thus, the relative sensitivity of spliceosome assembly to modification at each purine and pyrimidine of the RNA (and, by extrapolation, the contribution to assembly of each unmodified nucleotide) can be determined in a single experiment.

A purine (DEPC) modification sequence ladder, in the vicinity of the 5' splice junction, is shown in Fig. 10. Several points can be drawn. First, it is readily apparent that modification of either position 3 (A) or 5 (G) of the 5' splice site consensus sequence (GU$\underline{A}$U$\underline{G}$U) inhibits entry of the pre-mRNA into complex III, the first splicing complex. Second, the position 3 modification has stronger inhibitory effect than the position 5 modification. Third, relative to complex III formation, complex I formation is much more sensitive to 5' splice junction modifications. Similar observations have been made with pyrimidine modifications (hydrazine) in the 5' splice junction region as well as the TACTAAC box region (see Summary, Fig. 11). In conjunction with assembly assays on substrates containing point mutations, the data indicate that all of the consensus sequence nucleotides play a role, and perhaps multiple roles, in spliceosome assembly. By these criteria, only the yeast 3' junction appears to play no early role in spliceosome assembly, as originally proposed.[35]

These *in vitro* data can now be considered in the context of the *in vivo* observations made on genes containing mutations in the 5' splice junction and TACTAAC box sequences. The decreased splicing efficiency observed most likely results from effects of these mutations on spliceosome assembly. A decreased rate of assembly leads to an increased transit time through the presplicing pathway, the net (observable) result of which is an increase in pre-mRNA levels. With relatively mild mutants, the pre-mRNA is still well spliced but more slowly. Thus, it is likely that the splicing efficiency of these mild mutants is (almost) indistinguishable from a wild-type substrate; what we have been calling splicing efficiency (M/P) is likely to reflect spliceosome assembly efficiency and not splicing efficiency *per se*. Our very recent approach to spliceosome assembly is consistent with this hypothesis and suggests further that transport to the cytoplasm may be a competing process.[18a] More sophisticated *in vivo* assays will be required to examine these problems in more detail.

Materials

Oligonucleotides were prepared using a Biosearch Cyclone DNA synthesizer. Reverse transcriptase was purchased from Life Sciences (St. Petersburg, FL), T4 polynucleotide kinase from either U.S. Biochemical

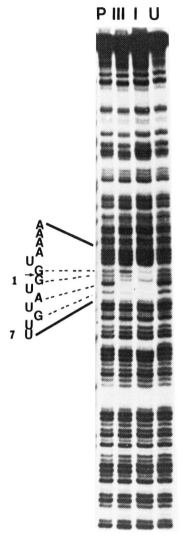

FIG. 10. Modification/interference pattern of splicing complex assembly surrounding the 5' splice site. DEPC-modified pre-mRNA was used as the substrate for spliceosome assembly. The pre-mRNA was recovered from the complexes, and cleavage was induced at the sites of modification. Interfering modifications are indicated by dashed lines between the partial nucleotide sequence (bounded by solid lines) and the corresponding cleavage product. The arrow indicates the exon I–intron border; the numbers refer to the corresponding intron positions. Lane P, Input pre-mRNA; lane III, complex III pre-mRNA; lane I, complex I pre-mRNA; lane U, unassembled pre-mRNA.

Corporation (Cleveland, OH) or Pharmacia (Piscataway, NJ) and T4 RNA ligase from Pharmacia. The enzymes are diluted to the working concentrations (indicated below) in 1× reaction buffer immediately prior to use. Actinomycin D and diethyl pyrocarbonate (DEPC) were from Sigma (St. Louis, MO), aniline and hydrazine from Aldrich (Milwaukee, WI) and 3'-CMP from Pharmacia. [γ-^{32}P]ATP was obtained from New England Nuclear (Boston, MA).

Methods

Primer Extension

Oligonucleotide Purification. For primer extension we routinely use oligonucleotides 17–22 nucleotides long and have noticed no consistent correlation between oligonucleotide length and efficacy. The DNA oligonucleotides should be size-selected by HPLC chromatography or polyacrylamide gel electrophoresis prior to labeling. This precaution removes the shorter, incompletely synthesized molecules that compete for labeling and may lead to background problems during primer extension. We commonly purify oligonucleotides on denaturing 10% polyacrylamide gels (~50 μg DNA in a lane 2 cm wide with 0.4–0.8 mm spacers). The proper band is recognized by UV shadowing, eluted from the polyacrylamide overnight at 23° in TE [10 mM tris(hydroxymethyl)aminomethane (Tris) (pH 7.5), 1 mM ethylenediaminetetraacetic acid (EDTA)], concentrated by ethanol precipitation, washed in 100% ethanol, and resuspended in TE. The A_{260} is determined, and the DNA concentration adjusted to 0.5 mg/ml with TE. The DNA is stored at −20°.

Kinase Reaction for 5'-End-Labeling Synthetic Single-Stranded DNA Oligonucleotides. Mix together 0.5 μl (0.25 μg) of oligonucleotide, 1.0 μl of 500 mM Tris-HCl (pH 7.6), 100 mM MgCl$_2$, 50 mM dithiothreitol (DTT), and 2.0 μl of [γ-^{32}P]ATP (~6000 Ci/mmol, 180 μCi/μl) and add distilled water to 9 μl. Add 1.0 μl of T4 polynucleotide kinase (~10 units) and incubate for 30 min at 37°. Add 40 μl of TE and inactivate the enzyme at 100° for 3 min. Store at −20° until used. In general, while the denatured protein and much of the unincorporated labeled nucleotide can be removed by phenol extraction followed by ethanol precipitation, these steps are not necessary and do not lead to significantly cleaner results.

Hybridization. Mix 10 μg of yeast total RNA,[17] 4 μl of 5× buffer [250 mM Tris-HCl (pH 8.0), 200 mM KCl, 2.5 mM EDTA], and 1 μl labeled single-stranded DNA oligonucleotide (~5 ng) and add distilled water to 15 μl. Incubate at 42° for 30 min.

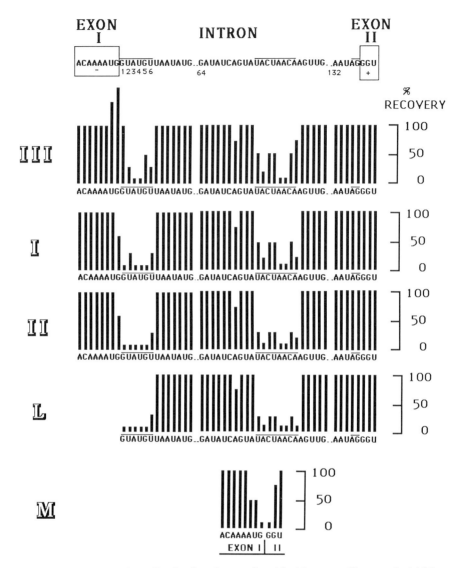

FIG. 11. Summary of modification/interference data. The histograms illustrate the inhibitory effects of certain exon and intron sequences on splicing complex formation (III, I, and II), lariat formation (L), and mRNA formation (M). The data represent average values obtained from several repetitions (between 2 and 5) of the modification/interference assay on complexes III and I, complex II, lariat intermediate, and mRNA. Where no inhibition (or enhancement) was reproducibly visible by eye, the percentage of recovery was set at 100. In the other cases, the average percentage of recovery was calculated from scans of autoradiograms and then set at the closest value, 150, 125, 75, 50, 25, or 10. The figure was designed to present a qualitative summary of the total modification pattern. The terminal two nucleotides of exon I exceed the 100% value in complex III, presumably owing to inhibition of the complex III → I transition. The numbers under the partial nucleotide sequence at top

Extension. Add 5 μl of extension mix to hybridization [extension mix is composed of 1 μl of 2.5 mM dNTP* (2.5 mM for each nucleotide), 1 μl of 200 mM MgCl$_2$, 1 μl of 20 mM DTT, 1 μl of actinomycin D (in ethanol at 1 mg/ml), and 1 μl of AMV (avian myeloblastosis virus) reverse transcriptase (2–5 units)]. Incubate at 37° for 30 min. Terminate the extension reaction by adding 2.5 μl of 3 M sodium acetate and 60 μl of 100% ethanol. Chill on dry ice for 5 min and collect by centrifugation in a refrigerated microcentrifuge. Wash the pellet with 1 ml of 80% ethanol and dry under vacuum. Resuspend the pellet in 15 μl of loading dye (20 mM Tris-HCl, 8 M urea, 1 mM EDTA, 0.05% bromphenol blue, 0.05% xylene cyanol). Heat denature (100°, 3 min) immediately prior to loading the sample on a denaturing (7 M urea) polyacrylamide gel (5–8%).

It is occasionally useful to treat the primer extension products with DNase-free RNase A (5 μl of a 0.2 mg/ml solution at 37° for 5 min) prior to adding the gel loading buffer. This is especially useful when larger amounts of RNA are necessary to generate an extension signal or when the extension product comigrates with one of the abundant, stable RNAs.

Notes: (1) The amount of RNA used in an extension reaction can be variable and depends on the abundance of the transcript in question. For rare transcripts it is helpful to use poly(A)$^+$ RNA, which effects an enrichment of 20- to 50-fold over total RNA if the transcript in question is polyadenylated. (2) The signal-to-noise ratio is superior with end-labeled primer than with cold primer and labeled dNTP incorporation. This is presumably due to the lack of a contribution from "self-priming" in the former case. (3) Different primers are successful to different extents. Our experience is that sometimes the failure of a primer to work well is comprehensible (e.g., intramolecular RNA–RNA pairing which "competes" with primer–RNA hybridization) while at other times it is not. Our rule of thumb is to try two or three different oligonucleotides for a particular RNA before entertaining seriously more elaborate hypotheses about the state of the RNA.

Modification/Interference Assay

3'-End-Labeling RNA. For the preparation of pCp, mix together 3 μl of 10× kinase salts [500 mM Tris-HCl (pH 7.6), 100 mM MgCl$_2$, 50 mM

---

* In the case where a dideoxynucleotide is to be incorporated, use a stock solution consisting of 2.5 mM for the three dNTPs and 100 $\mu$$M$ for the ddNTP.

---

represent the corresponding intron positions. The pairs of dots represent segments of the intron omitted from the diagram in which no inhibitory lesions were found. Overlined from left to right are the 5' splice site, TACTAAC box, and 3' splice site consensus sequences.

DTT, 1 mM spermidine, 1 mM EDTA], 1 μl of 10 mM 3'-CMP, 15 μl of distilled water, 10 μl of [γ-^{32}P]ATP ($\sim$6000 Ci/mmol, 180 μCi/μl), and 1 μl of T4 polynucleotide kinase (10 units). Incubate at 37° for 30 min. Heat inactivate the kinase at 100° for 3 min. Store at $-$20° until needed.

For the labeling reaction,[38] mix 1 μl of pCp, 2 μl of RNA ($\sim$0.5 μg of gel-purified transcript),[38a] and 2 μl of 2.5$\times$ buffer [25% dimethyl sulfoxide (DMSO), 125 mM N-2-hydroxyethylpiperazine-N'-2-ethanesulfonic acid (HEPES) (pH 7.5), 7.5 mM DTT, 50 mM MgCl$_2$, 12.5 μM ATP, 5 units T4 RNA ligase]. Incubate at 0° for 12–18 hr. The labeled RNA is extracted with phenol and precipitated with ethanol prior to use in the modification experiments.

Chemical Modifications. The base-specific modification of nucleic acids with hydrazine and DEPC has been previously described.[39–41] The reactions described below, with minor variations, represent those previously used for RNA sequencing.[42] As noted,[42] the reagents employed in the modification reactions are hazardous and should be handled appropriately (i.e., wear protective clothing and dispense reagents in a fume hood). Treat hydrazine wastes with 5 volumes of 2 M ferric chloride prior to disposal. All reactions are carried out in sterile, 1.5-ml microcentrifuge tubes.

DEPC–purine reactions. Mix 1 μl tRNA (5 mg/ml) and 10 μl 3'-end-labeled pre-mRNA ($\sim$10^7 cpm as described above). Freeze; lyophilize to dryness. Resuspend the RNA in 200 μl of 50 mM sodium acetate (pH 4.0), 1 mM EDTA. Add 1 μl DEPC and mix. Incubate at 90°C for 1 min. Precipitate the RNA with 100 μl of 3 M sodium acetate and 1 ml of 100% ethanol; place on dry ice for 5 min. Spin in a refrigerated microcentrifuge for 10 min. Discard the supernatant and wash the pellet once with 1 ml of 80% ethanol. Resuspend the pellet in 200 μl of distilled water, add 100 μl of 3 M sodium acetate, 1 ml of 100% ethanol, and precipitate as above. Wash and then dry under vacuum.

Hydrazine–pyrimidine modifications. For the C + U reaction, mix 1 μl tRNA (5 mg/ml) and 10 μl 3'-end-labeled pre-mRNA ($\sim$10^7 cpm). Freeze; lyophilize to dryness. Resuspend the RNA pellet in 10 μl of fresh, anhydrous hydrazine–NaCl (17.5 mg NaCl/100 μl hydrazine; at room temperature this solution is saturated with NaCl). Incubate on ice for

---

[38] T. E. England, A. G. Bruce, and O. C. Uhlenbeck, this series, Vol. 65, p. 65.
[38a] J. K. Yisraeli and D. A. Melton, this series, Vol. 180, p. 42.
[39] A. R. Cashmore and G. B. Peterson, *Nucleic Acids Res.* **5**, 2485 (1978).
[40] D. H. Hayes and F. Hayes-Baron, *J. Chem. Soc. C*, 1528 (1967).
[41] A. Vincze, R. E. L. Henderson, J. J. McDonald, and W. J. Leonard, *J. Am. Chem. Soc.* **95**, 2677 (1973).
[42] D. A. Peattie, *Proc. Natl. Acad. Sci. U.S.A.* **76**, 1760 (1979).

15 min. Add 1 ml of 100% ethanol and place on dry ice for 5 min. Spin for 10 min in a refrigerated microcentrifuge. Wash with 80% ethanol. Reprecipitate as described above, wash pellet, and dry.

For the U reaction, mix 1 μl tRNA (5 mg/ml) and 10 μl 3'-end-labeled pre-mRNA ($\sim$10[7] cpm). Freeze; lyophilize to dryness. Resuspend the RNA pellet in 10 μl of fresh, 50% hydrazine (diluted 1:1 in distilled water). Incubate on ice for 2 min. Add 1 ml of 100% ethanol and 100 μl of 3 M sodium acetate. Place on dry ice for 5 min. Spin for 10 min in a refrigerated microcentrifuge, wash with 80% ethanol, and then reprecipitate as described above. Wash pellet and dry.

The labeled, modified pre-mRNA is resuspended in urea dye and repurified on a denaturing, 5% acrylamide gel. This second gel purification of the pre-mRNA assures that the full-length transcripts are separated from nicked molecules generated during the labeling and modification reactions and further assures that no modifying agent remains in the RNA sample. The labeled RNAs are localized by autoradiography of the wet gel, the relevant areas excised, and the RNA eluted overnight at 37° in 300 μl of 100 mM Tris-HCl (pH 7.5), 12.5 mM EDTA, 150 mM NaCl, 1% sodium dodecyl sulfate (SDS), 10 μg/ml tRNA. The RNA is extracted once with phenol–chloroform–isoamyl alcohol (50:49:1, v/v) precipitated with 150 μl of 7.5 M ammonium acetate (pH 6.2) and 1 ml of ethanol, washed 2 times with 80% ethanol, and dried.

Interference Assay. 3'-End-labeled, modified pre-mRNA (1–2 × 10[6] cpm; either A + G modifications, or U modifications and C + U modifications mixed 1:1 to balance the pyrimidine lesions) is incubated in a 50-μl (5×) standard splicing reaction.[14] After 1 min of incubation (for the isolation of complexes III and I) or 20 min of incubation (for the isolation of complex II) at 25°, the reaction is stopped by the addition of an equal volume of Q buffer [400 mM KCl, 50 mM EDTA, 2 mM magnesium acetate, 50 mM HEPES (pH 7.5)].

A portion (10%) of the sample is deproteinized by proteinase K treatment [add 10 μl of sample to 150 μl of 100 mM Tris-HCl (pH 7.5), 12.5 mM EDTA, 150 mM NaCl, 1% SDS (w/v) 0.4 mg/ml proteinase K; incubate at 37° for 10 min] followed by phenol extraction. Ammonium acetate (pH 6.2) is added to 2 M, and the RNA is precipitated with 2 volumes of 100% ethanol, washed twice with 80% ethanol, dried, and put aside as the "input" RNA control.

To the remainder of the splicing reaction is added 100 μg of total yeast RNA, and the mixture is incubated on ice for 10 min. This additional RNA competes for nonspecific associations of extract factors with the pre-mRNA and improves the resolution of the splicing complexes.[23] Thirty microliters of glycerol dye is added (1.65× TBE, 73.5% glycerol,

0.01% bromphenol blue), and the sample is loaded into a preparative well of a horizontal 3% acrylamide–0.5% agarose gel in 0.5× TBE [0.5× TBE is 50 mM Tris–borate (pH 8.5), 1 mM EDTA] and run at 4° for 900–1200 V hr (not exceeding 150 V).

Recipe for native gel. Mix 109 ml of distilled water, 0.65 g agarose, and 6.5 ml of 10× TBE (see above). Heat in a microwave oven to dissolve the agarose. Be careful to make up with additional water any loss in volume caused by evaporation. Cool to approximately 65° and add 13 ml of a 30% solution of acrylamide (acrylamide : bisacrylamide 59 : 1, w/w). Carefully degas. Add 1.3 ml of 10% ammonium persulfate and 0.130 ml N, N, N', N'-tetramethylethylenediamine (TEMED). Pour into a horizontal gel casting tray[43] and cover for at least 1 hr. Remove gel from its casting tray and soak in 0.5× TBE for 30 min. Prerun the gel (submerged in 0.5× TBE) at 100 v for 15 min prior to loading sample.

Transfer of RNA from native gel to DEAE membrane. Prepare the DEAE membrane (NA45, Schleicher & Schuell, Keene, NH), cut to the size of the gel, by soaking, with gentle agitation, in 100 ml of 0.5 N NaOH, followed by 5 successive 5-min rinses (100 ml each) in distilled water. Finally, equilibrate the membrane and two like-sized pieces of Whatman 3MM paper in the transfer buffer (0.5× TBE, 0.1% SDS).

Place the DEAE membrane against the gel and sandwich the membrane–gel between the pieces of Whatman paper. Transfer at 250 mA in an electroblot apparatus (the cross section of our electroblotter is approximately 30 × 20 cm), filled with 0.5× TBE, 0.1% SDS, for 3 hr. The membrane is then removed, wrapped in cellophane, and exposed to X-ray film with a Du Pont Cronex intensifying screen for 15 min at −70°. The relevant regions of the membrane (corresponding to complexes I, II, and III and the unassembled pool) are excised and the RNA eluted by 3 serial washes (65°, 5 min) of the membrane with 150 μl of 55% formamide, 1.8 M sodium acetate (pH 6.5), 0.02 mg/ml tRNA, 2 mM EDTA, 0.2% SDS. To remove any fragments of DEAE membrane, the eluates are pooled and spun in a microcentrifuge for 10 min at room temperature. The solution is carefully transferred to two fresh microcentrifuge tubes and an equal volume of sterile water added to each. The RNA is extracted with 1 volume of phenol–chloroform–isoamyl alcohol (50 : 49 : 1, v/v), precipitated with 2 volumes of ethanol, washed with 80% ethanol, and dried under vacuum. Typically 50–70% of the RNA bound to the DEAE membrane is recovered.

Cleavage of modified pre-mRNA. Cleavage of the pre-mRNA at the site of modification is induced by aniline treatment. Complexes III and I

---

43 W. Hendrickson and R. F. Schleif, *J. Mol. Biol.* **178**, 611 (1984).

and the unassembled pool contain only one labeled species, the full-length pre-mRNA. Complex II, however, also contains lariat intermediate and mRNA. These molecules must be separated from each other on a denaturing polyacrylamide gel prior to aniline treatment.

Aniline reaction. Aniline is light-sensitive and rapidly oxidized. We have found that Aldrich Gold Seal aniline can be used without additional purification. The aniline should be stored in small (1-ml) aliquots, frozen ($-70°$) in a light-tight container.

To the dry RNA pellet add 20 μl of 1 M aniline [diluted in 0.6 M sodium acetate (pH 4.0)]. Mix and incubate in the dark at 60° for 20 min. To remove the aniline and precipitate the RNA, add 1 ml of n-butanol, mix, and spin for 3 min at room temperature in a microcentrifuge. Remove the butanol, resuspend the pellet in 150 μl of 1% SDS, and reprecipitate the RNA with 1.2 ml of butanol. Resuspend the RNA pellet in 150 μl of distilled water and precipitate once again with butanol as described above. Wash the pellet 2 times with 80% ethanol, dry, and resuspend in 10 μl of urea dye. Determine the relative radioactivity of each sample by Cerenkov counting. Adjust to the lowest value by diluting the samples with additional urea dye. Run the samples on a denaturing sequencing gel [6–8% polyacrylamide; 19 : 1 acrylamide : bisacrylamide). In addition to complex-derived RNA, each gel should include lanes containing the input RNA (nonfractionated) and unassembled RNA controls. To determine unambiguously the locations of interfering nucleotides, an RNA sequence of the substrate RNA, produced using the modifying techniques described above plus a guanosine-specific reaction,[42] should be run in adjacent lanes.

Acknowledgments

We would like to thank former members of our laboratory, Jose Rodriguez, Leo Kretzner, and John Teem, for significant contributions to this work and current members of our laboratory, Nadja Abovich, and Hildur Colot for assistance in preparing the manuscript. This work was supported by National Institutes of Health Grant GM23549 to M.R., an NIH postdoctoral fellowship to B.R.C., and a Fogarty International Fellowship to P.L.

[11] Analysis of Pre-mRNA Processing in Transfected Plant Protoplasts

By GREGORY J. GOODALL, KARIN WIEBAUER, and WITOLD FILIPOWICZ

Introduction

Spliceosome-mediated pre-mRNA processing arose before the divergence of fungi, plants, and animals. By studying the mechanism of splicing in all three kingdoms we can gain a better understanding of the general principles of mRNA splicing, as well as its evolutionary history. Plant and animal introns differ to the extent that if animal pre-mRNAs are expressed in plants, splicing is either absent[1-3] or inaccurate.[3] A better understanding of intron processing in plants could also provide some insight into the reported differences in pre-mRNA processing between monocots and dicots[4] or into the role of introns in increasing the level of expression of genes, reported for transformed maize cells.[5]

The analysis of RNAs transiently expressed in transfected plant protoplasts, as summarized below, appears to be at present the most suitable approach for addressing the problems of plant mRNA processing. An *in vitro* splicing system from plants has yet to be developed, and the transient expression procedure is much more rapid than the analysis of RNAs in stably transformed plants. Transient expression experiments have been reported with protoplasts originating from a number of different tissues or suspension cultures from both dicotyledonous and monocotyledonous plants.[6-13]

[1] A. Barta, K. Sommergruber, D. Thompson, K. Hartmuth, M. A. Matzke, and A. J. M. Matzke, *Plant Mol. Biol.* **6**, 347 (1986).
[2] V. L. van Santan and R. A. Spritz, *Gene* **56**, 253 (1987).
[3] K. Wiebauer, J. J. Herrero, and W. Filipowicz, *Mol. Cell. Biol.* **8**, 2042 (1988).
[4] B. Keith and N.-H. Chua, *EMBO J.* **5**, 2419 (1986).
[5] J. Callis, M. Fromm, and V. Walbot, *Genes Dev.* **1**, 1183 (1987).
[6] M. Fromm, L. P. Taylor, and V. Walbot, *Proc. Natl. Acad. Sci. U.S.A.* **82**, 5824 (1985).
[7] W. Werr and H. Lörz, *Mol. Gen. Genet.* **202**, 471 (1986).
[8] E. A. Howard, J. C. Walker, E. S. Dennis, and W. J. Peacock, *Planta* **170**, 535 (1987).
[9] T.-M. Ou-Lee, R. Turgen, and R. Wu, *Proc. Natl. Acad. Sci. U.S.A.* **83**, 6815 (1986).
[10] B. Junker, J. Zimny, R. Lühts, and H. Lörz, *Plant Cell Rep.* **6**, 329 (1987).
[11] R. M. Hauptmann, P. Ozias-Akins, V. Vasil, T. Tabaeizadeh, S. G. Rogers, R. B. Horsh, I. K. Vasil, and R. T. Fraley, *Plant Cell Rep.* **6**, 265 (1987).
[12] D. W. Ow, J. D. Jacobs, and S. H. Howell, *Proc. Natl. Acad. Sci. U.S.A.* **84**, 4870 (1987).
[13] P. Vankan and W. Filipowicz, *EMBO J.* **7**, 791 (1988).

We have used *Orychophragmus violaceous* and *Nicotiana tabacum* suspension cultures, along with HeLa cells, to compare the processing of animal and plant pre-mRNAs.[3] We have also used leaf protoplasts from *Nicotiana plumbaginifolia* and maize to study the cis-acting elements required for pre-mRNA splicing in dicots and monocots. In dicots we find the required elements to be appropriate splice site sequences and a high A + U content within the intron.[14] Transient expression systems using electroporated protoplasts have been described in a recent article in this series.[15] (For extensive discussion of protoplast preparation see Potrykus and Shillito.[16])

In most instances the expression of introduced plasmids has been determined by assaying the activity of enzymes encoded by reporter genes. In a few cases the expressed RNAs were also directly analyzed by RNase A/T1[3,6,14,16a] or nuclease S1[3,9,14] mapping. Below we briefly outline the protocols used in our laboratory for preparation and transformation of protoplasts prepared from suspension cultures of *O. violaceus* and *N. tabacum,* and from leaves of *N. plumbaginifolia* and *Zea mays.* This part is followed by procedures describing RNA isolation from protoplasts and RNA mapping with the use of ^{32}P-labeled complementary RNA probes.

Methods

Abbreviations. PIPES, 1,4-Piperazinediethanesulfonic acid; HEPES, *N*-2-hydroxyethylpiperazine-*N'*-2-ethanesulfonic acid; MES, 4-morpholinoethanesulfonic acid; Tris, tris(hydroxymethyl)aminomethane; EDTA, ethylenediaminetetraacetic acid; 2,4-D, 2,4-dichlorophenoxyacetic acid; NAA, naphthaleneacetic acid; BAP, 6-benzylaminopurine; GA3, gibberellin; DEPC, diethyl pyrocarbonate.

Materials

Stainless steel filters (Saulas, 16 rue du Buisson St-Louis, Paris 10, France)

Cellulolytic enzymes: Cellulase RS, Cellulase Onozuka R10, and Macerozyme R10 (Yakult Honsha Co., Takarazuka, Japan), Pectolyase Y23 (Seishin Pharmaceutical Co., Noda, Japan), Driselase (Fluka, Buchs, Switzerland)

---

[14] G. J. Goodall and W. Filipowicz, *Cell* **58,** 473 (1989).
[15] M. Fromm, J. Callis, L. P. Taylor, and V. Walbot, this series, Vol. 153, p. 351.
[16] I. Potrykus and R. Shillito, this series, Vol. 118, p. 549.
[16a] P. Vankan, D. Edoh, and W. Filipowicz, *Nucleic Acids Res.* **16,** 10425 (1988).

Enzymes for RNA analysis: SP6 polymerase and RNasin (Promega Biotec), RNase H (Gibco-BRL), RNase A (Boehringer-Mannheim), RNase T1 [Calbiochem (Sankyo)]; enzymes from other sources should also be suitable; for DNase treatment we recommend RQ1 RNase-free DNase from Promega Biotec

All other chemicals are of the highest purity available from different commercial sources.

Media, Solutions, and Buffers

Media A, AA, and MS as described in Table I.[17–19]

Supplemented medium AA: medium AA containing 0.35 M sorbitol, 0.1 M sucrose (instead of 20 mg/liter), 1 mM glucose, 1 mM xylose, 1 mM arabinose, and 0.6 mM inositol (instead of 20 mg/liter)

Medium RMP2: MS medium but with one-fifth the concentration of KNO_3, NH_4NO_3, and $MgSO_4$ and with 0.2 mg/liter indoleacetic acid (IAA) in place of 2,4-D

1/2MS salts: inorganic components of MS medium at one-half concentration

Solution EP: 10 mM HEPES, 150 mM NaCl, 5 mM $CaCl_2$, and 0.2 M mannitol, adjusted to pH 7.2

Solution MC: 5 mM MES, 20 mM $CaCl_2$, and 0.5 M mannitol (pH 5.7)

Solution W5: 154 mM NaCl, 5 mM KCl, 125 mM $CaCl_2$, and 5 mM glucose, adjusted to pH 6.0

40% Polyethylene glycol (PEG): 40 g of PEG MW 4000 (Merck) is dissolved in 70 ml of a solution containing 0.1 M $Ca(NO_3)_2$ and 0.4 M mannitol; after dissolving, the solution is brought to pH 10 with KOH, filter-sterilized, and stored frozen in aliquots

Solution of carrier DNA: Calf thymus DNA (Sigma, St. Louis, MO) is sonicated to yield fragments of average length 3–5 kb, precipitated with ethanol for sterilization, and redissolved in sterile water at 2 mg/ml

Buffer TE: 10 mM Tris-HCl (pH 7.5) containing 0.1 mM EDTA

Buffer TBE: 90 mM Tris–borate, 10 mM EDTA (pH 8.3)

Formamide loading solution for polyacrylamide gels: 90% formamide, 10 mM EDTA, 0.1% (w/v) bromphenol blue, 0.1% (w/v) xylene cyanol FF

[17] K. N. Kao and M. R. Michayluk, *In Vitro* **17**, 645 (1981).
[18] A. J. Müller and R. Grafe, *Mol. Gen. Genet.* **161**, 67 (1978).
[19] T. Murashige and K. Skoog, *Physiol. Plant.* **15**, 473 (1962).

TABLE I
COMPOSITION OF PLANT TISSUE CULTURE MEDIA[a]

Component	A medium[b] (mg/liter)	AA medium[c] (mg/liter)	MS medium[d] (mg/liter)	K3M medium (mg/liter)
KCl	—	2950	—	—
KNO$_3$	2200	—	1900	2500
KH$_2$PO$_4$	170	170	170	—
NH$_4$NO$_3$	600	—	1650	250
NaH$_2$PO$_4$ · H$_2$O	75	—	—	150
CaCl$_2$ · 2H$_2$O	295	440	440	900
MgSO$_4$ · 7H$_2$O	310	370	370	250
(NH$_4$)$_2$SO$_4$	75	—	—	134
Na$_2$EDTA	74.6	74.6	74.6	74.6
FeCl$_3$ · 6H$_2$O	27.0	27.0	27.0	27.0
H$_3$BO$_3$	3.0	6.2	6.2	1.5
KI	0.75	0.83	0.83	0.375
MnSO$_4$ · H$_2$O	10.0	16.9	16.9	5.0
ZnSO$_4$ · 7H$_2$O	2.0	8.6	8.6	1.0
CuSO$_4$ · 5H$_2$O	0.025	0.025	0.025	0.125
Na$_2$MoO$_4$ · 2H$_2$O	0.25	0.25	0.25	0.0125
CoCl$_2$ · 6H$_2$O	0.025	—	—	—
CoSO$_4$ · 7H$_2$O	0.025	0.025	0.03	—
m-Inositol	100	20	100	100
L-Glycine	—	7.5	2.0	—
Pyridoxine-HCl	1.0	0.1	0.5	1.0
Thiamin-HCl	10.0	0.4	0.1	10.0
Nicotinic acid	1.0	0.5	0.5	1.0
Glucose	72,000	—	—	83,000
Xylose	250	—	—	38
Sucrose	—	20,000	20,000	—
GA3	—	0.1	—	—
2,4-D	1.0	1.0	0.1	0.1
NAA	0.1	—	—	1.0
BAP	0.5	—	—	0.2
Kinetin	—	0.2	—	—
L-Glutamic acid	—	876	—	—
L-Aspartic acid	—	266	—	—
L-Arginine-HCl	—	174.2	—	—
Sodium pyruvate	5	—	—	—
Citric acid	10	—	—	—
Malic acid	10	—	—	—
Fumaric acid	10	—	—	—
Casein hydrolyzate	500	—	—	—
pH	5.7–5.8	5.8	5.8	5.7

[a] Media A, AA, and K3M are sterilized by filtration. Medium MS is sterilized by autoclaving.
[b] Medium A is from Kao and Michayluk[17] but contains the additional components 2,4-D, BAP, and NAA and glucose in place of sucrose.
[c] Medium AA is from Müller and Grafe.[18]
[d] Medium MS is from Murashige and Skoog[19] but additionally contains sucrose, inositol, and 2,4-D.

Isolation of Protoplasts from Suspension Cultures

Suspension cultures of *O. violaceus* (a member of the family Cruciferae, obtained from C. Matsui, Nagoya University, Nagoya, Japan) and *N. tabacum nia*-115 (nitrate reductase-deficient line, obtained from I. Potrykus, ETH, Zürich, Switzerland) are grown at 26° on a rotatory shaker (150 rpm) in MS and supplemented AA medium (see Materials), respectively, in 250-ml Erlenmeyer flasks (40 ml/flask). The cultures are diluted every 4 (*O. violaceus*) or 7 (*N. tabacum*) days with 3 volumes of the appropriate fresh medium. It is important that cells used for protoplast preparation are in the exponential phase of growth. Four days after subculturing *O. violaceus*, the cells from a 40-ml suspension culture are sedimented and resuspended in 25 ml of a solution containing 5 mM MES, 0.4 M mannitol, 0.1% Pectolyase Y23, and 1% cellulase R10 (pH 5.7).[20] Aliquots of 12–15 ml are transferred to 10-cm petri dishes and left on a rocking platform in the dark at 26° for 1 hr. The protoplasts are filtered through 100- and 50-μm mesh stainless steel sieves, sedimented at 100 g for 5 min, washed twice with EP solution, and finally resuspended in EP solution at 3 × 10⁶ protoplasts/ml. The yield is 4–5 × 10⁷ protoplasts per 40 ml of suspension culture.

Nicotiana tabacum protoplasts are isolated from 40 ml of a 4-day-old culture by incubation of the cells in 50 ml of a solution containing 5 mM MES, 0.6 M sucrose, 2% cellulase R10, 1% Macerozyme R10, and 0.13% Driselase, distributed into petri dishes as described above. After 3 hr at 26° the protoplasts are filtered through a 100-μm sieve. When protoplasts are to be transfected by electroporation, the filtrate is distributed into 12-ml polystyrene round-bottomed Falcon tubes (10 ml/tube), and the protoplast suspension in each tube is overlayed with 1 ml of EP solution. After centrifugation at 100 g for 10 min the floating protoplasts are collected, washed 2 times with EP, resuspended in EP (3 × 10⁶ protoplasts/ml), and used immediately for electroporation. When protoplasts are to be used for the PEG-mediated transformation, the protocol of Negrutiu *et al.*[21,22] is followed. Ten-milliliter filtrate aliquots are overlayed with 1 ml of W5 solution. After centrifugation at 100 g for 10 min, the floating protoplasts are collected, washed twice with W5 solution, and resuspended in W5 at 2 × 10⁶ protoplasts/ml. The protoplast suspension is stored in this form for 0.5–4 hr at 5–8°. Immediately before transformation, protoplasts are sedimented by centrifugation and resuspended at the same concentration in solution MC. The yield of *N. tabacum* protoplasts is 20–25 × 10⁶ per 40 ml of suspension culture.

[20] C. Matsui, personal communication (1987).
[21] I. Negrutiu, R. Shillito, I. Potrykus, G. Biasini, and F. Sala, *Plant Mol. Biol.* **8,** 363 (1987).
[22] I. Negrutiu, personal communication (1987).

Isolation of Leaf Protoplasts

Nicotiana plumbaginifolia. Sterile shoot cultures grown in 0.7% agar containing RMP2 medium are maintained as a source of sterile leaves. Leaves are best taken 6–8 weeks after subculturing. About 6 leaves are cut into strips of approximately 1 mm in a 10-cm petri dish and bathed overnight in the dark at 25° in 12 ml of freshly prepared enzyme solution. This solution, containing 0.5% Driselase, 0.25% polyvinylpyrrolidone, 0.5 M sucrose, 3.85 mM CaCl$_2$, 6 mg/ml NAA, and 2 mg/ml BAP, is stirred for 30 min and then sterile-filtered before use. After the overnight incubation the dish is gently shaken by hand and the cell suspension poured through a 100-μm sieve. The subsequent steps are as already described for preparation of *N. tabacum* protoplasts for PEG-mediated transformation.

Maize. Maize seeds are sterilized by stirring for 30 min in 0.45% Ca(OCl)$_2$, 0.1% Tween 80, followed by three 15-min washes in sterile water. A 2-cm layer of porous material such as Perlite granules is sterilized in a 1-liter beaker by autoclaving, and about 80 ml of 1/2MS salts containing 0.1% MES (pH 5.7) is added. The surface of the Perlite layer should be moist but not wet. About 12 seeds, only half of which may germinate, are placed on the Perlite. A crystallizing dish of diameter similar to that of the beaker is sealed upside down on top of the beaker with Parafilm. After 12 or 13 days the second and third leaves are cut into diagonal 1-mm strips. The strips are placed in 10-cm Petri dishes, 4 leaves per dish, and bathed overnight at 22° in freshly prepared enzyme solution. This solution, containing 1% cellulase RS, 0.5% Macerozyme R10, 0.05% Pectolyase Y23, 5 mM CaCl$_2$, 0.6 M mannitol, and 0.05% MES (pH 5.6) is stirred for 30 min and then sterile-filtered before use. After the overnight incubation the dishes are gently shaken by hand and left a further 30 min before the protoplasts are passed through a 100-μm sieve. Ten-milliliter aliquots of the filtrate are centrifuged for 10 min at 100 g, and the pelleted protoplasts are washed twice with W5 solution.

Transfection of Protoplasts

Two alternative procedures can be used for transfecting protoplasts to study transient gene expression. Electroporation involves the use of an electric field to stimulate the uptake of exogenous DNA by the protoplasts.[6,9,11,15] In the chemical method DNA uptake is mediated by high concentrations of polyethylene glycol.[7,10,21,22]

Electroporation Procedure. Electroporation is carried out essentially as described by Fromm *et al.*[15] Prechilled protoplast suspension in EP solution (0.7 ml, 2 × 10^6 protoplasts) is mixed with 0.1 ml of DNA solution in EP. Supercoiled plasmid DNA (50–100 μg) or 10–20 μg of plasmid

DNA supplemented by 100 μg of carrier DNA (see Materials) is used per electroporation. The addition of carrier increases the level of expression from lower concentrations of plasmid, but the highest levels of expression are achieved with a large amount of the plasmid DNA.

After preincubation for 10 min at 0° the suspension is transferred into the prechilled electroporation chamber, which is made from a 1-ml disposable plastic cuvette (Grainer). This has a piece of aluminum foil (0.025 mm thick, 0.9 cm wide, and 6 cm long) attached to each of the two wide inner walls of the cuvette. The distance separating the two electrodes is 0.4 cm. The chambers are sterilized in 70% ethanol and can be reused many times. Electroporation is carried out by discharging an 820-μF capacitor charged at 200 V (*O. violaceus* protoplasts) or 300 V (*N. tabacum* protoplasts), using a homemade electroporation apparatus, similar to that described by Fromm *et al.*,[15] and an LKB Model 2103 power supply for charging of the capacitor. The electric parameters important for protoplast electroporation are discussed in detail by Fromm *et al.*[15] Following the electroporation, samples are left on ice for an additional 10 min and then diluted with 10 ml of supplemented medium A (*O. violaceus* protoplasts) or supplemented medium AA (*N. tabacum* protoplasts). Protoplasts are incubated at 26° in the dark for 24–36 hr before RNA isolation.

PEG-Mediated Transformation of N. tabacum, N. plumbaginifolia, and Maize Protoplasts. Chemical transformation is carried out essentially according to the protocol of Negrutiu *et al.*[21,22] Supercoiled plasmid DNA (5–20 μg, in 10 μl of TE buffer) is mixed at room temperature with 6 × 10⁵ protoplasts in 0.3 ml of MC solution. The use of larger amounts of plasmid does not result in a substantial increase in gene expression. Three-tenths milliliter of 40% PEG 6000 solution is then added, and the samples are gently mixed and incubated at room temperature for 1–5 min. Samples are finally diluted with 4 ml of supplemented medium AA (*N. tabacum*), or medium K3M containing 0.001% Tween 80 plus 0.45 *M* glucose (*N. plumbaginifolia*) or 0.6 *M* glucose (*Zea mays*) and transferred to 15-ml plastic Falcon tubes, which are incubated in a horizontal position in the dark at 26° for 24 hr.

Expression Vectors for Protoplast Transformation

For strong constitutive expression in a broad range of plant species, the 35 S promoter from cauliflower mosaic virus is the best promoter described so far. Tandem duplication of upstream sequences may further increase the efficiency of the promoter.[12,23] In the expression vector

[23] R. Kay, A. Chan, M. Daly, and J. McPherson, *Science* **236**, 1299 (1987).

pDH51,[24] a polylinker with multiple cloning sites separates the 35 S promoter and the terminator region of cauliflower mosaic virus. We use either pDH51 or a modified form of this vector in which part of the promoter has been duplicated.[14] With the latter vector we find that sequences upstream of the RNA start site are also expressed to some extent[25] (this has not been tested for pDH51). We have not determined whether this is due to incomplete termination and transcription all the way around the plasmid or to initiation either within the promoter or elsewhere within the vector. The plasmids used for protoplast transfection are purified by CsCl density-gradient centrifugation following standard protocols.

Isolation of RNA

We isolate RNA from protoplasts either by centrifugation of a guanidium thiocyanate extract through CsCl[26] or by extraction of a guanidinium thiocyanate extract with phenol–chloroform. Centrifugation through CsCl gives clean, easily dissolvable RNA. The phenol–chloroform extraction method is more convenient for handling many small samples, but the resulting RNA may contain contaminants (probably polysaccharides) that can make it more difficult to dissolve but that do not interfere with the RNase protection assay. Both methods give a yield of 5–25 μg of RNA from 6×10^5 protoplasts. A slightly different phenol–chloroform extraction method for isolating RNA from plant tissues with high polysaccharide content has been described by Logemann *et al.*[27]

CsCl Method. The protoplasts are pelleted by centrifugation and lysed by the addition of 2.5 ml of 4 *M* guanidinium thiocyanate in 0.1 *M* Tris-HCl (pH 7.5) and 0.1 *M* 2-mercaptoethanol with vigorous vortexing. The resulting solution is overlayed on a 1.3-ml cushion of 5.7 *M* CsCl in 25 m*M* sodium citrate (pH 5.0) in an SW60 (Beckman) centrifuge tube and centrifuged at 36,000 rpm for 20 hr at 20°. The supernatant is slowly aspirated off, leaving 100–200 μl covering the pellet. The tube is then cut about 1 cm from the bottom with a sterile blade and inverted to drain off the remaining supernatant. The RNA pellet is dissolved in 300 μl of TE buffer and precipitated with ethanol.

Guanidinium Thiocyanate–Phenol–Chloroform Extraction Method. This method is adapted from that of Chomczynski and Sacchi.[28] Pelleted

[24] M. Pietrzak, R. Shillito, T. Hohn, and I. Potrykus, *Nucleic Acids Res.* **14**, 5857 (1986).
[25] G. Goodall, unpublished observation (1988).
[26] R. J. MacDonald, G. H. Swift, A. E. Przybyla, and J. M. Chirgwin, this series, Vol. 152, p. 219.
[27] J. Logemann, J. Schell, and L. Willmitzer, *Anal. Biochem.* **163**, 16 (1987).
[28] P. Chomczynski and N. Sacchi, *Anal. Biochem.* **162**, 156 (1987).

protoplasts from a single transfection are lysed in 0.4 ml of solution D [4 M guanidinium thiocyanate, 25 mM sodium citrate (pH 7.0), 0.5% sarcosyl, 0.1 M 2-mercaptoethanol] in an Eppendorf tube with vigorous vortexing. To this is added 40 μl of 2 M sodium acetate (pH 4.0), 400 μl of water-saturated phenol, and 80 μl of chloroform–isoamyl alcohol (49 : 1, v/v). After mixing by vortexing, the sample is kept on ice for 15 min, then centrifuged at 4° for 20 min. The supernatant aqueous phase, avoiding material near the interphase, is transferred to an Eppendorf tube and mixed with 0.4 ml of 2-propanol. After 15 min at -20° the RNA is pelleted by centrifugation for 15 min. The pellet is dissolved in 0.3 ml of solution D, then mixed with 0.3 ml of 2-propanol and left at -20° for 15 min. After centrifugation for 15 min, the pellet is washed with cold 75% ethanol, briefly air dried, and dissolved in 6 μl of high-quality formamide, which improves solubility, and 50 μl of DEPC-treated water. An aliquot of 4 μl can be removed to check the RNA concentration by measuring the A_{260}.

DNase Treatment. If one is to measure accurately the proportion of expressed RNA which remains unspliced, it is important that none of the transfected plasmid contaminates the isolated RNA, since this would protect the same size probe fragment as would unspliced transcript. To eliminate this possibility, we include a DNase treatment of the RNA. An alternative is RNase H treatment after hybridization of the RNA with the complementary probe.[13] Labeled probe hybridized to the contaminating DNA should be degraded by RNase H. RNase H treatment has the advantage of not requiring an extra step, and, since it is done after the probe is hybridized with the target RNA, potential contamination of the enzyme with single-strand-specific ribonucleases is not a problem. However, if the amount of contaminating plasmid is of the order of 1 ng or more, then DNase treatment should be performed. Otherwise the plasmid will compete with the mRNA for the 1 fmol of probe. RNase H treatment is described under RNase mapping. The extent of DNase treatment required will, of course, depend on the amount of contaminating plasmid in the RNA preparations. This can be minimized by washing the protoplasts prior to lysis and, if using the phenol–chloroform extraction method, by being careful to avoid material at the interphase during the extraction step.

For DNase treatment, 6 μl of 100 mM Tris-HCl (pH 8.0) with 100 mM $MgCl_2$, 1 μl of 10 units/μl RNasin, and 1 μl of 1 unit/μl RNase-free DNase (RQ1 DNase from Promega Biotec) is added to the 50 μl of RNA solution. After incubation at 37° for 30 min, 100 μl of TE and 160 μl of phenol–chloroform (1 : 1, v/v) is added, the solution is thoroughly mixed, and the aqueous phase is precipitated with ethanol. RNA is dissolved in 50 μl of DEPC-treated water.

Analysis of RNA Processing

For the analysis of splicing patterns in protoplasts we use RNase A/T1 or nuclease S1 protection assays and primer extension with reverse transcriptase. The S1 nuclease and primer extension assays are used to map accurately the location of individual splice sites. These methods are described elsewhere in this series[29,30] and are not described here. RNase mapping,[31] which we describe in detail, is used to quantitatively analyze the splicing pattern. In this assay the RNA isolated from transfected protoplasts is hybridized with complementary uniformly ^{32}P-labeled RNA probes and the mixture digested with RNase A and RNase T1. The spliced RNA fails to protect the regions in the probe which are complementary to introns, thus giving rise to shorter labeled fragments which correspond to exon sequences (see Fig. 1).

Preparation of ^{32}P-Labeled Complementary RNA Probes. Various systems for the *in vitro* synthesis of RNA are described elsewhere in this series.[32] We use SP6 RNA polymerase under conditions similar to those described by Melton *et al.*,[31] except that we gel-purify the RNA product, rather than use DNase treatment, to remove the template. This ensures that the probe is of uniform length and free of traces of complementary transcripts which may arise by unspecific end to end transcription of plasmid DNA.

The transcribed RNA is ethanol precipitated, redissolved in 5 μl of formamide loading solution, and electrophoresed in a small (16 × 16 cm) 0.4-mm-thick 6% polyacrylamide denaturing gel. After electrophoresis the upper glass plate is removed, and the gel, still attached to the lower plate, is wrapped in plastic wrap. Marker strips with radioactive or fluorescent spots are taped to the plastic wrap, and the gel is exposed to X-ray film for 30 sec. The labeled RNA band is located and cut from the gel. The gel fragment is placed in an Eppendorf tube and thoroughly mashed using a device made by briefly igniting the end of a 200-μl plastic pipet tip so that a molten blob forms. Two hundred fifty microliters of elution buffer (0.5 *M* ammonium acetate, 0.1 m*M* EDTA, 0.1% SDS) containing 10 μg of tRNA is then added, and the tube is placed in a shaker at room temperature for 1 hr. Polyacrylamide pieces are pelleted or removed by centrifugation of the slurry through a yellow Eppendorf tip clogged with glasswool and placed in an Eppendorf tube. Two hundred microliters of the

[29] A. J. Berk, this series, Vol. 180, p. 334.
[30] W. R. Boorstein and E. A. Craig, this series, Vol. 180, p. 347.
[31] D. A. Melton, P. A. Krieg, M. R. Rebagliatti, T. Maniatis, K. Zinn, and M. R. Green, *Nucleic Acids Res.* **12**, 7035 (1984).
[32] J. K. Yisraeli and D. A. Melton, this series, Vol. 180, p. 42.

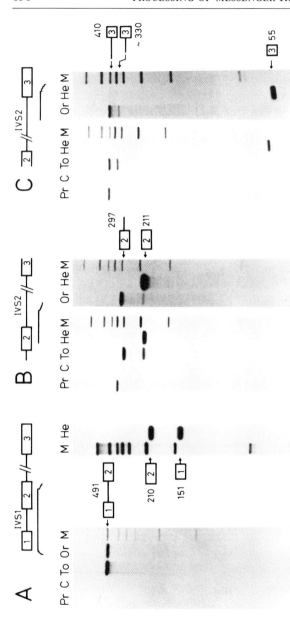

FIG. 1. RNase A/T1 mapping of the human β-globin pre-mRNA expressed in electroporated *N. tabacum* (lanes To) and *O. violaceus* (lanes Or) protoplasts and in HeLa cells (lanes He). Total cellular RNA (10–20 μg) was hybridized with labeled RNA probes complementary to different regions of globin pre-mRNA (depicted by lines below the diagrams of the gene). After digestion with RNases, the protected probe fragments were separated on a 6% polyacrylamide gel. Analyses were also carried out on RNA from nontransfected tobacco protoplasts (lanes C). Lanes Pr, Undigested probe; lanes M, DNA size markers (3'-end-labeled *Hin*fI digest of pBR322). All probes contain additional vector sequences, which are indicated by wavy lines. The structure and lengths of RNase-resistant products are diagrammed beside each autoradiogram. The results indicate quantitative processing of both introns in HeLa cells, no splicing of intron 1 (IVS1) in protoplasts, and aberrant, inefficient splicing of IVS2 (involving an authentic 5' and a cryptic 3' splice site) in protoplasts. Only one-tenth of each reaction with HeLa RNA was loaded on the gel. (Adapted from Wiebauer *et al.*,[3] with permission.)

supernatant is collected, and the RNA is precipitated by the addition of 85 μl of 5 M ammonium acetate and 700 μl of ethanol. After centrifugation the pellet is washed with 70% ethanol, dried, and dissolved in 50 μl of DEPC-treated water.

RNase A/T1 Mapping. Some adjustment of the conditions described for RNase mapping may be required to optimize signal strength while keeping background to a minimum. For example, the hybridization temperature and RNase concentrations will depend to some extent on the length and sequence of the probe being used.

The RNA (10–20 μg) and probe [1 fmol, ~20,000 Cerenkov counts per minute (cpm) for a 200-nucleotide probe] are mixed and dried in an Eppendorf tube, redissolved in 10 μl of hybridization solution [40 mM PIPES (pH 6.7), 400 mM NaCl, 1 mM EDTA, 80% formamide] and incubated in an oven at 45° for at least 16 hr. If the RNA had been treated with DNase before hybridization, then RNase A/T1 digestion is carried out directly by the addition of 100 μl of RNase buffer [10 mM Tris-HCl (pH 7.5), 0.1 M LiCl, 0.2 M NaCl, 5 mM EDTA] containing 8 μg/ml RNase A and 0.4 μg/ml RNase T1. If the RNA was not DNase treated, then the hybridized RNA is first incubated with RNase H. Fifty microliters of RNase buffer (see above) containing in addition 10 mM MgCl$_2$ and 10 units/ml of RNase H is added, and the sample is incubated at 37° for 30 min, followed by the addition of 50 μl of RNase buffer containing 16 μg/ml RNase A and 0.8 μg/ml RNase T1.

The digestion with RNase A and T1 is carried out at 26° for 40 min, after which 2 μl of 10% SDS and 3 μl of 10 mg/ml proteinase K is added. The mixture is incubated at 37° for 15 min, and then 110 μl is transferred to a tube containing 10 μg of tRNA and 110 μl of phenol–chloroform (1 : 1, v/v). The transfer to a fresh tube is done to separate the digested probe fragments from any probe that may have avoided complete digestion through having been splashed up on the wall or lid of the tube. After thorough mixing, followed by centrifugation for 5 min, the aqueous phase is transferred to a tube containing 250 μl of ethanol. The ethanol precipitate is washed with 70% ethanol, dried, dissolved in 4 μl of formamide loading solution, and electrophoresed in a 7 M urea–polyacrylamide gel, along with control and marker samples. After electrophoresis the gel is dried and autoradiographed.

Results of RNase A/T1 protection analysis of the human β-globin pre-mRNA expressed in *O. violaceus* and *N. tabacum* protoplasts and in HeLa cells are shown, as an example, in Fig. 1.

Quantitation. To make comparisons of the splicing efficiencies of various mutant forms of a gene, one can calculate the percentage of pre-mRNA that has been spliced by scanning the autoradiographs with a scanning densitometer. However, care must be taken to calibrate the film response to different amounts of radioactivity. We use a calibration strip that has a series of bands of known relative amounts of radioactivity and preflash[33] the film to improve the linearity of response.

The calibration strip is made by performing a serial dilution of ^{32}P-labeled 40-mer oligonucleotide, beginning with 2500 cpm in 8 μl of diluted formamide loading solution (4% formamide loading solution, 48% formamide, 48% TBE). An aliquot of 5 μl is removed and mixed with 3 μl of diluted formamide loading solution in a second tube. Five microliters is removed from this and mixed with another 3 μl of diluted formamide loading solution in a third tube and so on, so that each tube has 1.6 times the relative radioactivity of the next. These are run in a gel of the same type as used to separate the fragments from the RNase protection assay, using a single lane and loading at 6-min intervals, beginning with the most diluted sample. This gives a series of evenly spaced bands, each with 1.6 times the radioactivity of the next. The lane is cut from the dried gel and taped next to each gel that is to be quantitated. By scanning the calibration strip one can plot a correction curve of peak area versus relative amount of radioactivity.

A more convenient method of quantitation is to scan the dried gel directly with a β-scanner such as the AMBIS Radioanalytic Imaging System.

Comments

Numerous reports demonstrating transient expression of exogenous genes in transfected plant protoplasts indicate that this approach can be applied to protoplasts from any plant tissue or suspension culture provided that good quality, metabolically active protoplasts are prepared and transfection parameters are carefully optimized. The quality of protoplasts and efficiency of the transformation protocol can be monitored by measuring chloramphenicol acetyltransferase (CAT) activity after transformation of protoplasts with the plasmid pDW2.[24] CAT activity in protoplast extracts can be conveniently measured, and nontransformed protoplasts usually have little or no endogenous background.[15]

Quantitative differences in the level of exogenous gene expression in different protoplasts exist. For example, we obtain higher levels of RNA expression in *N. tabacum* protoplasts than in *O. violaceus* protoplasts. With *N. tabacum* protoplasts, PEG-mediated transformation usually gives significantly better results than electroporation. PEG-mediated

---

[33] R. A. Laskey, this series, Vol. 65, p. 363.

transformations of mesophyll protoplasts from *N. plumbaginifolia* and maize give even better expression as indicated by measurements of CAT activity[22] or of RNA levels.[25] Even under optimal conditions, however, we find that the levels of expression from the cauliflower mosaic virus 35 S promoter achieved in protoplasts are about an order of magnitude lower than expression from the SV40 promoter in transfected HeLa cells. We prefer the PEG protocol for other reasons as well. It is simple and does not require extra equipment, and lower amounts of plasmid DNA are needed for maximal levels of expression. This last factor is of importance when protoplast RNA free of contaminating plasmid DNA has to be prepared for RNase mapping experiments.

Acknowledgments

We thank Jirina Petruska and Juan-Jose Herrero for help in establishing transient expression systems, and Drs. C. Matsui and I. Negrutiu for making available the *O. violaceus* line and PEG protocols, respectively.

[12] Polyadenylate Polymerases

By MARY EDMONDS

Introduction

Polyadenylate polymerases were known long before poly(A) sequences were identified on the 3' ends of both nuclear and messenger RNAs. This discovery stimulated many efforts to purify and characterize these enzymes in the hope of linking them to cellular polyadenylation reactions. Activities purified from many sources carry out the reaction shown below, but none have as yet been directly linked to polyadenylation *in vivo* or in nuclear extracts.

$$\text{Polyribonucleotide} + n\text{ATP} \xrightarrow{\text{Mg}^{2+} \text{ or Mn}^{2+}} \text{polyribonucleotide (A)}_n + n\text{PP}_i$$

The properties of polyadenylate polymerases have been reviewed extensively in the past.[1–3] All have shown a nearly complete specificity for ATP and a requirement for a polyribonucleotide with a 3'-hydroxyl group to which AMP residues are attached. None have shown much specificity

[1] M. Edmonds and M. A. Winters, *Prog. Nucleic Acid Res. Mol. Biol.* **17**, 149 (1976).
[2] M. Edmonds, *in* "The Enzymes" (P. Boyer, ed.), Vol. 15, p. 217. Academic Press, New York, 1982.
[3] S. T. Jacob, and K. Rose, *in* "Enzymes of Nucleic Acid Synthesis and Modification" (S. T. Jacob, ed.), Vol. 2, p. 135. CRC Press, Boca Raton, Florida, 1983.

for either the length or the sequence of the primer. Oligonucleotides of a few AMP residues were usually as effective as poly(A) sequences of 50 or more, while tRNAs and other small RNAs were also effective primers. Some insight into this lack of primer specificity was gained by the more recent discovery of a polyadenylation recognition sequence, AAUAAA, present within 10–30 nucleotides of the poly(A) addition site of most mRNAs. This sequence has been shown to be essential for the correct 3'-end-processing and polyadenylation of pre-mRNAs *in vivo*[4–6] and *in vitro*.[7,8] The recent detection of a specific protein that binds to short regions of 3' untranslated sequences which contain this AAUAAA and that is chromatographically separable from a poly(A) polymerase activity in the extract suggests that sequence specificity for polyadenylation does not reside in the polymerase.[9]

Several investigators have identified a complex of several proteins that include a U-type small nuclear ribonucleoproteins (snRNPs) that are needed for the correct polyadenylation of pre-mRNAs *in vitro*.[10]

Recently a purified poly(A) polymerase from calf thymus has been shown to replace a poly(A) polymerase activity in HeLa nuclear extracts that carry out site-specific polyadenylation when combined with a specificity factor from these same extracts.[10a] In addition to providing an activity for such reconstitution experiments, at least one purified poly(A) polymerase has been used to generate anti-poly(A) polymerase antibodies.[10b] The purified enzymes will continue to be used for the addition of poly(A) tails to RNA molecules and for the labeling of the 3' ends of RNAs.

Purification of Poly(A) Polymerases

Source. Poly(A) polymerases have been purified from many animal and plant tissues, from cultured animal cells, bacteria, and several viruses.[1,2] Detailed descriptions of the most purified and best characterized preparations from calf thymus,[11,12] rat liver nuclei,[13] HeLa cells,[14] and

---

[4] M. Fitzgerald and T. Shenk, *Cell* **24**, 251 (1981).

[5] D. R. Higgs, S. E. Goodbourn, J. Lamb, J. B. Clegg, D. J. Weatheral, and N. J. Proudfoot, *Nature (London)* **306**, 398 (1983).

[6] M. Wickens and P. Stephenson, *Science* **226**, 2801 (1984).

[7] D. Zarkower, P. Stephenson, M. Sheets, and M. Wickens, *Mol. Cell. Biol.* **6**, 2317 (1986).

[8] J. L. Manley, H. Yu, and L. Ryner, *Mol. Cell. Biol.* **5**, 373 (1985).

[9] J. Wilusz and T. Shenk, *Cell* **52**, 221 (1988).

[10] C. L. Moore, this volume [6].

[10a] V. J. Bardwell, D. Zarkower, M. Edmonds, and M. Wickens, *Mol. Cell. Biol.* **10** (in press).

[10b] M. P. Terns and S. T. Jacob, *Mol. Cell. Biol.* **9**, 1435 (1989).

[11] M. A. Winters and M. Edmonds, *J. Biol. Chem.* **248**, 4756 (1973).

[12] C. Tsiapalis, J. Dorson, and F. Bollum, *J. Biol. Chem.* **250**, 4486 (1975).

[13] K. Rose and S. T. Jacob, *Eur. J. Biochem.* **67**, 11 (1976).

Escherichia coli[15] have been published. The *E. coli* enzyme is currently available from Bethesda Research Laboratories [(BRL) Gaithersburg, MD].

Purification Strategies. Most purifications have relied on essentially similar protocols. In contrast to RNA polymerases, poly(A) polymerases are readily extracted from cells or isolated nuclei by homogenization in neutral buffers of low ionic strength. A typical purification begins with ammonium sulfate fractionation of the homogenate or extract to concentrate the proteins for application to a DEAE-cellulose or DEAE-Sephadex column. Here, the weak affinity of the polymerase for anionic sites is exploited by binding in low ionic strength to prevent it but not most proteins, including RNA polymerases, from binding. This approach was recently used to separate a poly(A) polymerase activity from a 64K protein that recognizes the AAUAAA polyadenylation signal which apparently controls the specificity of polyadenylation.[9]

Primer dependence develops at this stage since polynucleotides bind tightly to DEAE groups. The unbound poly(A) polymerase after concentration is then applied to a negatively charged ion-exchange column (most often phosphocellulose, but CM-Sepharose columns have also been used).

Purifications of several hundred to more than a thousandfold are characteristically achieved after the two chromatography steps, particularly if either one or both are repeated. At this stage the enzyme is 10–20% pure if the reported specific activities for several preparations are assumed to be close to the maximum obtainable. For several preparations these fall within a rather narrow range of 5–10 μmol of AMP polymerized/mg of protein/hr. The *E. coli* poly(A) polymerase, however, was reported to have a specific activity more than 5-fold greater.[15]

Removal of remaining proteins has been done with standard techniques such as gel filtration, hydroxyapatite binding, and Blue or heparin-Sepharose chromatography. DNA linked to Sepharose has also been used effectively, especially where amounts of protein are limited.[16]

Affinity chromatography with either ATP or poly(A) bound to Sepharose resins has been used in a few cases, although protein binding was most likely the result of ionic interactions between the bound ligand and the polymerases since the enzyme was released with high concentrations of salt. In general, these materials have not enhanced the purity of poly(A) polymerases beyond that achieved with conventional protocols. The similarity in specific activity previously noted for most of the highly

[14] J. R. Nevins and W. K. Joklik, *J. Biol. Chem.* **252**, 1477 (1977).
[15] A. Sippel, *Eur. J. Biochem.* **37**, 31 (1973).
[16] D. A. Stetler and S. T. Jacob, *Biochemistry* **24**, 5163 (1985).

purified mammalian enzymes suggests that several preparations may be nearly pure.

The advent of fast performance liquid chromatography (FPLC) along with the array of high-capacity pressure-resistant ion-exchange and gel-filtration matrices such as Mono Q and Mono S (Pharmacia-LKB) provides alternative materials for purification. These should reduce the time required and enhance the resolution that has been obtained with the low-pressure conventional chromatography used in the purifications discussed above. A poly(A) polymerase has recently been purified from HeLa nuclear extracts on such columns.[17]

Contaminating Enzymes. Activities that may interfere with the assay of the enzyme such as RNA polymerases, transfer RNA nucleotidyltransferases, polynucleotide phosphorylases in bacterial extracts, and ATPases are apparently removed at early steps of the purification. Nucleases, however, are likely to remain a problem, which has not been adequately assessed for most preparations. Relatively stringent criteria were applied to those preparations that were reported to have undetectable ribonuclease activity.[11,12]

Properties of the Poly(A) Polymerase Protein

Aside from estimates of size, little is known of the structure of any poly(A) polymerase. A tabulation of molecular weights of both denatured and nondenatured enzymes purified from mammalian tissues, cells, and viruses has been published.[2] These data should be viewed cautiously since these enzymes were not protected from proteolysis during purification, nor were molecular weights obtained for a variety of conditions. Determinations of molecular weights of denatured proteins included the presumption that the enzyme was the major band seen on denaturing gels, since activities were not measured. Among these enzymes only the polymerase from cytoplasmic extracts of calf thymus gave close agreement for the sizes determined by nondenaturing gel filtration and denaturing gel electrophoresis (60 kDa).[12] More recently a poly(A) polymerase from *Artemia salinas* was estimated to be 70 kDa by both techniques.[18] There were several cases in the earlier compilation for which much higher molecular weights were reported for analysis by gel filtration or sedimentation velocity. The rat liver nuclear enzyme was also reported later to be 156 kDa by gel filtration but 38 kDa in denaturing gels.[19] The higher values

[17] G. Christofori and W. Keller, *Cell* **54**, 875 (1988).
[18] E. Roggen and H. Sleegers, *Eur. J. Biochem.* **147**, 225 (1985).
[19] D. A. Stetler and S. T. Jacob, *J. Biol. Chem.* **259**, 7239 (1984).

observed for gel filtration were assumed to represent either some multi-meric form of the enzyme or simply protein aggregation. It now seems possible that they are related to the large protein complexes recently found in nuclear extracts that function in the specific polyadenylation of nuclear transcripts.[10]

Isoelectric points of about pH 6 have been reported for several prepa-rations, although that of the enzyme from the cytoplasm of calf thymus was 7.3.[12] The nuclear poly(A) polymerases of rat liver and rat hepatoma have been reported to contain zinc ions, which are required for catalytic activity.[20] Certain evidence suggested that zinc was involved in the inter-action of the enzyme with the polynucleotide primer.

Requirements for Poly(A) Synthesis

Components needed for optimal poly(A) synthesis have been re-viewed in detail[1-3] and are briefly described here.

Substrates. Purified poly(A) polymerases are specific for ATP. The low levels of other nucleotides incorporated ($\leq 1\%$ of the ATP) are most likely the result of contaminating activities. Neither ADP nor 2'-dATP are substrates, but a single 3'-dATP (cordycepin triphosphate) can be incor-porated at the 3' end of the primer to terminate elongation.[21] 3'-dATP has been used to label 3' ends of RNAs.[22] It has also been used for the quantitative measurement of poly(A)$^+$ mRNAs in complex RNA mix-tures, which depends on its incorporation by the *E. coli* poly(A) poly-merase to terminate elongation.[23]

Primers. All poly(A) polymerases have shown a lack of specificity for the polyribonucleotide primer that has been amply documented.[1,2] Aside from the necessary 3'-hydroxyl group, the structure and size of RNAs that can prime poly(A) synthesis vary widely. However, some marked differences in the efficiency of utilization of different primers by poly(A) polymerases from any one source have also been reported. The signifi-cance of these variations are unclear since contaminating nuclease activi-ties and/or the number and availability of 3'-hydroxyl groups on the primer were seldom reported. The enzyme from *E. coli* polyadenylates a wide variety of natural RNAs as well as synthetic homopolynu-cleotides.[15,23]

Deacylated tRNAs have often been used to assay poly(A) poly-

[20] K. M. Rose, M. Allen, I. Crawford, and S. T. Jacob, *Eur. J. Biochem.* **88**, 29 (1978).
[21] W. Muller, G. Seibert, R. Beyer, H. Breter, A. Maidhof, and R. Zahn, *Cancer Res.* **37**, 3824 (1977).
[22] W. R. Beltz and S. H. Ashton, *Fed. Proc., Fed. Am. Soc. Exp. Biol.* **41**, 1450 (1982).
[23] M. S. Krug and S. L. Berger, *Anal. Biochem.* **153**, 315 (1986).

merases because they are among the most active primers of a well-defined size that are readily available and have common 3' ends. Polyadenylates of 10–50 nucleotides are even more effective primers but in the past were not as readily available in discrete size classes.

Divalent Cations. Different levels of Mg^{2+} or Mn^{2+} have been reported to give maximal poly(A) synthesis with poly(A) polymerases from different sources. Some preparations use only Mn^{2+} and not Mg^{2+}, whereas others are most active with Mn^{2+} but also use Mg^{2+}. Some of the nuclear polymerases are more active with Mg^{2+} but retain activity with Mn^{2+}. The meaning of these differences need not be related to differences in the catalytic properties of polymerase proteins from different sources but could be effects on primer structure or on interactions between primer and polymerase. As noted above, different primers were usually used for preparations from different sources, and a variety of primers have not been compared with Mg^{2+} and Mn^{2+} for any one poly(A) polymerase. Differential effects of Mg^{2+} and Mn^{2+} on nucleases present in crude preparations have also been reported and need to be considered.[1,2]

However, the most important consideration may be the well-known effects of the replacement of Mg^{2+} with Mn^{2+} on the stringency of the template requirements for both DNA and RNA polymerases and on the sequence specificity for restriction endonucleases. Evidence has suggested that this extends to poly(A) polymerases as well. An enzyme in a chromatin fraction of rat liver nuclei that was active with Mg^{2+} used only endogenous RNAs as primers but became inactive with Mg^{2+} when released from chromatin.[24] If Mn^{2+} replaced Mg^{2+}, however, a variety of small RNAs could be used by the solubilized enzyme to prime poly(A) synthesis. This finding may relate to the recent demonstration that the sequence specificity required for polyadenylation of a specific pre-mRNA in a nuclear extract was diminished when Mn^{2+} replaced Mg^{2+} as the divalent cation.[25] In this case, polyadenylation was not observed with Mg^{2+} when the poly(A) recognition sequence, AAUAAA, was mutated to AAUCAA. However, when Mn^{2+} replaced Mg^{2+} in the extract, both the wild-type and the mutant pre-mRNAs were polyadenylated to the same extent.

From the considerations noted above, it is to be expected that Mg^{2+} and Mn^{2+} are likely to have complex effects on poly(A) polymerases that will need to be identified to develop the appropriate assays for a particular set of conditions.

[24] K. M. Rose, F. J. Roe, and S. T. Jacob, *Biochim. Biophys. Acta* **478**, 180 (1977).
[25] Y. Takagaki, L. C. Ryner, and J. L. Manley, *Cell* **52**, 731 (1988).

Monovalent Cations. Purified poly(A) polymerases are activated by salt concentrations of 50 mM but can be inhibited by levels above 100 mM (NaCl or KCl).

Hydrogen Ions. Poly(A) polymerases are active in the range of pH 7–9 with the optimum somewhat above pH 8.

Inhibitors. In contrast to the eukaryotic RNA polymerases, a specific inhibitor of poly(A) polymerase such as α-amanitin has not been reported. Cordycepin apparently inhibits as the triphosphate (3'-dATP) by chain termination. Substrate levels of CTP, GTP, or UTP inhibit significantly, and the effects are additive when combined. Substrate levels of pyrophosphate and inorganic phosphate are also inhibitory. Enzyme reactivity in phosphate-buffered reactions of greater than 10 mM salt at pH 8 was greatly reduced relative to that in tris(hydroxymethyl)aminomethane hydrochloride (Tris-HCl) at 50 mM at this pH.[11]

Kinetics of Reaction. Many poly(A) polymerase preparations have shown a pronounced lag at 37° that precedes the linear phase of ATP uptake by 5–10 min.[1,2] The lag was not abolished for the calf thymus nuclear enzyme by increasing the enzyme concentration nor by preincubating enzyme with the primer. The linear phase of the reaction can be sustained for 30–60 min or longer with most preparations. Concentrations of the purified calf thymus nuclear poly(A) polymerase below 0.1 mg/ml resulted in a reduced reaction rate that was not overcome by including bovine serum albumin (BSA) in the reaction.[11]

Assay of Poly(A) Polymerases

The variations in the requirements for divalent cations and primers, as well as in the optimum levels of the reactants that have been reported for polymerases from different sources, make it unlikely that a single set of conditions will be satisfactory for the assay of any one preparation. Each component may need to be varied to optimize activity. This is likely to be necessary for Mg^{2+} and Mn^{2+} levels, as well as levels of the specific primer selected for the assay. The molar concentration of divalent cations should be adjusted to equal or surpass the ATP levels. Table I serves as a guide for selection of conditions that should detect poly(A) polymerase activity.

Analysis of Reaction Product. Assays based on radioactivity incorporated from labeled ATP into an acid-insoluble product give rapid and accurate measurements of enzyme activity. Milder procedures that also remove labeled ATP effectively but preserve the polyadenylated product

TABLE I
GENERALIZED REACTION CONDITIONS FOR
POLY(A) POLYMERASES

Parameter	Value
Reaction volume	5–100 μl
Temperature	37°
Time	30–60 min
Buffer	50 mM Tris-chloride (pH 8)
[α-^{32}P]ATP	0.4–4 mM
MgCl$_2$ (or MnCl$_2$)	1–10 mM (depends on ATP levels)
RNA primer	1–50 μM [tRNAs, poly(A)$_{(10-50)}$]
Dithiothreitol	0.4 mM
Enzyme[a]	0.1–1 mg/ml

[a] Specific activity of 1–10 μmol ATP/mg/hr.

for further analysis can provide information on the number and size of the reaction products.

Assay for Production of Acid-Insoluble Radioactivity. Reactions are terminated by the sequential addition with mixing of 1 ml of cold 100 mM NaCl followed by 0.1 ml of 0.5% sodium dodecyl sulfate (SDS) containing 1 mg of BSA and then 1 ml of cold 10% trichloroacetic acid (TCA). After 15 min on ice, the entire reaction is poured onto a Whatman GF/B glass fiber filter maintained under a vacuum produced by a water aspirator. After removal of liquid, the filter is washed 5 or 6 times with 2-ml portions of cold 5% TCA. The damp filter is transferred to a scintillation vial. ^{32}P-Labeled products are counted directly for Cerenkov radiation. Five or ten milliliters of an all-purpose scintillation solvent is added to vials that contain filters with ^{3}H-, ^{14}C-, or ^{35}S-labeled products. After shaking gently for 15–30 min, radioactivity is measured.

Assay of ^{32}P-Labeled Products Hybridized to Oligo(dT)-Cellulose or Poly(U)-Sephadex. To each reaction is added 5–20 mg of oligo(dT)-cellulose (Type 7, Pharmacia-LKB) suspended in 0.5–1.0 ml of 250 mM NaCl, 10 mM Tris (pH 7.5), 2 mM ethylenediaminetetraacetic acid (EDTA), and 0.2% SDS (2.5× NETS). After gentle shaking at room temperature for 15 min, the suspension is transferred to a 2-ml Spin-X filter containing a removable 0.22-μm cellulose acetate membrane filter (Costar, Cambridge, MA). After centrifuging for about 1 min at low speed to transfer liquid from the filter insert to the centrifuge tube, the filter is washed 3 or 4 times with 0.5 ml of 2.5× NETS. When washing is complete, the filter insert can be transferred to a vial appropriate for detection of Cerenkov radiation. It is important that the amount of ^{32}P measured in this assay is comparable to that recovered by the acid precipitation method. A decreased amount may mean that some poly(A) chains were too short to form a hybrid with oligo(dT) which was stable under these conditions. Binding in higher salt concentrations or use of celluloses with longer oligo(dT) chains may remedy this. However, the currently available commercial celluloses have maximum chain lengths of 18 to 30 TMPs.

Poly(U)-Sephadex (BRL), which is reported to have lengths of more than 300 UMPs per molecule, can provide an alternative. However, poly(U)-Sephadex has less than 10% of the binding capacity of commercial oligo(dT)-celluloses [2–3 mg poly(A) cellulose]. The same is true for poly(U)-Sepharose resins (Pharmacia-LKB), which may not behave well in filtration–centrifugation methods because of their collapsible matrices.

Recovery of Polyadenylated Products from Filters. After the radioactivity of the filter has been measured, the filter is transferred to a 2-ml centrifuge tube identical to that used in the Spin-X filter unit. Polyadenylated species are eluted at room temperature with three or four 0.40-ml washes of 0.2% SDS. The temperature of the eluting solution can be raised to 60–70° in cases of incomplete or inefficient elution. Quantitative elutions from poly(U)-Sephadex will usually require washing of the filters with 90% formamide. Polyadenylated products can be recovered from the eluates by adding tRNA carrier (10 μg/ml) and adjusting to 0.2 M with potassium acetate before precipitation in 70% ethanol.

Gel Electrophoresis. The ethanol-precipitated RNAs are dissolved in 90% formamide (v/v) and heated at 70° for 1 min before applying to a 10% polyacrylamide gel containing 7 M urea.[26]

Concluding Remarks

The poly(A) polymerases purified from many sources catalyze the same overall reaction. However, enzymes purified from different sources display marked differences in their activity in the presence of Mg^{2+} and Mn^{2+} as well as in their use of specific RNAs as primers. These variations along with some evidence for the actual separation of activities from single cells have led to the view that multiple poly(A) polymerases may exist in the cell.[2] Although satisfactory evidence has yet to be found for this conclusion, such a hypothesis is compatible with long-standing observations that, in addition to the nucleus, poly(A) synthesis occurs in other cellular compartments. The elongation of the poly(A) tails of cytoplasmic mRNAs *in vivo* reported in the past[27,28] may involve a cytoplasmic poly-

[26] T. Maniatis and A. Efstratiadis, this series, Vol. 65, p. 299.
[27] G. Brawerman and J. Diez, *Cell* **5,** 271 (1975).
[28] S. G. Sawicki, W. Jelinek, and J. E. Darnell, *J. Mol. Biol.* **113,** 219 (1977).

merase different from the nuclear enzyme. The recent demonstration that correct polyadenylation requires specific recognition signals within the nuclear transcript as well as proteins in addition to the polymerase may soon clarify the role these purified poly(A) polymerases have in cellular polyadenylation reactions.

[13] Purification and Use of Vaccinia Virus Messenger RNA Capping Enzyme

By STEWART SHUMAN and BERNARD MOSS

As interest in the role of the 5'-RNA cap in mRNA biogenesis and function has increased, so too has the need for enzymatic reagents that allow investigators to manipulate the 5'-terminal structure of RNA. The properties of the capping and methylating enzymes from vaccinia virus[1] have proved particularly useful for 5'-RNA modification, including radioactive labeling of RNA ends.[2] This chapter provides an overview of the properties of the vaccinia capping enzyme, detailed protocols for enzyme assay and purification, and examples of how the purified enzyme may be of use to biochemists and molecular biologists.

Cap Synthesis

The cap structure of eukaryotic viral and cellular mRNAs consists of a 7-methylguanosine residue linked to the penultimate nucleoside of the RNA via a 5'–5' triphosphate bridge. The enzymatic pathway of cap synthesis has been determined to proceed by the following sequence of reactions:

$$pppN(pN)_n \rightarrow ppN(pN)_n + P_i \tag{1}$$
$$ppN(pN)_n + pppG \rightleftharpoons G(5')pppN(pN)_n + PP_i \tag{2}$$
$$G(5')pppN(pN)_n + AdoMet \rightarrow m^7G(5')pppN(pN)_n + AdoHcy \tag{3}$$

The enzymes responsible for these reactions are, respectively, RNA triphosphatase (1), RNA guanylyltransferase (2), and RNA (guanine-7)-methyltransferase (3). The structure $m^7G(5')pppN$ is referred to as cap 0. Additional modification reactions, e.g., methylation at the 2'-O position

[1] S. Shuman and J. Hurwitz, in "The Enzymes" (P. Boyer, ed.), Vol. 15, p. 245. Academic Press, New York, 1982.
[2] B. Moss, in "Gene Amplification and Analysis" (J. G. Chirikjian and T. S. Papas, eds.), Vol. 2, p. 253. Elsevier/North-Holland, New York, 1981.

of the ribose sugars to generate cap 1 and cap 2 structures, are catalyzed by separate 2'-O-methyltransferases (not discussed further herein). The enzymes catalyzing reactions (1)–(3) are present in purified virus particles from vaccinia virus,[3] reovirus,[4] and cytoplasmic polyhedrosis virus,[5] as well as in the nuclei of eukaryotic cells. Among the viral enzymes, only the one from vaccinia has proved amenable to purification from virus particles.[6] An especially attractive feature of the purified vaccinia capping enzyme is its ability to catalyze all three of the reactions leading to the synthesis of the cap 0 structure.[7] Cellular capping enzymes have been isolated from HeLa cells,[8–10] rat liver,[11] wheat germ,[12] brine shrimp,[13] and yeast[14]; however, in these cases it appears that separate capping and methylating enzymes are dissociated during purification, as are, in some instances, the capping and RNA triphosphatase activities. For these reasons, the capping and methylating activities of vaccinia virus are the only ones that have achieved general use as RNA-modifying reagents.

Structure of Vaccinia Capping Enzyme

Vaccinia virus capping enzyme has been purified to apparent homogeneity and shown to consist of two subunits of M_r 95,000 and 31,000.[6] Sedimentation and gel filtration reveal an M_r for the native enzyme of 127,000, indicating that the enzyme is a heterodimer. The purified enzyme is a multifunctional complex containing RNA triphosphatase, RNA guanylyltransferase, and RNA (guanine-7)-methyltransferase activities, collectively referred to as capping enzyme.[7,15] Each functional domain is independently active, and individual reactions are not obligatorily coupled.

The mechanism of transguanylylation has been studied in detail, and

[3] B. Moss, A. Gershowitz, C-M. Wei, and R. Boone, *Virology* **72**, 341 (1976).
[4] Y. Furuichi, S. Muthukrishnan, J. Tomasz, and A. Shatkin, *J. Biol. Chem.* **251**, 5043 (1976).
[5] Y. Furuichi and K. Miura, *Nature (London)* **253**, 374 (1975).
[6] S. A. Martin, E. Paoletti, and B. Moss, *J. Biol. Chem.* **250**, 9322 (1975).
[7] S. Venkatesan, A. Gershowitz, and B. Moss, *J. Biol. Chem.* **255**, 903 (1980).
[8] S. Venkatesan, A. Gershowitz, and B. Moss, *J. Biol. Chem.* **255**, 2829 (1980).
[9] S. Shuman, *J. Biol. Chem.* **257**, 7237 (1982).
[10] D. Wang, Y. Furuichi, and A. Shatkin, *Mol. Cell. Biol.* **2**, 993 (1982).
[11] K. Mizumoto and F. Lipmann, *Proc. Natl. Acad. Sci. U.S.A.* **76**, 4961 (1979).
[12] J. M. Keith, S. Venkatesan, and B. Moss, *Biochemistry* **21**, 327 (1982).
[13] Y. Yagi, K. Mizumoto, and Y. Kaziro, *J. Biol. Chem.* **259**, 4695 (1984).
[14] N. Itoh, H. Yamada, Y. Kaziro, and K. Mizumoto, *J. Biol. Chem.* **262**, 1989 (1987).
[15] S. Shuman, M. Surks, H. Furneaux, and J. Hurwitz, *J. Biol. Chem.* **255**, 11588 (1980).

the reaction was shown to involve a covalent enzyme–guanylate interme-diate, as follows:

$$pppG + E \rightleftharpoons EpG + PP_i \qquad (4)$$
$$EpG + ppRNA \rightleftharpoons G(5')pppRNA + E \qquad (5)$$

The intermediate consists of a GMP residue covalently attached to the M_r 95,000 enzyme subunit via a phosphoramidate bond[16] to the ε-amino group of a lysine.[17] Formation of the EpG occurs in the absence of an RNA acceptor. In the presence of GTP and [^{32}P]PP$_i$, the enzyme cata-lyzes PP$_i$ exchange with GTP.[15] Both partial reactions are readily revers-ible, yet pyrophosphorolysis of capped RNA is prevented by guanine-7 methylation,[18] accounting perhaps for the intimate association of the cap-ping and methylating functions. Though it is evident that the guanyl-yltransferase domain resides on the large subunit, it is not clear whether the small subunit is required for enzymatic activity. Nor is it known where the triphosphatase and methyltransferase domains are located. At-tempts to denature and renature the subunits have resulted in loss of enzymatic activity.

Capping Reactions and Enzyme Assays

General Considerations. Since the vaccinia capping enzyme exists as a complex of RNA triphosphatase, RNA guanylyltransferase, and RNA (guanine-7)-methyltransferase activities, any one of the three functions may be used to monitor enzyme activity during purification. Numerous assay systems have been developed for capping enzyme; these methods vary in their need for highly specialized substrates and hence in their accessibility to the investigator. The assay methods and the characteris-tics of the enzymatic reactions involved are described below.

RNA Guanylyltransferase. Guanylyltransferase catalyzes the transfer of GMP from GTP to the 5' terminus of RNA to form the cap structure G(5')pppN. Activity is generally determined by the incorporation of la-beled GMP from [^{3}H]GTP (or [α-^{32}P]GTP) into acid-insoluble material, i.e., RNA. Only RNAs containing a 5'-triphosphate or -diphosphate end will act as a cap acceptor.[18] RNAs that have 5'-monophosphate or 5'-hydroxyl ends are not utilized. Polyadenylic acid with a 5'-triphosphate terminus is generally used as the cap acceptor for assay purposes, and it may be synthesized by *Escherichia coli* RNA polymerase using the fol-lowing protocol.

[16] S. Shuman and J. Hurwitz, *Proc. Natl. Acad. Sci. U.S.A.* **78,** 187 (1981).
[17] M. J. Roth and J. Hurwitz, *J. Biol. Chem.* **259,** 13488 (1984).
[18] S. A. Martin and B. Moss, *J. Biol. Chem.* **251,** 7313 (1976).

A 1-ml reaction mixture is prepared containing 50 mM tris(hydroxymethyl)aminomethane hydrochloride (Tris-HCl) (pH 7.5), 2.5 mM dithiothreitol (DTT), 3 mM MnCl$_2$, 1 mM ATP, 20 μM cordycepin triphosphate, 0.9 A_{260} M13 single-stranded DNA, and 100 units $E.$ $coli$ RNA polymerase. [γ-^{32}P]ATP is included to quantitate the extent of synthesis and the concentration of 5' termini in the final product. For the synthesis of high specific activity 5'-triphosphate poly(A) (suitable for use as a substrate in the RNA triphosphatase assay), we typically add 0.4 mCi of [γ-^{32}P]ATP to the reaction. When lower specific activity poly(A) will suffice (as in the RNA guanylyltransferase assay) we use 20–40 μCi of labeled ATP. Cordycepin triphosphate is included as a chain terminator to favor the reinitiation of RNA chains, and, at the concentration used, it results in the synthesis of poly(A) of 150–200 nucleotides in length. The reaction mixture is incubated for 4–5 hr at 37° and precipitated by the addition of trichloroacetic acid (TCA) to 5%. Insoluble material is recovered by centrifugation and resuspended in 0.2 M Tris-HCl (pH 8). This procedure is repeated at least 3 times to remove all the unincorporated ATP. The material is then digested with DNase I, extracted twice with phenol and once with CHCl$_3$, reprecipitated with TCA, then ethanol precipitated and dissolved in 10 mM Tris-HCl (pH 8), 1 mM ethylenediaminetetraacetic acid (EDTA).

Though poly(A) is the standard cap acceptor we use, any RNA with an appropriate phosphorylation state should be able to be capped. There is no apparent base specificity for the penultimate nucleotide as a variety of homoribopolymers and naturally occurring RNAs are effective substrates.[18] An alternative substrate to poly(A) may most easily be prepared using commercially available bacteriophage RNA polymerases (T7 or Sp6) to synthesize RNAs of defined length and sequence from truncated plasmid vectors containing the appropriate promoters. Here, too, we recommend using the appropriate [γ-^{32}P]NTP to quantitate the concentration of 5'-triphosphate termini.

In addition to an appropriate cap acceptor, guanylyltransferase requires a divalent cation, preferably Mg^{2+}. Mn^{2+} is less effective and Ca^{2+} is inactive. Only guanine NTPs serve as donors in the capping reaction, including GTP, dGTP, ITP, and GTPγS. The K_m for GTP is 15 μM. 7-Methyl-GTP is inactive,[18] confirming that guanine methylation occurs after capping as indicated above. Activity is inhibited by NaCl concentrations in excess of 0.1 M. At 0.3 M NaCl the enzyme is inhibited by 90%. Guanylyltransferase is inactivated completely by heating at 55° for 5 min. Most importantly, the enzyme is exquisitely sensitive to inhibition by PP$_i$, a product of the capping reaction. In the presence of 9 μM GTP donor, capping activity is inhibited 50% by 4 μM PP$_i$. Phosphate is without effect up to 20 mM.

Guanylyltransferase activity is assayed in 50-μl reactions containing 25 mM Tris-HCl (pH 7.8), 2 mM MgCl$_2$, 1 mM DTT, 25 μM [^{3}H]GTP (10 Ci/mmol, and 10–20 pmol (of termini) of triphosphate-terminated poly(A). After 30 min at 37°, the reaction is terminated by the addition of 5% TCA, and insoluble material is collected by filtration. Alternatively, samples are spotted onto DEAE (DE-81) filters which are then washed successively with 5% NaH$_2$PO$_4$, distilled water, ethanol–diethyl ether (1 : 1), and ether. Using either method, radioactivity is then determined in a liquid scintillation counter. One unit catalyzes the capping of 1 pmol of RNA. We recommend the inclusion of 25–100 μM S-adenosylmethionine (AdoMet) in the reaction. AdoMet, though not required for activity, stimulates capping 2-fold,[19] apparently by preventing reversal of the reaction.

GTP–PP$_i$ Exchange Reaction. The advantage of the GTP–PP$_i$ exchange assay is the lack of requirement for special RNA substrates. Reaction mixtures (50 μl) containing 60 mM Tris-HCl (pH 8.4), 10 mM DTT, 5 mM MgCl$_2$, 0.2 mM GTP, and 1 mM [^{32}P]PP$_i$ (20–50 cpm/pmol) are incubated for 30 min at 37°, then halted by addition of 0.5 ml cold 10% TCA, 0.1 ml of 0.1 M NaPP$_i$, and 0.2 mg bovine serum albumin. Acid-insoluble material is removed by centrifugation in a microcentrifuge, and the supernatants are added to 0.1 ml of a 30% suspension of Norit A charcoal. Norit-absorbed material is collected on glass fiber filters, washed with TCA, and quantitated by Cerenkov counting. One unit catalyzes the incorporation of 1 nmol of ^{32}P from PP$_i$ into Norit-absorbable material (i.e., into GTP).

Enzyme–Guanylate Complex Formation. Formation of the covalent enzyme–GMP reaction intermediate is the simplest and therefore the preferred method for the assay of capping enzyme. It involves no RNA substrate and requires only [α-^{32}P]GTP (or dGTP), which is readily available at high specific activity and low cost. Reaction mixtures (10 μl) containing 50 mM Tris-HCl (pH 8.2), 2 mM DTT, 5 mM MgCl$_2$, and 2 μCi [α-^{32}P]GTP (3000 Ci/mmol) are incubated for 10 min at 37°, then halted by the addition of sodium dodecyl sulfate (SDS) to 1%. For precise quantitation of the labeling of the M_r 95,000 subunit of capping enzyme, the sample is analyzed directly by polyacrylamide gel electrophoresis. Electrophoresis must be carried out until the dye front is well out of the bottom of the gel (so as to eliminate free GTP). The gel may be autoradiographed without drying and the labeled protein cut out of the gel and quantitated by Cerenkov counting. By knowing the specific activity of the label, one can determine, using this assay, the precise molar concentration of active enzyme molecules.

An abbreviated version of this assay, which we suggest be employed as the assay of choice for the monitoring of capping enzyme during purifi-

[19] S. A. Martin and B. Moss, *J. Biol. Chem.* **250**, 9330 (1975).

cation, omits the electrophoretic analysis. The samples may be precipitated with TCA after denaturation with SDS and the acid-insoluble radioactivity determined by collection on glass fiber filters.

RNA (Guanine-7)-methyltransferase. Methyltransferase activity is assayed by the transfer of [^{3}H]methyl groups from Ado[*methyl*-^{3}H]Met to an
appropriate acceptor. Initial studies of this activity were performed using
unmethylated, capped RNA as the acceptor.[6] Such RNAs do not occur
naturally but can be synthesized *in vitro* by permeabilized vaccinia virions
when transcription reactions are carried out in the presence of S-adeno-
sylhomocysteine (AdoHcy). Alternatively, capped unmethylated poly(A)
can be synthesized *in vitro,* but this requires prior isolation of the vaccinia
capping enzyme. With capped RNA substrate, the enzyme methylates
only the 5'-guanine nucleoside of the cap and does so exclusively at the
N-7 position. The K_m for capped poly(A) is 0.21 μM.[18] Activity requires
neither Mg^{2+} nor GTP and is optimally active over a broad pH range near
neutrality. As expected, methyltransferase is inhibited by AdoHcy, a
product of the reaction.[19]

The need to synthesize and purify capped, unmethylated RNA substrate has been eliminated by the introduction of a far more convenient
assay that utilizes GTP as the methyl acceptor.[20] The K_m for GTP is quite
high (0.53 mM) compared to capped RNA.[18] The reaction mixture (50 μl)
contains 50 mM Tris-HCl (pH 7.6), 1 mM DTT, 10 mM GTP, and 1 μM
Ado[*methyl*-^{3}H]Met (11.6 Ci/mmol). After incubation with enzyme at 37°
for 30 min, 40 μl is applied to a DE-81 filter. Unincorporated AdoMet is
removed by washing the filters sequentially with 25 mM ammonium formate, water, ethanol–ether (1 : 1), and ether. Washing of the filters may be
performed using a vacuum filtration apparatus or by soaking in the respective solutions.

RNA Triphosphatase. RNA triphosphatase is assayed by the libera-
tion of [^{32}P]P$_i$ from ^{32}P-labeled triphosphate-terminated poly(A). The K_m
of the enzyme for RNA is 0.6 μM with respect to 5' ends.[15] Reactions are
performed in 25-μl mixtures containing 50 mM Tris-HCl (pH 8.4), 2.5 mM
MgCl$_2$, 2 mM DTT, and 1 pmol (of termini) of [γ-^{32}P]poly(A). After incubation at 37° for 5 min, the samples are spotted on PEI-cellulose thin-layer
sheets and chromatographed in 0.75 M KH$_2$PO$_4$ (pH 3.4). P$_i$ is located by
autoradiography, and the spot is cut out and counted.

Purification of Vaccinia Virus

The major hurdle for biochemists or molecular biologists in need of
capping enzyme is not the purification of the enzyme per se, but rather the

---

[20] E. Barbosa and B. Moss, *J. Biol. Chem.* **253,** 7692 (1978).

growth and purification of sufficient quantities of virus. The procedure used in our laboratory, derived from that of Joklik,[21] is described below.

Safety Precautions. Vaccinia is stable at ambient temperature even when dried, and parenteral inoculation and droplet or aerosol exposure of mucous membranes or broken skin with infectious virus are the primary hazards to laboratory personnel. The U.S. Department of Health and Human Services has provided safety precautions for working with vaccinia virus.[22] Biosafety Level 2 practices and containment equipment (e.g., Class I or II biological safety cabinets) are recommended for all activities involving the use or manipulation of vaccinia virus. The U.S. Public Health Service further recommends that all persons working in or entering laboratory areas where vaccinia is being used have documented evidence of satisfactory smallpox vaccination within the preceding 3 years.

Procedures. Large-scale preparations (typically 10–20 liters) of vaccinia virus strain WR are conveniently grown in HeLa cell suspension cultures. The virus inoculum is prepared by incubating a crude lysate of vaccinia-infected cells (containing 2×10^9 to 2×10^{10} plaque forming units/ml in Eagle's medium supplemented with 2% horse serum) with 0.12 mg/ml of crystallized trypsin for 30 min at 37°, with frequent vortexing to disperse aggregates. Growing HeLa cells are then collected by centrifugation and resuspended at a concentration of 2×10^7/ml in Eagle's spinner medium containing 5% horse serum. The cells are inoculated with virus at a multiplicity of 3 and incubated at 37° for 30 min with constant stirring. The cells are then diluted 40-fold with fresh medium containing 5% horse serum and incubated at 37° for 48 hr. The cells are harvested by low-speed centrifugation and resuspended at 5×10^7/ml in cold 10 mM Tris-HCl (pH 9).

Subsequent procedures are carried out at 4°. The cells are disrupted using a Dounce homogenizer, and nuclei are removed by centrifugation at 200 g for 5 min. The supernatant is saved, and the nuclear pellet is resuspended in a minimal volume of Tris buffer and recentrifuged. The supernatants are combined and dispersed by sonication for eight 15-sec intervals at maximum power using a Branson sonifier. Approximately 17 ml of this cytoplasmic fraction is layered over an equal volume of 36% (w/v) sucrose in 10 mM Tris-HCl (pH 9) and centrifuged in an SW27 rotor at 13,500 rpm for 80 min. The pellet is resuspended by sonication in 5 ml of 1 mM Tris-HCl (pH 9), layered over a linear 25–40% sucrose gradient in 1

[21] W. K. Joklik, *Biochim. Biophys. Acta* **61**, 290 (1962).
[22] HHS Publication No. (CDC) 84-8395, p. 66. U.S. Government Printing Office, Washington, D.C., 1984.

mM Tris buffer, and centrifuged in an SW27 rotor at 12,000 rpm for 50 min. The thick white virus band is collected by puncturing the side of the tube with a needle and syringe, then diluted with 2 volumes of 1 mM Tris and centrifuged at 13,500 rpm for 60 min. The pellet is resuspended by sonication in 3 ml of 1 mM Tris and subjected to a second round of sucrose-gradient sedimentation.

Since a significant amount of virus aggregates during sedimentation, the yield can be increased by resuspending the viral pellets from several tubes and reapplying them to sucrose gradients. An average yield is approximately 40 mg (1 A_{260} is taken to equal 64 μg of protein) of twice-banded virus from 6 liters of cells. Virus is stored at $-70°$.

Purification of Capping Enzyme

Step 1: Preparation of Virus Cores. Purified virus (100 mg) is incubated in 10 ml of 50 mM Tris-HCl (pH 8.4), 0.5% Nonidet P-40 (NP-40), 50 mM DTT for 30 min at 37° with frequent shaking. Cores are pelleted by centrifugation at 10,000 rpm for 10 min in an SS34 rotor. The supernatant, containing the viral envelope fraction, may be discarded.

Step 2: Disruption of Viral Cores. All subsequent steps are carried out at 4°. Cores are resuspended in 10 ml of 0.3 M Tris-HCl (pH 8.0), 50 mM DTT, 0.25 M NaCl, then made 0.2% in sodium deoxycholate. The mixture is homogenized in a Dounce homogenizer and allowed to sit on ice for 60 min. An increase in viscosity will be apparent owing to the release of the viral DNA from the cores. The mixture is centrifuged at 10,000 rpm for 10 min and the supernatant (containing soluble enzyme) removed. The recovery of enzyme can be enhanced by resuspending the pellet in buffer A [50 mM Tris-HCl (pH 8.0), 2.5 mM DTT, 1 mM EDTA, 10% glycerol, 0.01% NP-40] containing 0.2 M NaCl, adding deoxycholate to 0.2%, homogenizing vigorously, and repeating the centrifugation. The deoxycholate supernatants are then combined and adjusted to contain 0.2 M NaCl in buffer A.

Step 3: DEAE-Cellulose Chromatography. The deoxycholate supernatant is applied to a 15-ml column of DEAE cellulose (DE-51) equilibrated with 0.2 M NaCl in buffer A. After washing with the same buffer, the protein-containing flow-through material (now free of nucleic acid) is pooled and diluted with buffer A to adjust the NaCl concentration to 50 mM. This fraction is applied to a second 15-ml column of DEAE-cellulose equilibrated with 50 mM NaCl in buffer A. The bulk of the capping enzyme is not retained on this column and is recovered in the flow-through. The viral DNA-dependent RNA polymerase is retained quantitatively on the column and can be recovered by step elution with 0.25 M NaCl if so desired.

The DEAE-cellulose II flow-through fraction, though still relatively crude, can be used to cap and/or 2'-O-methylate RNA. However, this fraction contains numerous other enzymes including poly(A) polymerase, single-strand-specific DNase, protein kinase, DNA topoisomerase, DNA-dependent ATPase, and 5'-phosphate polynucleotide kinase that may complicate the use of the capping enzyme. Thus, further purification is suggested.

Step 4: Heparin-Agarose Chromatography. The DEAE II flow-through is applied to a 4-ml column of heparin-agarose equilibrated with 0.1 M NaCl in buffer A, washed with the same buffer, and eluted with a 100-ml linear gradient of 0.1–1.0 M NaCl in buffer A. Capping enzyme elutes as a discrete peak between 0.17 and 0.3 M NaCl and is resolved from poly(A) polymerase and DNA topoisomerase at this step.

Step 5: Phosphocellulose Chromatography. The pooled heparin-agarose fraction is diluted 4-fold with buffer A and applied to a 6-ml column of phosphocellulose equilibrated with 80 mM NaCl in buffer A, washed with the same buffer and eluted with an 80-ml gradient of 0.08–0.5 M NaCl in buffer A. Capping enzyme elutes as a discrete peak at 0.2 M NaCl.

Step 6: DNA-Agarose Chromatography. The phosphocellulose enzyme is diluted with buffer A to adjust the NaCl concentration to 40 mM, then applied to a 6-ml column of single-strand DNA-agarose equilibrated with 30 mM NaCl in buffer A. It is advisable to verify the salt concentration of the enzyme prior to application to the column, as the enzyme does not absorb tightly to this resin and may be apt to partition between the retained and flow-through fractions. After washing with the same buffer, the column is developed with a 90-ml linear gradient of 0.03–0.45 M NaCl in buffer A. The capping enzyme elutes as a discrete peak at 0.12 M NaCl and is resolved from DNA-dependent ATPase at this stage. The active enzyme fractions are routinely stored at $-80°$.

Comments on Purification. The protocol described above provides highly purified capping enzyme devoid of interfering activities, including 2'-O-methyltransferase. The yield at the DNA-agarose step is about 75% relative to the DEAE-cellulose II fraction.[23] We find this procedure particularly well suited to the simultaneous purification of several of the enzymes associated with vaccinia virus cores. Alternative schemes may be developed based on the needs of the individual investigator. At the very least, we would suggest that the DNA-agarose step be included in such protocols.

---

[23] S. Shuman, S. Broyles, and B. Moss, *J. Biol. Chem.* **262,** 12372 (1987).

Application of Vaccinia Capping Enzyme

Labeling of RNA. Vaccinia capping enzyme is useful as a highly specific reagent for 5′-end-labeling of RNAs that contain (or can be converted to a form that contains) either 5′-triphosphate or 5′-diphosphate termini. The relatively low K_m for GTP and the availability of high specific activity [α-^{32}P]GTP make the capping enzyme valuable in analytical studies of RNA. The weight of current evidence would indicate that any RNA which can be cap labeled with GTP without prior decapping can be construed as constituting a primary transcript.

RNAs that originally contain capped termini may be decapped by periodate oxidation of the 2′,3′-*cis*-diol on the blocking 7-methylguanosine followed by β-elimination with aniline.[24] This is carried out in 0.1–0.2 ml of freshly prepared 0.9 M sodium periodate (>100-fold molar excess), 0.15 M sodium acetate (pH 5.3) in the dark at 0°. After 30 min, the RNA is precipitated twice with ethanol. If carrier RNA is needed, alkaline phosphatase-treated tRNA may be added prior to precipitation. Approximately 0.1 to 0.2 ml of a solution containing 0.3 M distilled aniline and 10 mM acetic acid adjusted to pH 5 with concentrated HCl is used to dissolve the RNA. Aniline is known to be carcinogenic and should be used only in a fume hood. After 3 hr in the dark at room temperature, the RNA is either precipitated repeatedly with ethanol or filtered through a Sephadex G-50 column. The product of this reaction is a 5′-triphosphate-terminated RNA from which the 3′-terminal nucleoside has been removed.

Chemically decapped RNA can be recapped and methylated by the vaccinia capping enzyme, with radioactive label introduced either in the blocking guanosine or the N-7 methyl group. The conditions for capping are similar to those described in the assay for RNA guanylyltransferase. RNA in picomole amounts can be capped quantitatively in the presence of high enzyme concentrations and GTP in excess of the K_m. However, the need to achieve high specific labeling of small quantities of substrate may necessitate using large amounts of undiluted [α-^{32}P]GTP (in the range of 1 μM). Under such conditions, capping is less than quantitative, but the addition of unlabeled AdoMet will increase the yield. AdoMet is unstable at even mildly alkaline pH and must be prepared as a stock solution in 5 mM H_2SO_4. In cases where 5′-RNA labeling is undertaken for RNA sequencing, it is preferable to omit AdoMet (and to include AdoHcy) since incomplete methylation or alkaline degradation (ring opening) of the m^{7}G can lead to artificial heterogeneity. Labeling can also be performed with [^{3}H]GTP. However, since ^{3}H is commonly at the 8 position of

24 H. Fraenkel-Conrat and A. Steinschneider, this series, Vol. 12B, p. 243.

guanine, AdoMet must not be used since methylation at the 7 position will lead to loss of the isotope. When Ado[*methyl*-³H]Met is used to label RNA, a high concentration of unlabeled GTP (1 mM) can be used to attain nearly quantitative capping. Optimal enzyme concentrations are best determined empirically.

Verification that the label is in the cap structure is obtained by digestion of a sample with nuclease P1 ($+/-$ alkaline phosphatase). The released cap structures, e.g., G(5′)pppN, m⁷G(5′)pppN, and m⁷G(5′)pppNm, can be analyzed by high-voltage electrophoresis at pH 3.5[18] or (more simply) by PEI-cellulose thin-layer chromatography[16,18] with appropriate marker dinucleotides.

Other Applications. In *vitro* manipulation of the RNA cap structure using the vaccinia capping enzyme system has been used to study the important function of the cap in promoting RNA binding to ribosomes in the initiation of protein synthesis.[25] The enzyme has also been utilized to determine the role of the methylated RNA cap in the priming of influenza virus transcription by exogenous RNAs.[26] Cap labeling can be applied to great advantage in the genomic mapping of transcription initiation sites (either by blotting techniques or S1 nuclease protection). In addition, cap labeling may be used to sequence RNA in the same manner as polynucleotide kinase.

[25] S. Muthukrishnan, B. Moss, J. Cooper, and E. Maxwell, *J. Biol. Chem.* **253**, 1710 (1978).
[26] M. Bouloy, S. Plotch, and R. Krug, *Proc. Natl. Acad. Sci. U.S.A.* **77**, 3952 (1980).

[14] RNA Lariat Debranching Enzyme as Tool for Analyzing RNA Structure

By BARBARA RUSKIN and MICHAEL R. GREEN

The primary transcripts (pre-mRNAs) of most structural genes in higher eukaryotes contain intervening sequences (introns) that are removed by splicing. Splicing occurs in a two-step reaction[1,2] (reviewed in Refs. 3 and 4). In the first stage, the pre-mRNA is cleaved at the 5′ splice

[1] B. Ruskin, A. R. Krainer, T. Maniatis, and M. R. Green, *Cell* **38**, 317 (1984).
[2] R. A. Padgett, M. M. Konarska, P. J. Grabowski, S. F. Hardy, and P. A. Sharp, *Science* **225**, 898 (1984).
[3] R. A. Padgett, P. J. Grabowski, M. M. Konarska, and P. A. Sharp, *Annu. Rev. Biochem.* **55**, 1119 (1986).
[4] M. R. Green, *Annu. Rev. Genet.* **20**, 671 (1986).

FIG. 1. Debranching reaction. The 2'–5' phosphodiester bond (RNA branch) between the A residue at the branch point and the G residue at the 5' end of the excised intron is indicated. Hydrolysis of the bond catalyzed by the lariat debranching enzyme results in a linear RNA with a 2'-hydroxyl at the A residue of the branch point.

site to generate two splicing intermediates, the first exon RNA species and an RNA species composed of the intron and second exon (IVS–exon 2 RNA species). In the second stage, cleavage at the 3' splice site and ligation of the exons occurs, resulting in excision of the intact intron. The IVS–exon 2 splicing intermediate and the excised IVS are in the form of a lariat in which the 5' end of the intron is joined to an adenosine residue near the 3' end of the intron by a 2'–5' phosphodiester bond. Thus, the two products of the splicing reaction are the linear spliced mRNA and the excised intron, which is in a lariat configuration. The mRNA is transported to the cytoplasm whereas the excised intron is degraded in the nucleus. This implies that there is a nuclear RNA processing activity that hydrolyzes the 2'–5' phosphodiester bond (RNA branch) of the excised intron.

The lariat debranching enzyme is an RNA processing activity, originally identified in mammalian cell extracts, that specifically hydrolyzes the 2'–5' phosphodiester bond in RNA lariats[5] (Fig. 1). Analysis of yeast pre-mRNA mutants supports a role for the enzyme in the intron degradation pathway.[6] The 2'–5' phosphodiester bond is resistant to digestion by most nucleases. The lariat debranching enzyme displays high substrate specificity; the single 2'–5' phosphodiester bond in the RNA lariat is degraded whereas all of the 3'–5' phosphodiester bonds are left intact. This result distinguishes the lariat debranching enzyme from 2'→5'-phosphodiesterases such as snake venom phosphodiesterase, which hydrolyzes both 2'–5' and 3'–5' phosphodiester bonds. The lariat debranching enzyme will not hydrolyze the 2'–5' phosphodiester bonds in 2'–5' oligoisoadenylate[5] (a linear RNA consisting of adenosine residues joined by 2'–5' phosphodiester bonds), a standard substrate used for assaying 2'→5'-phosphodiesterases. This distinguishes the lariat debranching en-

[5] B. Ruskin and M. R. Green, Science 229, 135 (1985).
[6] A. Jacquier and M. Rosbash, Proc. Natl. Acad. Sci. U.S.A. 83, 5835 (1986).

zyme from other mammalian 2'→5'-phosphodiesterases such as the inter-feron-inducible 2'→5'-phosphodiesterase.[7-9] Finally, the RNA debranch-ing enzyme cannot remove a 2'-phosphomonoester,[10] distinguishing it from many phosphatases. Taken together, these results indicate that the minimal substrate for the lariat debranching enzyme is a branched nucleo-tide containing both a 2'–5' and a 3'–5' phosphodiester bond.[5,10] Although most naturally occurring RNA lariats contain an A-(2'–5')-G bond, the lariat debranching enzyme is able to hydrolyze 2'–5' bonds composed of other nucleotides.[5,6,11]

The high substrate specificity of the lariat debranching enzyme has made it a powerful tool for the analysis of RNA structure. Accordingly, the lariat debranching enzyme has been used (1) to verify the presence of RNA lariats in cellular RNA populations,[11-14] (2) to determine whether particular RNA species are in a lariat or a circular configuration,[15,16] (3) to aid in mapping the positions of RNA branches generated during *in vitro* pre-mRNA splicing,[17-20] (4) to demonstrate the existence of a Y structure RNA intermediate generated during trans-splicing in trypanosomes,[21,22] and (5) to help deduce the structure of a novel branched RNA that is covalently linked to the 5' end of single-stranded DNA in the bacterium *Stigmatella aurantiaca*.[23]

[7] I. M. Kerr and R. E. Brown, *Proc. Natl. Acad. Sci. U.S.A.* **75**, 256 (1978).

[8] A. Schmidt, Y. Chernajovsky, L. Shulman, P. Federman, H. Berissi, and M. Revel, *Proc. Natl. Acad. Sci. U.S.A.* **76**, 4788 (1979).

[9] M. A. Minks, S. Benvin, P. A. Maroney, and C. Baglioni, *Nucleic Acids Res.* **6**, 767 (1979).

[10] J. Arenas and J. Hurwitz, *J. Biol. Chem.* **262**, 4274 (1987).

[11] L. A. Fouser and J. D. Friesen, *Cell* **45**, 81 (1986).

[12] S. Zeitlin and A. Efstratiadis, *Cell* **39**, 589 (1984).

[13] J. R. Rodriguez, C. W. Pikielny, and M. Rosbash, *Cell* **39**, 603 (1984)

[14] A. Cellini, R. Parker, J. McMahon, C. Guthrie, and J. Rossi, *Mol. Cell. Biol.* **6**, 1571 (1986).

[15] C. L. Peebles, P. S. Perlman, K. L. Mecklenburg, M. L. Petrillo, J. H. Tabor, K. A. Jarrell, and H. L. Cheng, *Cell* **44**, 213 (1986).

[16] K. Ehrenman, J. Pedersen-Lane, D. West, R. Herman, F. Maley, and M. Belfort, *Proc. Natl. Acad. Sci. U.S.A.* **83**, 5875 (1986).

[17] B. Ruskin, J. M. Greene, and M. R. Green, *Cell* **41**, 833 (1985).

[18] B. Ruskin, C. W. Pikielny, M. Rosbash, and M. R. Green, *Proc. Natl. Acad. Sci. U.S.A.* **83**, 2022 (1986).

[19] R. Reed and T. Maniatis, *Cell* **41**, 95 (1984).

[20] D. Rio, *Proc. Natl. Acad. Sci. U.S.A.* **85**, 2904 (1988).

[21] R. E. Sutton and J. C. Boothroyd, *Cell* **47**, 527 (1986).

[22] W. J. Murphy, K. P. Watkins, and N. Agabian, *Cell* **47**, 517 (1986).

[23] T. Furuichi, A. Dhundale, M. Inouye, and S. Inouye, *Cell* **48**, 47 (1987).

Preparation of Extracts Containing RNA Lariat Debranching Activity

HeLa cell nuclear extracts are prepared according to published procedures.[24] A majority of the RNA lariat debranching activity is present in the "cytoplasmic" S100 fraction, presumably owing to leakage from the nucleus during extract preparation. The S100 fraction can be stored in Buffer D[24] containing 20% glycerol at $-70°$ for several months with little or no loss of debranching activity. Following thawing, enzyme activity is retained for at least several days at $4°$.

Various experiments suggest that the lariat debranching enzyme is composed of protein without an associated RNA component.[5,10] If desired, the lariat debranching activity can be partially purified. The enzyme is relatively stable and maintains activity following fractionation over a number of conventional chromatographic resins.[10,25] In all instances, the activity fractionates as a single component. In our experience, there has been no need to use the partially purified enzyme since enzymatic debranching can be performed in the crude S100 fraction in the presence of ethylenediaminetetraacetic acid (EDTA) (see below). Under these conditions we have not observed interfering nuclease or other RNA processing activities.

Biochemical Requirements for Enzymatic Debranching

The biochemical requirements for the lariat debranching enzyme are extraordinarily simple; in the crude S100 fraction, the reaction proceeds in the absence of added divalent or monovalent cation or a nucleoside cofactor.[5] The lack of a divalent cation requirement enables the reaction to be performed in the presence of relatively high concentrations of EDTA (8–50 mM).[5] High concentrations of salt (400 mM) inhibit debranching activity.[5,10] It has been reported, in contrast to results using S100, that the partially purified enzyme requires divalent cations.[10] The enzyme displays a relatively broad pH range with an optimum at pH 7.0.[10]

Determining if RNA Contains RNA Branch

RNA lariats (and branched linear RNAs) have anomalous electrophoretic mobilities on denaturing polyacrylamide gels.[1,26] Following enzymatic debranching, the linear RNA product migrates more rapidly than the lariat.[5] RNA lariats within a population of RNA molecules can thus be

[24] J. D. Dignam, R. M. Lebovitz, and R. G. Roeder, *Nucleic Acids Res.* **11**, 1475 (1983).
[25] A. Kramer and W. Keller, *EMBO J.* **4**, 3571 (1985).
[26] P. J. Grabowski, R. A. Padgett, and P. A. Sharp, *Cell* **37**, 415 (1984).

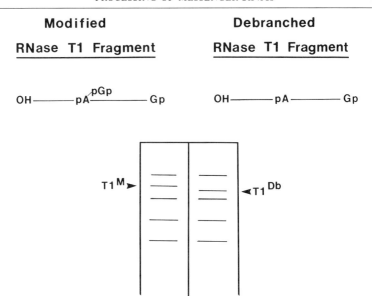

FIG. 2. Localization of the RNA branch to an RNase T1 fragment. The structures of the RNase T1 fragment containing the RNA branch (Modified RNase T1 Fragment) and the same debranched RNase T1 fragment (Debranched RNase T1 Fragment) are shown. Below is a schematic diagram of the RNase T1 digestion products of an intact and enzymatically debranched RNA lariat fractionated on a denaturing polyacrylamide gel. The expected positions of the RNase T1 fragment containing the RNA branch (T1^M) and the same debranched RNase T1 fragment (T1^{Db}) are indicated.

Procedure 3: RNase T1 Digestion of RNA Lariats

1. The ^{32}P-labeled RNA lariat is first purified from a polyacrylamide gel. The ^{32}P-labeled RNA lariat is detected by autoradiography and eluted from a gel slice by soaking overnight at room temperature in 200–400 μl of 2× proteinase K buffer containing 10 μg of tRNA.

2. The purified RNA lariat is precipitated with ethanol, and a portion is removed and enzymatically debranched (Procedure 1).

3. Both the intact and debranched RNAs are resuspended in a small volume of water and added to an RNase T1 digestion reaction. The RNase T1 digestion reaction contains 1 μl RNA (containing 10 μg tRNA carrier), 1 μl of 10 mM Tris-HCl (pH 7.4), 1 mM EDTA, and 1 μl RNase T1 (2 units/μl).

4. The reaction proceeds at 37° for 30 min and is terminated by addition of formamide loading dye (90% formamide, 1× TBE, 0.05% bromphenol blue, 0.05% xylene cyanole FF).

5. The sample is heated at 90° for 3 min and loaded directly on a 20–25% polyacrylamide gel that has been preelectrophoresed for at least 1 hr. Electrophoresis is stopped when bromphenol blue migrates one-third the gel length.

6. A complete RNase T1 digest of a ^{32}P-labeled RNA substrate is analyzed in parallel to provide size standards.

RNase A Secondary Digestion Analysis of the Modified
RNase T1 Fragment

The RNA branch can be mapped within the modified RNase T1 fragment by secondary digestion analysis with RNase A. The 2'–5' phosphodiester bond renders the adjacent 3'–5' phosphodiester resistant to RNase A digestion.[1] Therefore, the RNase A digestion products of the modified RNase T1 fragment will differ from those of the corresponding unmodified RNase T1 fragment.

Procedure 4: RNase A Secondary Digestion of RNase T1 Fragments

1. The modified and unmodified RNase T1 fragments are isolated from a polyacrylamide gel (see Procedure 3, Step 1).

2. The small size of some RNase T1 fragments prevents efficient ethanol precipitation; the addition of magnesium increases the precipitation efficiency of these small RNA fragments. To 200 μl of 2× proteinase K buffer containing the ^{32}P-labeled RNase T1 fragment and carrier tRNA are added 20 μl of 0.1 M magnesium acetate and 2.5 volumes of ethanol.

3. The ethanol precipitate containing the RNase T1 fragment is collected and digested to completion with RNase A in a reaction mixture containing 1 μl RNase T1 fragment (containing 10 μg tRNA carrier), 1 μl of 30 mM Tris-HCl (pH 7.5), 1 mM EDTA, and 1 μl of 10 mg/ml RNase A.

4. The reaction is allowed to proceed at 37° for 60 min.

5. The RNase A digestion products are fractionated by two-dimensional thin-layer chromatography on polyethyleneimine (PEI) plates.[1,31] By comparison to the RNase A digestion pattern of the unmodified RNase T1 fragment, the position of the RNA branch can be deduced (see, for example, Ref. 1).

Composition of Branched Trinucleotide Core

As a final step, the trinucleotide core composition of the RNA branch can be determined by nuclease P1 and RNase T2 digestion analyses.[1]

---

[31] G. Volckaert, W. M. Jou, and W. Fiers, *Anal. Biochem.* **72**, 433 (1976).

These sequence-nonspecific nucleases are also unable to cleave the 3′–5′ phosphodiester bond adjacent to the RNA branch.[1,2,29,30] Nuclease P1 clevage results in 5′-phosphate/3′-hydroxyl termini whereas RNase T2 clevage results in 5′-hydroxyl/3′-phosphate termini. First, four RNA substrates are synthesized, each of which is labeled with one of the four α-^{32}P-labeled ribonucleoside triphosphates. The RNA lariats resulting from processing of each of these RNA substrates is then digested to completion with (1) nuclease P1, (2) RNase T2, and (3) nuclease P1 plus RNase T2. In each instance the nuclease-resistant component (trinucleotide) migrates less rapidly than the mononucleotide when analyzed by thin-layer chromatography. By determining whether an RNA lariat labeled with a particular ribonucleoside triphosphate gives rise to a labeled nuclease-resistant component, the composition of the RNA branch can be deduced (see, for example, Ref. 1).

Procedure 5: Nuclease P1 and RNase T2 Digestion Analysis

1. The nuclease P1 digestion reaction contains 1 μl RNA (containing 10 μg tRNA carrier), 1 μl of 60 mM sodium acetate (pH 5.5), and 1 μl of 2 mg/ml nuclease P1. The reaction is allowed to proceed at 37° for 2 hr. The nuclease P1 digestion products are fractionated by one-dimensional chromatography on cellulose plates using as a solvent 0.1 M sodium phosphate (pH 6.8)–ammonium sulfate–n-propanol (100 : 60 : 2, v/w/v).[32]

2. The RNase T2 digestion reaction contains 1 μl RNA (containing 10 μg tRNA carrier), 1 μl of 60 mM sodium acetate (pH 5.5), and 1 μL of 2000 units/ml RNase T2. The reaction is allowed to proceed at 37° for 2 hr. The RNase T2 digestion products are fractionated on PEI plates using as a solvent 1 M formic acid adjusted with pyridine to pH 4.3.[31]

3. For digestion with nuclease P1 plus RNase T2, the nuclease T2 digestion is conducted first, followed by the addition of nuclease P1 and further incubation (2 hr) at 37°. The nuclease P1 plus RNase T2 digestion products are fractionated by one-dimensional chromatography on cellulose plates using as a solvent 0.1 M sodium phosphate (pH 6.8)–ammonium sulfate–n-propanol (100 : 60 : 2, v/w/v).[32]

---

[32] M. Silberklang, A. M. Gillum, and U. L. RajBhandary, this series, Vol. 59, p. 58.

[15] *Escherichia coli* Ribonuclease III

By Hugh D. Robertson

Introduction

Escherichia coli RNase III was discovered in 1967,[1] purified the following year,[2] and identified as the RNA processing enzyme responsible for maturation of *E. coli* rRNA and phage T7 mRNA several years later.[3-5] Its role in RNA processing has been analyzed in reviews.[6,7] The enzyme is also useful in identifying regions of 20 base pairs or more of perfect RNA–RNA duplex structure[2,8] (which it prefers to cleave despite convincing data that its intracellular targets are not composed of such perfect RNA–RNA structures) and in introducing specific, limited cleavages into long RNAs under nonphysiological, cell-free conditions.[9,10] In this chapter we consider methods for the assay of RNase III, preparation of the substrates required, and a purification scheme for the enzyme that has been tested on strains of *E. coli* carrying cloned copies of the gene for RNase III (the *rnc* gene), which has recently been identified and sequenced.[11]

[1] H. D. Robertson, R. E. Webster, and N. D. Zinder, *Virology* **11**, 718 (1967).
[2] H. D. Robertson, R. E. Webster, and N. D. Zinder, *J. Biol. Chem.* **243**, 82 (1968).
[3] J. J. Dunn and F. W. Studier, *Proc. Natl. Acad. Sci. U.S.A.* **70**, 1559 (1973).
[4] N. Nikolaev, L. Silengo, and D. Schlessinger, *Proc. Natl. Acad. Sci. U.S.A.* **70**, 3361 (1973).
[5] H. D. Robertson and J. J. Dunn, *J. Biol. Chem.* **250**, 3050 (1975).
[6] H. D. Robertson, *in* "Columbia P & S Symposia: Protein–Nucleic Acid Interactions" (H. Vogel, ed.), p. 549. Academic Press, New York, 1977.
[7] H. D. Robertson, *Cell* **30**, 669 (1982).
[8] H. D. Robertson and T. Hunter, *J. Biol. Chem.* **250**, 418 (1975).
[9] G. Paddock, K. Fukada, J. Abelson, and H. D. Robertson, *Nucleic Acids Res.* **3**, 1351 (1976).
[10] J. J. Dunn, *J. Biol. Chem.* **251**, 3807 (1976).
[11] N. Watson and D. Apirion, *Proc. Natl. Acad. Sci. U.S.A.* **82**, 849 (1985).

METHODS IN ENZYMOLOGY, VOL. 181

Materials and Methods

RNase III processing of various RNA precursors is carried out by polyacrylamide gel electrophoresis followed by characterization of the cleavage products.[3-5] For enzyme purification, however, a synthetic double-helical RNA, poly(A-U), is customarily used together with precipitation of reactions in 5% trichloroacetic acid (TCA). Since RNase III copurifies with a principal RNase H activity *E. coli*,[2] substrates for this activity of equal specific radioactivity can be conveniently prepared at the same time and used to monitor this second activity.

Test Substrates for RNase III and RNase H

Conditions are described below for the synthesis *in vitro* of ^{3}H-labeled RNA preparations that can be used as test substrates for detection of RNase III or RNase H activity. The synthesis conditions are designed to produce about 10^5 counts per minute (cpm) per nanomole of total polymer phosphorus of either double-stranded RNA or DNA–RNA hybrid substrate. Of course, the exact specific activity must be determined experimentally for each batch synthesized since it depends on such variables as tritium efficiency of various scintillation counters, amount of quenching of tritium on glass fiber filters caused by the bovine serum albumin (BSA) carrier, and degree of precipitability of the particular preparation in TCA. The double-stranded RNA test substrate is ^{3}H-labeled poly(A-U)n; the DNA–RNA hybrid test substrate consists of RNA synthesized *in vitro* on the single-stranded phage DNA obtained from the filamentous coliphage M13.

Reagents. Ingredients for the two syntheses are the same and are prepared or purchased as follows:
   ^{3}H-Labeled UTP, ^{3}H-labeled uridine 5′-triphosphate, tetrasodium salt (New England Nuclear, Cat. No. NET-287)
   5× reaction buffer: 0.1 M tris(hydroxymethyl)aminomethane hydrochloride (Tris-HCl) (pH 7.9), 50 mM MgCl$_2$, 0.75 M KCl, 0.5 mM ethylenediaminetetraacetic acid (EDTA); dithiothreitol (DTT) is added separately from an 0.1 M stock solution to give a final concentration of 1 mM
   Unlabeled ribonucleoside triphosphates (P-L Pharmaceutical)
   Poly[d(A-T)$_n$] template for poly(A-U)$_n$ (Miles)
   Single-stranded bacteriophage DNA (Biogenics Research Corp., Chagrin Falls, OH)
   Escherichia coli RNA polymerase (Promega Biotec), 10,000 μ/ml
   Reactions for Double-Stranded RNA and Hybrid Synthesis. For synthesis of dsRNA, the reaction mixture contains the following:

Ingredient	Volume added (µl)
5× buffer	60
DTT (0.1 M)	3
ATP (30 mM)	2
UTP (30 mM)	2
[^{3}H]UTP (100 µCi)	Dried in reaction tube
RNA polymerase	2
Poly[d(A-T)$_n$](1 mg/ml)	20
Water (glass distilled)	211
Total	300

The reaction is incubated for 1 hr at 37° and then heated for 2 min at 75° and allowed to cool to ambient temperature. The reaction is tested for uptake of radiolabel by spotting 1 µl onto a GF/A filter while precipitating another 1 µl aliquot in 5% TCA in the presence of 50 µg of BSA carrier. The TCA precipitate is collected by filtering through a GF/A filter under suction. Both filters are then dried and counted. At least 5% of the ^{3}H-labeled UTP should have been taken up. Purification of the RNA is described below.

For the synthesis of DNA–RNA hybrids, the reaction mixture contains the following:

Ingredient	Volume added (µl)
5× buffer	100
DTT	5
UTP (30 mM)	2.5
ATP (30 mM)	5
GTP (4 mM)	37.5
CTP (4 mM)	37.5
[^{3}H]UTP (850 µCi)	Dried in tube
M13 DNA (1 mg/ml)	67.5
RNA polymerase	4
Water	241
Total	500

The reaction is incubated for 2 hr at 37°, heated for 2 min at 65°, chilled on ice, and assayed as above (goal, 3% incorporation).

Purification of Synthesized RNAs. The volume of each reaction is increased to 1 ml by addition of TSE buffer [50 mM Tris-HCl (pH 7.0), 0.1 M NaCl, 1 mM EDTA]. Then 0.5 ml of phenol saturated with 10 mM Tris-HCl (pH 7.6), 1 mM EDTA is added; the mixture is vortexed for 3 min at ambient temperature and then centrifuged for 5 min at 3000 rpm in a swinging-bucket rotor at ambient temperature. The 1-ml aqueous layer is removed, and 1 ml of absolute ethanol is added to it. The sample is then subjected to cellulose–ethanol chromatography on a column of Whatman CF-II cellulose (Franklin column). Columns are poured in 5-ml disposable syringes packed with glass wool and then a slurry of cellulose CF-II, washed first in 3 changes of TSE buffer and then allowed to settle so that the fines could be removed.

The 2-ml columns are allowed to settle and then washed with water (4 times, 3 ml each), TSE buffer alone (4 times, 3 ml), and water (4 times, 3 ml). Shortly before use, they are equilibrated with a buffer containing 50% (v/v) ethanol and TSE buffer, which is thoroughly degassed by aspiration before being poured onto the column. After washes with the buffer containing 50% ethanol (4 times, 3 ml), the 2-ml sample is added, the column is washed with 12 ml of buffer containing 50% ethanol, and 4 fractions of 3 ml each are collected which contain all of the labeled triphosphates and much of the small RNA fragments. In the case of the poly(A-U) synthesis, the column is then washed (4 times, 3 ml) with a mixture made by adding 65 ml of TSE buffer to 35 ml of absolute ethanol followed by extensive degassing. The column is next washed with 3 ml of absolute ethanol, then an aliquot of 12 ml of water is added, and 12 fractions of 1 ml each are collected. In the case of the DNA–RNA hybrids, identical procedures are followed except that the 35% ethanol step is omitted. Aliquots of the water fractions are counted. In each case, the peak of radioactive RNA eluted in fractions 3–5, which are pooled, tested, and kept as substrate.

Assay Conditions. Standard assay conditions for RNase III and accompanying RNase H are 20 mM Tris-HCl (pH 7.6), 10 mM magnesium acetate, 0.13 M NH$_4$Cl, substrate, enzyme, and 5% sucrose (particularly for dilute protein fractions). This assay buffer is called TM13. Incubation is carried out at 37° for the time chosen, followed by addition of 5% TCA and filtration through GF/A filters as described above. One unit of RNase III is defined as the amount that solubilizes 1 nmol of substrate per hour at 37°.

Enzyme Purification

Several approaches to RNase III purification were worked out in earlier papers, and the one suggested here is principally based on that of Dunn[10] as modified by our experience. Briefly, to obtain reasonable quantities of RNase III from an *E. coli* strain such as BL-15[10] or one carrying cloned enzyme,[11] at least 50 g of packed, frozen cells are needed. Cells are warmed to 0° in a cold room and mixed with 2.5 times their weight of washed alumina[2,8] plus 2.5 times their weight of extraction buffer [50 mM Tris-HCl (pH 7.9), 0.25 M KCl, 2 mM EDTA, 1 mM 2-mercaptoethanol,

0.1 mM DTT]. In a particular example, 55 g of frozen cells was mixed with 127.5 g alumina and 127.5 ml of extraction buffer, homogenized in a chilled mortar and pestle until a "creamy" consistence was reached, and then blended briefly with a Virtis Omnimixer. At that point, 0.75 mg of Worthington crystalline DNase (DPFF) was added and the mixture stirred 20 min.

Centrifugation at 10,000 rpm in the SS34 rotor of the Sorvall RC-5B centrifuge was followed by isolation of the supernatant and rehomogenization of the pellets in 20 ml of additional extraction buffer. After a second 10,000 rpm spin, the supernatants (totalling 144 ml) were pooled, spun at 15,000 rpm for 20 min, recovered, and centrifuged for 2.5 hr in the Type 40 rotor of a Beckman ultracentrifuge at 39,000 rpm. The supernatant of this preparation is recovered (totalling 125.5 ml). The rationale of the above steps is that the RNase III activity, which sticks to ribosomes at salt concentrations below 0.2 M,[2] is released under these conditions into the postribosomal supernatant obtained in the type 40 rotor. Ammonium sulfate fractionation is then carried out, first at 22 g/100 ml of original volume (thus, 27.45 g here), slowly added with stirring on ice. The pH must be adjusted to 7.0 by adding 1 N NaOH (about 20 μl/g ammonium sulfate) as needed. After 15 min of stirring, the mixture is centrifuged at 15,000 rpm for 20 min and the supernatant, now totalling 135 ml, is saved. The pellets, which do not contain RNase III, are saved for future assays and controls. To the 135 ml of supernatant is added an additional 14 g/100 ml of solid ammonium sulfate (18.9 g here) with stirring on ice (total time 15 min). Adjustment of pH is carried out by addition of 0.38 ml of 1 M NaOH. Centrifugation at 15,000 rpm in the SS34 rotor for 10 min is followed by resuspension of the pellets (which contain most or all of the RNase III activity) in 15 ml of buffer B plus 1 M KCl per 100 g of starting *E. coli* cell weight. Buffer B is 20 mM Tris (pH 7.9), 0.1 mM EDTA, 0.1 mM DTT, 5% glycerol, to which various amounts of KCl are added as indicated. Here, 8.25 ml of buffer B plus 0.1 M KCl is added to the pellets, giving a total volume after dissolving the pellets of 14.5 ml.

The next step is dialysis overnight against 1 liter of buffer B plus 0.1 M KCl, followed by 1 hr of further dialysis against a fresh 1 liter. Dialysis tubing is prepared by boiling for 5 min in buffer B plus 0.1 M KCl without DTT or glycerol. The dialysis sample is transferred to a 100-ml graduated cylinder prechilled on ice (total volume 16 ml). The conductivity of 50 μl of dialyzed enzyme diluted to 10 ml is determined and compared to standards (all prepared by adding 50 μl to 9.95 ml of water). The objective here is to reduce the conductivity below the level of buffer B plus 50 mM KCl, which is the starting buffer of the DEAE column to be run as the next step. In this case, distilled water had a conductivity of 1.6; buffer B (no KCl), 3.4; buffer B plus 25 mM KCl, 15.0; buffer B plus 50 mM KCl, 28.0; dialyzed enzyme sample, 47. The enzyme sample was diluted with enough buffer B (no KCl) to give a final conductivity reading of 22.

   DEAE-Cellulose Chromatography. RNase III exhibits strange behavior on DEAE-cellulose[2] in that some enzyme flows through in low salt (i.e., buffer B alone) although sometimes lagging behind the front by several column volumes. The bulk of the enzyme is removed from the column by addition of buffer B plus 50 mM KCl. The practice of adding enzyme sample containing buffer B and 30–50 mM KCl to a column equilibrated in buffer B alone probably explains this behavior.

   We use Whatman microgranular preswollen DE-52 DEAE-cellulose (Cat. No. 2452751), which is prepared by acid and base washing as follows: 50 g of DE-52 is rinsed with 15 times its volume (~750 ml) of 0.5 N HCl, then water, then 0.5 N NaOH, then water again. The pH is adjusted to 8.5–9.0. The washed DE-52, 600 ml in water, is then allowed to settle in a graduated cylinder, collected in a 1-liter beaker, and washed twice with 500-ml rinses of 10× buffer B [0.2 M Tris-HCl (pH 7.9), 1.0 mM each of EDTA and DTT], after which the resin is washed twice with 500-ml rinses of 1× buffer B then chilled for column preparation. A column 2.4 × 18 cm is poured and rinsed (during which the column shrinks to 2.5 × 15 cm). The column is prewashed with buffer B (1 liter) in the cold room (4°) and the pH of input and output read (8.1 versus 8.05 in this case). The sample (~45 ml) is then loaded, and 15-ml fractions are collected. A 0.1-ml portion of the diluted sample is saved for later assays. The column is washed in buffer B and 20 fractions of 15 ml collected. Fractions 3–6 showed significant RNase III activity when assayed with ^{3}H-labeled poly(A-U)$_n$ as follows: A cocktail of 10× TM13 (0.17 ml), ^{3}H-labeled poly(A-U) (1200 cpm/μl, 0.17 ml), and 1.36 ml water is prepared, and in the assay 20 μl of this cocktail and 1–5 μl of the sample or column fraction to be assayed are mixed, incubated for 20 min at 37°, stopped with 5% TCA and 50 μl of 0.5% BSA, and filtered on Whatman GF/A filters, after which the radioactivity is determined as remaining cpm.

   At this point, the DE-52 column is rinsed with buffer B plus 50 mM KCl, and 27 fractions of 15 ml collected. Activity appears starting in fraction 24 and continues until the buffer is changed to buffer B plus 0.1 M KCl and 18 more 15-ml fractions collected. Fractions 25–60 are pooled, some stored in aliquots, and the rest subjected to chromatography on phosphocellulose.

   Phosphocellulose Chromatography. RNase III binds to phosphocellulose and comes off at about 0.3–0.35 M KCl. Whatman phosphocellulose P-11 (Cat. No. 211181, 20 g) is washed with 15 times its volume (thus 300 ml here) of 0.5 *N* NaOH, then water, then 0.5 *N* HCl, then water again, and the pH is adjusted to 7.0 with 1 *N* KOH. The slurry in water is stirred, and 100 ml is poured into a graduated cylinder and allowed to settle, after which the P11 phosphocellulose is collected in a 1-liter beaker. The resin is then washed with buffer B exactly as described above for the DEAE-cellulose and a column poured. Because of low flow rates, a column 2.5 × 3 cm was poured in this case and equilibrated in the cold room by running through 200 ml of buffer B plus 50 m*M* KCl. The flow rate was 2 ml/hr, the input pH was 8.15, and the output pH 8.1. Pooled fractions 25–60 from the DEAE-cellulose column above are loaded and the flow-through collected. Then the column is washed overnight with buffer B plus 50 m*M* KCl, then a linear gradient of buffer B plus 50 m*M* KCl–buffer B plus 0.5 *M* KCl is applied (total volume 100 ml) and 47 fractions of 2 ml collected. Fractions are assayed for RNase III, and the peak of activity occurs between fractions 20 and 35 of the KCl gradient. The active fractions are pooled and stored in liquid nitrogen or prepared for affinity chromatography.

Chromatography on Poly(I)-poly(C)-Agarose.

Dunn[10] worked out an affinity step for RNase III purification based on the ability of the enzyme to bind, but not cleave, double-stranded RNA in the absence of magnesium ions. Agarose-hexane-poly(I)-poly(C) is from P-L/Pharmacia. In the particular preparation considered here, 2 ml of column material (Cat. No. 355381) was packed into a 5-ml disposable pipet and equilibrated overnight with a buffer containing 20 m*M* Tris–acetate (pH 7.5), 0.1 m*M* DTT, 1.0 m*M* EDTA, 0.5 *M* NH₄Cl, and 5% glycerol (IC buffer with 0.5 *M* NH₄Cl) at 4°. Then the column was run dry and loaded with 34.5 ml of phosphocellulose-purified enzyme diluted. 1 : 1 (total 69 ml) with a buffer designed to remove magnesium ions from the enzyme on the column[10]: 5% glycerol, 20 m*M* Tris–acetate (pH 7.5), 1 *M* NH₄Cl, 0.1 m*M* DTT, and 11 m*M* EDTA. The column was then run stepwise as follows: flow-through, 4 20-ml fractions; wash with IC buffer plus 0.5 *M* NH₄Cl, 13 3-ml fractions; wash with IC buffer plus 1.0 *M* NH₄Cl, 12 1-ml fractions; and elution with IC buffer plus 2.0 *M* NH₄Cl, 12 1-ml fractions. The majority of activity binds to the column, with the principal peak eluting with the buffer containing 2.0 *M* NH₄Cl.

Active enzyme fractions can be stored frozen in small aliquots in liquid nitrogen or at −70°, where they are stable for more than 5 years and more than 2 years, respectively. After thawing, activity is best preserved by storing the tube unfrozen or on ice in a cold room.

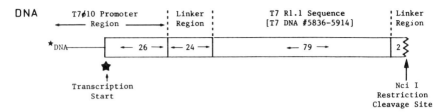

FIG. 1. DNA map of a region permitting *in vitro* transcription of an RNase III cleavage site from bacteriophage T7. The cloned DNA segment diagrammed here was generously provided by Dr. J. J. Dunn (Brookhaven National Laboratories). It contains the sequences for the RNase III cleavage site separating bacteriophage T7 genes 1.0 and 1.1 as part of a 79-base segment (residues 5836–5914 in the T7 DNA sequence[1]). These bases are joined by commercially available DNA linkers to the T7 RNA polymerase promoter T7ϕ10. In preparation for *in vitro* synthesis of a 131-base RNA containing the RNase III cleavage site, the cloned DNA is restricted at the position indicated by incubation with the enzyme *Nci*I.

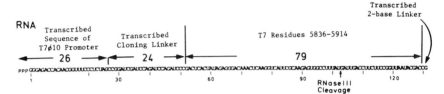

FIG. 2. Model RNase III substrate synthesized *in vitro* from cloned T7 DNA. A DNA clone containing an RNase III cleavage site from bacteriophage T7 (diagrammed in Fig. 1) was cleaved with *Nci*I and transcribed *in vitro* by bacteriophage T7 RNA polymerase (kindly supplied by Dr. J. J. Dunn) in the presence of [α-32P]GTP. RNA was purified by phenol extraction, chromatography on cellulose CF11, and fractionation in 10% polyacrylamide gels containing 7 *M* urea. The anticipated primary structure (based on the sequence of the DNA template) is diagrammed. The runoff RNA transcript contains 26 bases from the phage T7ϕ10 promoter region, 24 bases of linker sequence, 79 bases containing the T7 R1.1 RNase III cleavage site, and two additional bases of linker sequence at the 3′ end.

Studies on a Cloned RNase III Processing Site Using Cloned Enzyme

Mapping and Synthesis of T7 R1.1 RNA. When Dunn and Studier completed their sequence analysis of bacteriophage T7[12] they were able to propose primary and secondary structures for each of the encoded RNase III processing sites in the early mRNA region of the phage genome. Figure 1 shows the DNA map of a cloned segment of T7 DNA (residues 5836–5914) that maps between genes 1.0 and 1.1 and contains the RNase III clevage site which separates the mRNAs for these two genes. This cloned DNA was placed adjacent to a phage T7 RNA polymerase promoter and a linker region, as shown in Fig. 1, leading to a template for a

[12] J. J. Dunn and F. W. Studier, *J. Mol. Biol.* **166**, 2/7 (1983).

131-base RNA molecule, whose sequence is diagrammed in Fig. 2. As can be seen, RNase III cleavage occurs at a unique site, at least in this region of the intact T7 early mRNA precursor, which is near the center of the 79-base T7-specific portion of the 131-base RNA molecule. It is widely believed that this RNA folds into a characteristic structure, the Watson–Crick base pairs of which are shown in Fig. 3. Here the 79 bases of T7-specific sequence are folded into a hairpin loop–stem structure, with the stem interrupted in the middle by nine unpaired bases. Such a configuration is characteristic of four of the five RNase III cleavage sites in T7 early mRNA precursor,[1] and these four sites have extensive sequence homology as well. It is noteworthy that RNase III cleaves in a region of apparently unpaired bases. The real structure in this region of such RNase III cleavage sites is one of the key questions which we are planning to study in the experiments proposed here.

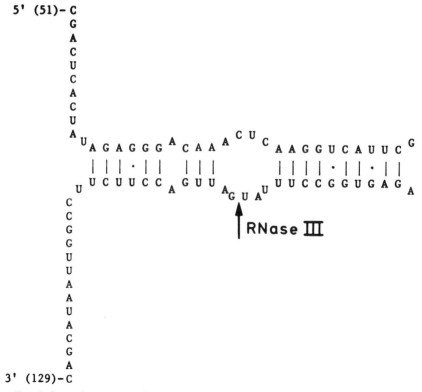

FIG. 3. Secondary structural map depicting the phage T7 R1.1 RNase III cleavage site. The map shows the potential for base pairing in the 79 bases surrounding the RNase III cleavage site (arrow) that separates the bacteriophage T7 genes 1.0 and 1.1.

TABLE I
RELATIVE YIELDS OF RNase III FROM *E. coli* BL-15 AND *E. coli* BL-107 TRANSFORMED
BY RNase III-ENCODING PLASMID pNA9[a]

Enzyme fraction	units/μl	μg protein/μl	units/μg protein
BL-15 DE-52 peak	0.102	0.29	0.351
pNA9/BL-107 DE-52 peak	1.40	0.198	7.07
BL-15 P11 peak	0.306	0.02	15.3
pNA9/BL-107 P11 peak	1.49	0.134	11.12

[a] In order to prepare large quantities of RNase III for processing studies, the enzyme was purified from *E. coli* cells transformed by the plasmid pNA9, kindly supplied by Dr. D. Apirion, who demonstrated that it contains the *rnc* gene.[11] Sixty-gram aliquots of various strains of *E. coli* were analyzed in parallel to determine the levels of RNase III, using procedures described previously for the purification and quantitation of RNase III activity.[3,4] Identical volumes were collected from DE-52 DEAE-cellulose ion-exchange columns and from P11 phosphocellulose columns. Units of enzyme were determined in 70-μl reactions, under standard conditions,[10] containing 10,500 cpm of poly[r(A-U)], specific activity 3.5×10^5 cpm/nmol. Kinetic time points of 10 μl were collected at 10-min intervals, and residual radioactivity precipitable in ambient 5% TCA on Whatman GF/A glass fiber filters determined. The yields of the two preparations, at both the De-52 and P11 stages, are best compared by analyzing the units/μl column: the cloned enzyme is in 14-fold excess at the first stage and about 5-fold excess at the second. Protein determinations were by the Bradford procedure.[13]

Purification and Properties of Cloned RNase III. The plasmid pNA9 from the laboratory of D. Apirion (Washington University, St. Louis, MO) was used to transform the RNase III of *E. coli* strain BL-107. The resulting bacterial strain was grown and subjected to the normal steps of RNase III purification in parallel with *E. coli* BL-15, its wild-type RNase III[+] counterpart. Outlined in Table I,[11,12] the usual steps of the RNase III purification procedure were carried out on 60 g of both strains[3–5] and their yields compared. Using standard units relating to acid solubilization of double-stranded RNA, we found that the amount of cloned enzyme exceeded the wild-type control by 5- to 14-fold. It was also important to compare the purified RNase III protein obtained from cloned enzyme with that previously obtained by Dunn.[10] In Fig. 4,[14] a number of fractions produced during the purification of RNase III from pNA9-transformed *E. coli* BL-107 are analyzed for their protein content. It is clear that about 50% of the phosphocellulose peak of enzyme activity, and greater than 95% of the poly(I).poly(C)-agarose affinity column peak, consists of a

---

[13] M. M. Bradford, *Anal. Biochem.* **72**, 248 (1976).
[14] U. K. Laemmli, *Nature (London)* **227**, 680 (1970).

polypeptide of molecular weight 25,000, in agreement with Dunn's earlier findings.[10] The amount of enzyme which could be purified using this technique, and its great stability, can be attributed to the increased amounts of RNase III protein present. Thus, it is possible to purify thousands of units at a time over the poly(I).poly(C)-agarose affinity column without loss of stability.

Cleavage of R1.1 RNA with Cloned and Wild-Type RNase III. Two critical questions are addressed in this section. First, will the truncated substrate molecule undergo the same cleavage as the 7000-nucleotide intact T7 early mRNA precursor? And, second, will cloned enzyme exhibit the same specificity and activity in processing that we would expect from its behavior during purification and when assayed using acid solubilization of double-stranded RNA? The answers to both questions are yes as shown in Figs. 5 and 6. In Fig. 5, we see polyacrylamide gel electrophoresis analysis of R1.1 RNA with and without incubation in RNase III reaction buffer, and with either cloned or wild-type RNase III. While buffer has no effect, both RNase III preparations cleave the substrate quantitatively to yield a prominent band about 100 bases long (labeled 2 in Fig. 5) and further products in the 30-base size class (labeled 3). The dimer

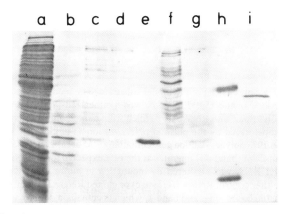

a   b   c   d   e   f   g   h   i

FIG. 4. SDS–polyacrylamide gel analysis of column fractions showing the purification of RNase III from *E. coli* BL-107 transformed by pNA9. RNase III preparations from various stages of the purification process were fractionated in Laemmli 17.5% polyacrylamide gels containing 0.1% sodium dodecyl sulfate (SDS)[14] along with protein markers of known size. Lane a, ammonium sulfate fraction of RNase III; lane b, flow-through from a DE-52 column; lane c, peak from a DE-52 column; lane d, flow-through from a poly[r(I-C)]-agarose affinity column; lane e, peak from a poly[r(I-C)]-agarose affinity column; lane f, flow-through from a P11 phosphocellulose column; lane g, peak from a P11 phosphocellulose column; lane h, protein markers of 45,000 and 16,000; lane i, protein marker of 35,000. The prominent band in lane e represents the RNase III protein subunit purified as described previously by Dunn.[10]

Conclusion

With the availability of increased amounts of RNase III enzyme through genetic engineering techniques, it should be possible to use the enzyme with increased efficiency in studies of its own *in vivo* processing reactions as well as in studies in which RNase III is used as a reagent to detect double-stranded RNA and to introduce specific cleavages into RNA molecules. As more RNA-level studies are carried out on mRNA and on RNAs that contain enzymatic activities within their structure, it will be more important than ever to have a variety of specific cleavage agents available for their study. Furthermore, the characteristic structures formed by RNase III substrates which specify cleavage within long regions of single-stranded RNA may themselves provide clues to the function of RNA secondary and tertiary structure in a variety of reactions.[7] We have recently begun to use ultraviolet light-induced RNA–RNA cross-linking[16] in the study of cleavage of R1.1 RNA by RNase III and have found that a specific photo-cross-link involving bases near the cleavage site depicted in Fig. 3 can strongly inhibit the reaction. Further studies of this sort should provide evidence for the specific local tertiary structural features involved in RNA processing by RNase III.

Acknowledgments

Thanks are due to a number of colleagues who over the years have supported my continuing interest in RNase III, most especially Drs. A. D. Branch, J. J. Dunn, J. E. Dahlberg, J. A. Steitz, and N. D. Zinder. Financial support for continued studies on RNA processing has come principally from several grants from the National Institute of General Medical Sciences of the National Institutes of Health, with support for the substrate structural studies coming also from the National Science Foundation and the McKnight Foundation.

[16] A. D. Branch, B. J. Benenfeld, and H. D. Robertson, *Proc. Natl. Acad. Sci. U.S.A.* **82**, 6590 (1985).

[16] Messenger RNA Turnover in Cell-Free Extracts

By GARY BREWER and JEFFREY ROSS

The steady-state level of a messenger RNA (mRNA) depends not only on the rate at which it is synthesized, processed, and transported, but also on the rate at which it decays. The half-life of a mRNA, in turn, determines the length of time that molecule can function to synthesize protein. Therefore, mRNA turnover is linked directly to "gene expression," and it is important to know why some mRNAs are more stable than others.

There are several advantages to using an *in vitro* system for studying mRNA turnover.[1-4] First, mRNA turnover is difficult to study in whole cells, in large part because of the difficulties inherent in inhibiting transcription specifically or in "chasing" radioactive nucleotides. With a cell-free system containing only cytoplasmic components, cytoplasmic events can be studied independently of nuclear events. Second, a cell-free system allows biochemical dissection of the mRNA turnover machinery, including mRNA ribonucleases (mRNases) as well as auxiliary factors that modulate mRNA decay rates. Third, site-directed mutations that affect mRNA turnover rates can be mapped rapidly, using radioactive mRNA substrates transcribed *in vitro* with bacteriophage RNA polymerases.

The polysome is the most important component of the cell-free mRNA decay system we have described. It contains both the mRNA to be degraded and the enzymes and cofactors that degrade the mRNA. The major criterion used to assess the reliability and fidelity of the system is the assay for relative decay rates of different mRNAs. Each mRNA should have a characteristic half-life or lifetime under a specific set of conditions. Therefore, in a system in which mRNases and their cofactors are functioning properly, different mRNAs should be degraded at different rates. If the system is not functioning properly, all RNAs should be degraded at similar rates. The *in vitro* mRNA decay system is thus judged by assaying the relative turnover rates of mRNAs that are known to decay at different rates in whole cells. In the system described below, the mRNAs we have assayed are degraded *in vitro* in the same rank order as in whole cells.[1]

Preparation of Cell-Free Extracts

Source of Extracts. We have used cell lines of human and mouse origin but have no reason to doubt that primary animal tissues could also be used, provided that nonspecific RNase activity does not interfere with polysome recovery. We typically use K562 human cells, a suspension line of hemopoietic origin derived from a patient with chronic myeloid leuke-

[1] J. Ross and G. Kobs, *J. Mol. Biol.* **188,** 579 (1986).
[2] J. Ross, S. W. Peltz, G. Kobs, and G. Brewer, *Mol. Cell. Biol.* **6,** 4362 (1986).
[3] S. W. Peltz and J. Ross, *Mol. Cell. Biol.* **7,** 4345 (1987).
[4] G. Brewer and J. Ross, *Mol. Cell. Biol.* **8,** 1697 (1988).

mia.[5] We have also had success with several mouse fibroblast cell lines, Ltk⁻ and NIH/3T3, and a mouse lymphoma line, S49.

Solutions for Cell Extracts. Solutions are made with ultrapure reagents using double-distilled water. Buffers containing tris(hydroxymethyl)aminomethane (Tris) are adjusted to the indicated pH at room temperature with HCl. Dithiothreitol (DTT) is made as a 1M stock which is stored in small aliquots at −20° and diluted into the buffers before use.

Buffer A: 10 mM Tris-HCl (pH 7.6), 1 mM potassium acetate, 1.5 mM magnesium acetate, 2 mM DTT

Buffer B: 30% (w/v) sucrose, 10 mM Tris-HCl (pH 7.6), 1 mM potassium acetate, 1.5 mM magnesium acetate, 2 mM DTT

Extract Preparation. Cells are harvested in mid-log phase. All subsequent manipulations are performed at 4°. Cells are washed 3 times in cold Ham's F12 medium without serum and are resuspended in buffer A by gentle pipetting. (We use 3.5 ml of buffer A per 2 × 10⁸ cells.) Cells are lysed by 20 strokes (5 sec down, 5 sec up) with a tight-fitting, Teflon pestle homogenizer (clearance ~0.1 mm). The crude lysate is centrifuged at 12,000 g for 10 min to remove nuclei. The supernatant is removed in such a way as to avoid lipid at the top of the tube and nuclei at the bottom. The supernatant is gently layered over a 1.5-ml cushion of buffer B in an SW60 Ultraclear tube (Beckman) and is centrifuged in the SW60 rotor at 36,000 rpm (130,000 g) for 2.5 hr, 4°. The high-speed supernatanat (S130) is removed without disturbing the sucrose–S130 interface and is stored at −70° in small aliquots (see below). The sucrose pad is removed by aspiration and discarded.

The centrifuge tubes are inverted for several minutes (in a cold room), to allow buffer to drain away from the polysomal pellet. (At this point, the upper two-thirds or so of the tubes may be cut off with a clean razor blade and discarded, to allow easier access to the pellet.) The polysomal pellet is rinsed twice with 0.5-ml portions of buffer A, which are then discarded. An additional 0.5 ml of buffer A is added, and the pellet is resuspended by repeated up and down pipetting with a micropipet. We cut off 2–3 mm from the end of the pipet tip with a clean razor blade, to permit easier pipetting of the polysome pellet. Undissolved chunks of poorly dissolved pellet are broken with several gentle strokes in a 1-ml Dounce glass homogenizer (A-type pestle; clearance 0.06 mm). For K562 cells we typically recover 2.5 A_{260} units of polysomes per 10⁷ cells at a concentration of 150–200 A_{260} units of polysomes/ml (determined in buffer A). Polysomes are stored in small aliquots at −70°. We have observed no detectable loss of activity with up to 3 thawing and freezing cycles.

⁵ C. B. Lozzio and B. B. Lozzio, *Blood* **45**, 321 (1975).

Optional Extract Preparation: Salt Extraction of Polysomes

High-salt extraction of polysomes elutes loosely bound material.[6] The high-salt eluate or ribosomal salt wash (RSW) is free of polysomes and mRNA but includes translation initiation factors and a $3' \rightarrow 5'$-exonuclease that degrades histone mRNA. mRNA turnover in RSW-containing reactions can be assayed using radioactive *in vitro*-sythesized mRNA substrates.

To isolate RSW, polysomes from 6×10^8 cells (prepared as described above) are resuspended in 1.3 ml of buffer A at 4°. Subsequent steps are also performed at 4°. The polysomes are transferred to a small tube or beaker containing a stirring bar. Sufficient 4 M KCl is added dropwise, and with stirring, to bring the final concentration to 0.3 M KCl. After stirring for an additional 15 min, polysomes are pelleted through a cushion of buffer B in the SW60 rotor, as described above. Smaller or larger rotors and tubes can be used, depending on the amount of material. The polysomal pellet is saved. The supernatant above the cushion is the RSW, which is removed and stored in small aliquots at $-70°$. The RSW typically contains approximately 3.3 mg of protein per 6×10^8 cells, and approximately 80% of the polysome-associated $3' \rightarrow 5'$-exonuclease activity originally present in the polysomes is recovered. We have also prepared RSW by bringing the KCl concentration to 0.5 M. This procedure results in the recovery of 85–95% of the exonuclease activity, but additional proteins are also eluted.

Cell-Free mRNA Decay Reactions

Our cell-free mRNA decay system is based on the protein synthesis conditions originally described by Atkins *et al.*[7] However, the incorporation of radiolabeled amino acids is low or nil as a result of our modifications of the system.[8] The reaction mixture contains mono- and divalent cations, ATP and GTP, an energy-regenerating system, spermine, a nonspecific ribonuclease inhibitor, and polysomes or RSW. The polysomes serve as a source of both mRNAs and the mRNA decay machinery. Radioactively labeled, capped, *in vitro*-synthesized mRNA substrates can also be assayed, using either polysomes or RSW. Since the RSW contains no mRNAs, exogenous substrates must be used with RSW-containing reactions.

[6] J. Ross, G. Kobs, G. Brewer, and S. W. Peltz, *J. Biol. Chem.* **262,** 9374 (1987).
[7] J. F. Atkins, J. B. Lewis, C. W. Anderson, and R. F. Gesteland, *J. Biol. Chem.* **250,** 5688 (1975).
[8] S. W. Peltz and J. Ross, unpublished observations.

Solutions. All solutions are made with enzyme-grade reagents using double-distilled water. Glassware is treated with undiluted diethyl pyro-carbonate (DEPC) for 5 min, rinsed several times with double-distilled water, and baked in an oven at 200° for at least 1 hr. A stock solution of 2 *M* potassium acetate is adjusted to pH 7.6 at room temperature with acetic acid. A stock solution of 2 *M* Tris-HCl is adjusted to pH 7.5 at room temperature with HCl. A stock solution of 1 *M* magnesium acetate is not pH-adjusted. The three stock solutions are filtered through 0.22-μm Millipore filters. DTT is stored in small aliquots at $-20°$ as a 1 *M* stock solution. A stock solution of 2.5 m*M* spermine is adjusted to pH 7.6 at room temperature with KOH and stored in small aliquots at $-20°$. A stock solution of 1 *M* creatine phosphate (dipotassium salt) is stored in aliquots at $-20°$. A stock solution of 50 m*M* ATP, 20 m*M* GTP is made from the Tris salt of each triphosphate and is stored in small aliquots at $-20°$. Each ATP–GTP aliquot is thawed only once. A stock of creatine phosphokinase is made in 50% glycerol at 675 units/ml (4 mg/ml protein) and stored in small aliquots at $-20°$. Urea lysis buffer [ULB; 7 *M* urea, 2% sodium dodecyl sulfate (SDS), 0.35 *M* NaCl, 10 m*M* ethylenediaminetetraacetic acid (EDTA), 10 m*M* Tris-HCl (pH 7.5)] is used to stop the reactions and is also stored at $-20°$.

Decay Reactions. Decay reactions are carried out in 1.5-ml microcentrifuge tubes, with one tube per time point. All manipulations prior to the 37° incubation are performed on ice. The components described above are added at the following final concentrations in a reaction volume of 25 μl: 100 m*M* potassium acetate, 2 m*M* magnesium acetate, 2 m*M* DTT, 10 m*M* creatine phosphate, 1 μg of creatine phosphokinase, 1 m*M* ATP, 0.4 m*M* GTP, 0.1 m*M* spermine, 10 m*M* Tris-HCl (pH 7.6), 80 units/ml of placental ribonuclease inhibitor, and 0.7 A_{260} units of K562 cell polysomes. Polysomes are added last. Reactions are incubated at 37° for various times, after which 0.4 ml of ULB is added. The tube is then vortexed briefly, and the reactions are extracted twice with phenol–chloroform and twice with chloroform alone, using standard procedures. RNA is precipitated with ethanol. The RNA pellet is dried and resuspended in a small volume of double-distilled water or standard Tris–EDTA buffer. The ribosomal RNA from the reactions should be intact, since ribosomal RNA is stable in the growing cell. RNA integrity is assayed by electrophoresing a small aliquot in a 1% agarose minigel (nondenaturing) containing 0.5 μg/ml ethidium bromide. The 28 and 18 S ribosomal RNAs should appear as discrete bands in the gel.

There are many possible variations of the above protocol. Radioactively labeled, *in vitro*-synthesized, capped mRNA substrates may be added to the reaction mixtures just prior to adding the polysomes. Incubations and RNA extractions are performed as described above. Alternatively, polysomes may be replaced with RSW. We use 18 μg of RSW protein per 25-μl reaction. Exogenous mRNA substrates must be used in these RSW reactions. We have not tested all possible combinations of the various ions and cofactors, although several general properties of the system have been described.[6,9]

We have also found that some regulatory aspects of mRNA turnover can be reconstituted *in vitro* by adding the S130 fraction. For example, H4 histone mRNA turnover is accelerated when both histone proteins and S130 are included in the reaction mixture.[3] c-*myc* mRNA turnover is accelerated in reactions supplemented with S130 from growing cells but not with S130 from cycloheximide-treated cells.[10] These S130-mediated changes in mRNA turnover rates correlate with observations in intact cells under different growth conditions.[11–14] Therefore, the S130 seems to contain important regulating factors that can be assayed *in vitro*. We typically add equal cell equivalents of polysomes and S130 to standard reaction mixtures, e.g., 2×10^6 cells of both. Incubations and RNA extractions are performed as described above.

Assay for mRNA Turnover. mRNA degradation is assayed by a variety of methods. If radioactive substrates are used, the mRNAs are electrophoresed in standard denaturing polyacrylamide gels and are visualized by autoradiography. To assay polysome-bound, endogenous mRNAs, equal amounts of RNA from each time point are analyzed by either RNA blot hybridization or nuclease protection assays, using standard techniques.

Although RNA blot hybridization and nuclease protection assays are not as easy or as fast as direct analysis of radioactive substrates by electrophoresis, it may not be possible to synthesize radiolabeled substrates, if appropriate cDNA clones are unavailable. To increase the ease and speed of nuclease protection assays for endogenous mRNAs, our laboratory has developed a rapid hybridization protocol using ^{32}P-labeled RNA probes.

We make stock solutions of 3.6 M NaCl, 30 mM EDTA (pH 7.4) (buffer C), which is stored at 4°, and 5 mM EDTA (pH 7.4), 30 μg/ml

[9] S. W. Peltz, G. Brewer, G. Kobs, and J. Ross, *J. Biol. Chem.* **262**, 9382 (1987).
[10] G. Brewer and J. Ross, *Mol. Cell. Biol.* **9**, 1996 (1989).
[11] J.-M. Blanchard, M. Piechaczyk, C. Dani, J.-C. Chambard, A. Franchi, J. Pouyssegur, and P. Jeanteur, *Nature (London)* **317**, 443 (1985).
[12] M. Dean, R. A. Levine, and J. Campisi, *Mol. Cell. Biol.* **6**, 518 (1986).
[13] M. S. Kindy and G. E. Sonenshein, *J. Biol. Chem.* **261**, 12865 (1986).
[14] C. B. Thompson, P. B. Challoner, P. E. Nieman, and M. Groudine, *Nature (London)* **319**, 374 (1986).

RNase P1 (Calbiochem, San Diego, CA; 500 units/mg), 10 μg/ml RNase T1 (Sigma, St. Louis, MO; 300 units/μg) (buffer D), which is stored in 1-ml aliquots at −20°. EDTA is adjusted to pH 7.4 at room temperature with NaOH. Cell-free decay reactions are scaled down to a reaction volume of 10 μl per time point, but the scale-down is not required. After incubation at 37° for the appropriate time, 5 μl of buffer C is added, and the tube is vortexed and stored on dry ice until all reactions are complete.

Tubes are thawed, and ³²P-labeled RNA probe is added to a final volume of 20 μl. (Because of the EDTA and the NaCl in buffer C, ribonucleases associated with the polysomes or in the RSW are inactive, and the RNA and probe are not degraded during subsequent steps.) Tubes are incubated under appropriate conditions for RNA–RNA hybridization. Conditions will vary depending on the probe. For example, using a ³²P-labeled RNA probe which protects the 3′-terminal 200 nucleotides (nt) of human c-*myc* mRNA (A + U content 80%), 70° for 60 min is optimum. At the end of the incubation period, 105 μl of buffer D is added, the tube is mixed gently, and the reactions are further incubated at 20° for 60 min. RNase digestion is stopped by adding 180 μl of ULB (described above), followed by one extraction with phenol–chloroform and one with chloroform. RNA is precipitated with ethanol. The protected fragments are electrophoresed in denaturing polyacrylamide minigels and visualized by autoradiography. Using this method, it is possible to assay a decay reaction in 5–6 hr. We have found that the phenol step is necessary; when it is omitted, the RNA smears in the gel.

Assay for Poly(A) Shortening. As a polyadenylated mRNA ages, its poly(A) tract is progressively shortened. An early degradative step in c-*myc* mRNA turnover is shortening and removal of the poly(A) tract, prior to nucleolytic cleavage in the 3′ untranslated region.[4] For large mRNAs, such as c-*myc* (~2000 nt), poly(A) shortening results in a relatively small size change, which is difficult to detect using standard RNA blotting techniques. To circumvent this problem, we have developed a modified blotting technique we call H-mapping. A deoxyoligonucleotide complementary to a unique region is annealed to the mRNA. The annealing site is chosen to be within 400 nt from the poly(A) addition site. The hybrids are treated with RNase H, to cleave the mRNA at the RNA–DNA duplex, and the RNA fragments are separated by electrophoresis, blotted, and hybridized to a ³²P-labeled probe specific for the 3′ end fragment.

The RNase H step is a modification of a published method for poly(A) removal.[15] We make a stock solution of 4 *M* KCl and a stock solution of buffer TM [40 m*M* Tris-HCl (pH 7.5), 60 m*M* MgCl₂]. RNAs from decay

¹⁵ F. B. Mercer and S. A. Wake, *Nucleic Acids Res.* **13,** 7929 (1985).

reactions are precipitated with ethanol, resuspended in 20 μl of 1 mM EDTA (pH 7.4), and heated at 70° for 10 min, to denature the RNA. Deoxyoligonucleotide (0.5 μg; 14–20 nt in length) is added, and the reaction mixtures are incubated at 20° for 15 min. One microliter of 4 M KCl is added, and the mixtures are incubated at 20° for 15 min, followed by addition of 20 μl of buffer TM. RNase H (Bethesda Research Laboratories, Gaithersburg, MD; 2000 units/ml) is then added to a final concentration of 20 units/ml. Digestion is performed at 37° for 30 min. Reactions are extracted once with phenol–chloroform and once with chloroform. RNA is precipitated with ethanol and assayed either by blot hybridization or by electrophoresis in denaturing polyacrylamide gels (for ^{32}P-labeled mRNAs).

In principle, a deoxyoligonucleotide complementary to any relevant portion of the mRNA could be used. However, it might be advantageous to try several that are complementary to different regions of the mRNA, in order to ensure specificity.

[17] Nuclear Cap Binding Protein from HeLa Cells

By MUTSUHITO OHNO and YOSHIRO SHIMURA

Introduction

It has been shown that a cap structure, m[7]G(5')ppp(5')N, blocking the 5' termini of many eukaryotic mRNAs, plays important roles in cellular functions. The cap structure has been shown to enhance translational efficiency by facilitating ribosome binding to mRNA. A 25K cytoplasmic cap binding protein (CBP-I) binds to the cap structure during an early step in the initiation of protein synthesis and stimulates translation of capped but not uncapped mRNAs.[1] CBP-I was purified to near homogeneity by various chromatographic methods.[2-5] Its cDNA has also been isolated and sequenced.[6]

[1] A. J. Shatkin, *Cell* **40**, 223 (1985).

[2] N. Sonenberg, K. M. Rupprecht, S. M. Hecht, and A. J. Shatkin, *Proc. Natl. Acad. Sci. U.S.A.* **76**, 4345 (1979).

[3] N. R. Webb, R. V. J. Chari, G. DePillis, J. W. Kozarich, and R. E. Rhoads, *Biochemistry* **23**, 177 (1984).

[4] H. Trachsel, N. Sonenberg, A. J. Shatkin, J. K. Rose, K. Leong, J. E. Bergmann, J. Gordon, and D. Baltimore, *Proc. Natl. Acad. Sci. U.S.A.* **77**, 770 (1980).

[5] G. M. Hellmann, L.-E. Chu, and R. E. Rhoads, *J. Biol. Chem.* **257**, 4056 (1982).

[6] W. Rychlik, L. L. Domier, P. R. Gardner, G. M. Hellmann, and R. E. Rhoads, *Proc. Natl. Acad. Sci. U.S.A.* **84**, 945 (1987).

There has been cumulative evidence indicating that the cap structure also plays an important role in mRNA splicing[7-10] and 3'-terminal processing of mRNA.[11,12] It is likely that a cap binding protein(s) is involved in these nuclear events as is the case with translation. However, very little is known about nuclear cap binding proteins. Patzelt *et al.* previously identified in HeLa cells, by photoaffinity labeling, three nuclear cap binding proteins having molecular weights of 120K, 89K, and 80K, respectively, which are associated with the nuclear matrix.[13] More recently, Rozen and Sonenberg identified 115K and 20K cap binding proteins in a HeLa nuclear extract by UV cross-linking.[14] However, the biochemical properties and functions of the cap binding proteins remains to be elucidated.

We developed a novel gel mobility shift assay to detect cap binding activity. Using this method, we have purified an 80K cap binding protein to homogeneity from a HeLa nuclear extract.[15] Purification of this protein may provide a vital clue to the functions of the cap structure in nuclei. Described herein are the purification procedures and some properties of the 80K nuclear cap binding protein.

Assay for Cap Binding Activity

Rationale of Assay

The rationale behind the assay we developed is as follows. If a cap binding protein binds to the cap structure of an RNA probe, the electrophoretic mobility of the bound RNA in a native gel becomes slower than that of the unbound RNA, and the mobility shift of the RNA owing to the cap binding protein is sensitive to the presence of cap analogs such as m^7GpppG or m^7GTP in the binding mixture. Although binding of other proteins, specific or nonspecific, to the RNA may also shift the electrophoretic mobility, the shift of the RNA band would not be sensitive to cap analogs and should occur even with an uncapped counterpart used as a probe.

---

[7] M. M. Konarska, R. A. Padgett, and P. A. Sharp, *Cell* **38**, 731 (1984).
[8] I. Edery and N. Sonenberg, *Proc. Natl. Acad. Sci. U.S.A.* **82**, 7590 (1985).
[9] M. Ohno, H. Sakamoto, and Y. Shimura, *Proc. Natl. Acad. Sci. U.S.A.* **84**, 5187 (1987).
[10] H. Sakamoto, M. Ohno, and Y. Shimura, *J. Biochem.* **102**, 1289 (1987).
[11] O. Georgiev, J. Mous, and M. Birnstiel, *Nucleic Acids Res.* **12**, 8539 (1984).
[12] R. P. Hart, M. A. McDevitt, and J. R. Nevins, *Cell* **43**, 677 (1985).
[13] E. Patzelt, D. Blaas, and E. Kuechler, *Nucleic Acids Res.* **11**, 5821 (1983).
[14] F. Rozen and N. Sonenberg, *Nucleic Acids Res.* **15**, 6489 (1987).
[15] M. Ohno and Y. Shimura, manuscript in preparation.

Preparation of Probes

A capped RNA probe (57 nucleotides long) is synthesized by m⁷GpppG-primed transcription of the plasmid pSP64[16] linearized with *Eco*RI (Takara Shuzo Co., Japan) using SP6 RNA polymerase (Takara Shuzo Co.) in the presence of [α-^{32}P]GTP (410 Ci/mmol, Amersham) as described previously.[7] An uncapped probe is synthesized similarly without the dinucleotide primer. The specific activity of the probes should be 2–10×10^6 counts per minute (cpm)/μg. The RNA probes are purified on a 5% polyacrylamide gel containing 8 M urea, dissolved in water, and stored at $-20°$.

Assay Procedure

The RNA probe (2–5×10^4 cpm) is mixed with a subnuclear fraction in 10 μl of 10 mM N-2-hydroxyethylpiperazine-N$'$-2-ethanesulfonic acid (HEPES) (pH 7.9), 50–500 mM KCl, 50–1000 μg/ml yeast RNA, 5% glycerol, and incubated at 20° for 30 min. Then the reaction mixture is directly loaded on a 6% polyacrylamide (1:30 bisacrylamide) gel in one-quarter strength TAE buffer [10 mM tris(hydroxymethyl)aminomethane (Tris)–acetate, 0.5 mM ethylenediaminetetraacetic acid (EDTA) (pH 8.3)] which has been prerun at 10 V/cm for 1 hr. A dye marker (0.05% bromphenol blue, 0.05% xylene cyanol in the same buffer as the reaction mixture) is applied to another lane of the same gel to determine the electrophoretic mobility. The gel is run at 10 V/cm with buffer circulation until the bromphenol blue migrates about 8 cm, which usually takes about 1.5 hr. The gel is then dried and autoradiographed with an intensifying screen.

Comments

(1) The concentrations of protein and yeast RNA in the reaction mixture should be properly chosen. When a HeLa nuclear extract described below is used in the assay, 1–2 mg/ml protein and 1000 μg/ml yeast RNA are recommended. (2) The assay should be performed with both capped and uncapped probes, because several proteins in a HeLa nuclear extract bind to the RNA moieties of the probes irrespective of the presence of the cap structure and specifically shift the gel mobilities of the probes. Some of these bands disappear with higher concentrations of yeast RNA in the reaction mixture. (3) With a HeLa nuclear extract, only a single cap-

[16] D. A. Melton, P. A. Krieg, M. R. Rabagliati, T. Maniatis, K. Zinn, and M. R. Green, *Nucleic Acids Res.* **12**, 7035 (1984).

specific band is observed. As described later, this band corresponds to the 80K nuclear cap binding protein (see Fig. 1).

Purification Procedures

The 80K cap binding protein can be purified to near homogeneity from a HeLa nuclear extract by the following chromatographic procedures.

Preparation of Capped RNA-Sepharose

The capped RNA, which is the same as the one used in the gel mobility shift assay, is prepared as described above on a larger scale and covalently attached to cyanogen bromide-activated Sepharose CL-2B in the same manner as the preparation of sequence-specific DNA affinity resin described previously.[17] In short, Sepharose CL-2B (Pharmacia, 5 ml settled volume) is activated by cyanogen bromide, and 300 μg of the capped RNA is coupled to the resin. The resin is then treated with 1 M ethanol-amine-HCl (pH 8.0) to inactivate the unreacted Sepharose, washed extensively, and stored as described previously.[17]

Preparation of HeLa Nuclear Extract

HeLa cells are grown in spinner flasks at 37° in Eagle's minimal essential medium (MEM) No. 4 (Nissui Co., Japan) supplemented with 0.02% glutamine and 5% horse serum. The cells are grown to 4–6 × 10^5 cells/ml and then harvested. A nuclear extract is prepared from 2 × 10^{10} HeLa cells as described by Dignam et al.[18] with the following modifications. When crude nuclei are suspended in buffer C, the buffer containing 0.6 M KCl instead of 0.42 M NaCl is employed, so that the KCl concentration of the mixture is approximately 0.25 M, as determined by the conductivity of the mixture. The amount of buffer C is usually about 1.5 ml/10^9 cells.

DEAE-Cellulose Chromatography

Unless otherwise stated, all operations are performed at 4°. A nuclear extract prepared from 2 × 10^{10} HeLa cells (20 ml, 23.4 mg/ml protein) is loaded onto a DEAE-cellulose column (Whatman, 2.5 × 18 cm) previously equilibrated with buffer A [10% glycerol, 20 mM HEPES (pH 7.9), 0.1 mM EDTA, 1 mM dithiothreitol (DTT)] containing 0.1 M KCl. The

[17] J. T. Kadonaga and R. Tjian, Proc. Natl. Acad. Sci. U.S.A. 83, 5889 (1986).
[18] J. D. Dignam, R. M. Levovitz, and R. G. Roeder, Nucleic Acids Res. 11, 1475 (1983).

column is washed with 300 ml of buffer A containing 0.15 M KCl. The 0.15 M KCl eluate need not be collected. The column is then washed with buffer A containing 0.3 M KCl, and 4-ml fractions are collected. Fractions having an $A_{280 \, nm}$ greater than 1.0 are pooled (36 ml, 3.6 mg/ml protein). Most of the cap binding activity is in this fraction.

BioGel A 1.5m Chromatography

Solid ammonium sulfate (47.6 g/100 ml, 75% saturation at 0°) is added to the DEAE-cellulose active fraction, and the mixture is gently stirred on ice for 1 hr. The precipitate is collected by centrifugation for 20 min at 25,000 rpm in the type 60 rotor (Beckman) and then dissolved in 4 ml of buffer A containing 0.1 M KCl and 0.2 mM (p-amidinophenyl)methane-sulfonyl fluoride hydrochloride (APMSF, Wako Chemical Co., Japan). This protein solution is mixed with 2 ml of 3 M KCl and 0.5 ml of 50% glycerol and fractionated by gel filtration on a BioGel A-1.5m column (Bio-Rad, Richmond, CA; 2.5 × 92.5 cm) previously equilibrated with buffer A containing 1 M KCl and 0.2 mM APMSF. One hundred and forty 4-ml fractions are collected, and every other fraction between fractions 48 and 130 is assayed for cap binding activity. Active fractions are pooled (27 ml, 414 μg/ml protein). The cap binding activity is eluted from the column in the vicinity of approximately 100K as calibrated by protein size markers.

DEAE-Toyopearl 650 M Chromatography

The pooled BioGel active fraction is dialyzed for 5 hr against 1 liter of buffer A containing 50 mM KCl with one change and applied to a DEAE-Toyopearl 650 M column (Toyo Soda Co., 1.8 ml of resin in a Bio-Rad polypropylene column) previously equilibrated with buffer A containing 50 mM KCl. The column is washed with 10 ml of the same buffer, then eluted with buffer A containing 0.2 M KCl. Ten 0.5-ml fractions are collected, and fractions having an $A_{280 \, nm}$ greater than 0.8 are pooled (2 ml, 3.5 mg/ml protein).

Capped RNA-Sepharose Chromatography

The pooled DEAE-Toyopearl fraction is adjusted to 0.1% Nonidet P-40 (NP-40) and 0.2 mM APMSF with 10% NP-40 and 0.2 M APMSF, respectively, and mixed with 3 ml of capped RNA-Sepharose previously equilibrated with buffer A containing 0.2 M KCl, 0.1% NP-40, and 0.2 mM APMSF in a screw-capped polypropylene tube. The mixture is incubated at 4° for 1 hr with gentle mixing on a rotating platform, then trans-

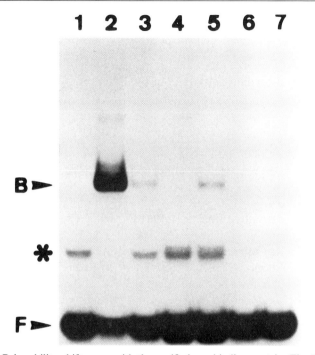

FIG. 1. Gel mobility shift assay with the purified cap binding protein. The RNA bands, unshifted (free RNA) and shifted specifically by the cap binding protein, are indicated by arrowheads designated F and B, respectively. The bands indicated by an asterisk correspond to the free RNA probe of an alternative secondary structure. Lane 1, m⁷GpppG-primed probe alone; lane 2, m⁷GpppG-primed probe plus purified nuclear cap binding protein (70 ng); lane 3, same as lane 2 but with the addition of 300 μM m⁷GpppG; lane 4, GpppG (unmethylated)-primed probe alone; lane 5, GpppG-primed probe plus purified cap binding protein (70 ng); lane 6, uncapped probe alone; lane 7, uncapped probe plus purified cap binding protein (70 ng).

ferred to a polypropylene column (Bio-Rad). The column is washed with 40 ml of buffer A containing 0.2 M KCl and 0.2 mM APMSF, then with 40 ml of buffer A containing 0.7 M KCl and 0.2 mM APMSF. Most of the 80K cap binding protein remains bound to the resin even after washing with 0.7 M KCl, whereas proteins that bind nonspecifically to the RNA moiety are eluted. The 80K cap binding protein is then eluted with buffer A containing 4 M urea and 0.2 mM APMSF, and 0.5-ml fractions are collected. Active fractions are pooled and dialyzed for 5 hr against 1 liter of buffer A containing 20% glycerol, 0.1 M KCl, and 0.2 mM APMSF with two changes. This preparation is designated the purified cap binding protein (1.5 ml, 14.4 μg/ml protein).

Properties of Purified Cap Binding Protein

If the purified cap binding protein is subjected to SDS–polyacrylamide gel electrophoresis[19] followed by silver staining, a single protein band of approximately 80K is visible. The native form of this protein is yet to be clarified. ATP and Mg^{2+} are not required for the cap binding activity. KCl in the concentration range of 50–500 mM does not significantly affect the activity, and substantial binding occurs even in the presence of 1 M KCl or 2 M urea. The binding specificity of this protein is shown in Fig. 1. The purified protein is very active in binding to the cap structure (lane 2), and its binding is strongly inhibited by m^7GpppG. The protein binds weakly to the unmethylated cap structure (GpppG-primed transcript, lane 5) and not at all to uncapped RNA (lane 7), indicating that not only the methyl residue at the N-7 position of the blocking guanosine but the blocking structure itself is important for the binding of this protein. The biological function of the cap binding protein remains to be clarified. The protein is barely detectable in cytoplasmic fractions such as the S100 and ribosome wash fractions by the assay procedure described above.

Acknowledgments

This work was supported by grants from the Ministry of Education, Science, and Culture of Japan.

[19] U. K. Laemmli, *Nature* (*London*) **227**, 680 (1970).

[18] Purification of Small Nuclear Ribonucleoprotein Particles Active in RNA Processing

By Angela Krämer

Introduction

The U-type small nuclear ribonucleoprotein particles (snRNPs) have gained considerable attention over the last decade when a function for several of these particles in the processing of primary transcripts became evident (for recent reviews, see Ref. 1). To date snRNPs have been found in a large variety of animal and plant cells, in unicellular organisms (for

[1] M. L. Birnstiel (ed.), "Structure and Function of Major and Minor Small Nuclear Ribonucleoprotein Particles." Springer-Verlag, Berlin and New York, 1988.

references, see Reddy and Busch[2]), and in *Herpesvirus saimiri*-infected cells.[3] In this case the snRNAs are encoded in the viral genome. In addition to U1–U6 snRNPs which are present at approximately 10^6 copies per mammalian cell, 20–30 species of less abundant U-type snRNPs have been resolved from yeast, fungi, plants, and vertebrates.[4–10]

The snRNPs are localized in the nucleoplasm,[11,12] except for U3 and possibly U8 snRNPs which reside in the nucleolus.[6,13,14] Each particle consists of a uridine-rich RNA[2,15,16] which is associated with several proteins.[17,18] U4 and U6 RNAs are organized into one snRNP[19–21]; however, a small portion of these RNAs has also been detected in separate particles.[22–26] With the exception of U6 RNA, which contains a cap structure of unknown composition, the 5′ ends of the U RNAs are characterized by

[2] R. Reddy and H. Busch, *in* "Structure and Function of Major and Minor Small Nuclear Ribonucleoprotein Particles" (M. L. Birnstiel, ed.), p. 1. Springer-Verlag, Berlin and New York, 1988.

[3] S. I. Lee, S. C. S. Murthy, J. J. Trimble, R. C. Desrosiers, and J. A. Steitz, *Cell* **54**, 599 (1988).

[4] J. A. Wise, D. Tollervey, D. Maloney, H. Swerdlow, E. J. Dunn, and C. Guthrie, *Cell* **35**, 743 (1983).

[5] K. Strub, G. Galli, M. Busslinger, and M. L. Birnstiel, *EMBO J.* **3**, 2801 (1984).

[6] R. Reddy, D. Henning, and H. Busch, *J. Biol. Chem.* **260**, 10930 (1985).

[7] N. Riedel, J. A. Wise, H. Swerdlow, A. Mac, and C. Guthrie, *Proc. Natl. Acad. Sci. U.S.A.* **83**, 8097 (1986).

[8] D. Tollervey, *J. Mol. Biol.* **196**, 355 (1987).

[9] A. Krämer, *Proc. Natl. Acad. Sci. U.S.A.* **84**, 8408 (1987).

[10] K. A. Montzka and J. A. Steitz, *Proc. Natl. Acad. Sci. U.S.A.* **85**, 8885 (1988).

[11] M. Mattioli and M. Reichlin, *J. Immunol.* **107**, 1281 (1971).

[12] J. O. Northway and E. M. Tan, *Clin. Immunol. Immunopathol.* **1**, 140 (1972).

[13] R. A. Weinberg and S. Penman, *J. Mol. Biol.* **38**, 289 (1968).

[14] A. W. Prestayko, M. Tonato, and H. Busch, *J. Mol. Biol.* **47**, 505 (1970).

[15] A. Krol and P. Carbon, this series, Vol. 180, p. 212.

[16] R. Reddy, this series, Vol. 180, p. 521.

[17] R. Lührmann, *in* "Structure and Function of Major and Minor Small Nuclear Ribonucleoprotein Particles" (M. L. Birnstiel, ed.), p. 71. Springer-Verlag, Berlin and New York, 1988.

[18] S. O. Hoch, this volume [20].

[19] P. Bringmann, B. Appel, J. Rinke, R. Reuter, H. Theissen, and R. Lührmann, *EMBO J.* **3**, 1357 (1984).

[20] C. Hashimoto and J. A. Steitz, *Nucleic Acids Res.* **12**, 3283 (1984).

[21] J. Rinke, B. Appel, M. Digweed, and R. Lührmann, *J. Mol. Biol.* **185**, 721 (1985).

[22] C. W. Pikielny, B. C. Rymond, and M. Rosbash, *Nature (London)* **324**, 341 (1986).

[23] M. M. Konarska and P. A. Sharp, *Cell* **49**, 763 (1987).

[24] S.-C. Cheng and J. N. Abelson, *Genes Dev.* **1**, 1014 (1987).

[25] A. I. Lamond, M. M. Konarska, P. J. Grabowski, and P. A. Sharp, *Proc. Natl. Acad. Sci. U.S.A.* **85**, 411 (1988).

[26] D. A. Brow and C. Guthrie, *Nature (London)* **334**, 213 (1988).

the presence of a 2,2,7-trimethylguanosine residue.[2] (Details on the structure of snRNPs are given elsewhere in this volume.[15,18,27])

Except for U3, the snRNPs are immunoprecipitated with antibodies from patients suffering from connective tissue diseases. Anti-Sm antibodies recognize proteins that are common to all particles,[28] whereas proteins characteristic for U1 or U2 snRNPs are recognized either by patient sera or by monoclonal antibodies raised against U1 snRNP (reviewed by Lührmann[17]). In addition, the particles can be precipitated by antibodies to the trimethylated cap structure.[29,30] These and other properties have allowed the purification of individual snRNPs by immunological methods.[31–34]

The function of snRNPs in the processing of primary transcripts is now well documented. U1, U2, U4, U5, and U6 snRNPs participate in nuclear pre-mRNA splicing[35,36]; U7 snRNP acts in histone pre-mRNA 3' end formation[37–40]; and U11 snRNP has been implicated in the polyadenylation of mRNA precursors.[40,41] The function of U3 snRNP is not quite clear. U3 RNA is partially complementary to ribosomal RNA, which suggests a role in the processing of rRNA precursors, in ribosome assembly, or in the transport of the ribosome to the cytoplasm.[36,42] Thus, other snRNPs of yet unknown function may participate in the maturation of primary transcripts as well.

For a detailed understanding of the function of snRNPs in one or the other processing reaction it is desirable that these particles be available as purified components. Several procedures for the purification of U

[27] C. Brunel and G. Cathala, this volume [21].

[28] M. R. Lerner and J. A. Steitz, *Proc. Natl. Acad. Sci. U.S.A.* **76**, 5495 (1979).

[29] R. Lührmann, B. Appel, P. Bringmann, J. Rinke, R. Reuter, S. Rothe, and R. Bald, *Nucleic Acids Res.* **10**, 7103 (1982).

[30] J. H. Smith and G. L. Eliceiri, *J. Biol. Chem.* **258**, 4636 (1983).

[31] P. Bringmann, J. Rinke, B. Appel, R. Reuter, and R. Lührmann, *EMBO J.* **2**, 1129 (1983).

[32] P. Bringmann and R. Lührmann, *EMBO J.* **5**, 3509 (1986).

[33] A. R. Krainer, *Nucleic Acids Res.* **16**, 9415 (1988).

[34] M. Bach, P. Bringmann, and R. Lührmann, this volume [19].

[35] T. Maniatis and R. Reed, *Nature (London)* **325**, 673 (1987).

[36] J. A. Steitz, D. L. Black, V. Gerke, K. A. Parker, A. Krämer, D. Frendewey, and W. Keller, *in* "Structure and Function of Major and Minor Small Nuclear Ribonucleoprotein Particles" (M. L. Birnstiel, ed.), p. 115. Springer-Verlag, Berlin and New York, 1988.

[37] G. Galli, H. Hofstetter, H. G. Stunnenberg, and M. L. Birnstiel, *Cell* **34**, 823 (1983).

[38] K. L. Mowry and J. A. Steitz, *Science* **238**, 1682 (1987).

[39] O. Gick, A. Krämer, A. Vasserot, and M. L. Birnstiel, *Proc. Natl. Acad. Sci. U.S.A.* **84**, 8937 (1987).

[40] K. L. Mowry and J. A. Steitz, *Trends Biochem. Sci.* **13**, 447 (1988).

[41] G. Christofori and W. Keller, *Cell* **54**, 875 (1988).

[42] K. Parker and J. A. Steitz, *Mol. Cell. Biol.* **7**, 2899 (1987).

snRNPs by biochemical means have been described.[43,44] The method presented below was first established for the fractionation of splicing factors from HeLa cell nuclear extracts.[45] The major snRNPs U1–U6 can be considerably purified through four chromatographic steps. In addition, this scheme allows the separation of the minor snRNPs U7 and U11 from the major snRNP particles.[9]

Methods

Analytical Procedures

Analysis of snRNAs. Fractions (20 μl) are treated with proteinase K (300 μg/ml; Boehringer Mannheim) in a total volume of 150 μl containing 2% sarkosyl (w/v). 0.1 M tris(hydroxymethyl)aminomethane hydrochloride (Tris-HCl) (pH 7.5), and 20 mM ethylenediaminetetraacetic acid (EDTA) for 30 min at 37°. Following phenol–chloroform extraction and ethanol precipitation, RNAs are separated in denaturing 10 or 12% polyacrylamide–8.3 M urea sequencing gels in 1× TBE (0.089 M Tris-OH, 0.089 M boric acid, 1 mM EDTA).[46] RNAs are visualized by staining with ethidium bromide (0.5 μg/ml) and illumination with UV light.

3′-End-Labeling of snRNAs. Fractions are treated with proteinase K as described above. After phenol–chloroform extraction the RNA is precipitated in the presence of 3 M ammonium acetate, 8 mM magnesium acetate, and 1 μl of glycogen carrier (Renner, Dannstadt, FRG). RNA is 3′-end-labeled overnight at 4° with 10 μCi of [α-^{32}P]pCp (3000 Ci/mmol; Amersham) in a total volume of 14 μl in the presence of 50 mM *N*-2-hydroxyethylpiperazine-*N*′-2-ethanesulfonic acid (HEPES–KOH (pH 8.3), 10 μM ATP, 3.3 mM dithiothreitol, 10% (v/v) dimethyl sulfoxide, 10% (v/v) glycerol, and 5 units of T4 RNA ligase (P-L Biochemicals).[47] Labeled RNA is extracted with phenol–chloroform, ethanol precipitated, separated in a 10% polyacrylamide–8.3 M urea gel, and detected by autoradiography.

SDS–Gel Electrophoresis. Proteins are prepared for electrophoresis by precipitation with 2 volumes of 20% trichloroacetic acid followed by 2 washes with acetone. They are dissolved in sodium dodecyl sulfate (SDS)

[43] M. Hinterberger, I. Pettersson, and J. A. Steitz, *J. Biol. Chem.* **258**, 2604 (1983).
[44] C. S. Kinlaw, B. L. Robberson, and S. M. Berget, *J. Biol. Chem.* **258**, 7181 (1983).
[45] A. Krämer, M. Frick, and W. Keller, *J. Biol. Chem.* **262**, 17630 (1987).
[46] T. Maniatis, E. F. Fritsch, and J. Sambrook, "Molecular Cloning: A Laboratory Manual." Cold Spring Harbor Laboratory, Cold Spring Harbor, New York, 1982.
[47] T. E. England and O. C. Uhlenbeck, *Nature (London)* **275**, 560 (1978).

sample solution and separated in SDS–polyacrylamide gels.[48] Proteins are either stained with Coomassie blue or with silver.[49] Molecular weight markers (Sigma, St. Louis, MO) are as follows: myosin (M_r 205,000), β-galactosidase (M_r 116,000), phosphorylase b (M_r 97,400), bovine serum albumin (M_r 66,000), ovalbumin (M_r 45,000), carbonate dehydratase (M_r 29,000), and lysozyme (M_r 14,300).

Immunoblotting of snRNP Proteins. Proteins are transferred from SDS–polyacrylamide gels to nitrocellulose (Schleicher & Schüll) for 1 hr at 150 mA as described by Kyhse-Anderson.[50] After transfer the filters are incubated in TN–Tween [20 mM Tris (pH 7.5), 150 mM NaCl, 0.05% (v/v) Tween 20 (Serva, Heidelberg, FRG)] for 60 min at room temperature. [The efficiency of the transfer can be monitored by staining of the proteins on the filter with Ponceau S (Sigma). The filter is briefly soaked in a 0.2% Ponceau S solution and rinsed with water. The stain can be completely removed by washing in TN–Tween.] Incubation with antibodies (diluted in TN–Tween) is performed in a sealed plastic bag for 1–2 hr at room temperature, followed by 5 washes with TN–Tween for 5 min each. For detection of antigen-bound antibodies, the filters are incubated with ^{125}I-labeled protein A in TN–Tween for 1 hr at room temperature. The filters are then washed as above, dried, and exposed on X-ray film. ^{14}C-Methylated proteins (Amersham) are used as molecular weight markers: myosin (M_r 200,000), phosphorylase b (M_r 92,500), bovine serum albumin (M_r 69,000), ovalbumin (M_r 46,000), carbonate dehydratase (M_r 30,000), and lysozyme (M_r 14,300).

Assay Systems for snRNP Activity

Antibody-Depleted Extracts. Anti-Sm antibodies [Y12 monoclonals[51] (as mouse ascites fluid) or human autoantibodies[52] (as antiserum)] are added to 5 μl of packed protein A-Sepharose (PAS) and incubated in a total volume of 200 μl in IPP [10 mM Tris-HCl (pH 7.9)], 0.5 M NaCl 0.1% Nonidet P-40 (NP-40)][52] for 1 hr in the cold (on a rotating wheel). Unbound proteins are removed by 2 washes with IPP, followed by 3 washes with extract buffer (buffer C, see below). After the last wash the buffer is carefully removed, and 20 μl of nuclear extract is added to the anti-Sm PAS. Following a 30-min incubation at 4°, anti-Sm PAS is removed by

[48] U. K. Laemmli, *Nature* (*London*) **277**, 680 (1970).
[49] R. C. Switzer, C. R. Merril, and S. Shifrin, *Anal. Biochem.* **98**, 231 (1979).
[50] J. Kyhse-Anderson, *J. Biochem. Biophys. Methods* **10**, 203 (1984).
[51] E. A. Lerner, M. R. Lerner, L. A. Janeway, and J. A. Steitz, *Proc. Natl. Acad. Sci. U.S.A.* **78**, 2737 (1981).
[52] A. Fritz, R. Parisot, D. Newmeyer, and E. De Robertis, *J. Mol. Biol.* **178**, 273 (1984).

centrifugation, and the resulting supernatant (depleted nuclear extract) is used in a processing reaction. If multiple samples of depleted extract are needed, the reaction is scaled up.

Micrococcal Nuclease-Treated Extracts. Nuclear extract (15 μl) is incubated for 10 min at 30° in the presence of 1 mM CaCl$_2$ and 500 units/ml of micrococcal nuclease (P-L Biochemicals). The reaction is terminated by chelating Ca^{2+} with the addition of ethylene glycol bis(β-aminoethyl ether)-N,N,N',N'-tetraacetic acid (EGTA) to a final concentration of 2 mM.[53] Fractions derived from a nuclear extract are treated accordingly.

In Vitro Splicing Reactions. In vitro assays for splicing complex formation or splicing of mRNA precursors are described by Krämer and Keller.[54] Other methods that evaluate functional characteristics of snRNPs are detailed elsewhere in this volume.[55–57]

Chromatographic Procedures

Buffers. All stock solutions and buffers are made with sterile components.

Buffer A: 10% (v/v) glycerol, 20 mM HEPES–KOH (pH 7.9), 3 mM MgCl$_2$, 0.1 mM EDTA, 0.5 mM dithiothreitol, 0.5 mM phenylmethylsulfonyl fluoride, and KCl at the concentrations indicated in the text

Buffer B: equivalent to buffer A, but containing 50 mM Tris-HCl (pH 7.9)

Buffer C: equivalent to buffer A, but containing 20% (v/v) glycerol

Buffer D: equivalent to buffer B, but containing 20% (v/v) glycerol

General Remarks. All chromatographic resins used were purchased from Pharmacia LKB Biotechnology. Prior to use they are treated according to the manufacturer's instructions and equilibrated with the appropriate buffer.

Fractionation is carried out in a cold room at 4–8°. The purification steps are schematically represented in Fig. 1. The procedures for DEAE-Sepharose, heparin-Sepharose, and Mono Q chromatography are the same as those employed in the fractionation of splicing activities from HeLa cell nuclear extracts.[45] These methods are described in detail in this volume by Krämer and Keller[54]; therefore, only a brief description for the initial steps will be given here.

[53] A. R. Krainer and T. Maniatis, *Cell* **42,** 725 (1985).
[54] A. Krämer and W. Keller, this volume [1].
[55] K. A. Parker and J. A. Steitz, this series, Vol. 180, p. 454.
[56] J. A. Steitz, this series, Vol. 180, p. 468.
[57] A. Krämer, this volume [23].

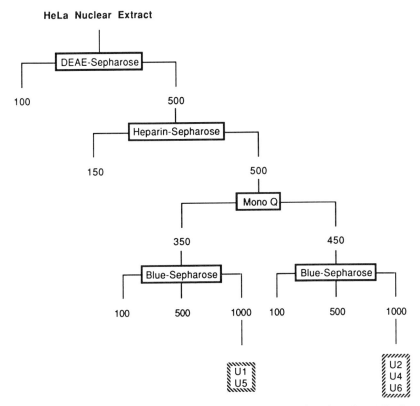

FIG. 1. Schematic representation of the chromatographic fractionation of snRNPs from HeLa nuclear extracts. The different chromatographic resins used are shown in boxes. Numbers designate the millimolar KCl concentration at which the snRNPs elute from a resin. For the Mono Q column, where elution was performed with a linear salt gradient, the approximate KCl concentration required for desorption is given.

DEAE-Sepharose Chromatography. Nuclear extract (in buffer C plus 100 mM KCl) is loaded onto a DEAE-Sepharose column which has been equilibrated with buffer A plus 100 mM KCl. The column is washed with the same buffer to remove unbound material. Bound components are eluted with buffer A plus 500 mM KCl. The resulting fractions are designated DS100 (flow-through fraction) and DS500 (eluted material). The DS500 fraction is dialyzed against several changes of buffer D plus 100 mM KCl.

Heparin-Sepharose Chromatography. The DS500 fraction is applied to a heparin-Sepharose column equilibrated with buffer B plus 150 mM KCl. After removal of unbound material, elution is performed with buffer

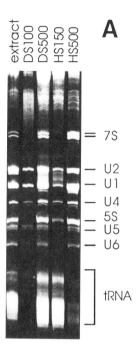

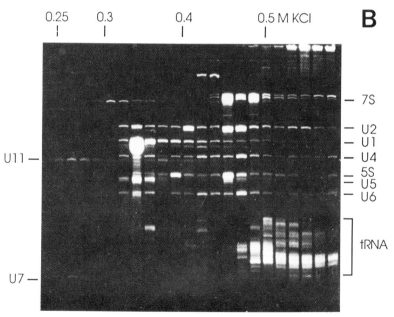

FIG. 2. Analysis of small RNAs present in DEAE-Sepharose, heparin-Sepharose, and Mono Q fractions. (A) RNA from 20 μl of the fractions indicated above each lane was extracted as described in the methods section and electrophoresed in a denaturing 12%

B plus 500 mM KCl. The resulting fractions, designated HS150 and HS500, are dialyzed against buffer D plus 100 mM KCl.

Mono Q Chromatography. HS500 is loaded onto a Mono Q FPLC column that has been equilibrated with buffer B plus 100 mM KCl. The column is washed with the same buffer and developed with a linear gradient of 100–500 mM KCl in buffer B. Fractions are individually dialyzed against buffer D plus 100 mM KCl. Fractions enriched in U1 and U5 snRNPs are designated Q-U1, those enriched in U2, U4, and U6 snRNPs are designated Q-U2.

Blue-Sepharose Chromatography. Blue-Sepharose is equilibrated with buffer B plus 100 mM KCl. Pooled Mono Q fractions enriched in either U1 or U2 snRNPs are loaded onto the column at 1–2 column volumes/hr. After washing with 6 column volumes of buffer B plus 100 mM KCl, bound material is eluted in two steps of 500 and 1000 mM KCl in buffer B. The protein peaks of each step are pooled and dialyzed against buffer C plus 100 mM KCl. The fractions are frozen at −80°. Activity of the snRNPs in *in vitro* splicing assays is still detectable in fractions that had been stored for more than 2 years.

Gel Filtration of snRNPs. Two hundred microliters of a fraction enriched in snRNPs is loaded onto a Superose 12 FPLC column in buffer B plus 100 mM KCl at 30 ml/hr. Fractions of 0.5 ml are collected. The column is calibrated with the following molecular weight markers under the conditions used for the gel filtration of snRNPs: Blue dextran (M_r 2,000,000), thyroglobulin (M_r 669,000), ferritin (M_r 440,000), aldolase (M_r 158,000), bovine serum albumin (M_r 67,000), ovalbumin (M_r 43,000), and chymotrypsinogen A (M_r 25,000) (Pharmacia/LKB). Alternatively, a Superose 6 column can be used, which gives a more efficient resolution in the high molecular weight range.

Results

The distribution of small RNAs in the chromatographic fractions described above is shown in Figs. 2 and 3. It is apparent that the bulk of the snRNPs bind to DEAE- as well as heparin-Sepharose (Fig. 2A). The flow-through fractions of both columns (DS100 and HS150) contain only small

---

polyacrylamide gel. The designation of the different RNA species is shown at right. [From Krämer *et al.*, *J. Biol. Chem.* **262**, 17630 (1987).[45]] (B) RNA from 20 µl of every second fraction obtained after Mono Q chromatography was extracted and electrophoresed in a denaturing 10% polyacrylamide gel. The approximate molar KCl concentration at which the fractions eluted from the column is indicated at top. RNA species are indicated at right; the positions of U7 and U11 RNAs are shown at left.

amounts of snRNPs. Most of the tRNA and part of the high molecular weight RNA do not bind to heparin-Sepharose and are thus removed from the snRNP-enriched HS500 fraction.

Chromatography of HS500 on the anion-exchange resin Mono Q results in a partial separation of individual snRNPs (Fig. 2B). U1 and U5 snRNPs are enriched in fractions eluting at 350 mM KCl (Q-U1), whereas U2, U4, and U6 snRNPs are found in fractions that elute above 400 mM KCl (Q-U2). However, low concentrations of the snRNPs are present in all fractions eluting above 300 mM KCl. U2 snRNP (and to a lesser extent U4, U5, and U6 snRNPs) exhibits a rather peculiar behavior in that it reproducibly elutes in several peaks (e.g., at ~0.4 and ~0.45 M KCl; Fig. 2B). Although the reason for this is unknown, it may indicate a heterogeneity in a particular snRNP population, which could be explained by the following observations. First, U2 and U5 snRNPs have been shown to exist in a nuclear extract in two forms that can be separated by nondenaturing gel electrophoresis.[23,58] For example, one form of U2 snRNP migrates in the gel as expected for a particle of 8–10 S. In addition, U2 RNA is detected in a slower migrating complex which appears to turn over to the fast migrating form in the presence of ATP. Preliminary experiments from our laboratory indicate that U2 RNA is present in the fast migrating form in most of the Mono Q fractions, whereas it is found in slow migrating complexes in the late eluting fractions.[59] Second, on Mono Q chromatography snRNPs can be separated that either contain the entire complement of snRNP proteins or lack characteristic or common proteins.[34]

In addition to the major snRNPs, two new minor species were detected on Mono Q chromatography (Fig. 2B).[9] U7 and U11 snRNPs contain RNAs of approximately 63 nucleotides[37,60,61] and 131/134 nucleotides,[10,62] respectively, and coelute at 250–300 mM KCl. U11 is present in minor quantities also in later fractions, as examined by Northern blot analysis.[63] Whether this is also true for U7 snRNP has not been determined. Gel filtration of the particles on Superose 12 revealed an M_r of about 150,000 for U11 and an M_r of around 115,000 for U7 snRNP.[9] In glycerol gradients U11 sediments with 10–12 S, whereas U7 snRNP shows a slightly lower sedimentation value.[63]

U11 snRNP can be further purified by centrifugation in CsCl gradi-

[58] M. M. Konarska and P. A. Sharp, *Cell* **46**, 845 (1986).
[59] D. Frendewey, A. Krämer, and W. Keller, unpublished observation (1987).
[60] M. Cotton, O. Gick, A. Vasserot, G. Schaffner, and M. L. Birnstiel, *EMBO J.* **7**, 801 (1988).
[61] D. Soldati and D. Schümperli, *Mol. Cell. Biol.* **8**, 1518 (1988).
[62] C. Suter and W. Keller, personal communication (1988).
[63] G. Christofori and W. Keller, personal communication (1988).

ents,[41] where it sediments with a buoyant density of 1.40–1.49 g/cm³, characteristic for ribonucleoproteins.[64] Additional fractionation steps such as chromatography on spermine-agarose and Blue Sepharose have also been used. Details of these procedures are described by Christofori and Keller.[41]

RNAs characteristic for U8, U9, U10, and U12 snRNPs with sizes of 140, ~130, ~60, and ~150 nucleotides, respectively,[6,10] have not been detected in any of the fractions obtained by the procedure described here. These particles could be present in very low concentrations such that they cannot be visualized by ethidium bromide staining, or they could be obscured by the mass of the major snRNAs present in these fractions. Alternatively, U8–U10 and U12 snRNPs may exhibit chromatographic properties that do not allow their purification by the procedures employed here.

High molecular weight RNA elutes from Mono Q at KCl concentrations above 450 mM and is thus efficiently separated from the snRNPs. Residual tRNA also elutes at high salt concentrations, and only minor amounts remain in the U2 snRNP-enriched fractions. The 7S RNA that coelutes with U2 snRNP from Mono Q is removed by Blue Sepharose chromatography.

Fractionation on Blue Sepharose yields snRNP preparations that are more than 70% pure. U1 snRNP is recovered from this resin in the BS500 and BS1000 fractions (Fig. 3). Most of the proteins as well as residual U2, U4, U5, and U6 snRNPs present in Q-U1 are found in the 500 mM KCl step. Analysis of the RNA contents of the BS1000 fraction shows that only U1 RNA and a small amount of U5 RNA are present (Fig. 3A). The RNA species migrating just ahead of U1 RNA probably represents a degradation product of U1 RNA that is missing a few nucleotides from the 5' end.[65] End-labeling of the RNAs present in this fraction with [α-³²P]pCp confirmed the absence of U2, U4, and U6 RNAs.[66] Apart from several high molecular weight polypeptides, only snRNP proteins are detected in the BS1000 fraction. The U1-specific polypeptides (70K and A) as well as the polypeptides common to all snRNPs (B', B, and D) are clearly visible (Fig. 3B). The U1-specific C protein has not been detected in the stained gel, which may indicate that this protein is underrepresented in the purified U1 snRNPs (cf. Refs. 31 and 32). The polypeptides

[64] C. Brunel, J. Sri-Widada, M. N. Lelay, P. Jeanteur, and J. P. Liautard, *Nucleic Acids Res.* **9**, 815 (1981).

[65] M. R. Lerner, J. A. Boyle, S. M. Mount, S. L. Wolin, and J. A. Steitz, *Nature (London)* **283**, 220 (1980).

[66] A. Krämer, unpublished result (1988).

that are smaller than the D protein probably represent the common poly-peptides E, F, and G. Thus, the BS1000 fraction contains highly purified U1 snRNP together with a small amount of U5 snRNP.

All snRNPs present in Q-U2 bind to Blue Sepharose and are eluted with 1 *M* KCl (BS1000). In addition to U2, U4, and U6 snRNPs this fraction contains very low amounts of U1 and U5 snRNPs (Fig. 3A) which

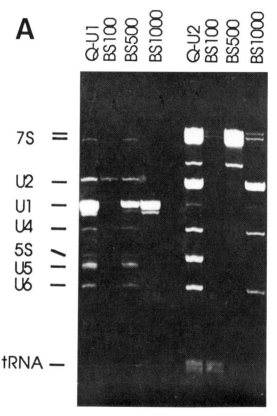

FIG. 3. Analysis of RNA and protein contents of the Blue Sepharose fractions. (A) RNA from 25 µl of the fractions indicated above each lane was extracted and electrophoresed in a denaturing 12% polyacrylamide gel. RNA species are indicated at left. (B) Proteins from 20 µl of the fractions indicated above each lane were precipitated as outlined in the methods section, separated in a 12% polyacrylamide gel in the presence of SDS, and stained with silver. The sizes of molecular weight markers (lane M) are given at left. Characteristic snRNP proteins are indicated at the right of each gel. (C) Proteins were transferred to nitrocellulose from a gel identical to that shown in (B). The filter was successively incubated with antiserum V26 and ^{125}I-labeled protein A. The snRNP proteins recognized by antiserum V26 are indicated at the right of each gel; molecular weight markers (lane M) are indicated at left.

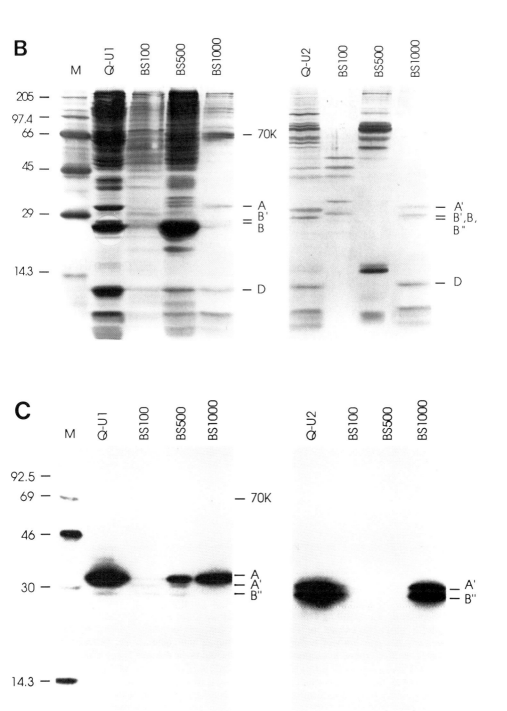

Fig. 3. (*continued*)

are hardly visible in the ethidium bromide stained gel but which can be detected after 3'-end-labeling.[66] Most of the contaminating 7 S RNA, 5 S RNA, and proteins present in Q-U2 are removed by washing the resin with 500 mM KCl. Analysis of the protein content of the BS1000 fraction reveals the common polypeptides B', B, D, E, F, and G, as well as protein A' which is unique to U2 snRNP (Fig. 3B).[32,43,44,67–69] Polypeptide B″, a second U2-specific protein, is detected in this fraction after immunoblotting (Fig. 3C). It cannot be identified unambiguously in the silver-stained gel because it comigrates with B and/or B'.

Inspection of the distribution of U2, U4, and U6 RNAs in the Blue Sepharose fractions (Fig. 3A) again hints to a certain degree of heterogeneity within these particles. Whereas U2, U4, and U6 snRNPs that are derived from Q-U1 fractions are eluted with 500 mM KCl, the same snRNPs derived from Q-U2 are eluted only with 1 M KCl. An immunoblot of the Blue Sepharose fractions performed with an antibody which recognizes the U1-specific 70K and A proteins as well as the U2-specific A' and B″ proteins (V26)[69] shows that these proteins are present in all fractions that contain the corresponding snRNAs (Fig. 3C). (The V26 antiserum cross-reacts only weakly with the U1-specific 70K protein.[69] A band corresponding to the 70K protein was detected in the original autoradiograph but could not be reproduced in Fig. 3C.) Also, the snRNP proteins D, E, F, and G, which are common constituents of the snRNPs, are found in all fractions that contain snRNAs (Fig. 3B). It therefore appears that the heterogeneity in the elution properties of U1 and U2 snRNPs is not a result of the absence or the presence of U1- or U2-specific proteins, with the possible exception of the U1-specific C protein (see above).

Figure 4 shows two examples for assays of snRNP function in splicing. An extract depleted of snRNPs after incubation with anti-Sm antibodies coupled to protein A-Sepharose is inactive in the processing of pre-mRNA (Fig. 4A). Addition of DS500 and HS500, which are enriched in snRNPs, to the depleted extract restores splicing activity, whereas the concentration of residual snRNPs in DS100 and HS150 does not suffice to generate processed RNA. Similar results have been obtained for snRNPs present in Mono Q and Blue Sepharose fractions.[45,66]

The function of U1 and U2 snRNPs in splicing complex formation is demonstrated in Fig. 4B. It has recently been shown that, in addition to

[67] T. Mimori, M. Hinterberger, I. Pettersson, and J. A. Steitz, *J. Biol. Chem.* **259,** 560 (1984).

[68] P. B. Billings and S. O. Hoch, *J. Biol. Chem.* **259,** 12850 (1984).

[69] W. Habets, M. Hoet, P. Bringmann, R. Lührmann, and W. van Venrooij, *EMBO J.* **4,** 1545 (1985).

these particles, two protein factors (SF1 and SF3) are necessary for the assembly of the pre-mRNA into a presplicing complex.[70] To test an involvement of the snRNPs in this process, the particles present in the SF1-containing fraction were subjected to treatment with micrococcal nuclease (MN). A similar treatment of the SF3-containing fraction is not necessary because this fraction is free of snRNPs. In combination, MN-treated SF1 and SF3 are inactive in presplicing complex formation (Fig. 4B, lane 3). Addition of the BS1000 fraction (which contains U1 snRNP but no U2), cannot restore the ability to form a complex (lane 4), whereas addition of the BS1000-U2 fraction (which is contaminated by a very small amount of U1 snRNP) leads to the assembly of a presplicing complex (lane 5). When both U1- and U2-containing Blue Sepharose fractions are combined with MN-treated SF1 and SF3, complex formation is considerably enhanced (lane 6). This indicates that U1 and U2 snRNPs are essential elements in the early events of the splicing reaction. It also shows that very low amounts of U1 snRNP are sufficient to obtain a low level of complex formation.

Conclusions

The procedure described here for the purification of snRNPs was initially developed for the fractionation of factors that function in pre-mRNA splicing[45] The snRNPs that are at least partially resolved after Mono Q chromatography can be purified to near homogeneity by Blue Sepharose chromatography. Although a complete separation of U2, U4, and U6 snRNPs has not been achieved, this might be possible by employing additional fractionation steps, such as gel filtration procedures. A combination of the biochemical methods outlined in this chapter with immunological procedures[32-34] might also help in the separation of individual snRNPs.

In addition to the major snRNPs U1–U6, two minor species, U7 and U11, were discovered during the fractionation.[9] The rather low abundance of these snRNPs had previously precluded their detection in nuclear extracts. More refined purification schemes may also allow the detection of snRNPs that are even less abundant than U7 and U11.

The purified snRNPs described here function in a number of processing reactions. Assay systems for snRNP activity, such as antibody-depleted or micrococcal nuclease-treated extracts or chromatographic fractions, have been employed to demonstrate directly an involvement of U1–U6 snRNPs in pre-mRNA splicing and a function of U1 and U2

[70] A. Krämer, *Genes Dev.* **2**, 1155 (1988).

snRNPs in presplicing complex formation.[33,45,70] In addition, the activity of U7 snRNP in histone pre-mRNA 3' end formation could be monitored throughout the fractionation steps described.[39] The extensive secondary structure of U11 RNA has precluded its degradation by micrococcal nuclease,[10,63] although a participation of U11 snRNP in the polyadenylation of pre-mRNAs was inferred from its cofractionation through a number of chromatographic steps with one of the activities involved in this process.[41]

Acknowledgments

I thank Walter Keller for support, advice, and stimulating discussions; Marion Frick for excellent and dedicated assistance throughout this work; Joan Steitz for the Y12 cell line; Eddy DeRobertis and Walter van Venrooij for gifts of antibodies; and Walter Keller and Graeme Bilbe for comments on the manuscript. This work was supported by grants from the Deutsche Forschungsgemeinschaft and the Schweizerischer Nationalfonds.

[19] Purification of Small Nuclear Ribonucleoprotein Particles with Antibodies against Modified Nucleosides of Small Nuclear RNAs

By MONTSERRAT BACH, PETER BRINGMANN, and
REINHARD LÜHRMANN

Introduction

The small nuclear (sn)RNAs U1, U2, U4, U5, and U6 are the most abundant members of the class of snRNAs, which are found in the nucleoplasm of all eukaryotic cells (for reviews, see Birnstiel[1]). A hallmark of these RNAs is the presence of the 5'-terminal cap structure containing the base 2,2,7-trimethylguanosine (m_3G) at the 5' end, with the single exception of U6 RNA, which also appears to be capped but which lacks the nucleoside m_3G. The snRNAs also contain a number of modified nucleosides such as pseudouridines and those containing methylated bases such

---

[1] M. L. Birnstiel (ed.), "Structure and Function of Major and Minor Small Nuclear Ribonucleoprotein Particles." Springer-Verlag, Berlin and New York, 1988.

as N^6-methyladenosine (m^6A). One m^6A residue is present in each of the snRNAs U2, U4, and U6.[2,3]

The snRNAs exist in the cell as ribonucleoprotein (RNP) complexes.[4] While the snRNAs U1, U2, and U5 are organized in separate RNP particles, the majority of the snRNAs U4 and U6 reside in a single RNP complex.[5-7] The protein constituents of the isolated snRNPs from human cells have recently been characterized. Seven proteins of apparent molecular weights 29K (B'), 28K (B), 16K (D), 15.5K (D'), 12K (E), 11K (F), and 9K (G) are common to the snRNPs U1, U2, U5, and U4/U6. U1 snRNPs contain in addition to these common proteins three unique polypeptides of apparent molecular weights 70K, 34K (A), and 22K (C). U2 snRNPs are characterized by the presence of a 33K and a 28.5K protein, denoted A' and B", respectively.[8-14] Several proteins in the molecular weight range of 25K to 200K were recently identified as possible U5-specific proteins[15-18] (see also below). Information about the primary structure of the U1- and U2-specific proteins as well as the common snRNP proteins E and D has recently been obtained by cDNA cloning.[19-28]

[2] R. Reddy and H. Busch, *in* "Structure and Function of Major and Minor Small Nuclear Ribonucleoprotein Particles" (M. L. Birnstiel, ed.), p. 1. Springer-Verlag, Berlin and New York, 1988.

[3] R. Reddy, this series, Vol. 180, p. 521.

[4] M. R. Lerner and J. A. Steitz, *Proc. Natl. Acad. Sci. U.S.A.* **76,** 5495 (1979).

[5] P. Bringmann, B. Appel, J. Rinke, R. Reuter, H. Theissen, and R. Lührmann, *EMBO J.* **3,** 1357 (1984).

[6] C. Hashimoto and J. A. Steitz, *Nucleic Acids Res.* **12,** 3283 (1984).

[7] J. Rinke, B. Appel, M. Digweed, and R. Lührmann, *J. Mol. Biol.* **185,** 721 (1985).

[8] M. Hinterberger, I. Pettersson, and J. A. Steitz, *J. Biol. Chem.* **258,** 2604 (1983).

[9] P. Bringmann, J. Rinke, B. Appel, R. Reuter, and R. Lührmann, *EMBO J.* **2,** 1129 (1983).

[10] P. B. Billings and S. O. Hoch, *J. Biol. Chem.* **259,** 12850 (1984).

[11] C. S. Kinlaw, B. L. Robberson, and S. M. Berget, *J. Biol. Chem.* **258,** 7181 (1983).

[12] P. Bringmann and R. Lührmann, *EMBO J.* **5,** 3509 (1986).

[13] R. Lührmann, *in* "Structure and Function of Major and Minor Small Nuclear Ribonucleoprotein Particles" (M. L. Birnstiel, ed.), p. 71. Springer, Verlag, Berlin and New York, 1988.

[14] C. Brunel and G. Cathala, this volume [21].

[15] M.-N. Lelay-Taha, I. Reveillaud, J. Sri-Widada, C. Brunel, and P. Jeanteur, *J. Mol. Biol.* **189,** 519 (1986).

[16] V. Gerke and J. A. Steitz, *Cell* **47,** 973 (1986).

[17] J. Tazi, C. Alibert, J. Temsamani, I. Reveillaud, G. Cathala, C. Brunel, and P. Jeanteur, *Cell* **47,** 755 (1986).

[18] M. Bach, G. Winkelmann, and R. Lührmann, *Proc. Natl. Acad. Sci. U.S.A.* **86,** 6038 (1989).

[19] H. Theissen, M. Etzerodt, R. Reuter, C. Schneider, F. Lottspeich, P. Argos, R. Lührmann, and L. Philipson, *EMBO J.* **5,** 3209 (1986).

The major U snRNPs have received much attention in the past, mainly for two reasons. One is the role that the snRNPs may play in the splicing of premessenger RNA (pre-mRNA) molecules in the cell nucleus. This was originally proposed for U1 snRNP,[29,30] and experimental evidence was soon provided for the participation of all major nucleoplasmic snRNPs in the splicing of pre-mRNAs. One of the tasks of the snRNPs appears to be the recognition of important signal structures contained in the intron such as the 5' and 3' splice sites and the branch point.[31–33]

The second reason for interest in the snRNPs, and in particular their proteins, is their relevance from a clinical and immunological point of view. Patients with systemic lupus erythematosus (SLE) or related connected tissue diseases often develop autoantibodies that react with particular snRNP proteins.[4,34] Anti-U1 RNP autoantibodies precipitate only U1 snRNPs, and it has been shown that they react with the polypeptides characteristic of U1 snRNP, namely, 70K, A, and C. In contrast, anti-Sm autoantibodies precipitate all the nucleoplasmic snRNPs U1 to U6. The

[20] R. A. Spritz, K. Strunk, C. S. Surowy, S. O. Hoch, D. E. Barton, and U. Francke, *Nucleic Acids Res.* **15,** 10373 (1987).

[21] C. C. Query and J. D. Keene, *Cell* **51,** 211 (1987).

[22] W. J. Habets, P. T. G. Sillekens, M. H. Hoet, J. A. Schalken, A. J. M. Roebroek, J. A. M. Leunissen, W. J. M. van de Ven, and W. J. van Venrooij, *Proc. Natl. Acad. Sci. U.S.A.* **84,** 2421 (1987).

[23] P. T. G. Sillekens, W. J. Habets, R. P. Beijer, and W. J. van Venrooij, *EMBO J.* **6,** 3841 (1987).

[24] K. Yamamoto, H. Miura, Y. Moroi, S. Yoshinoya, M. Goto, K. Nishioka, and T. Miyamoto, *J. Immunol.* **140,** 311 (1988).

[25] P. T. G. Sillekens, R. P. Beijer, W. J. Habets, and W. J. Van Venrooij, *Nucleic Acids Res.* **16,** 8307 (1988).

[26] D. R. Stanford, A. Rohleder, K. Neiswanger, and E. D. Wieben, *J. Biol. Chem.* **262,** 9931 (1987).

[27] D. R. Stanford, M. Kehl, C. A. Perry, E. L. Holicky, S. E. Harvey, A. M. Rohleder, R. Lührmann, and E. D. Wieben, *Nucleic Acids Res.* **16,** 10593 (1988).

[28] L. A. Rokeach, J. A. Haselby, and S. O. Hoch, *Proc. Natl. Acad. Sci. U.S.A.* **85,** 4832 (1988).

[29] M. R. Lerner, J. A. Boyle, S. M. Mount, S. L. Wolin, and J. A. Steitz, *Nature (London)* **283,** 220 (1980).

[30] J. Rogers and R. Wall, *Proc. Natl. Acad. Sci. U.S.A.* **77,** 1877 (1980).

[31] T. Maniatis and R. Reed, *Nature (London)* **325,** 673 (1987).

[32] P. A. Sharp, *Science* **235,** 766 (1987).

[33] J. A. Steitz, D. L. Black, V. Gerke, K. A. Parker, A. Krämer, D. Frendewey, and W. Keller, *in* "Structure and Function of Major and Minor Small Nuclear Ribonucleoprotein Particles" (M. L. Birnstiel, ed.), p. 115. Springer-Verlag, Berlin and New York, 1988.

[34] E. M. Tan, *Adv. Immunol.* **33,** 167 (1982).

major immunoreactive Sm proteins are B', B, and D.[4,9,13,35–43] The further analysis of the U snRNPs is therefore important not only because they are agents in the splicing of nuclear pre-mRNAs but also because they are antigens and immunogens, so that a better understanding of the mechanisms leading to anti-Sm and anti-RNP autoantibody production in SLE patients may be reached. One of the prerequisites for this is the availability of quick and efficient procedures for the purification of all individual snRNPs (U1, U2, U5, and U4/U6) from nuclear extracts. Ideally, the procedure should yield native snRNP particles with retention of their functions in the splicing of pre-mRNAs as demonstrated by assays *in vitro*.

Several purification protocols for U snRNPs have been developed during recent years, some using strictly biochemical approaches,[8,11,14,15,44–46] others using immunological methods or a mixture of both.[9,10,12,13,47] We describe here purification procedures in which antibodies raised against modified nucleosides in the snRNAs, such as m_3G and m^6A, play an important part. Antibodies against m_3G and m^6A were first raised in rabbits,[48–51] and we have demonstrated that both the m_3G caps

[35] E. A. Lerner, M. R. Lerner, J. A. Hardin, C. A. Janeway, and J. A. Steitz, *Arthritis Rheum.* **25**, 761 (1982).

[36] P. B. Billings, R. W. Allen, F. C. Jensen, and S. O. Hoch, *J. Immunol.* **128**, 1176 (1982).

[37] G. E. Conner, D. Nelson, R. Wisniewolski, R. G. Lahita, G. Blobel, and H. G. Kunkel, *J. Exp. Med.* **156**, 1475 (1982).

[38] L. Matter, K. Schopfer, J. A. Wilhelm, T. Nyffenegger, R. F. Parisot, and E. M. De Robertis, *Arthritis Rheum.* **25**, 1278 (1982).

[39] D. E. Fisher, W. H. Reeves, G. E. Conner, G. Blobel, and H. G. Kunkel, *Proc. Natl. Acad. Sci. U.S.A.* **81**, 3185 (1984).

[40] I. Pettersson, M. Hinterberger, T. Mimori, E. Gottlieb, and J. A. Steitz, *J. Biol. Chem.* **259**, 5907 (1984).

[41] W. J. Habets, J. H. M. Berden, S. O. Hoch, and W. J. van Venrooij, *Eur. J. Immunol.* **15**, 992 (1985).

[42] I. Pettersson, G. Wang, E. I. Smith, H. Wigzell, E. Hefors, J. Horn, and G. C. Sharp, *Arthritis Rheum.* **29**, 986 (1986).

[43] R. Lührmann and R. Reuter, in "Modern Methods in Protein Chemistry" (H. Tschesche, ed.), Vol. 3, p. 61. de Gruyter, Berlin and New York, 1988.

[44] C. Brunel, J. Sri-Widada, M. N. Lelay, P. Jeanteur, and J. P. Liautard, *Nucleic Acids Res.* **9**, 815 (1981).

[45] J. Sri-Widada, C. Assens, J. P. Liautard, P. Jeanteur, and C. Brunel, *Biochem. Biophys. Res. Commun.* **104**, 457 (1982).

[46] I. Reveillaud, M. N. Lelay-Taha, J. Sri-Widada, C. Brunel, and P. Jeanteur, *Mol. Cell. Biol.* **4**, 1890 (1984).

[47] C. Brunel, J. Sri-Widada, and P. Jeanteur, *Prog. Mol. Subcell. Biol.* **9**, 1 (1985).

[48] R. Lührmann, B. Appel, P. Bringmann, J. Rinke, R. Reuter, S. Rothe, and R. Bald, *Nucleic Acids Res.* **10**, 7103 (1982).

and the m^6A bases of the respective snRNAs were available for antibody binding in the snRNP particles.[51,52] This allowed the use of these antibody systems for the one-step purification of all snRNPs U1 to U6 (in the case of the anti-m_3G antibody[9,12,52]) or of a mixture of the snRNPs U1 and U4/U6 (in the case of anti-m^6A antibodies[51]). The main advantage of these procedures is that they offer the possibility of desorbing the antibody-bound U snRNPs by elution with an excess of nucleoside m_3G or m^6A, respectively, i.e., under nondenaturing conditions. A further advantage is that these antibodies may be used for the purification of U snRNPs from any organism.

The combined use of the antibody systems with DEAE ion-exchange chromatography allows fractionation of individual snRNP types. As an alternative, the mixture of snRNPs purified by anti-m_3G or anti-m^6A affinity columns may be fractionated by ion-exchange chromatography on Mono-Q columns under FPLC conditions. The power of Mono-Q columns to separate U snRNP particles was first demonstrated by Krämer *et al.* during studies on the fractionation of individual splicing factors contained in total nuclear extracts.[53,54] The high resolution of Mono-Q columns allows even the separation of snRNP populations of the same class, for example U1 RNPs, which differ by the presence or absence of single snRNP proteins.[55] The availability of such protein-deficient snRNP populations is expected to be of substantial value for the investigation of the roles of individual snRNP proteins in the functioning of the snRNPs during splicing.

Purification of Anti-m_3G and Anti-m^6A Antibodies and
Preparation of Immunoaffinity Columns

The critical aspects of the major steps necessary for the production of antinucleoside antibodies in general, such as synthesis of the immunogen, immunization, and characterization of immune sera, have already been

---

[49] J. H. Smith and G. L. Eliceiri, *J. Biol. Chem.* **258**, 4636 (1983).
[50] T. W. Munns, M. K. Liszewski, and H. F. Sims, *Biochemistry* **16**, 2163 (1977).
[51] P. Bringmann and R. Lührmann, *FEBS Lett.* **213**, 309 (1987).
[52] P. Bringmann, R. Reuter, J. Rinke, B. Appel, R. Bald, and R. Lührmann, *J. Biol. Chem.* **258**, 2745 (1983).
[53] A. Krämer, M. Frick, and W. Keller, *J. Biol. Chem.* **262**, 17630 (1987).
[54] A. Krämer, this volume [18].
[55] M. Bach, B. Kastner, and R. Lührmann, manuscript in preparation.

extensively reviewed[56–64] and are not discussed in detail here. For the isolation of snRNPs from nuclear extracts, it is important that the anti-m_3G antibodies used are highly specific for the m_3G base and, in particular, do not cross-react with the m^7G caps of mRNA. This is of critical importance when anti-m_3G antibodies are used for the intracellular location of snRNAs by immunohistochemical methods.

In our experience, the specificity of antibodies against m_3G caps is determined largely by two parameters. One of these is the three methyl groups at the 5′-terminal m_3G. The greater the contribution of the interaction between the N^2-methyl group and the antibody, the better is the discrimination between m_3G and m^7G cap structures. There is much variation in this respect between different sera: we have characterized rabbit sera whose affinities for these two nucleosides differ by three orders of magnitude. At the other extreme, some anti-m_3G antibodies show a much lower ability to discriminate between the two nucleosides (only one order of magnitude difference), and those appear first and foremost to recognize the N^7-methyl group.[62]

Another factor that can be decisive for the selectivity of an antibody for m_3G caps is the presence of negatively charged phosphate groups in the cap. For some rabbit sera, we have observed a lowering of the affinity of the binding to m_3G by three orders of magnitude when the m_3G cap was replaced by the single nucleoside.[62] This explains why the anti-m_3G antibodies cross-react with m^7G at the nucleoside level while failing to do so at the level of intact cap structures.

Purification of Antibodies from Sera

The choice of a particular method for the purification of anti-m_3G and anti-m^6A antibodies is determined not only by the efficiency of the purification but also by the requirement that the resulting purified antibody preparation be as free as possible from contamination by RNases and

---

[56] B. D. Stollar, in "The Antigens" (M. Sela, ed.), Academic Press, New York, 1973.
[57] T. W. Munns and M. K. Liszweski, Prog. Nucleic Acid Res. Mol. Biol. 24, 109 (1980).
[58] W. L. Highes, M. Christine, and B. D. Stollar, Anal. Biochem. 55, 468 (1973).
[59] L. Levine and H. Gjika, ABB 164, 583 (1974).
[60] M. Becker, Scand. J. Immunol. Suppl. 3, 11 (1978).
[61] W. J. Herbert, in "Handbook of Experimental Immunology" (D. M. Weir, ed.), p. A3.1. Alden Press, Oxford, 1978.
[62] B. D. Stollar, this series, Vol. 70, p. 70.
[63] B. F. Erlanger, this series, Vol. 70, p. 85.
[64] B. A. L. Hurn and S. M. Chantler, this series, Vol. 70, p. 104.

proteases. For the preparation of total IgG from serum, we generally use fractional precipitation by ammonium sulfate[65] followed by separation of the antibodies by gel chromatography on Sephadex G-150 or an equivalent gel. Enrichment in antibodies specific against m_3G or m^6A can then be performed by immunoaffinity chromatography on affinity columns with bound nucleoside. These columns are made by periodate oxidation of the nucleoside m^6A or m^7G (the latter for anti-m_3G antibodies) and subsequent coupling to aminohexyl-Sepharose 4B (Pharmacia) by Schiff base reduction.[66] The nucleoside-bound antibodies can be desorbed from the column either by the use of denaturing reagents (urea, chaotropic salts, or acidic buffers) or, better, by elution with an excess of the cognate nucleoside. We use the same method for the purification of monoclonal anti-m_3G antibodies from hybridoma supernatants.[67]

Immobilization of Purified IgGs to Sepharose Matrices

We use two methods for the production of immunoaffinity columns; the choice depends on the ability of the antibody to react with protein A. In general, we couple purified rabbit anti-m_3G or anti-m^6A IgG with dimethyl suberimidate (DMS) as a cross-linking agent, essentially as described by Gersten and Marchalonis.[68] About 20 mg IgG is bound per milliliter of gel, so that this method gives columns with a binding capacity for snRNPs that is some 3 times higher than that of columns made from CNBr-activated Sepharose.

In brief, protein A-Sepharose CL-4B (Pharmacia) is allowed to swell overnight by end-over-end rotation at 4° in phosphate-buffered saline (PBS) (10 mM KH_2PO_4 (pH 8), 150 mM NaCl] containing 0.02% NaN_3. After extensive washing of the beads on a sintered glass filter (with ~50 bed volumes of PBS, pH 8), the gel is resuspended, packed in a glass column, and finally equilibrated by the passage of 5 bed volumes of PBS, pH 8. Purified IgGs in PBS, pH 8 (~10 mg/ml), are loaded onto the protein A-Sepharose column at a flow rate of 0.2 ml/min. The IgGs that leave the column are recirculated twice over the column, to maximize the yield of bound IgG. After a wash with 5 bed volumes of PBS, pH 8, the column is equilibrated by washing with 5 bed volumes of 0.2 M triethanolamine (pH 8.5). Cross-linking between the Fc part of the IgG molecule

[65] P. Minden and R. S. Farr, "Handbook of Experimental Immunology" (D. M. Weir, ed.), p. 13.1. Alden Press, Oxford, 1978.

[66] B. F. Erlanger and S. M. Beiser, *Proc. Natl. Acad. Sci. U.S.A.* **52,** 68 (1964).

[67] P. Bochnig, R. Reuter, P. Bringmann, and R. Lührmann, *Eur. J. Biochem.* **168,** 461 (1987).

[68] D. M. Gersten and J. J. Marchalonis, *J. Immunol. Methods* **24,** 305 (1978).

and protein A is carried out by recirculation (for 4 hr at 4°) of 3 bed volumes of dimethyl suberimidate [10 mg/ml in 0.2 M triethanolamine (pH 8.5)]. To desorb the non-cross-linked material from the matrix, the column is then washed with 5 bed volumes of PBS, pH 8, and 3 bed volumes of 6 M urea in PBS, pH 8. Finally, the column is washed and equilibrated with 5 bed volumes of anti-m_3G chromatography buffer (see below).

For the preparation of immunoaffinity columns with antibodies that do not bind efficiently to protein A, such as our monoclonal anti-m_3G antibody H-20 (IgG$_1$ isotype), purified antibodies are immobilized on CNBr-activated Sepharose 4B (Pharmacia) according to the manufacturer's protocol. Affinity columns prepared in this way generally contain 6 mg antibody bound per milliliter of packed Sepharose gel.

Preparation of Nuclear Extracts from HeLa Cells

We use two kinds of nuclear extracts from HeLa cells as a source of U snRNPs. In the first procedure, nucleoplasmic ribonucleoprotein particles are obtained from highly purified nuclei by extraction with high-salt buffer containing 50 mM MgCl$_2$ as described by Weinberg and Penman.[69,70] This extract is here termed NX-50. The second type of extract, denoted splicing extract, is prepared essentially by the method of Dignam et al.[71] as modified by Krainer et al.[72]

The principal difference between the snRNPs produced by anti-m_3G affinity chromatography of the two kinds of extract described above lies in the protein composition of the U5 snRNP particles.[18] U5 snRNPs isolated from NX-50 no longer possess the U5-specific proteins, whereas these proteins are present in at least a fraction of U5 snRNPs isolated from splicing extracts. The extraction method is chosen according to the use to which the snRNPs will be put. For example, NX-50 extracts give a higher yield of U4/U6 particles; they also have the important advantage of giving a mixture of snRNPs U1 to U6 after the anti-m_3G chromatography step from which the U1 and U2 snRNP particles are more easily obtained in a pure form than when the splicing extract is used as starting material. The reason for this is that the U5 snRNP particles containing their complement of U5-specific proteins, as obtained from splicing extract but not from NX-50 extract, cofractionate on ion-exchange columns with U1 snRNPs (see below).

[69] R. Weinberg and S. Penman, *Biochim. Biophys. Acta* **190**, 10 (1968).
[70] G. W. Zieve and S. Penman, *J. Mol. Biol.* **145**, 501 (1981).
[71] J. D. Dignam, R. M. Lebovitz, and R. G. Roeder, *Nucleic Acids Res.* **11**, 1475 (1983).
[72] A. R. Krainer, T. Maniatis, B. Ruskin, and M. R. Green, *Cell* **36**, 993 (1984).

Cell Culture

HeLa S3 cells are maintained in suspension at 37° in Eagle's minimum essential medium supplemented with 5% (v/v) newborn calf serum, 50 μg/ml penicillin, and 0.1 mg/ml streptomycin. Cells are grown logarithmically by daily dilution and are harvested at a density of 5–6 × 10^5 cells/ml.

Preparation of Nuclear Extracts (NX-50)

Isolation of Nuclei. In a standard preparation, 5 × 10^9 cells in mid-log phase (5–6 × 10^5 cells/ml) are harvested by centrifugation at 552 g for 10 min and are washed by careful resuspension with a glass pipet in 100 ml of PBS, pH 7.4 [20 mM KH$_2$PO$_4$ (pH 7.4), 130 mM NaCl]. After centrifugation at 175 g for 5 min, the cells are resuspended in 167 ml of RSB [10 mM tris(hydroxymethyl)aminomethane hydrochloride (Tris-HCl) (pH 7.4), 10 mM NaCl, 1.5 mM MgCl$_2$, 0.5 mM dithiothreitol (DTT), and 1 mM phenylmethylsulfonyl fluoride (PMSF) containing 0.5% (v/v) Nonidet P-40 (NP-40)] and agitated vigorously on a vortex mixer for 1 min (rate 8). The nuclei are pelleted by centrifugation at 175 g for 5 min and washed 3 times, first with the same volume of RSB, second with the same volume of RSB containing additionally 1% (v/v) Tween 40 and 0.5% (w/v) sodium deoxycholate, and then with an additional one-half volume of RSB. Each washing is carried out by successive agitation on a vortex mixer (rate 10) and centrifugation as described above.

Extraction of Purified Nuclei. For preparation of nuclear extracts, isolated nuclei are suspended in buffer HSB [10 mM Tris-HCl (pH 7.4), 500 mM NaCl, 50 mM MgCl$_2$, 0.5 mM dithioerythritol (DTE) and 1 mM PMSF]. The cells, suspended at a density of 4 × 10^7 nuclei/ml, are then agitated vigorously on a vortex mixer (rate 10) for 2 min. Structures of higher molecular weight are pelleted by centrifugation at 113,308 g for 90 min, and the supernatant, termed NX-50, is stored in liquid nitrogen.

All the procedures are performed at 0–4°, and the protease inhibitor (PMSF) (100 mM stock solution in 2-propanol), DTT, and detergents are always added to the buffers immediately before use. Buffers (without PMSF, DTE, and detergents), glassware, and other apparatus must be RNase free.

Preparation of Splicing Extracts

All the following steps are performed on ice, and the buffers and the homogenizer are precooled. Harvesting and washing of the HeLa cells with PBS–Earle's medium is performed as described above for the preparation of NX-50 extracts. The washed cells are resuspended in 5 packed cell volumes of buffer A [10 mM N-2-hydroxyethylpiperazine-N-ethane-sulfonic acid (HEPES)–KOH (pH 7.9 at 4°), 1.5 mM MgCl$_2$, 10 mM KCl, 0.5 mM DTT] and left to swell for 10 min. After centrifugation of the cells for 10 min at 1000 g, the cells are resuspended again in 2 packed cell volumes of buffer A. The cell membranes are then disrupted by 10 strokes with a homogenizer (Dounce B). This treatment should leave the nuclei intact. The nuclei are centrifuged down for 10 min at 1000 g. The supernatant is then carefully removed with a pipet.

The nuclear pellet is centrifuged again at 25,000 g for 20 min. After removal of the supernatant, the nuclear pellet is resuspended in buffer C [20 mM HEPES–KOH (pH set to 7.9 at 4°), 25% (v/v) glycerol, 420 mM NaCl, 1.5 mM MgCl$_2$, 0.2 mM ethylenediaminetetraacetic acid (EDTA)–NaOH (pH 8), 0.5 mM DTT, and 0.5 mM PMSF; 3 ml per 10^9 nuclei], transferred to the Dounce homogenizer, and homogenized with 10 strokes of the pestle. The resulting nuclear suspension is stirred slowly for 30 min (on ice). This is followed by centrifugation at 25,000 g for 30 in. The resulting supernatant is the source of U snRNPs and is kept frozen in liquid nitrogen.

One-Step Purification of Nucleoplasmic snRNPs U1, U2, U4/U6, and U5 by Anti-m$_3$G Affinity Chromatography

Isolation of snRNPs U1–U6 from NX-50

All purification steps are performed at 4°. NX-50 from 5 × 10^9 cells (i.e., 125 ml) is passed over a 5-ml affinity column carrying 30 mg monoclonal antibody (MAb) H-20 equilibrated with 10 bed volumes of TMK 300 buffer [300 mM KCl, 10 mM MgCl$_2$, 20 mM Tris-Cl (pH 8), 0.5 mM DTT, and 1 mM PMSF] at a flow rate of 3 ml/hr. (Alternatively, 10 ml of Sepharose containing 200 mg of rabbit anti-m$_3$G IgG may be used.) Any thawed nuclear extracts showing precipitation are clarified by centrifugation at 16,500 g for 10 min before loading onto the affinity column. At this flow rate a single passage of the nuclear extract over the anti-m$_3$G column (usually overnight) suffices for quantitative adsorption to the antibodies of the snRNPs from the extract. The column is then washed with TMK 300 buffer (30 ml/hr) until the absorbance of the eluate at 280 nm reaches zero. Usually, 4 bed volumes of buffer is sufficient. U snRNPs are desorbed from the column by elution with 1 bed volume of TMK 300 buffer containing 15 mM 7-methylguanosine in the following way. After two-thirds of the m^{7}G solution has penetrated the affinity gel matrix (at a flow rate of 6 ml/hr), the elution process is interrupted for 30 min in order to allow the nucleoside m^{7}G to compete with snRNPs for binding to the anti-m$_3$G antibodies. The U snRNPs are finally eluted with the remaining one-third bed volume of the m^7G solution and then with one additional bed volume of TMK 300 buffer. During the whole procedure, 1-ml fractions are collected. The presence of the U snRNPs in the various fractions is followed by protein determination and/or by gel electrophoretic analysis of proteins and snRNAs, respectively. Purified U snRNPs are frozen in liquid nitrogen and stored at $-80°$. The column is regenerated by washing with 3 bed volumes of TMK 300 buffer containing 6 M urea and then with 3 bed volumes of TMK 300–0.02% NaN_3 without urea.

Isolation of snRNPs U1 to U6 from Splicing Extracts

Anti-m_3G affinity chromatography of snRNPs U1 to U6 from splicing extracts is carried out essentially as described above for NX-50, but with the following modifications. The anti-m_3G column is equilibrated with 10 bed volumes of buffer C-5, which is essentially buffer C used for preparation and storage of splicing extract (see Preparation of Splicing Extracts, above) but which contains only 5% glycerol instead of 25%. It is our experience that adsorption of snRNPs to the anti-m_3G column is more efficient at lower glycerol concentrations.

Splicing extracts from 5×10^9 cells (i.e., ~15 ml) are thawed and centrifuged at 16,500 g for 10 min. The supernatant is diluted 5-fold with buffer C containing no glycerol and passed over the anti-m_3G affinity column overnight. The washing of the column and specific desorption of antibody-bound snRNPs by elution with m^7G are carried out essentially as described for NX-50 (see above), except that buffer C-5 is used in all steps.

Comparison of Protein and RNA Composition of snRNPs Isolated from NX-50 and from Splicing Extracts

The protein and RNA compositions of snRNPs from both extracts are displayed in Fig. 1. U snRNPs from NX-50 generally contain more U4/U6 snRNPs than those from splicing extracts. This is sometimes also true of U5 snRNP.

The most striking difference between the two U snRNPs preparations becomes apparent when the protein compositions are compared. Several proteins of molecular weights between 100K and 200K are contained only in the U snRNPs isolated from splicing extracts. They cofractionate on anion-exchange columns and on glycerol gradients with U5 snRNPs and are therefore possible U5-specific proteins[18] (see also below). At least five different proteins of apparent molecular weights 100K, 102K, 116K, and 200K (double band) may be distinguished. Two further proteins of appar-

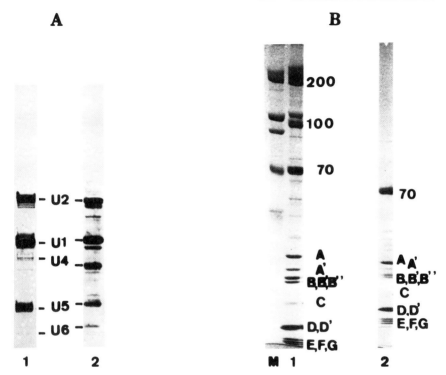

FIG. 1. One-step purification of U1 to U6 snRNPs by immunoaffinity chromatography with anti-m₃G IgG. Affinity chromatography of snRNPs from HeLa splicing extracts (lanes 1) or NX-50 (lanes 2) was performed as outlined in the text. (A) Silver-stained RNAs after fractionation in 10% denaturing (urea) polyacrylamide gels. (B) Coomassie-stained proteins of corresponding snRNP fractions. Lane M contains marker proteins (200K, 116.25K, 92.5K, 66.2K, and 45K from top to bottom).

ent molecular weights 40K and 52K are also enriched in purified U5 snRNP preparations and may therefore also be considered possible U5 snRNP proteins.[18] All the other proteins described so far, i.e., 70K, A, C, A', B'', and the common proteins B' to G are present in both snRNP preparations (Fig. 1B, lanes 1 and 2). In general, the U snRNPs isolated from NX-50 extracts contain fewer proteins in the background than do those purified from splicing extracts.

The yields in snRNPs from splicing extracts are generally about 5–7 mg per 1 × 10¹⁰ cells, whereas NX-50 extracts from the same number of cells produce about 3–5 mg in terms of protein mass. The differences in yield may be due in part to the presence of the high molecular weight proteins in the U5 snRNPs from splicing extracts.

Fractionation of U snRNPs on Anti-m₃G Affinity Columns

It is possible to fractionate the U snRNPs on the anti-m_3G affinity columns by sequential elution of the particles with m^7G at different, controlled salt concentrations. Thus, fractions of snRNPs may be obtained which contain the snRNP species U1, U2, and U5 if the affinity column is eluted with m^7G at low salt concentrations (100 mM KCl or less). U4/U6 RNP particles are desorbed from the column, alongside a further fraction of snRNPs U1, U2, and U5, only at higher KCl concentrations in the m^7G elution buffer. The particular KCl concentrations at which the differential desorption of U4/U6 RNPs can be carried out differ when individual anti-m_3G IgG batches are compared.

With affinity columns prepared with rabbit anti-m_3G IgG preparations, fractionation of snRNPs is achieved by sequential elution with m^7G in the presence of 100 mM KCl (low-salt m^7G eluate) (Fig. 2A, lane 1) and 300 mM KCl (high-salt eluate) (Fig. 2A, lane 2). With MAb H-20 affinity columns, snRNPs U1, U2, and U5 are selectively adsorbed only at 20–50 mM KCl, while U4/U6 particles are already desorbed with m^7G when the salt concentration in the elution buffer is raised to 0.1 M. This chromatographic behavior was observed for snRNPs from human and mouse cells alike.[12] The physical basis for the particular strength of complex formation of the anti-m_3G IgGs with the 5′-terminal cap structure of U4 RNA is not known.

Isolation of snRNPs U2 and U4/U6 by Affinity Chromatography with
  Antibodies Specific for m^6A

The presence of the modified nucleoside m^6A in the snRNPs, U2, U4, and U6, along with the accessibility of the m^6A residues for antibody binding when the snRNAs are part of intact RNP particles,[51] permits the use of anti-m^6A affinity columns for a one-step isolation of the snRNPs U2 and U4/U6 from nuclear extracts.[12] Anti-m^6A immunoaffinity chromatography of snRNPs is carried out along the same lines as described above for the one-step purification of snRNPs with anti-m_3G affinity columns. Since the efficiency of binding to snRNPs of our rabbit anti-m^6A IgG preparations is lower than that of the anti-m_3G IgG, columns with 10 ml of Sepharose carrying about 200 mg of anti-m^6A are generally used for the isolation of snRNPs from nuclear extracts (either NX-50 or splicing extracts) equivalent to 0.5 × 10^9 cells. The antibody-bound snRNPs are recovered in this case by elution with 10 mM m^6A in the appropriate elution buffer (see above). Although the m^6A eluate is generally free of U5

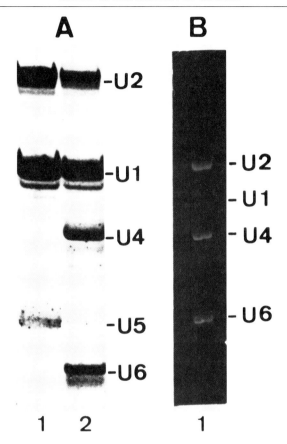

FIG. 2. Fractionation of snRNPs by immunoaffinity chromatography with anti-m₃G IgG (A) or anti-m⁶A IgG (B). (A) Anti-m₃G-bound snRNPs from NX-50 extracts were eluted sequentially with m⁷G in buffer containing 100 mM KCl (lane 1) and 300 mM KCl (lane 2) as described in the text. (B) Affinity chromatography of snRNPs with anti-m⁶A IgG from NX-50. The photographs show the RNAs of corresponding snRNP fractions after fractionation in 10% denaturing polyacrylamide gels stained with silver (A) or ethidium bromide (B).

snRNPs (Fig. 2B), it is important to note that a certain fraction of U1 snRNPs (~10–20% of the U1 snRNPs originally contained in the nuclear extract) is reproducibly coeluted from the anti-m⁶A affinity column.[12] This is true despite the fact that U1 RNA, like U5 RNA, does not contain m⁶A, and it may therefore be an indication of intermolecular interactions between U1 and U2 RNPs.[51,73]

[73] I. W. Mattaj, W. J. Habets, and W. J. van Venrooij, *EMBO J.* **5**, 997 (1986).

Isolation of Individual snRNPs U1, U2, and U5 by Differential
Immunoaffinity Chromatography

The differential immunoaffinity procedure uses antibodies specific for
the nucleosides m_3G and m^6A and antibodies against selected snRNP
proteins together with ion-exchange chromatography.[12] A schematic representation of the fractionation procedure used for the isolation of
snRNPs U1, U2, and U5 from NX-50 is shown in Fig. 3. All solutions
used for the extraction of snRNPs from HeLa cell nuclei and for immunoaffinity chromatography of the snRNPs are autoclaved and contain the
following protease inhibitors: 0.5 mM PMSF, 10 μg/ml aprotinin, and 5
μg/ml leupeptin.

To begin, snRNPs are isolated from NX-50 nuclear extracts equivalent
to 1.2×10^{10} HeLa cell by immunoaffinity chromatography with anti-m_3G
antibodies using the two-step m^7G elution procedure described above.
The detailed procedure is as follows. After overnight loading with the
nuclear extract, the affinity column (bed volume 25 ml of Sepharose,
carrying 500 mg of rabbit anti-m_3G IgG) is washed with 2 bed volumes of
TMK 300 buffer followed by 2 volumes of TMK 100 buffer (same as TMK
300 except for the presence of 100 mM KCl). snRNPs U1, U2, and U5 are
desorbed from the anti-m_3G column by elution with 25 ml of TMK 100
buffer containing 15 mM m^7G (flow rate 0.5 ml/min). After a wash with 25
ml of TMK 100, the affinity column is eluted a second time with 25 ml of
TMK 300 buffer containing 15 mM m^7G. This high-salt m^7G eluate contains the U4/U6 complex and a further fraction of snRNPs U1, U2, and
U5.

Further fractionation of snRNPs U1, U2, and U5 contained in the low-salt m^7G eluate is carried out by ion-exchange chromatography on DEAE-Sepharose. For this purpose, the snRNPs U1, U2, and U5 are dialyzed for
5 hr against 80 volumes of DE-50 buffer [20 mM Tris-HCl (pH 7), 15 mM
MgCl$_2$, 50 mM KCl, 0.2 mM DTT] with a change of dialysis buffer after
2.5 hr. The resulting dialyzate, containing U1, U2, and U5, is applied to a
DEAE-Sepharose (Pharmacia) column (1 ml bed volume/4 mg protein
applied, flow rate 1 ml/min) which has been preequilibrated with DE-50
buffer. Under these conditions, all snRNPs bind to the DEAE matrix.
Some basic high molecular weight proteins contaminating the snRNPs are
removed here and in the following DE-75 wash without loss of the other
snRNPs proteins. After a wash with 3 bed volumes of DE-75 buffer (composition as for DE50 but containing 75 mM KCl), snRNP U1 is eluted
with 3 volumes of DE-175 buffer (DE50 but containing 175 mM KCl), and
0.5-ml fractions are collected. The peak fraction contains snRNP U1, with
a protein concentration in the range of 0.6–0.8 mg/ml (Fig. 4, lane 3).

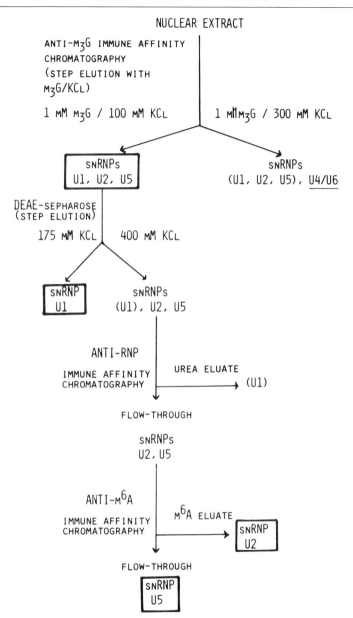

FIG. 3. Fractionation protocol used for the isolation of U1, U2, and U5 snRNPs. Fractions containing the individual snRNPs are boxed.

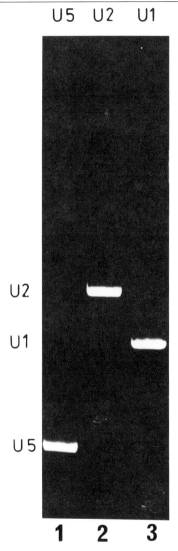

FIG. 4. RNA composition of isolated U1, U2, and U5 snRNPs. Individual U1, U2, and U5 snRNPs were purified from NX-50 extracts as schematically shown in Fig. 3 and described in detail in the text. The RNA moieties from the U snRNPs were fractionated in a 10% polyacrylamide gel containing urea. The photograph shows the ethidium bromide-stained proteins obtained as follows: lane 1, flow-through of the anti-m^6A column (isolated U5 snRNPs); lane 2, m^6A eluate of the anti-m^6A column (isolated U2 snRNPs); lane 3, DE-175 fraction (isolated U1 snRNPs).

Subsequently, snRNPs U2, U5, and a residual amount of U1 are eluted with DE-400 buffer (DE50 but with 400 mM KCl). Again, 0.5-ml fractions are collected, and the protein concentration is about 0.15 mg/ml. The pooled DE-400 eluate is directly applied to an anti-(U1)RNP column equilibrated in DE-400 buffer (bed volume 1 ml, containing 20 mg of purified IgGs from serum, obtained from a patient with mixed connective tissue disease, that contains exclusively autoantibodies against the U1-specific proteins 70K, A, and C). The U1 RNPs are quantitatively retained on the affinity column, and a flow-through fraction is obtained that contains only U2 and U5 snRNPs.

In the final step, the mixture of U2 and U5 RNPs is fractionated by anti-m^{6}A affinity chromatography. For this purpose the flow-through fraction is pooled with the first DE-400 wash fraction and loaded on a 2-ml anti-m^{6}A affinity column preequilibrated with TMK 300 buffer. After loading of the sample (flow rate 0.1 ml/min), the column is washed with 6 ml of TMK 300 buffer. Fractions of 2 ml are collected. The flow-through and the first wash fraction contain pure U5 snRNPs (Fig. 4, lane 1). Desorption of U2 snRNP is achieved by elution with 2 ml of 10 mM m^{6}A nucleoside in TMK 300 buffer followed by 4 ml of TMK 300 only. As can be seen in Fig. 4, lane 2, pure U2 snRNP is obtained in this step. The affinity column is regenerated with 6 M urea in TMK 300 and subsequent equilibration in TMK 300–0.02% (w/v) NaN$_3$. The fractions containing U2 and U5 usually exhibit protein concentrations of about 10 μg/ml and may be concentrated by successive dialysis and DEAE chromatography. The yields obtained with the fractionation are shown in Table I.

TABLE I
RECOVERIES OF snRNPs DURING ISOLATION[a]

snRNP-containing fraction	Protein content (μg)
1. 0.1 M KCl/m^{7}G eluate of anti-m$_3$G column (U1, U2, and U5)	1050
2. 0.3 M KCl/m^{7}G eluate of anti-m$_3$G column (U1–U6)	1920
3. DE-175 eluate (isolated U1 RNPs)	400
4. DE-400 eluate [U2, U5, (U1)]	200
5. Flow-through anti-(U1)RNP column (U2, U5)	140
6. Flow-through anti-m^{6}A column (purified U5 RNPs)	45
7. m^{6}A eluate (isolated U2 RNPs)	55

[a] Recoveries of snRNPs are expressed in micrograms snRNP protein. Nuclear extracts (NX-50) from 1.2×10^{10} HeLa cells were used as the starting material for the initial fractionation of snRNPs on anti-m$_3$G affinity columns (steps 1 and 2). The snRNPs contained in the low-salt m^{7}G eluate (1050 μg) were used for further fractionation (steps 3–7).

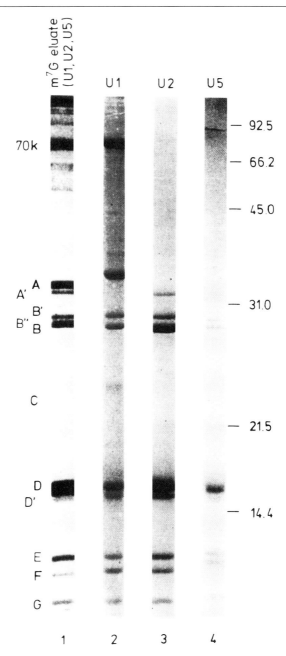

FIG. 5. Protein composition of isolated U1, U2, and U5 snRNPs. Individual U1, U2, and U5 snRNPs were purified from NX-50 extracts as schematically shown in Fig. 3. Proteins

If NX-50 extracts are used as a source of U snRNPs, this purification procedure produces pure snRNPs U1 and U2 containing the common snRNP proteins B' to G as well as the particle-specific proteins 70K, A, and C (for U1) (Fig. 5, lane 2) and B" and A' (for U2) (Fig. 5, lane 3). The U5 snRNPs obtained in this way contain only the common proteins B' to G (Fig. 5, lane 4).

The same purification protocol may also be applied to the snRNPs isolated from splicing extracts. In this case, however, U5 snRNP particles containing the U5-specific high molecular weight proteins (see next section) will cofractionate with U1 snRNPs on the DEAE-Sepharose column. The mixture of U1 and U5 snRNPs may then be fractionated by a centrifugation step on a glycerol gradient (see below). Those U5 RNPs contained in splicing extracts that lack the U5-specific proteins will fractionate as described above for the snRNPs from NX-50.

Fractionation of Anti-m$_3$G Affinity-Purified snRNPs by Anion-Exchange Chromatography on Mono-Q Columns

The advantage of a quick, one-step purification of the mixture of U1 to U6 snRNPs by anti-m$_3$G affinity chromatography may be combined with the advantages of carrying out subsequent anion-exchange chromatography on Mono-Q resin. In this section, Mono-Q fractionation is described for the mixture of U1, U2, U5, and U4/U6 snRNPs isolated from splicing extracts by anti-m$_3$G immunoaffinity chromatography. The same protocol may, of course, be used for the separation of any other snRNP mixture obtained by the various immunoaffinity chromatography steps described above.

A mixture of about 10 mg of U1 to U6 snRNPs from splicing extracts in approximately 30 ml of the m^{7}G elution buffer used in the anti-m$_3$G affinity chromatography (containing 420 mM NaCl, see also Isolation of snRNAs U1 to U6 from Splicing Extracts) is diluted 3-fold with 2 volumes of Mono-Q-0 buffer [20 mM Tris-HCl (pH 7), 1.5 mM MgCl$_2$, 0.5 mM DTT, 1 mM PMSF, 2 μg/ml leupeptin] in order to reduce the concentration of univalent electrolyte. The snRNPs are then allowed to adsorb to a Mono-Q column (1 ml bed volume) and equilibrated with Mono-Q-50 buffer (Mono-Q-0 buffer but containing 50 mM KCl) by passing the

---

were fractionated in a 15% polyacrylamide gel in the presence of SDS. The photograph shows the Coomassie-stained proteins obtained as follows: lane 1, 0.1 M KCl/m^{7}G eluate (U1, U2, and U5 snRNPs); lane 2, DE-175 fraction (isolated U1 snRNPs); lane 3, m^{6}A eluate of the anti-m^{6}A column (isolated U2 snRNPs); lane 4, flow-through of the anti-m^{6}A column (isolated U5 snRNPs).

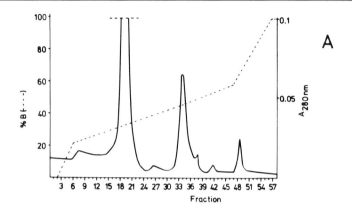

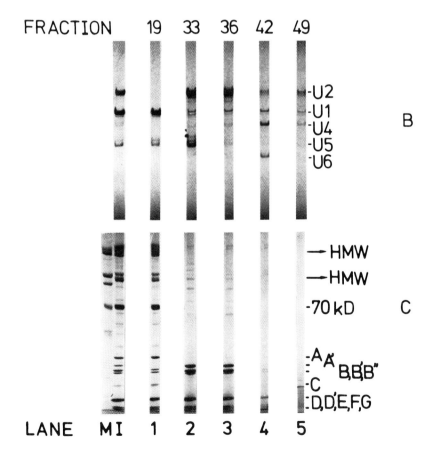

snRNP solution over the column at a flow rate of 2 ml/min (pressure ~3 MPa). All steps are carried out at 4°, and solutions are precooled on ice. The m^7G nucleoside is found in the flow-through of the column as well as contaminating basic proteins contained in the mixture of snRNPs U1 to U6. The column is washed with Mono-Q-50 buffer until the absorbance at 280 nm reaches zero, which normally requires about 4 bed volumes. The snRNPs are then fractionated by elution with a linear KCl gradient (50 mM–1 M KCl) in Mono-Q-0 buffer (~25 ml each, see the gradient shape in Fig. 6A) at a flow rate of 1 ml/min, with 1-ml fractions being taken. Elution of the snRNPs is monitored by measuring the absorbance at 280 nm. Figure 6B,C shows the RNA and protein compositions of the respective RNP peaks resolved in Fig. 6A.

The first peak, eluting at about 370 mM KCl (denoted Q-370 eluate), contains most of the total U1 snRNPs originally present in the snRNP mixture. In addition, this peak contains a fraction (~15–20%) of the total U5 snRNPs. The greater part of the U5 snRNP elutes at 480 mM KCl (denoted Q-480 eluate). The major fraction of U2 snRNP elutes slightly behind the U5 RNP peak at 490 mM KCl (Q-490 eluate). U4/U6 snRNPs are eluted last, at a KCl concentration of about 550 mM (Q-550 eluate).

The protein compositions of the respective snRNP fractions are shown in Fig. 6C. There are several high molecular weight proteins contained in the Q-370 eluate. When the proteins are fractionated on lower percentage polyacrylamide gels, at least five proteins with apparent molecular weights of about 100K, 102K, 116K, and a double band at 200K may be distinguished (Fig. 6C). These proteins are associated with the U5 snRNPs contained in this peak. This has been recently demonstrated by centrifugation in a 5–20% glycerol gradient, where the high molecular weight proteins cosediment with U5 RNA at about 20 S[18] (the glycerol gradient centrifugation step may also be used to purify the 20 S U5 snRNP

Fig. 6. Mono-Q chromatography of U1 to U6 snRNPs affinity-purified with anti-m_3G IgG from splicing extracts. Chromatography of snRNPs was performed at 4° as described in the text. (A) Elution profile of snRNPs from the Mono-Q column, where buffer B corresponds to the Mono-Q-1000 buffer (containing 1 M KCl). The segmented gradient was performed with a flow rate of 1 ml/min as follows: 4 min at 5.4% increase of buffer B/min, 28 min at 0.9% increase of buffer B/min, 13 min at 1% increase of buffer B/min, and 10 min at 4.2% increase of buffer B/min. (B) RNA from the snRNP fractions indicated at the top of each lane was extracted, electrophoresed in a denaturing 10% polyacrylamide gel, and stained with silver. (C) Proteins from the corresponding snRNP fractions as shown in (B) were extracted, separated in a 4–20% polyacrylamide gradient gel in the presence of SDS, and stained with Coomassie blue. Lane M contains molecular weight markers (200K, 116.25K, 92.5K, 66.2K, and 45K from top to bottom).

particle). Interestingly, the greater part of U5 snRNPs from splicing extracts contained in the Q-480 eluate lacks the U5-specific proteins.
Whether this heterogeneity in protein composition of U5 snRNPs reflects
functional heterogeneity or simply loose association of the U5-specific
proteins with the core U5 snRNP particle is not yet known.

The separation of the individual snRNP species from each other may
even be improved by applying shallower gradients in particular regions of
KCl concentration. If this is done between 320 and 400 mM KCl, for
example, U5 snRNPs containing the U5-specific proteins may be obtained, and these U5 snRNPs are 80–90% pure. Likewise, U1 snRNPs,
which elute at slightly higher salt concentrations than U5 RNPs, can also
be obtained at a purity of 80–90%.

The overall yield of snRNPs fractionated by Mono-Q chromatography
is generally higher than 50%. Furthermore, the snRNPs may be obtained
in high concentrations. For example, the Q-370 eluate (containing U1 and
U5 RNPs) gives protein concentrations of 2–3 mg/ml, if 10 mg of snRNPs
U1–U6 have originally been applied to the column. Most important, the
snRNPs obtained by these procedures are functionally active and restore
the splicing of pre-mRNA molecules *in vitro* on addition to HeLa nuclear
extracts that have been depleted of their endogenous snRNPs before-
hand.[74]

Isolation of U1 snRNP Particles by Mono-Q Chromatography
Lacking One of the U1-Specific Proteins A, C, and 70K or Defined Sets Thereof

The high power of resolution of the Mono-Q resin allows the separation and purification of U1 snRNP particles that differ in their protein
compositions, in particular with respect to the presence of the U1-specific
proteins C, A, and 70K.[55] Although such protein-deficient U1 snRNP
particles may already be observed to some extent when snRNPs are
fractionated on Mono-Q columns in the cold room (see the additional
peaks in Fig. 6A eluting between the major U1 and U2 snRNP-containing
peaks), the yield of these various U1 snRNP populations are considerably
increased when the snRNPs are fractionated on Mono-Q columns at
higher temperatures. Apparently, the U1-specific proteins are more
loosely associated with the U1 RNA than the set of common snRNP
proteins. This would explain the preferential dissociation of proteins C,
A, and 70K in the presence of the strong Mono-Q anion-exchange resin,

[74] G. Winkelmann, M. Bach, and R. Lührmann, *EMBO J.* in press (1989).

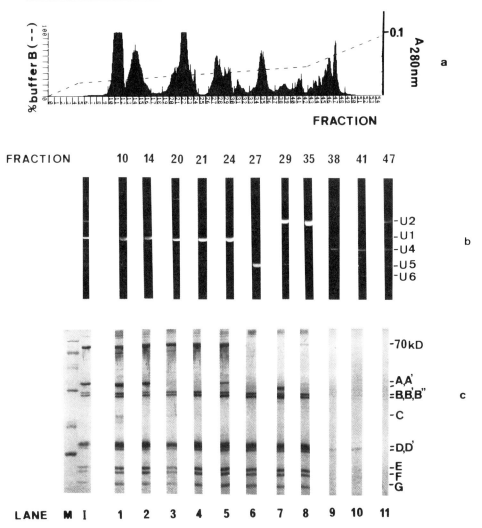

FIG. 7. Mono-Q chromatography of U1 to U6 snRNPs, affinity-purified from NX-50 extracts with anti-m₃G IgG. Chromatography of snRNPs was performed at room temperature as described in the text. (a) Elution profile of snRNPs from the Mono-Q column. The segmented gradient was performed with a flow rate of 1 ml/min as follows: 4 min at 5.4% increase of buffer B/min, 30 min at 1% increase of buffer B/min, and 10 min at 4.2% increase of buffer B/min. (b) RNA from the snRNP fractions indicated at the top of each lane was extracted, electrophoresed in a denaturing 10% polyacrylamide gel, and stained with ethidium bromide. (c) Proteins from the corresponding snRNP fractions as shown in (B) were extracted, separated in a 15% polyacrylamide gel in the presence of SDS, and stained with Coomassie blue. Lane M contains molecular weight markers (92.5K, 66.2K, 45K, 31K, 21.5K, and 14.4K from top to bottom).

which competes with the U1 snRNP proteins for binding to negatively charged regions of the U1 RNA. Figure 7 shows an example where a mixture of anti-m$_3$G affinity-purified snRNPs U1 to U6 from NX-50 (1 mg) has been subjected to anion-exchange chromatography on a 1-ml Mono-Q column under the same conditions as described above with the only exception that chromatography has been performed at room temperature instead of 4°.

On the basis of the protein gel electrophoretic pattern shown in Fig. 7c, the following U1 snRNP populations may be distinguished. Fraction 10 from the first peak contains U1 snRNPs which include all three U1-specific proteins (denoted full complement U1 snRNP) while fractions 14, 20, and 21 contain U1 snRNP species which all include the 70K protein but lack either proteins C (fraction 14), A (fraction 20), or both A and C (fraction 21). Fraction 24 contains a U1 snRNP species which lacks protein C and, by comparison of the intensity of Coomassie staining, displays less A protein than U1 snRNPs in fraction 14. It is tempting to suggest that this particle has lost only one of the two copies of A protein which are thought to be present in full complement U1 snRNPs.[11] Fraction 27 contains essentially pure U5 snRNPs lacking the high molecular weight U5 snRNP-specific proteins, and fractions 29 and 35 contain predominantly U2 snRNPs. Interestingly enough, the second U2 snRNP peak contains particles which lack completely the U2-specific proteins A' and B'' (Fig. 7c, lane 8). Thus, the same procedure can also be used to isolate protein-deficient U2 snRNPs.

It is further possible to produce U1 snRNP particles which lack all of the U1-specific proteins A, C, and 70K. For this purpose, full complement U1 RNPs isolated from the Mono-Q column at room temperature are rechromatographed on the Mono-Q column at 37°. From this second fractionation step, in addition to the A- and C-deficient U1 snRNP populations described above, a considerable amount (~20% of the original U1 RNPs applied to the column) of U1 snRNPs lacking proteins A, C, and 70K are recovered from the column at 0.47 M salt (not shown). As yet no U1 snRNP particles have been obtained which lack the 70K protein but retain the A and/or C protein(s). Qualitatively similar results are obtained when snRNPs U1 to U6, isolated by anti-m$_3$G affinity chromatography from splicing extracts, are fractionated on Mono-Q columns under the same conditions as described above.

The individual U1 snRNP particles lacking particular sets of U1-specific proteins should be ideally suited for studying the structure–function relationships of U1 snRNPs. In particular, they should be useful for investigating the contribution of the U1-specific proteins to the functioning of U1 snRNP during splicing of pre-mRNA molecules.

Acknowledgments

We thank Gaby Gutschmidt, Rosita Haupt, Irene Öechsner, Susanne Rothe, and Dorit Zeuske for excellent assistance throughout this work and Verena Buckow for help during the preparation of the manuscript. This work was supported by grants from the Deutsche Forschungsgemeinschaft. M. Bach was supported in part by a fellowship from the Alexander-von-Humboldt-Stiftung.

[20] Isolation of Small Nuclear Ribonucleoprotein Polypeptides

By SALLIE O. HOCH

Introduction

The preceding chapters have described the biochemical and immunochemical isolation of the snRNP (small nuclear ribonucleoprotein) particles involved in RNA processing. snRNP particles contain 5 major snRNA species (U1, U2, U4, U5, U6) and a minimum of 11 individual polypeptides designated as follows: (U1)RNP (70,000 MW), A (32,000 MW), A' (31,000 MW), B' (27,000 MW), B'' (26,500 MW), B (26,000 MW), C (18,500 MW), D (13,000 MW), E/F (11,000 MW doublet) and G (<10,000 MW).[1,2] These include the common or core polypeptides shared by all the snRNPs, namely, B', B, D, E, F, and G, and the so-called unique polypeptides associated with a particular snRNP, such as the (U1)snRNP-specific proteins 70K, A, and C and the (U2)snRNP proteins A' and B''. The molecular weights are estimates based on migration under standard Laemmli[3] conditions of sodium dodecyl sulfate–polyacrylamide gel electrophoresis (SDS–PAGE). At least one of the proteins may exhibit strikingly aberrant migration under different electrophoresis conditions. The 70K polypeptide exhibits molecular weights from 57,000 to 70,000 depending on the specific gel conditions;[4,5] recent cDNA sequence analysis relative to the protein suggested that its molecular weight is actually 52,000.[6] The snRNP polypeptides (with the possible exception of G) are

[1] M. R. Lerner and J. A. Steitz, *Proc. Natl. Acad. Sci. U.S.A.* **76,** 5495 (1979).
[2] P. Bringmann and R. Lührmann, *EMBO J.* **5,** 3509 (1986).
[3] U. K. Laemmli, *Nature (London)* **227,** 680 (1970).
[4] S. G. Harris, S. O. Hoch, and H. C. Smith, *Biochemistry* **27,** 4595 (1988).
[5] P. B. Billings and S. O. Hoch, *J. Biol. Chem.* **259,** 12850 (1984).
[6] R. A. Spritz, K. Strunk, C. S. Surowy, S. O. Hoch, D. E. Barton, and U. Francke, *Nucleic Acids Res.* **15,** 10373 (1987).

immunoreactive, using human autoantibodies designated anti-Sm or anti-RNP that are associated with the rheumatic diseases, systemic lupus erythematosus, and mixed connective tissue disease.[7]

While the snRNPs have been readily amenable to isolation, the individual protein species have not. The snRNP particles per se have proved quite resistant to complete disruption under a variety of chromatographic conditions, including the use of nucleases, salt, and denaturants to disrupt the complex. The approach that has been used successfully by several laboratories utilizes fractionation of these peptides under conditions of SDS-PAGE followed by electroelution.[8,9]

Isolation of Sm snRNPs

Extraction. Extracts are prepared from suspension cultures of HeLa cells.[5] Cells are harvested at a density of approximately $3-5 \times 10^5$ cells/ml and collected by centrifugation. Cell pellets are resuspended in one-tenth the original volume of the cell culture supernatant to which glycerol is added to a final concentration of 10% (v/v). The cells are stored in sealed plastic bags at $-70°$ until use. Approximately 2 liters of cells (original volume) are routinely used in making a cell extract. The frozen cells are thawed, collected by centrifugation, and washed once with 200 ml of phosphate-buffered saline [(PBS) 10 mM sodium phosphate (pH 7.4), 0.15 M NaCl]. The cells are then resuspended in extraction buffer [10 mM tris(hydroxymethyl)aminomethane hydrochloride (Tris-HCl) (pH 7.4), 0.35 M NaCl, 1.5 mM MgCl$_2$, 0.2 mM phenylmethylsulfonyl fluoride] at a ratio of 8–10 volumes buffer/volume cell pellet. The suspension is divided into four aliquots, and each aliquot is sonicated on ice for 15 sec, rested for 30 sec, and the cycle repeated [using a Branson sonicator with a microtip at output setting 4 (40 W)]. The cells are held on ice for 15 min, and an aliquot is checked for breakage under a light microscope. After a low-speed centrifugation (10,000 g for 15 min), extracts are spun at approximately 110,000 g for 60 min.

Immunoaffinity Chromatography. The clarified supernatant is passed slowly over an immunoaffinity column (2 $\times$ 5 cm) and washed through with the extraction buffer. The IgG, derived from a polyclonal human serum with anti-Sm reactivity, is coupled at 5–12 mg/g (wet weight) to

---

[7] I. Pettersson, G. Wang, E. I. Smith, H. Wigzell, E. Hedfors, J. Horn, and G. C. Sharp, *Arthritis Rheum.* **29**, 986 (1986).

[8] L. A. Rokeach, J. A. Haselby, and S. O. Hoch, *Proc. Natl. Acad. Sci. U.S.A.* **85**, 4832 (1988).

[9] R. Reuter, S. Rothe, and R. Lührmann, *Nucleic Acids Res.* **15**, 4021 (1987).

Sepharose CL-4B by the carbonyldiimidazole procedure.[10] After washing through nonbound material, the column is eluted with 6 M urea, 0.1 M NaCl, 10 mM Tris-HCl (pH 7.4), 0.2 mM phenylmethylsulfonyl fluoride. The peak tubes, based on A_{280}/A_{260} readings, are pooled. Carrier RNA (20–40 μg/ml) is added to the pooled eluate, which is concentrated by 70% ethanol precipitation at $-20°$ overnight.

This protocol has been designed such that the extraction and column chromatography are carried out in a matter of hours to permit snRNP isolation under conditions which minimize both protease and nuclease exposure. It can be noted that there is an alternative, abundant source of the snRNP polypeptides also in use in our laboratory, namely, acetone-extracted rabbit thymus (Pel-freez Biologicals, Rogers, AR; Immunovision, Springdale, AR). It has the disadvantage that there is some degradation of the snRNA during tissue extraction, but the snRNP polypeptides remain intact and can be isolated as a complex. The isolation procedure is quite similar to that described for HeLa cells.[11] The lyophilized rabbit thymus powder (1.5 g) is suspended in extraction buffer at 50 mg/ml. The suspension is immediately sonicated on ice for 30 sec, rested for 30 sec, and the cycle repeated 2 times [using a Branson sonicator with the macrotip at setting 5 (50 W)]. The cells are held on ice for 15 min before a low-speed centrifugation (10,000 g for 15 min). The clarified supernatant is then passed over the immunoaffinity column.

Isolation of Individual snRNP Polypeptides

Electroelution. The electroelution protocol that we use is essentially that of Hunkapiller *et al.*[12] It has proved ideal for our purposes because for each polypeptide band, one elutes microgram amounts of protein out of a large block of acrylamide (10–11 cm in width) into a microliter volume. The primary modification that we made was to change the staining conditions prior to excising gel bands.[8] A number of the snRNP polypeptides (e.g., B'/B) are quite closely spaced even on long gels such that we must be able to clearly differentiate them after the usual destaining period. Thus, the concentration of the Coomassie blue stain was decreased from 0.5 to 0.05%. The following is an abbreviated version of the electroelution protocol, as it has been described in detail in this series by the original authors.[12]

[10] M. T. W. Hearn, E. L. Harris, J. S. Bethell, W. S. Hancock, and J. A. Ayers, *J. Chromatogr.* **218**, 509 (1981).
[11] P. B. Billings and S. O. Hoch, *J. Immunol.* **131**, 347 (1983).
[12] M. W. Hunkapiller, E. Lujan, F. Ostrander, and L. E. Hood, this series, Vol. 91, p. 227.

Preparative Gel

Use sterile water, pipets, and glassware throughout. Chemicals utilized are reagent quality or better as described.[12] The SDS (Serva, Heidelberg, FRG) is recrystallized from ethanol prior to use. Gels are run using the standard Laemmli[3] procedure.

1. Pour an 11% lower gel (1.5 mm thickness) of 18 cm length. Polymerize the lower gel either overnight or several hours before use in order to reduce side chain destruction of the proteins.
2. Polymerize the upper gel with a 10-cm block well and a single molecular weight standard well.
3. To the cathode buffer reservoir (top chamber) add running buffer with 0.1 mM sodium thioglycolate.
4. Samples that are to be used on the gel should not be boiled. Heat them at 60° for 15 min in incubation solution [62.5 mM Tris-HCl (pH 6.8), 2.3% SDS, 5% 2-mercaptoethanol, 20% glycerol, 0.001% bromphenol blue].
5. The snRNP sample in SDS is loaded at a concentration of approximately 100 μg/cm.
6. The gel can be either run slowly overnight (6 mA for 19 hr) or quickly (40 mA for 5.5 hr).

Gel Stain

1. Stain the gel for 20 min in 0.05% Coomassie blue in acetic acid–2-propanol–water (1 : 3 : 6, by volume). Shake at room temperature.
2. Destain should be rapid (2 hr) at 4° using Kimwipes or plastic foam in the corner of the dish. Destain with acetic acid–methanol–water (50 : 165 : 785, by volume). Photocopy at this time to keep a record of what bands were cut out. The pattern of snRNP polypeptides (70K to G) is easily recognizable.
3. Slice stained bands from the gel with a sharp razor blade. Transfer each gel band to a plastic round-bottomed 15-ml tube and soak in sterile water for 2 hr total with several changes of water (approximately 3–4 changes).
4. Suction water off the gel and process immediately for elution or store at −20° (freezer with manual defrost) in a sealed plastic tube without water.

Preparation of Dialysis Membrane

1. Spectra/Por 6 tubular dialysis membrane with a molecular weight cutoff value of 10,000 is purchased from Spectrum Medical Industries (Los Angeles, CA).

2. The tubing is cut into 6-inch lengths, soaked in 1% NaHCO₃ solution at 60° for 1 hr, and then washed with water. It is soaked in 0.1% SDS solution at 60° for 1 hr, and washed again with water.
3. The clean tubing is spread onto a paper towel, and disks of membrane are cut using a sharp No. 8 cork borer. Use approximately 5–6 sheets of white folded paper towels or absorbent paper as the cutting surface.
4. The disks cut from the upper surface of the flat tubing are rinsed with water and stored in 0.1% SDS, 0.1% NaN₃ solution at room temperature. The disks cut from the lower surface of the tubing in contact with the paper towel are discarded.

Gel Electroelution[12]

The electroelution chamber was purchased from the Instrumentation Laboratory at the California Institute of Technology (Division of Biology).

1. Set up each elution cell with the prepared dialysis tubing and test for leaks by filling with sterile water and placing it on paper towels for 10 min. Then drain well.
2. Transfer the gel bands into petri dishes and cover with 10–20 ml water. Dice gel into 1-mm cubes with a new razor blade. Try not to mash gel.
3. Remove water and small gel fragments with a Pasteur pipet and blot with filter paper.
4. At this point, process each sample one at a time. Soak the gel in elution buffer (0.1% SDS in 50 mM NH₄HCO₃) for less than 5 min and then remove buffer. If the gel soaks too long, color starts leaking out of gel.
5. Use a spatula to transfer gel pieces into the large well of the electrophoretic eluter. Cover gel with 0.5 ml of soaking buffer (2% SDS in 0.4 M NH₄HCO₃) and add 5–50 μl of 10% dithiothreitol (DTT) to make the soaking buffer 0.1%[12]–1%[8] (v/v) in the reducing agent. With the addition of DTT, gently mix the gel pieces around in the buffer with the Eppendorf tip. Gently overlay the soaking buffer with elution buffer until the elution cell is filled on both sides to just above the cross passage. Remove any large bubbles in the cross passage by tilting the cell.
6. Insert the elution cell into the elution tank and add elution buffer (0.1% SDS in 50 mM NH₄HCO₃) to a level just above the drain ports in each electrode chamber. Add another 75 ml of elution buffer to the mixing chamber and use a two-channel peristaltic pump to move the buffer at 3 ml/min per line between the mixing

chamber and the electrode chamber. Buffer should drop rather than stream into and out of electrode chambers. Remove any bubbles under the elution cell with a bent-tip Pasteur pipet. It is easier to do this by drawing out the air below the gel rather than trying to force the bubble out with a stream of liquid.

7. Soak the gel pieces for 3–5 hr and then connect the electrode terminals. The cathode (black) should be near the well with the sliced gel pieces. Run at 50 V, constant voltage. Allow the elution to proceed for 12–19 hr.

8. Aspirate off the elution buffer. Rinse with water (~20 ml). Fill the tanks with dialysis buffer (0.02% SDS in 10 mM NH$_4$HCO$_3$) as before. Remove air bubbles under the cells. Dialyze at 80 V for 20–24 hr. At the end of this step, the Coomassie blue stain and protein should be concentrated in the small sample collection well.

9. Turn off the power supply. Remove the cell from the tank. Remove all buffer except that in the conical well with a Pasteur pipet. Start with removing buffer from the well with gel pieces, then the cross passage, and finally the buffer in the last well down to the blue color.

10. Use a Pasteur pipet to mix the remaining protein solution by filling and emptying the pipet. Then transfer the protein solution (0.2–0.4 ml) to a 1.5-ml Eppendorf tube. Rinse the collection well with 25–50 μl of fresh dialysis buffer and add this to the Eppendorf tube using a 50-μl Hamilton syringe.

11. Measure the total volume with a 500-μl Hamilton syringe. Transfer 10% to another Eppendorf tube. The bulk of the eluted sample is stored at $-70°$.

Polypeptide Characterization. The electroeluted proteins are each checked for homogeneity by both stain[13] and immunoblot.[14,15] Duplicate aliquots, equal to 5% of the total elution volume, are mixed with an equal volume of 2× SDS incubation solution, boiled for 3 min, and then run on an 11% polyacrylamide gel (18 cm) in the presence of SDS. One-half of the samples are silver stained following the Morrissey protocol[13] with the following changes. The gel is fixed in 25% 2-propanol, 10% acetic acid for 1 hr, followed by 10% acetic acid for 1 hr. The gel is washed extensively in distilled water for 45 min until the pH is between 4 and 5. The gel is then soaked in DTT (5 μg/ml) for 1 hr (step 4 of the original protocol); the

[13] J. H. Morrissey, *Anal. Biochem.* **117**, 307 (1981).
[14] H. Towbin, T. Staehelin, and J. Gordon, *Proc. Natl. Acad. Sci. U.S.A.* **76**, 4350 (1979).
[15] P. B. Billings, J. R. Barton, and S. O. Hoch, *J. Immunol.* **135**, 428 (1985).

remaining steps are unchanged. The remaining gel samples are transferred to nitrocellulose filters at 70 V for 2–3 hr in 12.5 mM Tris-OH–96 mM glycine buffer with 20% (v/v) methanol. The filter replicas are saturated for 30 min on a rocker platform in PBS containing 5% powdered milk. After the filters are rinsed briefly in PBS, they are incubated for 2–3 hr at room temperature with a pool of anti-Sm/RNP human sera, each at their appropriate dilution in 5% milk. Immune complexes are identified by reaction with iodinated protein A (2 × 10[5] cpm/ml) for 30 min. The immunoreactive species are visualized by autoradiography, using Kodak XAR film at −70° with an intensifying screen. The protein blot proves to be the most sensitive screen, at times identifying minor secondary bands not seen with silver staining.

Applications

snRNP polypeptides, thus isolated, have been used for a variety of applications, namely, microsequencing, generation of monospecific antibody probes, and biochemical characterization. Amino-terminal sequences derived from electroeluted proteins have been invaluable in designing oligonucleotide probes that were used successfully to screen cDNA libraries for the D[8] and B′/B[16] core snRNP polypeptides. Likewise, internal sequences derived from electroeluted 70K polypeptide were used to verify the cloning and expression of the (U1)snRNP-specific protein.[17] Rabbit antisera were generated against heterologous (human)[9] and homologous (S. Hoch, unpublished) snRNP polypeptides, encompassing the 70K, A, B′, B, and D species. In the former case, the proteins were subjected to ion-pair extraction prior to immunization, following the protocol of Konigsberg and Henderson[18] to remove SDS from small quantities of protein. These monospecific antibody probes were in turn utilized to discriminate structural homologies among the various snRNP proteins. To further discern shared components, the proteins themselves have been subjected to peptide mapping.[9] The applications to date have focused on structure rather than function as it is not yet clear whether biological activity is restored to these electroeluted proteins on removal of SDS.

[16] L. Rokeach, M. Jannatipour, J. Haselby, and S. Hoch, *J. Biol. Chem.* **264,** 5024 (1989).
[17] H. Thiessen, M. Etzerodt, R. Reuter, C. Schneider, F. Lottspeich, P. Argos, R. Lührmann, and L. Philipson, *EMBO J.* **5,** 3209 (1986).
[18] W. H. Konigsberg and L. Henderson, this series, Vol. 91, p. 254.

[21] Purification and Characterization of U Small Nuclear Ribonucleoproteins in Cesium Chloride Gradients

By CLAUDE BRUNEL and GUY CATHALA

Introduction

Since the discovery that antibodies from patients suffering from autoimmune diseases precipitate U small nuclear ribonucleoproteins (snRNPs) from Ehrlich ascites tumor and HeLa cells[1] and the hypothesis that U snRNPs could be involved in splicing,[2,3] isolation of native snRNPs has been attempted (see Ref. 4 for review). The procedures were based either on conventional fractionation methods[5,6] or on the utilization of antibodies against the unusual cap ($m_3^{2,2,7}G$) (see Refs. 7–9) which is found in all U snRNAs except U6.

In our laboratory, we have developed a protocol based on the stability of U snRNPs in high salt concentrations, allowing their purification by isopycnic centrifugation in cesium chloride[10–12] gradients. This technique has several advantages. First, it leads to the simultaneous isolation of functional U1 and U5 snRNPs starting either from nucleoplasmic extracts or from rapidly sedimenting structures (heterogeneous nuclear RNPs, hnRNPs) that contain a majority of U snRNPs. Second, it can be easily

[1] M. R. Lerner and J. A. Steitz, *Proc. Natl. Acad. Sci. U.S.A.* **76**, 5495 (1979).
[2] M. R. Lerner, J. Boyle, S. Mount, S. Wolin, and J. A. Steitz, *Nature (London)* **283**, 220 (1980).
[3] J. Rogers and R. Wall, *Proc. Natl. Acad. Sci. U.S.A.* **77**, 1877 (1980).
[4] C. Brunel, J. Sri-Widada, and P. Jeanteur, *in* "Progress in Molecular and Subcellular Biology" (F. E. Hahn, D. J. Kopecko, and W. E. G. Müller, eds.), Vol. 9, p. 1. Springer-Verlag, Berlin, Heidelberg, New York, and Tokyo, 1985.
[5] M. Hinterberger, I. Pettersson, and J. A. Steitz, *J. Biol. Chem.* **258**, 2604 (1983).
[6] C. S. Kinlaw, B. L. Robberson, and S. M. Berget, *J. Biol. Chem.* **258**, 7181 (1983).
[7] P. Bringmann, R. Reuter, J. Rinke, B. Appel, R. Bald, and R. Lührmann, *J. Biol. Chem.* **258**, 2745 (1983).
[8] P. Bringmann, J. Rinke, B. Appel, R. Reuter, and R. Lührmann, *EMBO J.* **2**, 1129 (1983).
[9] P. Bringmann and R. Lührmann, *EMBO J.* **5**, 3509 (1986).
[10] C. Brunel, J. Sri-Widada, M. N. Lelay, P. Jeanteur, and J. P. Liautard, *Nucleic Acids Res.* **9**, 815 (1981).
[11] J. Sri-Widada, C. Assens, J. P. Liautard, P. Jeanteur, and C. Brunel, *Biochem. Biophys. Res. Commun.* **104**, 457 (1982).
[12] M. N. Lelay-Taha, I. Reveillaud, J. Sri-Widada, C. Brunel, and P. Jeanteur, *J. Mol. Biol.* **189**, 519 (1986).

combined with other techniques of biochemical or immunological fractionation. Third, cesium chloride centrifugation in the presence of sarkosyl allows isolation of a core snRNP structure[10,11] whose protein complement consists exclusively of the small D–G proteins common to U1, U2, U4–U6, and U5 snRNPs, tightly bound to an $A(U)_nG$ sequence present in all but U6 snRNAs[13] and now referred to as the Sm binding site. Of course, centrifugation in cesium chloride–sarkosyl can be applied either to an isolated U snRNP (U1 or U5) or to a mixture containing all U1–U6 snRNPs. Fourth, isopycnic centrifugation in cesium chloride appears to be a useful procedure to monitor the *in vitro* assembly of U snRNPs starting with SP6 RNAs and S100 extracts.

General Procedure to Simultaneously Isolate U1 and U5 snRNPs in Native Form

The strong stabilizing effect of Mg^{2+} on RNA–protein interactions in U1 and U2 snRNPs,[14] and possibly in others, suggested that the snRNPs would withstand isopycnic density-gradient centrifugation in cesium chloride when carried out in high Mg^{2+} concentration (15 mM). Using this procedure, it has been possible to obtain fractions containing exclusively U1 and the less abundant U5 snRNPs which were then separated by anion-exchange chromatography.[12] U2 and U4–U6 snRNPs and incomplete forms of U1 and U5 comigrate; all attempts at separating them have been unsuccessful to date.

Buffers[5]

Buffer 1: 50 mM potassium phosphate (pH 7.5), 90 mM KCl, 10 mM NaCl

Buffer 2: 50 mM potassium phosphate (pH 7.5), 50 mM KCl, 10 mM NaCl, 1 mM ethylenediaminetetraacetic acid (EDTA), 0.1% (v/v) Triton X-100

Buffer 3: 5 mM potassium phosphate (pH 7.5), 25 mM KCl, 1 mM EDTA, 800 mM sucrose

Buffer 4: 5 mM potassium phosphate (pH 7.5), 25 mM KCl, 1 mM EDTA, 250 mM sucrose

Buffer 5: 20 mM potassium phosphate (pH 7.5), 100 mM KCl, 300 mM NaCl, 1 mM EDTA

snRNP buffer: 20 mM tris(hydroxymethyl)aminomethane hydrochlo-

---

[13] J. P. Liautard, J. Sri-Widada, C. Brunel, and P. Jeanteur, *J. Mol. Biol.* **162**, 623 (1982).
[14] I. Reveillaud, M. N. Lelay-Taha, J. Sri-Widada, C. Brunel, and P. Jeanteur, *Mol. Cell. Biol.* **4**, 1890 (1984).

ride (Tris-HCl) (pH 7.4), 100 mM NH$_4$Cl, 15 mM MgCl$_2$, 5 mM 2-mercaptoethanol

Equilibration buffer: 20 mM Tris-HCl (pH 7.4), 50 mM NH$_4$Cl, 15 mM MgCl$_2$, 5 mM 2-mercaptoethanol

Elution buffer: 20 mM Tris-HCl (pH 7.4), 450 mM NH$_4$Cl, 15 mM MgCl$_2$, 5 mM 2-mercaptoethanol

Step 1: Preparation of Nucleoplasmic Extract

Any cells growing in suspension culture or obtained from ascitic fluid, as well as tissues, can be used as starting material. Typically, we start with unlabeled 2 × 10^9 HeLa cells (S3 strain) grown in minimal essential medium supplemented with 5% heat-inactivated newborn calf serum and harvested at a density of 5 × 10^5 cells/ml. The nucleoplasmic extract is obtained essentially as described by Hinterberger *et al.*[5] All buffers contain protease inhibitors: 50 μM phenylmethylsulfonyl fluoride (Serva) and 5 mg/liter each of leupeptin, pepstatin, and chymostatin (Sigma). All manipulations are carried out at 0–4° unless otherwise stated. Harvested cells are washed twice with 40 ml of Buffer 1 and centrifuged for 5 min at 1,160 g. Cells are then resuspended in 40 ml of Buffer 2, transferred to a Dounce homogenizer, and broken by 10–20 strokes of pestle B. Cell breakage is always checked by phase-contrast microscopy, with more homogenization when necessary.

The disrupted cells are immediately layered on top of four 10-ml cushions of Buffer 3 and centrifuged for 20 min at 3,000 g. Material that does not enter the sucrose layer (cytoplasmic fraction) is discarded. The pellets of nuclei are washed twice with 5–10 ml of Buffer 4. Each pellet of clean nuclei is then resuspended in 9 ml of Buffer 5 and sonically disrupted by two 15-sec pulses at 50 W with a microtip-equipped Branson B12 sonifier. The sonicate is centrifuged for 2 hr at 26,500 rpm in a Beckman SW40 rotor. The supernatants are pooled and dialyzed for at least 5 hr against two changes of 500 ml snRNP buffer. Extending the dialysis overnight results in no significant degradation of snRNPs, and, in our hands, no differences were found when this and the subsequent steps were carried out with or without protease inhibitors. Once dialyzed, the nucleoplasmic extract is centrifuged at 10,000 g for 20 min to eliminate aggregates.

Step 2: Concentration of Nucleoplasmic Extract

At this stage, the final volume of the extract is about 35 ml. It must therefore be concentrated before centrifugation in a cesium chloride gradient. The most satisfactory procedure is to adsorb snRNPs onto a DEAE-Sepharose column from which they are step-eluted. The nucleoplasmic extract is applied to a 5-ml DEAE-Sepharose CL-6B (Pharmacia) column (0.8 cm diameter) in equilibration buffer. After washing with 10–20 ml of snRNP buffer, snRNPs are step-eluted with 10 ml of elution buffer. The first 2 ml correspond to the void volume of the column and are discarded.

Step 3: Centrifugation in Cesium Chloride Gradient

In two sterile polyallomer tubes for the Beckman SW40 rotor, dissolve 6.81 g of cesium chloride in water to a volume of about 7.5 ml. To each tube is added 160 μl of 1 M Tris-HCl (pH 7.4), 120 μl of 1 M MgCl$_2$, 2.5 μl of 2-mercaptoethanol, and 4 ml of the concentrated nucleoplasmic extract (layered above the cesium chloride solution), and the volume is brought to 12 ml with distilled water and mixed. Under these conditions, the cesium chloride concentration is 40% (w/w) at 20°, while the final concentrations of the other components are 20 mM Tris-HCl (pH 7.5), 15 mM MgCl$_2$, 5 mM 2-mercaptoethanol, and 150 mM NH$_4$Cl, the last derived from the extract. In a typical experiment, centrifugation is for 65 hr at 33,000 rpm and 20° in a Beckman SW40 rotor. Fractions of 0.8 ml are collected from the top, then 10- and 40-μl samples of each fraction are withdrawn for density measurement and RNA analysis, respectively. Fractions pooled from two gradients are dialyzed at 4° for 3 hr either against two 100 ml changes of snRNP buffer before purification or against any buffer appropriate to a particular use. The snRNPs are stable at 4° in snRNP buffer for several days or can be stored at $-20°$ indefinitely.

The distribution of the different RNA species across the gradient (Fig. 1A) serves to underline the following points: (1) U snRNAs are quantitatively recovered in buoyant structures with densities ranging from about 1.35 to 1.55 g/ml. (2) All other RNAs initially present in the nucleoplasmic extract, including 7 and 5 S RNAs, tRNAs, and some large RNA molecules, behave entirely as free RNA and migrate to the bottom. (3) As far as snRNPs are concerned, three regions of the gradient can be distinguished. Region I contains exclusively U1 and U5, region II mostly U2, a noticeable amount of U1, some U5, and traces of U4–U6, and region III essentially U4–U6, U2, and traces of U1. U1 and U5 snRNPs in region I can be separated by DEAE-Sepharose chromatography (see below). However, all attempts (using DEAE-Sepharose chromatography, second cycle of centrifugation in cesium chloride, hydrophobic chromatography, electrophoresis in nondenaturing gels, or combinations of these different

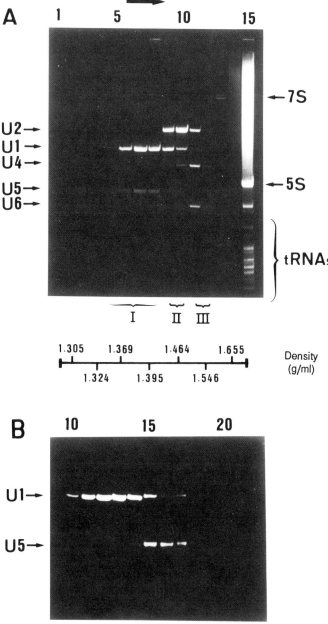

FIG. 1. Isolation of U1 and U5 snRNPs. (A) Distribution of U snRNAs in a cesium chloride density gradient (Step 3). RNAs contained in the indicated fractions (40-μl samples) from top to bottom (arrow) were extracted with phenol, precipitated with ethanol, and

procedures) at separating snRNPs contained in regions II and III into individual particles have been unsuccessful to date.

Step 4: Fractionation of U1 and U5 snRNPs by DEAE-Sepharose Chromatography

Step 4 follows the "DEAE-Sepharose II" procedure described by Hinterberger *et al.*[5] Briefly, the pooled fractions from two cesium chloride gradients corresponding to region I are dialyzed twice against 100 ml of snRNP buffer to eliminate cesium chloride and applied to a 5-ml DEAE-Sepharose (0.8 cm diameter) column in equilibration buffer. After washing with 10 ml of snRNP buffer, elution is carried out with a 80-ml linear gradient of 100 to 400 mM ammonium chloride in snRNP buffer. Fractions of 2.5 ml are collected with 100- and 400-μl aliquots of each fraction taken for RNA and protein analyses, respectively. When the isolated particles are to be used in splicing systems, they are dialyzed against the suitable buffer[15] and concentrated by centrifugation in Centricon 10 microconcentrators (Amicon), to a concentration estimated to be very similar to that of the corresponding snRNP in a splicing nuclear extract. This estimation is made from polyacrylamide gels stained with ethidium bromide.

Figure 1B shows the separation of the two particles. It is noteworthy that, in some preparations, U5 snRNP is noticeably contaminated by U1. On one hand, we now have evidence that cesium chloride fractions contain at least two forms of U1 snRNP differing by their protein composition, and, on the other hand, it appears that incomplete forms of U1 snRNP can be generated on storage which are more strongly adsorbed to the DEAE-Sepharose column and therefore contaminate U5 snRNP. To minimize these problems, we recommend using only the two first cesium chloride fractions containing U1 and U5 and performing the DEAE-Sepharose chromatography immediately after dialysis of the cesium chloride fractions.

---

[15] J. D. Dignam, R. M. Lebovitz, and R. G. Roeder, *Nucleic Acids Res.* **11**, 1475 (1983).

---

electrophoresed on a 12.5% denaturing polyacrylamide gel run in Tris–borate–EDTA buffer. Staining was with ethidium bromide. I, II, and III refer to the three regions described in the text. (B) Fractions from two cesium chloride gradients corresponding to region I were pooled and chromatographed on DEAE-Sepharose as described in the text. RNA analysis was as indicated above using 300-μl samples of each fraction. The photograph shows an ethidium bromide-stained 12.5% polyacrylamide gel exhibiting the U1 and U5 snRNA profiles. [From J. Tazi, C. Alibert, J. Temsamani, I. Reveillaud, G.Cathala, C. Brunel, and P. Jeanteur, *Cell* **47**, 755 (1986).]

Structural and Functional Criteria of Isolated U1 and U5 snRNPs

U1 snRNP

The U1 snRNP purified by the above method is entirely immunoprecipitable by anti-Sm and anti-RNP antibodies but not by the monoclonal antibody F78, an antibody recognizing a population of U snRNPs in which U1 has a different protein composition.[16] It contains all nine (68K and A to G) proteins previously known to be associated with U1 snRNA from immunoprecipitation data or other purification procedures. In micrococcal nuclease protection experiments,[12–14] it behaves as do U1 snRNPs prepared by other methods. According to all these criteria, this U1 snRNP seems to be structurally representative of the U1 snRNP population in the cell and is probably fully functional. As a matter of fact, it appears to recognize the 5′ splice site of a premessenger RNA.[17]

U5 snRNP

Only two procedures for the isolation of U5 snRNP have been described. They include our own protocol and another, using anticap and anti-N^6-methyladenosine antibodies, described by Bringmann *et al.*[9] While the latter authors concluded that U5 snRNP contains only the common BB′ and D–G proteins, we found that this particle is associated with a 100K intron-binding protein (IBP) which can specifically recognize intronic sequences at the 3′ splice site.[18] Recent results in our laboratory[17] now demonstrate that U5 snRNP purified by cesium chloride centrifugation can recognize the 3′ splice site of a mammalian pre-mRNA and that this recognition depends on the presence of IBP.

Versatility of Cesium Chloride Centrifugation Procedure

A major advantage of the cesium chloride centrifugation procedure is that it can be used at any stage of purification, from a crude nuclear extract to an already purified snRNP preparation. We have successfully applied it to splicing extracts,[15] hnRNPs,[19] or snRNPs prepared by affinity chromatography using anti-cap antibodies.[8] As described above, it leads to the simultaneous purification of U1 and U5 snRNPs. It is also perfectly

---

[16] A. Winkler, R. Watson-McKown, and K. S. Wise, *J. Immunol.* **140,** 69 (1988).

[17] J. Tazi, J. Temsamani, C. Alibert, W. Rhead, S. Khellil, G. Cathala, C. Brunel, and P. Jeanteur, *Nucleic Acids Res.* **17,** 5223 (1989).

[18] J. Tazi, C. Alibert, J. Temsamani, I. Reveillaud, G. Cathala, C. Brunel, and P. Jeanteur, *Cell* **47,** 755 (1986).

[19] V. M. Kish and T. Pederson, *J. Mol. Biol.* **95,** 227 (1975).

suited to the final purification of individual snRNPs obtained by more conventional techniques.[5,6] The only critical technical point is the presence of a high Mg^{2+} concentration (15 mM). For example, at low Mg^{2+}, U1 snRNP is nearly completely depleted of its specific proteins and is shifted to a density higher than that of the native form. In fact, a limited depletion of U1 snRNP proteins occurs even at high Mg^{2+} concentrations while U5 snRNP is much more stable. Interestingly, this situation can be exploited to obtain a U5 snRNP completely free of contaminating U1 snRNP. Indeed, the small amount of U1 snRNP still comigrating with U5 on DEAE-Sepharose clearly corresponds to incomplete forms, which are easily removed from U5 snRNP by a second cycle of centrifugation in cesium chloride.

Specific Utilizations of Isopycnic Centrifugation in Cesium Chloride

Isolation of Core snRNPs

When snRNPs are centrifuged in cesium chloride at low Mg^{2+} concentrations, specific proteins are released and shifted to the top of the gradient (see above). By introducing 0.5% sarkosyl and 5 mM Mg^{2+} in the cesium chloride, and after a final purification by velocity sedimentation in a sucrose gradient containing 0.5 M KCl and 0.5% sarkosyl, it is easy to obtain core snRNPs that contain only the small D–G common proteins[10,11] and an additional polypeptide[20] migrating between D and E, derived from BB′.[21]

Any type of nucleoplasmic fraction containing snRNPs (if necessary, concentrated in Centricon 10) is adjusted to 40% (w/w) cesium chloride, 0.5% sarkosyl, and 1 mM $MgCl_2$ in 20 mM Tris-HCl (pH 7.4) in a final volume of 12 ml. After centrifugation for 65 hr at 33,000 rpm in the SW40 Beckman rotor at 20°, 0.8-ml fractions are collected from the top. The SW60 Beckman rotor can also be used for analytical studies or when very small amounts of snRNPs are involved. With this rotor, centrifugation is for 48 hr at 50,000 rpm and 20°; 0.25-ml fractions are then collected from the top. At this stage, all but U5 snRNPs band at a density of 1.435 g/ml. The reason for the somewhat lower density of U5 snRNP is unknown but probably reflects a higher protein to RNA ratio.

At this stage, snRNPs are still contaminated by some free proteins that

[20] C. Assens, J. P. Liautard, J. Sri-Widada, C. Brunel, and P. Jeanteur, *Biochem. Biophys. Res. Commun.* **106,** 953 (1982).
[21] J. Temsamani, C. Alibert, J. Tazi, M. Rucheton, J. P. Capony, P. Jeanteur, G. Cathala, and C. Brunel, *J. Mol. Biol.* **206,** 439 (1989).

are easily removed by a final step of velocity centrifugation through a sucrose gradient. To this aim, fractions from the cesium chloride gradient containing snRNAs are pooled, dialyzed against a buffer containing 20 mM Tris-HCl (pH 7.4), 0.5% sarkosyl, 1 mM MgCl$_2$, and 500 mM salt (KCl or NH$_4$Cl are quite convenient for this purpose), concentrated to about 0.3 ml in a Centricon 10, and finally layered on top of a 12-ml 15–30% sucrose gradient in the same buffer. Centrifugation is for 23 hr at 38,000 rpm in the SW40 Beckman rotor at 15°. Under these conditions core snRNPs migrate as a sharp peak around 11–12 S while the remaining free proteins remain at the top of the gradient.[10,11] Collected fractions are dialyzed on ice against 20 mM Tris-HCl, 1 mM MgCl$_2$, 100–150 mM NH$_4$Cl, and the precipitated sarkosyl is eliminated by centrifugation. It is noteworthy that the snRNAs remain absolutely intact until sarkosyl is removed but are rapidly degraded at preferential positions[11,14] once sarkosyl is completely removed. We have now demonstrated that one of these small proteins has endonuclease activity.[21]

Preparation of Core snRNP Proteins

The core snRNPs contain D–G proteins that are tightly bound to the snRNA sequence at the Sm site.[13] It may be of interest to recover these proteins completely free of RNA. This can be accomplished by centrifuging core snRNPs in 40% (w/w) cesium chloride in the presence of 4 M guanidinium chloride under the same conditions as those described above. Free D–G proteins are recovered at the top and snRNAs at the bottom. Cesium chloride and guanidinium chloride are eliminated by dialysis against 20 mM Tris-HCl (pH 7.4), 1 mM MgCl$_2$, 100 mM NH$_4$Cl.

Monitoring snRNP Assembly

Until now, studies aimed at elucidating the function of the major U snRNPs led to relatively scant insight into the role of snRNP proteins in pre-mRNA splicing. The availability of *in vitro* assembled snRNPs with known composition will be extremely useful in testing their structural organization and function. Patton *et al.*[22] have assembled a U1 snRNP that withstands isopycnic banding in Cs$_2$SO$_4$ after incubating an SP6 U1 snRNA in HeLa S100 extracts. We[21] and others[23] have obtained similar results using the general procedure of centrifugation in cesium chloride described above. We have also used this general procedure to monitor the reconstitution of RNPs from small RNA fragments containing the A(U)$_n$G

[22] J. R. Patton, R. J. Patterson, and T. Pederson, *Mol. Cell. Biol.* **7,** 4030 (1987).
[23] C. Pikielny, A. Bindereif, and M. R. Green, *Genes Dev.* **3,** 479 (1989).

Sm binding site as the only snRNA sequence. Such an assembled particle has a buoyant density in cesium chloride of 1.4 g/ml, corresponding to the density of an authentic core snRNP (unpublished results).

[22] Structural Analysis of U Small Nuclear Ribonucleoproteins by *in Vitro* Assembly

By JÖRG HAMM and IAIN W. MATTAJ

Introduction

U small nuclear ribonucleoproteins (snRNPs) are essential cofactors in the splicing process of lower as well as higher eukaryotes. The major U snRNPs (U1, U2, U4–U6) of higher eukaryotes and the yeast counterparts are required to remove introns from messenger RNA precursors. In addition to their functional interest, U snRNPs are ideal systems to study the various types of interactions that occur in RNPs. The RNA component is relatively small [100–200 nucleotides (nt)], and up to 10 different proteins are present in a single particle. Some of the proteins recognize discrete RNA sequences whereas others are held in the RNP mainly by protein–protein interactions. There are subsets of proteins that probably enter the particle as a complex which forms in the absence of RNA. This complex contains some of the common U snRNP proteins (D, E, F, G) that are part of all major U snRNPs. The proteins specific for individual U snRNPs are able to interact with the U snRNA in the absence of the common proteins, although the complex is more stable when both groups of proteins are present.

The secondary structure and, in part, the primary structure of individual U snRNAs are highly conserved between species. Among the regions conserved are some that are required for interaction with proteins, as judged by the ability of U snRNAs from heterologous species (e.g., *Drosophila*, yeast) to associate specifically with vertebrate U snRNP proteins. cDNAs corresponding to several snRNP proteins have been sequenced, so the proteins can be synthesized *in vitro*. Monoclonal antibodies and monospecific or oligospecific sera against many of the proteins are available and enable a detailed analysis of snRNP composition.

An essential further requirement for structure–function studies is a simple, rapid, and efficient system to generate particles of defined composition in which the RNP components can be individually modified or

METHODS IN ENZYMOLOGY, VOL. 181

omitted. An assembly system was developed by combining U snRNAs synthesized *in vitro* with proteins from *Xenopus* egg extracts. Egg extracts were used because they contain a large stockpile of RNP proteins for use during the early stages of embryonic development. This system, in combination with a variety of conventional as well as novel assays, has enabled a detailed analysis of RNA–protein and protein–protein interactions in U snRNPs.

Methods

Extract Preparation

Mature *Xenopus* females are injected with 100 units of pregnant mare serum gonadotropin (Chrono-Gest PMSG 500, Intervet). Egg-laying is induced 2–10 days later by injecting 1000 units of human chorionic gonadotropin (HCG, Sigma, St. Louis, MO). After the injection of HCG, frogs are kept separately in 100 mM NaCl until eggs are layed (~6–8 hr). To dissolve the jelly coats, eggs are agitated gently in freshly prepared 2% cysteine–NaOH (pH 7.8) for 5–10 min. When the eggs are separated, the cysteine is removed, and the eggs are washed 5 times with tap water.

Next a whole cell extract[1] is prepared, where the packed cell volume (PCV) corresponds to two-thirds of the egg volume. The eggs are homogenized with 3 PCV of TE buffer [10 mM tris(hydroxymethyl)aminomethane (Tris) (pH 8.0), 1 mM ethylenediaminetetraacetic acid (EDTA)] and transferred to a 500-ml conical flask. Then 5 PCV of buffer L [38% glycerol, 20% sucrose, 50 mM Tris-HCl (pH 7.4), 2 mM dithiothreitol (DTT)] is added, and the MgCl$_2$ concentration is adjusted to 5 mM. Next, 1 PCV of a saturated (NH$_4$)$_2$SO$_4$ solution is added slowly with gentle shaking of the flask (~5 min). The lysed eggs are centrifuged for 5 hr at 4° in a 50.2 Ti rotor at 35,000 rpm. The run is stopped without using the brake. The supernatant is removed carefully (avoiding transfer of the lipid phase) and the volume determined. Powdered (NH$_4$)$_2$SO$_4$ is added (0.42 g/ml) over a period of 20 min while rotating the tube slowly at 4°, and then the solution is neutralized [20 μl of 1 M NaOH–1 g (NH$_4$)$_2$SO$_4$]. The tube is rotated for a further 1 hr at 4° and precipitated proteins pelleted (SW28, 26,000 rpm, 45 min, 4°). Proteins are resuspended in 0.4 PCV of TGMED-40 [17% glycerol, 25 mM Tris-HCl (pH 7.4), 2 mM MgCl$_2$, 0.1 mM EDTA, 0.25 mM DTT, 40 mM (NH$_4$)$_2$SO$_4$] on an agitator at 4°

[1] J. L. Manley, A. Fire, A. Cano, P. A. Sharp, and M. L. Gefter, *Proc. Natl. Acad. Sci. U.S.A.* **77**, 3855 (1980).

(~1 hr). The solution is dialyzed overnight at 4° against TGMED-200 [TGMED-40 but with 200 mM (NH$_4$)$_2$SO$_4$]. Precipitated material is removed by centrifugation (SW40, 36,000 rpm, 20 min, 4°), and the supernatant is dialyzed against TGMED-40 for 4 hr. The dialyzed extract is centrifuged for 5 min in an Eppendorf centrifuge at 4° for 5 min and is frozen in aliquots. The protein concentration is normally between 10 and 30 μg/μl.

Substrate Preparation

To synthesize RNA *in vitro* the templates must be connected to a phage RNA polymerase promoter (T7, T3, Sp6).[2] The available transcription vectors suffer from the limitation that insertion of a desired sequence into the restriction sites of the polylinker gives rise to an RNA with 5' and 3' extensions. This problem can be eliminated if templates are modified individually by inserting promoter sequences at the cap sites and restriction sites at their 3' end using site-directed mutagenesis.[3,4] It is often necessary to insert additional nucleotides at the 5' end because the consensus sequence of some promoters (e.g., T7) extend up to position +3, and not all 5' sequences allow efficient transcription. As a consequence RNAs synthesized by T7 polymerase normally carry three G residues at the 5' end. In order to be stable in extracts the RNAs require a cap structure, so transcription is performed in the presence of methylated or unmethylated cap analogs.

A typical T7 assay composition is as follows: 1 μg linearized template is incubated in a volume of 10 μl [40 mM Tris-HCl (pH 8.0), 8 mM MgCl$_2$, 2 mM spermidine, 50 mM NaCl, 30 mM DTT, 0.4 mM each ATP, UTP, and CTP, 10 units T7 RNA polymerase (Stratagene), 0.25 A_{260} units m[7] GpppG, 20 units RNasin, 10 μCi [α-^{32}P]GTP (10 μCi/μl, 400 Ci/mmol)] at 37° for 5 min. Unlabeled GTP is added (final concentration 0.4 mM) and incubation continued for 10 min. Ninety microliters of water is added, and proteins are extracted with 50 μl of phenol–chloroform (1 : 1, v/v). Unincorporated nucleotides are removed on spun columns[5] and RNA precipitated with 3 volumes of ethanol and 0.1 volume of 3 M sodium acetate. Two to five micrograms of RNA is normally synthesized by this method. An aliquot of the transcription product is analyzed on an 8% acrylamide–7 M urea gel.

[2] J. K. Yisraeli and D. A. Melton, this series, Vol. 180, p. 42.
[3] W. Kramer, V. Drutsa, H. W. Jansen, B. Kramer, M. Pflugfelder, and H. J. Fritz, *Nucleic Acids Res.* **12**, 9441 (1984).
[4] J. Hamm, M. Kazmaier, and I. W. Mattaj, *EMBO J.* **6**, 3479 (1987).
[5] T. Maniatis, E. F. Fritsch, and J. Sambrook, *in* "Molecular Cloning: A Laboratory Manual," p. 466. Cold Spring Harbor Laboratory, Cold Spring Harbor, New York.

In Vitro Assembly

The assembly reaction itself is very simple; it contains only extract, the synthetic T7 U snRNA, and tRNA. tRNA is required to trap RNA-binding proteins that would otherwise bind unspecifically to the synthetic RNA immediately and prevent assembly. The amount of tRNA required for efficient assembly must be titrated for every extract preparation, although a rough estimate can be made based on the protein concentration of the extract (Fig. 4a). It is important to combine the tRNA with the extract before addition of the synthetic RNA. In some extract preparations, addition of tRNA results in a precipitate if the reaction is performed in a small volume (e.g., 4 μl extract, 1 μl tRNA, 1 μl T7 RNA). This precipitation does not affect either the efficiency or the specificity of assembly. In 4 μl of extract (10 μg/μl protein), 20–50 ng of wild-type U1 T7 RNA can be assembled efficiently into particles within 30 min at room temperature.

Immunoprecipitation

Serum (1–100 μl, depending on the antibody titer) is mixed with 40 μl of protein A-Sepharose CL-4B beads (Pharmacia; 0.1 g/ml in IPP$_{500}$) and 400 μl of IPP$_{500}$ [10 mM Tris-HCl (pH 8.0), 500 mM NaCl, 0.1% Nonidet P-40, 0.1% sodium azide]. Binding takes place during incubation for 2–12 hr at room temperature in a rotary wheel. To remove unbound antibodies, the beads are washed 3 times with 1 ml of IPP$_{500}$.

IPP$_{500}$ is added to the assembly mix to a final volume of 1 ml, and this mixture is rotated with the washed beads for 2 hr at room temperature. The beads are then washed 3 times for 10 min in the rotary wheel with 1 ml of IPP$_{500}$. The precipitated RNA is released by incubating the beads in 400 μl of homomedium [50 mM Tris-HCl (pH 7.4), 5 mM EDTA, 1.5% sodium dodecyl sulfate (SDS), 300 mM NaCl, 1.5 mg/ml proteinase K] for 30 min at 37°. The beads are extracted once with phenol–chloroform, after which the RNA is precipitated with 3 volumes of ethanol and analyzed on 8% acrylamide–7 M urea gels.

Low-Stringency Immunoprecipitation

In order to precipitate particles of lower stability it is necessary to perform a low-stringency immunoprecipitation, which is carried out on ice and at a lower salt concentration than the standard assay. The assembly mix and the serum are incubated with 200 μl of IPP$_{150}$ (IPP$_{500}$ but with 150 mM NaCl) on ice for 15 min with occasional mixing. Then 40 μl of protein A-Sepharose beads (0.1 g/ml) is added, and incubation is continued for 30 min. The beads are washed 3 times with 1 ml of IPP_{150} at 0°, and the RNA is released as above. This type of assay suffers from the disadvantage of being extremely inefficient compared with the standard protocol.

Protein Sequestering Assay

The protein sequestering assay is very sensitive and enables the rapid detection of weaker interactions than does direct immunoprecipitation at high stringency. The principle is based on the template exclusion assay used in the analysis of transcription complexes. The ability of an RNA to bind a protein stably under binding conditions is measured rather than the ability of the formed particle to survive the immunoprecipitation procedure.

The RNA to be analyzed is unlabeled and is preincubated in molar excess over the proteins of interest under binding conditions. Subsequently, radioactively labeled wild-type RNA is added, and this RNA will bind the proteins that have not been sequestered by the preincubated RNA. If wild-type RNA is included in the preincubation, all proteins are sequestered and the labeled RNA added later will not bind any of the U snRNP proteins (Fig. 1a).

The proteins binding to the labeled RNA are identified by direct immunoprecipitation. Since this labeled RNA is wild type, the binding is stable and the RNP can be analyzed by immunoprecipitation at high stringency. For the assay, tRNA, 8 μl extract, and 300 ng unlabeled RNA are preincubated for 15 min at room temperature. Ten nanograms of ^{32}P-labeled wild-type RNA is added and incubation continued for 30 min at room temperature. Immunoprecipitation is then performed as described above.

An example is given in Fig. 1b. This experiment demonstrates that the 5' half of U1 RNA can efficiently bind to the U1-specific proteins A and 70 K in the absence of a binding site for the common (Sm antigenic) snRNP proteins, which has been removed by deletion of the 3' half of the molecule. Deletion of the 5'-most hairpin (in fA) results in loss of binding to proteins A and 70K.

Native Gel Analysis

Native gel systems have been used successfully to explore splicing and polyadenylation complexes.[6,7] The same gel system can be used to

---

[6] M. M. Konarska, this series, Vol. 180, p. 442.
[7] L. Conway and M. Wickens, this series, Vol. 180, p. 369.

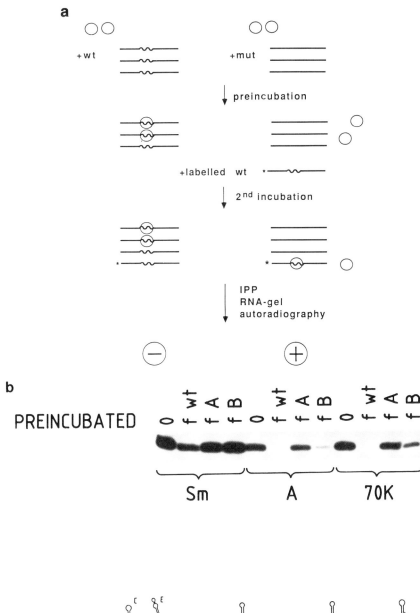

a

+wt   +mut

↓ preincubation

+labelled wt *———

↓ 2nd incubation

↓ IPP
RNA-gel
autoradiography

⊖   ⊕

b

PREINCUBATED   0 fwt fA fB   0 fwt fA fB   0 fwt fA fB

Sm   A   70K

U₁wt   fwt   fA   fB

analyze RNPs assembled *in vitro*. Unfortunately, every RNA, after incubation in an extract, will exhibit a reduced electrophoretic mobility so that at least small complexes, like single U snRNPs, have to be discriminated from nonspecific complexes by additional criteria.

One possibility is the differential response to added competitor RNA. Nonspecific complexes dissociate at lower competitor concentrations than specific complexes. Furthermore, specific complexes run as discrete bands whereas nonspecific complexes migrate as a broad smear (Fig. 2). Competitor RNA can be added either before or after the assembly reaction. If competitor and extract are mixed before assembly nonspecifically binding proteins are trapped and do not associate with the labeled RNA. If the competitor is added subsequently nonspecifically bound proteins are removed preferentially because their interaction with the specific RNA is weaker than the interaction of the specific proteins.

For the assay, tRNA, 4 μl extract, and 5 ng labeled RNA are incubated (see *In Vitro* Assembly, above). Five micrograms of total yeast RNA is added and incubated for 10 min on ice, $5\times$ sample buffer is added, and complexes are loaded on native gels. Gels are dried before exposure. The composition of the sample buffer and the native gels are given in the legend to Fig. 2.

Antibody-Induced Mobility Shifts

An elegant demonstration of the presence of a certain protein in a complex is made possible if monoclonal antibodies against the protein are available. The antibody is added after the assembly reaction and binds to complexes containing this protein. As a consequence, the electrophoretic mobility of these complexes is reduced. If antibodies against different proteins are added simultaneously, a second or third further mobility shift is introduced. These multiple shifts can be used to demonstrate that sev-

---

FIG. 1. Protein sequestering assay (PSA). (a) Principle of the PSA. The PSA measures the ability of RNAs (here wild type and mutant) to bind a certain protein. The RNA to be analyzed is unlabeled and is incubated in excess over the protein under binding conditions. A labeled wild-type RNA is added, and RNPs are then immunoprecipitated with antibodies against the protein of interest. Only if the preincubated RNA (here mutant) is unable to sequester a protein will the subsequently added labeled RNA be precipitated with an antibody recognizing this protein. (b) Example of a PSA. Different RNA fragments (derived from wild type, ΔA, or ΔB[4]) were preincubated, and a labeled wild-type RNA was added subsequently. None of the fragments is able to sequester the proteins recognized by anti-Sm antibodies, whereas fwt and, to a reduced extent, fB are able to bind the proteins A and 70K stably. Fragment fA is unable to sequester any of the proteins analyzed.

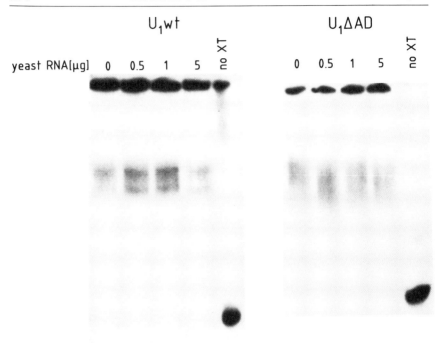

FIG. 2. Analysis of RNPs on native gels. U1 wild-type RNA and U1 ΔAD were incubated under assembly conditions (30 min, room temperature, 5 ng T7 RNA, 4 μl egg extract, 5 μg tRNA). Total yeast RNA was added as indicated, and the mixture was left for 10 min on ice. Then 5× loading buffer (250 mM Tris–glycine, 40% glycerol, 0.025% w/v bromphenol blue) was added, and samples were analyzed on native gels [4% acrylamide–bisacrylamide (60 : 1, v/v), 50 mM Tris–glycine, 10 V/cm, 4°]. U1 ΔAD gives rise to a smear, whereas wild-type U1 gives rise to two discrete bands.

eral proteins are present in the same particle. The extent of the shift is probably correlated with the number of proteins per particle that are recognized by the antibody.

For the assay, assembly is performed with 4 μl of extract (see Protein Sequestering Assay, above). Serum (1–10 μl, depending on the antibody titre) is added, and the reaction is incubated for 15 min at 4°. Then 5× sample buffer is added and complexes analyzed on native gels.

Comparison of Different Assembly Extracts

Different preparations of egg extract contain different amounts of protein. This has the consequence that each extract has a different optimum

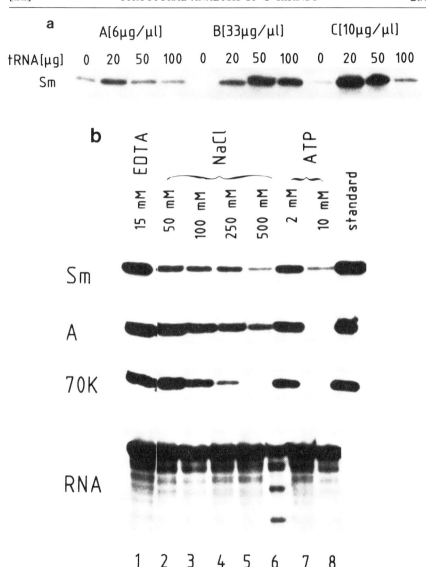

FIG. 3. Assembly conditions. (a) tRNA titration. To determine the optimal tRNA concentration for assembly of U snRNPs, egg extracts (A, B, and C) with different protein concentrations were used to assemble RNPs in the presence of increasing amounts of tRNA (4 μl extract plus tRNA as indicated), and complexes were immunoprecipitated with anti-Sm antibodies. (b) Factors effecting assembly. Standard conditions [25 mM Tris (pH 7.4), 40 mM (NH₄)₂SO₄, 0.25 mM DTT, 2 mM MgCl₂, 0.2 mM EDTA] were altered as indicated prior to assembly, and RNPs were immunoprecipitated after assembly with different antibodies (anti-Sm, anti-A, anti-70K). (From Ref. 4.)

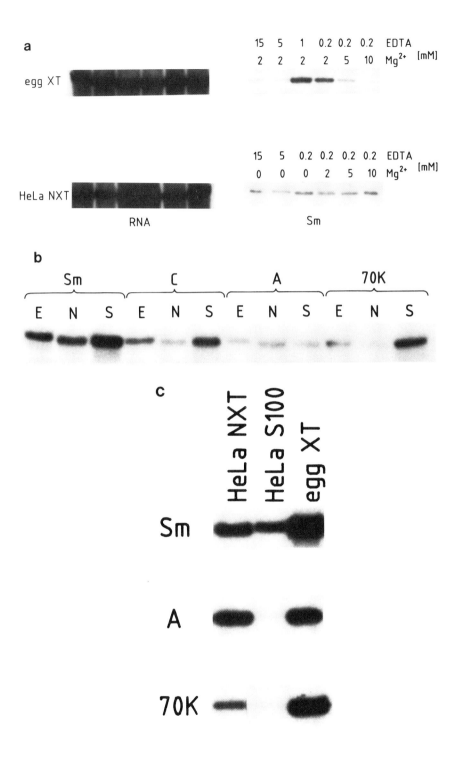

with respect to competitor RNA requirement. Although the amount of competitor for optimal assembly is roughly proportional to the amount of proteins in the assay (Fig. 3a), this parameter must be determined empirically for each new extract. It is also noteworthy that different egg extracts exhibit very different efficiencies of assembly per microgram protein (Fig. 3a).

Although our assembly studies were mainly performed with *Xenopus* egg extracts, many types of extracts can be used for assembly of U snRNPs. We have additionally tested HeLa nuclear extracts (NXT)[8] and HeLa S100 fractions.[8,9] They differ from egg extracts in some of the factors influencing assembly. Assembly does not require energy in the form of ATP in any of the extracts tested; in fact, 10 mM ATP inhibits assembly significantly in egg extracts (Fig. 3b) and in HeLa NXT (data not shown). In contrast, the response to Mg^{2+} and to EDTA is extract specific. In HeLa NXT and S100 assembly is not affected either by increasing the $MgCl_2$ concentration or by adding EDTA (Fig. 4a and data not shown), but in all egg extracts tested assembly is Mg^{2+} sensitive (Fig. 4a and data not shown). Increasing the EDTA concentration affects assembly in only some egg extracts (Figs. 3b and 4a). Thus, the Mg^{2+} sensitivity is a nonspecific effect rather than an assembly-specific one. Mg^{2+} probably stabilizes nonspecific interactions with RNA-binding proteins that are present in large amounts in egg extracts.

The highest assembly efficiency is routinely obtained in *Xenopus* egg extracts, but some HeLa S100 preparations are as efficient as egg extracts (Fig. 4b). However, note should be taken of the fact that S100 preparations can vary significantly in the content of U1-specific proteins. Some S100 extracts contain predominantly Sm proteins (Fig. 4c).[10] Each extract of this type must therefore be carefully checked before use.

[8] J. D. Dignam, R. M. Lebowitz, and R. G. Roeder, *Nucleic Acids Res.* **11**, 1475 (1983).
[9] J. R. Patton, R. J. Patterson, and T. Pederson, *Mol. Cell. Biol.* **7**, 4030 (1987).
[10] B. Ruskin, P. D. Zamore, and M. R. Green, *Cell* **52**, 207 (1988).

FIG. 4. Comparison of different assembly extracts. (a) Response to divalent cations. The concentration of Mg^{2+} or EDTA was altered as indicated prior to assembly of U1 snRNPs in *Xenopus* egg and HeLa nuclear extracts. Complexes formed were analyzed by immunoprecipitation with anti-Sm antibodies. (b) Assembly in different extracts. U1 wild-type RNA was assembled into RNPs in *Xenopus* egg, HeLa nuclear, and HeLa S100 extracts. The protein concentration of all extracts was about 10 $\mu g/\mu l$. Extract (16 μl) was incubated with 50 μg tRNA and 40 ng labeled U1 RNA for 30 min at room temperature. RNPs were analyzed by immunoprecipitation with different antibodies (anti-Sm, anti-C, anti-A, anti-70K). (c) As (b), but different extract preparations were used.

Output

These assays were developed in order to study the structure of U1 snRNPs. A set of U1 snRNA mutants lacking single secondary structure elements was designed and used to assemble particles *in vitro* and *in vivo*. The ability of various subsets of proteins to bind to these RNAs was analyzed by direct immunoprecipitation. The strength of interaction was determined by the protein sequestering assay. We were able to define two discrete protein binding sites on U1 snRNA. The common U snRNP proteins require only the conserved Sm binding site $A(U)_nG$. The essential contact points for the U1-specific proteins are located in the two 5'-most stem–loop structures of the RNA. Binding of these groups of proteins can occur independently, but the stability of the complex is greatly increased when both groups are present. One of the reasons for this higher stability is presumably protein–protein interactions between the two groups of proteins. Assembly is a very rapid process: complexes form within a few minutes. Assembly of U1 snRNPs does not require energy in the form of ATP, and assembly does not necessarily follow an ordered pathway, although, at least *in vitro*, there is a preferred order of addition of different proteins.[4]

Acknowledgments

Jörg Hamm is supported by a fellowship of the Boehringer Ingelheim Fond, Stuttgart. Heide Seifert typed the manuscript.

[23] Site-Specific Degradation of RNA of Small Nuclear Ribonucleoprotein Particles with Complementary Oligodeoxynucleotides and RNase H

By Angela Krämer

Small nuclear ribonucleoprotein particles (snRNPs) of the U type[1] have been found in the nuclei of a wide range of distantly related species.[2–11] These particles consist of small, stable, uridine-rich RNAs[12–14] and several proteins that are either common to all snRNPs or characteris-

[1] M. L. Birnstiel (ed.), "Structure and Function of Major and Minor Small Nuclear Ribonucleoprotein Particles." Springer-Verlag, Berlin and New York, 1988.
[2] G. Galli, H. Hofstetter, H. G. Stunnenberg, and M. L. Birnstiel, *Cell* **34,** 823 (1983).
[3] R. Reddy, D. Henning, and H. Busch, *J. Biol. Chem.* **260,** 10930 (1985).

tic for one particular type of particle.[15-17] In general only one RNA molecule associates with proteins to form a snRNP structure. An exception is the particle formed with U4 and U6 RNAs, most of which are found in one and the same snRNP.[18-22]

All U snRNPs of known function participate in the maturation of primary transcripts, such as nuclear pre-mRNA splicing and polyadenylation as well as histone pre-mRNA 3' end formation.[23-26] A number of methods have been used to probe the requirement for snRNPs in these reactions *in vivo* or *in vitro,* including antibody inhibition and depletion,[27-31] targeted degradation of individual snRNPs,[31-35] and reconstitu-

[4] N. Riedel, J. A. Wise, H. Swerdlow, A. Mac, and C. Guthrie, *Proc. Natl. Acad. Sci. U.S.A.* **83,** 8097 (1986).

[5] C. Tschudi, F. F. Richards, and E. Ullu, *Nucleic Acids Res.* **14,** 8893 (1986).

[6] D. Tollervey, *J. Mol. Biol.* **196,** 355 (1987).

[7] A. Krämer, *Proc. Natl. Acad. Sci. U.S.A.* **84,** 8408 (1987).

[8] S. I. Lee, S. C. S. Murthy, J. J. Trimble, R. C. Desrosiers, and J. A. Steitz, *Cell* **54,** 599 (1988).

[9] K. Van Doren and D. Hirsh, *Nature (London)* **335,** 556 (1988).

[10] J. P. Bruzik, K. Van Doren, D. Hirsh, and J. A. Steitz, *Nature (London)* **335,** 559 (1988).

[11] K. A. Montzka and J. A. Steitz, *Proc. Natl. Acad. Sci. U.S.A.* **85,** 8885 (1988).

[12] R. Reddy and H. Busch, *in* "Structure and Function of Major and Minor Small Nuclear Ribonucleoprotein Particles" (M. L. Birnstiel, ed.), p. 1. Springer-Verlag, Berlin and New York, 1988.

[13] A. Krol and P. Carbon, this series, Vol. 180, p. 212.

[14] R. Reddy, this series, Vol. 180, p. 521.

[15] R. Lührmann, *in* "Structure and Function of Major and Minor Small Nuclear Ribonucleoprotein Particles" (M. L. Birnstiel, ed.), p. 71. Springer-Verlag, Berlin and New York, 1988.

[16] S. O. Hoch, this volume [20].

[17] C. Brunel and G. Cathala, this volume [21].

[18] P. Bringmann, B. Appel, J. Rinke, R. Reuter, H. Theissen, and R. Lührmann, *EMBO J.* **3,** 1357 (1984).

[19] C. Hashimoto and J. A. Steitz, *Nucleic Acids Res.* **12,** 3283 (1984).

[20] J. Rinke, B. Appel, M. Digweed, and R. Lührmann, *J. Mol. Biol.* **185,** 721 (1985).

[21] M. M. Konarska and P. A. Sharp, *Cell* **49,** 763 (1987).

[22] D. A. Brow and C. Guthrie, *Nature (London)* **334,** 213 (1988).

[23] T. Maniatis and R. Reed, *Nature (London)* **325,** 673 (1987).

[24] J. A. Steitz, D. L. Black, V. Gerke, K. A. Parker, A. Krämer, D. Frendewey, and W. Keller, *in* "Structure and Function of Major and Minor Small Nuclear Ribonucleoprotein Particles" (M. L. Birnstiel, ed.), p. 115. Springer-Verlag, Berlin and New York, 1988.

[25] M. L. Birnstiel and F. J. Schaufele, *in* "Structure and Function of Major and Minor Small Nuclear Ribonucleoprotein Particles" (M. L. Birnstiel, ed.), p. 155. Springer-Verlag, Berlin and New York, 1988.

[26] K. L. Mowry and J. A. Steitz, *Trends Biochem. Sci.* **13,** 447 (1988).

[27] V. W. Yang, M. R. Lerner, J. A. Steitz, and S. J. C. Flint, *Proc. Natl. Acad. Sci. U.S.A.* **78,** 1371 (1981).

tion of a function by complementation assays.[2,36,37] Genetic procedures have helped in the elucidation of snRNP function in yeast and in mammalian cells.[38,39] (Several of these methods are reviewed in additional chapters in this series.[40-44]) In a situation where complex systems, such as living cells or cellular extracts, are investigated and genetic approaches cannot easily be applied, the use of antibody probes and the targeted degradation of snRNPs have been the most powerful methods for the analysis of a specific function.

Although antibodies directed against U snRNPs are useful reagents to demonstrate that a processing reaction is snRNP dependent,[24,26] most of these antibodies cannot be applied to analyze the function of individual snRNPs because they cross-react with antigenic determinants present on several particle types.[15,41,45] This becomes particularly relevant when a reaction involves more than one species of snRNP, such as pre-mRNA splicing. One way to circumvent this problem is site-directed cleavage of the RNA in a snRNP particle with a complementary oligodeoxynucleotide and RNase H.[46] This method was initially applied to the analysis of the secondary structure of U1 RNA,[47,48] and was later adopted to assay U1 snRNP function in an *in vitro* splicing system.[31] In brief, an oligodeoxynucleotide complementary to a specific sequence within a U RNA is added

[28] R. A. Padgett, S. M. Mount, J. A. Steitz, and P. A. Sharp, *Cell* **35,** 101 (1983).
[29] I. Bozzoni, F. Annesi, E. Beccari, P. Fragapane, P. Pierandrei-Amaldi, and F. Amaldi, *J. Mol. Biol.* **180,** 1173 (1984).
[30] A. Fradin, R. Jove, C. Hemenway, H. D. Keiser, J. C. Manley, and C. Prives, *Cell* **37,** 927 (1984).
[31] A. Krämer, W. Keller, B. Appel, and R. Lührmann, *Cell* **38,** 299 (1984).
[32] A. R. Krainer and T. Maniatis, *Cell* **42,** 725 (1985).
[33] D. L. Black, B. Chabot, and J. A. Steitz, *Cell* **42,** 737 (1985).
[34] S. M. Berget and B. L. Robberson, *Cell* **46,** 691 (1986).
[35] D. L. Black and J. A. Steitz, *Cell* **46,** 697 (1986).
[36] A. R. Krainer, *Nucleic Acids Res.* **16,** 9415 (1988).
[37] A. Krämer, *Genes Dev.* **2,** 1155 (1988).
[38] C. Guthrie and B. Patterson, *Annu. Rev. Genet.* **22,** 387 (1988).
[39] Y. Zhuang and A. M. Weiner, *Cell* **46,** 827 (1986).
[40] K. A. Parker and J. A. Steitz, this series, Vol. 180, p. 454.
[41] J. A. Steitz, this series, Vol. 180, p. 468.
[42] R. Parker, this series, Vol. 180, p. 510.
[43] S.-C. Cheng, A. Newman, R. J. Lin, G. McFarlane, and J. N. Abelson, this volume [8].
[44] A. Krämer, this volume [18].
[45] M. Bach, P. Bringmann, and R. Lührmann, this volume [19].
[46] H. Donis-Keller, *Nucleic Acids Res.* **7,** 179 (1979).
[47] E. Lazar and M. Jacob, *Nucleic Acids Res.* **10,** 1193 (1982).
[48] J. Rinke, B. Appel, H. Blöcker, R. Frank, and R. Lührmann, *Nucleic Acids Res.* **12,** 4111 (1984).

to a snRNP preparation (e.g., to a snRNP-containing cellular extract). RNase H will hydrolyze the RNA if a hybrid of at least 4 base pairs is formed with the oligodeoxynucleotide.[46] The formation of such a DNA–RNA hybrid requires that the RNA sequence complementary to the oligonucleotide be single stranded in the particle and that it not be obscured by bound proteins.[48] snRNPs treated this way can then be tested for secondary structure or processing activity *in vitro*. Similarly, oligodeoxynucleotides can be injected into the germinal vesicles of *Xenopus laevis* oocytes prior to the injection of mRNA precursors to analyze RNA processing *in situ*.[49]

Site-Specific Cleavage of snRNAs in Presence of Complementary Oligodeoxynucleotides and RNase H

A number of modifications of the initially published method for the targeted degradation of snRNAs in processing extracts have been described.[21,31–35,50–58] In the following, the procedures developed in our laboratory are presented. References for alternative protocols are given.

Preparation of Oligodeoxynucleotides. Oligodeoxynucleotides are prepared with a DNA synthesizer (Applied Biosystems 380B) and were supplied by the German Cancer Research Center (Heidelberg, FRG) or by the Biocenter (University of Basel, Switzerland). Synthesis follows the phosphotriester route,[59] the 3'-terminal trityl group being removed during synthesis. The oligodeoxynucleotides are purified by reversed-phase high-performance liquid chromatography.[60] Base deprotection is performed by heating the oligonucleotide (in ~25% ammonia) to 55° for 6 hr or overnight. The solution is then cooled and the ammonia is removed with a stream of compressed air or nitrogen. The sample is lyophilized, resuspended in 10 mM tris(hydroxymethyl)aminomethane hydrochloride

[49] Z.-Q. Pan and C. Prives, *Science* **241**, 1328 (1988).
[50] M. Cotton, O. Gick, A. Vasserot, G. Schaffner, and M. L. Birnstiel, *EMBO J.* **7**, 801 (1988).
[51] B. Chabot, D. L. Black, D. M. LeMaster, and J. A. Steitz, *Science* **230**, 1344 (1985).
[52] B. Chabot and J. A. Steitz, *Mol. Cell. Biol.* **7**, 281 (1987).
[53] D. Frendewey, A. Krämer, and W. Keller, *Cold Spring Harbor Symp. Quant. Biol.* **52**, 287 (1987).
[54] K. L. Mowry and J. A. Steitz, *Science* **238**, 1682 (1987).
[55] L. Kretzner, B. C. Rymond, and M. Rosbash, *Cell* **50**, 593 (1987).
[56] A. Krämer, *J. Mol. Biol.* **196**, 559 (1987).
[57] D. Soldati and D. Schümperli, *Mol. Cell. Biol.* **8**, 1518 (1988).
[58] P. Legrain, B. Seraphin, and M. Rosbash, *Mol. Cell. Biol.* **8**, 3755 (1988).
[59] E. Ohtsuka, M. Ikehara, and D. Söll, *Nucleic Acids Res.* **10**, 6553 (1982).
[60] H. J. Fritz, R. Belagaje, A. L. Brown, R. H. Fritz, R. A. Jones, R. G. Lees, and H. G. Khorana, *Biochemistry* **17**, 1257 (1978).

(Tris-HCl) (pH 7.4), 1 mM ethylenediaminetetraacetic acid (EDTA), and purified on a FPLC fast desalting column (HR10/10; Pharmacia/LKB Biotechnology) in distilled water according to the supplier's instructions. The sample is again lyophilized and the oligonucleotides are stored in distilled water at $-20°$. Alternatively, the oligomers can be purified by polyacrylamide gel electrophoresis and elution from the gel followed by a desalting step.[32,33,50,61]

Preparation of Extracts. HeLa cell nuclear extracts are prepared by a modification of the method of Dignam *et al.*,[62] as described by Krämer and Keller.[63] They are dialyzed against buffer D [20% (v/v) glycerol, 20 mM *N*-2-hydroxyethylpiperazine-*N'*-2-ethanesulfonic acid (HEPES)–KOH (pH 7.9), 100 mM KCl, 0.2 mM EDTA, and 0.5 mM dithiothreitol (DTT)]. Extracts prepared according to Heintz and Roeder[64] have been used as well.[35,51–54] Processing activity in these extracts appears to remain more stable after the preincubation period required to degrade the snRNAs.[53] Procedures for the preparation of nuclear extracts from cultured mouse cells and of yeast extracts have been described elsewhere.[43,61]

Site-Directed Hydrolysis of snRNAs in Nuclear Extracts. A standard reaction is carried out in a volume of 25 μl containing 15 μl of nuclear extract, 2.5 μl of buffer D, 3 mM MgCl$_2$, 0.4 mM ATP, 20 mM creatine phosphate, and purified oligodeoxynucleotides. The reaction is incubated for 60 min at $30°$ and is then transferred to an ice-water bath for further manipulation.

The oligonucleotide concentration needed to degrade a given snRNA should be titrated. Our laboratory uses oligonucleotides complementary to U2, U4, and U6 RNAs at 10 μM, whereas efficient degradation of U1 RNA usually requires higher concentrations, such as 40 μM.[53,56,58] The addition of ATP to the reaction is recommended because it has been observed in several cases that an oligonucleotide will hybridize to the RNA in the intact snRNP only under these conditions.[32–35,65] In contrast, the target sequences are usually available for base pairing in protein-free RNA. This ATP requirement could reflect conformational changes within the snRNP particle upon ATP binding or an ATP-dependent association or dissociation of some other component(s) present in the extract. The addition of RNase H is unnecessary in most cases because this nuclease is

[61] C. Stauber, D. Soldati, B. Lüscher, and D. Schümperli, this volume [7].
[62] J. D. Dignam, R. M. Lebovitz, and R. G. Roeder, *Nucleic Acids Res.* **11**, 1475 (1983).
[63] A. Krämer and W. Keller, this volume [1].
[64] N. Heintz and R. G. Roeder, *Proc. Natl. Acad. Sci. U.S.A.* **81**, 2713 (1984).
[65] M. Zillmann, S. Rose, and S. Berget, *Mol. Cell. Biol.* **7**, 2877 (1987).

present in sufficient concentrations in mammalian nuclear extracts[32,33,57] as well as in yeast extracts.[66] It has been reported, though, that the results are more consistent when exogenous RNase H is added.[33]

Purified snRNPs can be treated accordingly; however, here the addition of RNase H is required.[48,67] RNase H (*Escherichia coli*; Boehringer-Mannheim) is used at a concentration of 100 units/ml.

Functional Analysis of Oligonucleotide/RNase H-Treated snRNPs

Digestion of Oligodeoxynucleotides. In general, extracts treated as outlined above can be used directly in a standard assay after addition of the appropriate substrate RNA. In cases where oligodeoxynucleotides are partially complementary to the pre-mRNA, it is necessary to degrade the oligomer prior to the addition of the RNA substrate to prevent degradation of the latter.[32,34,53] This is achieved by incubating the oligonucleotide/RNase H reaction for 15 min at 30° in the presence of 4 μl (80 μg) of DNase I (Worthington DPFF; made RNase free by treatment with Macaloid[68]).[53]

Similarly, DNase I treatment is required when an oligomer/RNase H-treated extract is complemented with intact snRNPs to restore the processing activity in an inactivated extract (see below). Although it has been reported that most of the oligonucleotide is degraded during a 60-min incubation in a HeLa cell nuclear extract,[33] other workers have observed that the digestion of the oligonucleotide even after extensive DNase treatment is not always complete.[32,69] Thus, this reaction should be controlled carefully.

Analysis of snRNP Function in Vitro. The function of specifically cleaved snRNPs can be assessed in *in vitro* processing reactions. This procedure has allowed the demonstration of the involvement of U1, U2, U4, and U6 snRNPs in pre-mRNA splicing[31-35] as well as the participation of U7 snRNP in histone pre-mRNA 3' end formation.[50,54,57] *In vitro* systems for these reactions are detailed elsewhere in this volume.[61,63] As an example, the procedure used for the analysis of snRNP function in pre-mRNA splicing is given below.

The *in vitro* splicing reaction is performed in a total volume of 50 μl. After preincubation of the extract with oligodeoxynucleotides and RNase H (25 μl) and treatment with 4 μl of DNase I (see above), the splicing

---

[66] B. C. Rymond and M. Rosbash, *Nature (London)* **317**, 735 (1985).
[67] A. Krämer, unpublished results (1988).
[68] T. Maniatis, E. F. Fritsch, and J. Sambrook, "Molecular Cloning: A Laboratory Manual." Cold Spring Harbor Laboratory, Cold Spring Harbor, New York, 1982.
[69] J. Hamm, A. Krämer, and I. W. Mattaj, unpublished observation (1988).

reaction is initiated by the addition of 21 μl of a mixture containing radio-actively labeled pre-mRNA substrate,[53,63] 7.5 μg of *E. coli* tRNA (Boehringer-Mannheim), 0.4 mM MgATP, and 7.5 μl of buffer D. The reaction mixtures are incubated for 60 min at 30°, and reaction products are analyzed by appropriate methods.[53,63]

To exclude the possibility that the processing activity is inhibited un-specifically in the presence of an oligodeoxynucleotide, two types of con-trol reactions can be performed.[33,34] First, a small amount of nuclear extract (5 μl) is inactive when tested in the *in vitro* splicing reaction. However, the snRNPs are present in sufficient concentrations to generate spliced RNA when added to an extract that is deficient in intact snRNPs. Hence, the activity of an oligonucleotide/RNase H-treated extract should be restored to normal levels by the addition of exogenous snRNPs.[34]

The second control reaction can be applied to systems that require more than one snRNP, such as pre-mRNA splicing. Two extract samples treated with oligodeoxynucleotides complementary to different snRNAs (e.g., U1 and U2) are expected to complement each other for splicing activity.[33] On the other hand, when two oligomers complementary to different sequences in one and the same snRNA are used, splicing activity should not be restored. A positive result in this type of reaction requires that the snRNPs under investigation either exist as independent particles in the extract (at the start of the reaction) or that they readily exchange from any preexisting complex.

The predicted results were obtained in all cases tested,[33,34] confirming that the inhibition of splicing resulted from the specific degradation of the snRNP and not from unspecific effects caused by the presence of the oligomer or RNase H or by the degradation products themselves.

Analysis of Digestion Products

To interpret the results of functional studies with oligonucleotide/RNase H-treated snRNPs, the extent of degradation should be deter-mined. It has been observed that the digestion of a particular RNA se-quence is not always complete, thus resulting in varying levels of residual processing activity despite extensive degradation.[31–34,51,52,58,65,70] It should also be kept in mind that not all oligodeoxynucleotides direct the cleavage of their target RNA because of secondary structure or because the target sequence is protected from base pairing by bound proteins.[11,33–35,51,70] Two extreme cases, U5 and U11 snRNPs, have been reported, in which the

---

[70] M. Zillmann, M. L. Zapp, and S. M. Berget, *Mol. Cell. Biol.* **8**, 814 (1988).

RNAs are not available for base pairing with any complementary oligo-deoxynucleotide tested.[11,51,71]

One should also be aware that RNase H cleaves at different sites within the region of complementarity between oligodeoxynucleotide and target RNA, resulting in a heterogeneous population of digestion products.[35,46-49,72] Some of these may still contain sequences important for an interaction with a pre-mRNA and may be present in sufficient quantities for processing (or a certain step in a processing pathway) to occur.[35,72]

In the following section three methods are described that are suitable for the analysis, both quantitatively and qualitatively, of the products of the oligodeoxynucleotide/RNase H reaction.

Analysis of RNA by Gel Electrophoresis. To an oligodeoxynucleotide/RNase H reaction 120 μl of a solution containing 2% (w/v) sarkosyl, 0.1 M Tris-HCl (pH 7.5), 20 mM EDTA, and 40 μg of proteinase K (Boehringer-Mannheim) is added, and the reactions are incubated for 30 min at 37°. Following phenol–chloroform extraction and ethanol precipitation, RNAs are separated in a 10% polyacrylamide–8.3 M urea gel in 1× TBE.[68] RNAs are visualized by staining with ethidium bromide (0.5 μg/ml) and illumination with UV light.

This method allows a rather rapid examination of the main products of the degradation reaction and is suitable for an initial screen of the availability of snRNA sequences to hybridization with a particular oligonucleotide and cleavage by RNase H. The method is not very sensitive, however, and low concentrations of intact snRNPs may escape detection. For example, as little as 1% of residual intact U1 snRNP (which was detected only by Northern blotting) is sufficient in some cases to yield almost normal levels of splicing activity.[33]

Northern Blotting. RNAs are isolated and separated in denaturing polyacrylamide gels as described above. The gels are equilibrated in transfer buffer (see below) to remove the urea. The RNA is then transferred onto a nylon membrane (Gene Screen; New England Nuclear) by electroblotting at 200 mA in 0.1 M Tris-OH, 50 mM sodium acetate, and 5 mM EDTA (pH 7.8) for 16 hr at 4°.[21] The membrane is irradiated with UV light for 30 sec and baked at 80° for 1–2 hr. Prehybridization is performed for 1–3 hr at 42° in 10 ml of 50% formamide, 0.2% Ficoll, 0.2% bovine serum albumin, 0.2% poly(vinylpyrrolidone), 50 mM Tris-HCl (pH 7.5), 0.1% sodium pyrophosphate, 1% sodium dodecyl sulfate (SDS), 10% dextran sulfate, 1 M NaCl, and 0.15 mg yeast RNA/ml. Antisense RNA probes[21] (10^7 Cerenkov counts/min) are hybridized in the same buffer for

[71] G. Christofori and W. Keller, personal communication (1988).
[72] S. Bienroth and W. Keller, personal communication (1988).

16 hr at 42°. The membranes are washed at room temperature for 30 min each with four changes of 500 ml of 2× SSC plus 0.1% SDS (1× SSC is 0.15 M NaCl, 0.15 M sodium citrate) and finally with 500 ml of 0.1× SSC. The membranes are briefly dried and exposed for autoradiography.

With this procedure low concentrations of RNA can be detected, and the percentage of the RNA that is actually cleaved is easily determined.

Primer-Extension Analysis. The analysis of oligodeoxynucleotide/ RNase H-treated snRNAs by primer extension provides another method to calculate the ratio of intact versus cleaved RNA. In addition, the site of RNase H cleavage within the snRNA can be determined to the nucleotide when primer-extension products are compared to the sequence of the RNA under investigation.

Oligodeoxynucleotide primers complementary to snRNA sequences are 5′-end-labeled with polynucleotide kinase to a specific activity of 0.5– 10 mCi/nmol.[68] About 0.1–0.5 pmol of the primer is incubated in a total volume of 16 µl with RNA extracted from 5 µl of nuclear extract (either untreated or treated with oligonucleotides as described above), and final concentrations of 125 mM Tris-HCl (pH 7.5), 175 mM KCl, and 0.125 mM EDTA are added. After denaturation of the RNA for 5 min at 80°, the primer is annealed for 3 hr at 50°. Primer extension is carried out in a total volume of 20 µl after addition of final concentrations of 6 mM MgCl, 2.25 mM DTT, 0.5 mM each of dATP, dCTP, dGTP, and dTTP, and 7.5 units of Moloney murine leukemia virus (M-MuLV) reverse transcriptase (Pharmacia).[73,74] After incubation for 50 min at 42° and 10 min at 46°, the reaction is terminated by digestion of the RNA with RNase A (50 µg/ml; Boehringer-Mannheim) for 10 min at 37°. Primer-extension products are treated with proteinase K (see above), phenol–chloroform extracted, ethanol precipitated, and separated in a 18% polyacrylamide–8.3 M urea sequencing gel in 1× TBE, followed by autoradiography.

A sequence ladder of untreated RNA for reference is prepared by adding 0.6 µl of dideoxynucleotides (6.6 mM ddATP, 4 mM ddCTP, 6 mM ddGTP, or 6 mM ddTTP) to four separate primer extension reactions.[74]

Acknowledgments

I thank S. Bienroth, D. Frendewey, and W. Keller for providing details of the Northern blotting and primer-extension procedures and W. Keller for comments on the manuscript. This work was supported by grants from the Deutsche Forschungsgemeinschaft and the Schweizerischer Nationalfonds.

[73] D. Solnick, *Cell* **43**, 667 (1985).
[74] K. Parker and J. A. Steitz, *Mol. Cell. Biol.* **7**, 2899 (1987).

[24] Isolation and *in Vitro* Assembly of Nuclear Ribonucleoprotein Particles and Purification of Core Particle Proteins

By Stanley F. Barnett, Stephanie J. Northington, and Wallace M. LeStourgeon

During transcription in mammalian cells 700-nucleotide (nt) lengths of the elongating pre-mRNA molecule are packaged by multiple copies of a few abundant nuclear proteins into a repeating array of regular ribonucleoprotein particles [heterogeneous nuclear ribonucleoprotein (hnRNP) particles].[1] Monoparticles have a mass near 1.5 million, a sedimentation coefficient near 40 S, and a protein to RNA ratio near 6:1. They are primarily composed of six major "core particle" polypeptides which migrate in single-parameter sodium dodecyl sulfate–polyacrylamide gel electrophoresis (SDS–PAGE) as three groups of doublet bands [(A1, A2), (B1, B2), and (C1, C2)].[1-6] In monoparticles and in oligomeric hnRNP complexes isolated from actively growing HeLa cells, these proteins are present in the molar ratio 3A1, 3A2, 1B1, 1B2, 3C1, 1C2.[1,2,5,6] These proteins will spontaneously assemble on 700-nt lengths of single-stranded RNA or DNA to form 40 S particles with this stoichiometry even if the starting mix of protein is made deficient in one or more of the individual proteins. Dimers, trimers, and polyparticles assemble on integral multiples of 700 ± 20 nt.[1] In intact hnRNP particles, the major core particle proteins exist as tetramers of $(C1)_3C2$, $(A2)_3B1$, and $(A1)_3B2$.[5,6] Under appropriate conditions the $(A2)_3B1$ tetramers will package 700 nt of RNA into a 43 S particle,[6] or, in the absence of RNA, these tetramers will

[1] G. Conway, J. Wooley, T. Bibring, and W. M. LeStourgeon, *Mol. Cell. Biol.* **8**, 2884 (1988).

[2] A. L. Beyer, M. E. Christensen, B. W. Walker, and W. M. LeStourgeon, *Cell* **11**, 127 (1977).

[3] W. M. LeStourgeon, A. L. Beyer, M. E. Christensen, B. W. Walker, S. Poupore, and L. P. Daniels, *Cold Spring Harbor Symp. Quant. Biol.* **42**, 885 (1978).

[4] L. Lothstein, H. P. Arenstorf, S. Chung, B. W. Walker, J. C. Wooley, and W. M. LeStourgeon, *J. Cell Biol.* **100**, 1570 (1985).

[5] S. F. Barnett, D. Friedman, and W. M. LeStourgeon, *Mol. Cell. Biol.* **9**, 492 (1989).

[6] W. M. LeStourgeon, S. F. Barnett, and S. J. Northington, *in* "The Eukaryotic Nucleus: Molecular Structure and Macromolecular Assemblies" (P. Strauss and S. Wilson, eds.). Telford Press, Caldwell, New Jersey, in press.

METHODS IN ENZYMOLOGY, VOL. 181

assemble to form insoluble helical filaments of indefinite length.[4] For more information on hnRNP particles and individual particle proteins, see the reviews of Dreyfuss,[7] Chung and Wooley,[8] and LeStourgeon et al.[6]

Described below are (1) procedures for isolating 40 S hnRNPs from mammalian tissue culture cells, modified from the original methods of Samarina et al.[9]; (2) procedures for the dissociation and in vitro assembly of hnRNP complexes on defined substrates; (3) procedures for generating and isolating the 43 S and/or the filamentous rearrangement products of proteins A2–B1; and (4) procedures for purifying native tetramers of (A2)$_3$B1 and (C1)$_3$C2.

Isolating 40 S hnRNP Particles from Mammalian Cells in Culture: Nuclear Isolation

In order to obtain hnRNP complexes which are not contaminated with ribosomes or other cytoplasmic components it is very important to obtain rapidly nuclei free of cytoplasmic material. We have developed a one-solution procedure that yields nuclei from tissue culture cells free of other cellular components as monitored by phase-contrast microscopy at ×1000. In this procedure, fresh cells or cells frozen in 50% glycerol can be used. In our hands, log-phase HeLa cells, 3T3 cells, or MEL cells frozen at −60° for periods up to 3 years yield hnRNPs indistinguishable from those isolated from freshly harvested log-phase cells.

Suspend 1–3 × 10^9 cells in 360 ml of ice-cold nuclear isolation buffer [(NIB) 2.0 mM tris(hydroxymethyl)aminomethane (Tris) base, 0.1 mM MgCl$_2$, 0.002% w/v Triton X-100 (pH 7.5)]. This can be achieved by gentle agitation in an Erlenmeyer flask or in a hand-held loose-fitting Dounce homogenizer. Spin 45-ml aliquots of the suspension in eight 50-ml conical bottom polycarbonate tubes for 8 min at 300 g in a refrigerated centrifuge. Repeat this step by using a loose-fitting Dounce homogenizer to resuspend the cells or simply by swirling the cells in a small Erlenmeyer flask. Do not be overly aggressive. Some cells will begin to lyse at this time. Spin as above and again resuspend the cells in NIB.

Examine the preparation with the phase-contrast microscope. If more than 1–2% of the cells remain intact it may be necessary to homogenize a few times. Spin as above to pellet the purified nuclei. Note that the nuclei will appear quite swollen and of low contrast. This is a consequence of the very low [Mg^{2+}] in the NIB. Remember that hnRNPs do not require

---

[7] G. Dreyfuss, Annu. Rev. Cell Biol. **2**, 457 (1986).

[8] S. Y. Chung and J. Wooley, Proteins: Struct. Funct. Genet. **1**, 195 (1986).

[9] O. P. Samarina, E. M. Lukanidin, J. Molnar, and G. P. Georgiew, J. Mol. Biol. **33**, 251 (1968).

divalent cation for stability. The very low [Mg^{2+}] and the trace level of Triton aid in cell lysis, dispersion of cytoplasmic material, and hnRNP recovery. Note that if too many cells are used per volume of NIB, lysis will be inefficient and the nuclei will not be clean. Finally, be aware that fresh cells lyse more readily, and complete lysis may occur in the first suspension in NIB. If this occurs, a single wash in NIB usually achieves acceptable nuclear purity. Nuclear isolation should be complete in 20–30 min. Isolated nuclei may be frozen for later use, but hnRNP recovery always seems to be diminished somewhat.

Extracting hnRNP Particles from Isolated Nuclei

1. If the procedure used above is followed for nuclear isolation, one will have in hand 8 nuclear pellets. The pellets in groups of 4 should be washed with small volumes of freshly prepared ice-cold pH 7.2 STM [90 mM NaCl, 1.0 mM MgCl$_2$, 10 mM Tris-HCl (pH 7.2)] into a Dounce homogenizer and the volume of pH 7.2 STM brought to 45 ml. Disperse the nuclear clumps by homogenizing a few times. Spin this preparation in a 50-ml conical bottom polycarbonate tube for 8 min at 500 g in a refrigerated centrifuge. This washing step removes many nucleosolic and nuclear membrane-associated proteins. If the nuclei are examined with the phase-contrast microscope they will appear smaller and with more contrast owing to the higher Mg^{2+} concentration.

2. Combine the 2 washed nuclear pellet preparations into one 50-ml conical bottom centrifuge tube with pH 7.2 STM, pellet, and draw off the supernatant. At this stage 1 ml of packed nuclei represents about 1.2–1.5 × 10^9 nuclei.

3. At this point one may choose to use ultrasound to effect hnRNP extraction, or the preparation may be stirred at 37° for 1–2 hr or at room temperature for 2–4 hr. We routinely use ultrasound as follows: Add at least 2.0 ml of pH 8.0 STM for each 1.0 ml of nuclear pellet and subject this preparation to the equivalent of three 10-sec bursts from a Heat Systems-Ultrasonics, Inc., Model W140 Sonifier equipped with a microtip transducer set at 60–80 W. The 50-ml conical bottom centrifuge tube containing the nuclear suspension is held in ice during this procedure. Pause about 15 sec between bursts. The fragmented nuclear material is examined with the phase-contrast microscope using the 40× "high dry" objective. If more than a few intact nuclei appear per field, it may be necessary to sonicate more aggressively. Note here that the yield of hnRNP increases as the volume of pH 8.0 STM is increased. We usually

compromise on yield because when using Beckman SW28.1 centrifuge tubes for sucrose density gradients we do not want to load more than 2 ml of clarified sonicate on each gradient. If one chooses to use the "stir" method the volume of pH 8.0 STM should be increased to at least 4 : 1 (buffer to pellet volume). Evidence that stirring is too long can be seen in hnRNPs that are deficient in proteins A1 and the C-group polypeptides.

4. Incubate the nuclear sonicate at 37° for 15–20 min to allow endogenous nucleases to free the 40 S monoparticles. Swirl the tube occasionally during the incubation period. If particles are deficient in protein A1 and the C-proteins, either excessive nuclease action or protease action has occurred or the cells were not in log-phase growth. We do not routinely add protease inhibitors when working with HeLa cells, but this is sometimes necessary with some cell types. If the 40 S gradient peak appears as a light shoulder on a larger 45–50 S peak then insufficient nuclease action has not generated 40 S monoparticles.[4] To isolate polyparticles, then, this step should be avoided entirely and RNasin should be added prior to nuclear sonication. Note that only a small portion of nuclear hnRNP complexes will be recovered if the latter procedure is followed to obtain oligomeric complexes.

5. Centrifuge the incubated nuclear sonicate (or extract) in a 15-ml Corex centrifuge tube for 10 min at 8000 g (i.e., using the HB4 rotor of the Sorvall RC2-B centrifuge at 7000 rpm) to pellet the chromatin. After this spin the supernatant containing the hnRNP will appear somewhat opalescent.

6. Load 1–2 ml of the chromatin-clarified nuclear sonicate on one or more 15–30% sucrose density gradients containing pH 8.0 STM in SW28.1 centrifuge tubes and spin at 25,000 rpm for 16 hr in the Beckman SW28.1 rotor at 4°. These spin conditions will cause the 40 S monoparticles to sediment near the middle of the gradient.

7. Pump the gradients from bottom to top through the flow cell of a spectrophotometer set at 260 nm, fractionate into 0.5- to 1.0-ml aliquots, and analyze the gradient fractions on 8.75% polyacrylamide gels in the presence of SDS. We routinely add 3 volumes of ethanol to each fraction to precipitate the particles but trichloroacetic acid (TCA) also works well (add one-tenth the fraction volume of 100% TCA and wash the precipitates with 1–3 ml of 70% ethanol to remove the acid before adding electrophoresis sample buffer). Alternatively, one may simply dialyze the gradient fractions against the electrophoresis sample buffer.

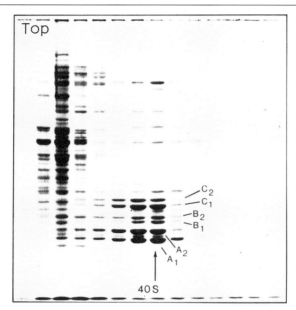

Fig. 1. Sedimentation and protein composition of HeLa 40 S hnRNP particles in 15–30% sucrose density gradients. The gradient was fractionated into 1.0-ml aliquots, and the material in each fraction was precipitated with 3 volumes of ethanol, dried, and solubilized in electrophoresis sample buffer. The samples were resolved by SDS–PAGE (lane dimensions 0.7 mm × 10 cm, 8.75% polyacrylamide). The top of the gradient is shown at left. The molar ratio of the major core particle proteins is 3A1, 3A2, 1B1, 1B2, 3C1, 1C2.

If the above procedure is followed for a preparation from 1.5×10^9 one should obtain two gradients, and in each gradient the peak absorbance should be between 1.5 and 2.0 and the peaks should be positioned near the center of the gradient tube. Since the protein to RNA ratio in 40 S hnRNP is near 6 : 1, adequate protein will be available to run many Coomassie blue-stained gels (Fig. 1) or for other experiments.

Isolating 43 S Rearrangement Product of (A2)₃B1 Tetramers

The core particle proteins A2 and B1 can easily be obtained in an almost pure state by inducing formation of the 43 S rearrangement product.[4] This is achieved by adding RNase A to the nuclear sonicate generated in Step 3 above. To the nuclear sonicate add 5 μg of RNase A per 10^8 nuclei and incubate for 15–20 min at 37°. Proceed with Step 5 above. Note that one should obtain a quantitative recovery of proteins A2 and B1 in the 43 S structure. The protein to RNA ratio in these structures is 12 : 1, so even though the OD_{260} peak in gradients is small and sharper, enough

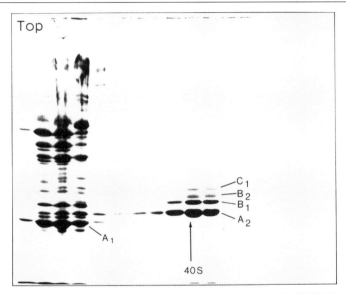

FIG. 2. Sedimentation of the 43 S rearrangement product composed of (A2)₃B1 tetramers in a 15–30% sucrose density gradient. Sample preparation and electrophoresis were as in Fig. 1. The top of the gradient is at the left. The trace quantities of proteins B2 and C1 which sediment with the 43 S rearrangement product vary in amount as a function of the extent of nuclease digestion.

protein will be obtained to run as many gels as above (Fig. 2). It is also possible to dissociate and reconstitute the 43 S rearrangement product on 700-nt lengths of RNA. Follow the procedures described below for dissociation and reconstitution of 40 S hnRNP particles except perform the nuclease digestion at 4° for 2 hr and use 12 μg synthetic RNA per OD unit of the original 43 S product. If the digestion is not performed at low temperatures, proteins A2 and B1 will form insoluble filaments as described below.

Generating Insoluble Helical Filaments Composed of (A2)₃B1 Tetramers

Pool the gradient fractions containing the 43 S rearrangement product described above and dialyze against 1.0 liter of one-tenth strength pH 8.0 STM overnight at 4° to remove sucrose and to lower the salt concentration. Digest the remaining endogenous RNA with ribonuclease A (10 μg/ml for 30 min at 0°), and add MgCl₂ to a final concentration of 5 mM and allow this preparation to come to room temperature. The filaments begin to form at temperatures above 10° and appear in electron micrographs as shown in Fig. 3.

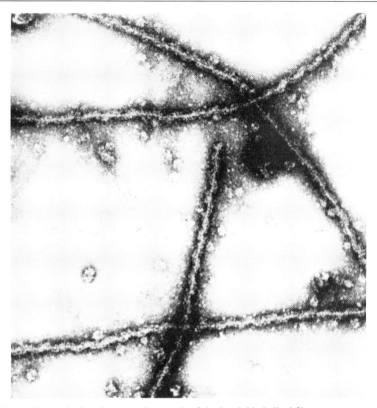

FIG. 3. Transmission electron micrograph of the insoluble helical filaments composed of $(A2)_3B1$ tetramers. The filaments have a width of 18 nm and a pitch near 60 nm. In most electron micrographs filament ends are not observed. [From L. Lothstein, H. P. Arenstorf, S.-Y. Chung, B. W. Walker, J. C. Wooley, and W. M. LeStourgeon, *J. Cell Biol.* **100,** 1570 (1985), courtesy of The Rockefeller University Press.]

Purification of $(A2)_3B1$ Tetramers

Soluble and native $(A2)_3B1$ tetramers can be purified using the 43 S rearrangement product as starting material. The sucrose gradient fractions containing the 43 S rearrangement product are pooled and dialyzed against 20 mM Tris (pH 8.0) and 50 mM NaCl. The RNA component is digested with ribonuclease A (10 μg/ml) or micrococcal nuclease (300 U/ml) for 2 hr at 0°. It is very important to keep this material below 10°, or the tetramers will assemble to form insoluble filaments as described above. The digested material is sedimented on 5–20% sucrose density gradients [20 mM Tris (pH 8.0), 50 mM NaCl] at 190,000 g at 2° for 35 hr.

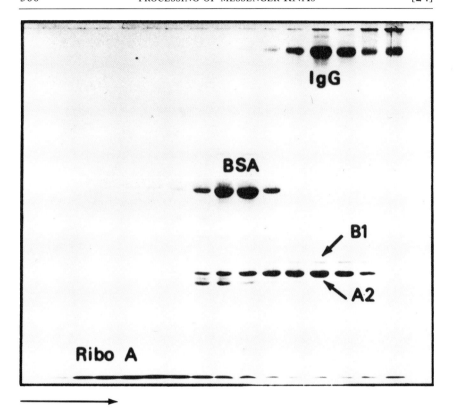

FIG. 4. Coomassie blue-stained 8.75% polyacrylamide gel showing sedimentation of (A2)₃B1 tetramers in 5–20% sucrose density gradients with internal sedimentation standards of ribonuclease A, bovine serum albumin (BSA), and immunoglobulin G (IgG). Sedimentation is from left to right. Some A2–A2 and A2–B1 dimers are seen near the BSA peak, but most of the material sediments with a mass near 140,000.

At this salt concentration the tetramers sediment mostly as one peak at 6.8 S (Fig. 4). At 90 mM NaCl the tetramers begin to dissociate, and about 50% of proteins A2 and B1 sediment as dimers at 3.2 S. At higher salt concentrations (200–400 mM) all of the A2 and B1 proteins sediment as dimers. This is also the salt concentration at which these proteins dissociate from endogenous RNA packaged in native 40 S hnRNP particles.

Purification of Native C-Protein Tetramer

Native C-protein tetramers can be purified from isolated 40 S hnRNP particles through a two-step procedure using gel filtration in 350 mM NaCl and anion-exchange chromatography. The C-protein tetramers obtained

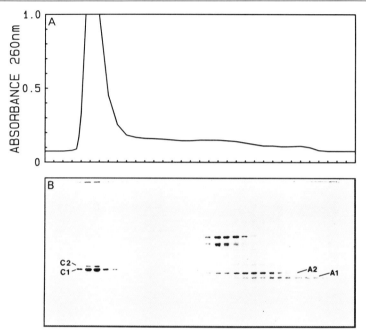

FIG. 5. BioGel A-0.5 m chromatography of isolated 40 S hnRNPs dissociated in 0.35 M NaCl. (A) Elution of the RNA (OD_{260} material) in the exclusion volume. (B) Protein elution pattern. Note that at a salt concentration of 0.35 M the C proteins elute in the exclusion volume associated with the RNA originally packaged in monoparticles.

as described below will bind RNA in the absence of other proteins and will participate in hnRNP particle reconstitution on defined RNA substrates.[10]

1. The peak sucrose gradient fractions containing the 40 S hnRNP particles isolated from 4.0×10^9 cells are pooled and adjusted to a final concentration of 350 mMNaCl and 1 mM dithiothreitol (DTT) and loaded onto a BioGel A-0.5 m (Bio-Rad Laboratories) column (1.6 i.d. × 75 cm) equilibrated with 20 mM Tris-HCl (pH 8.0), 1 mM MgCl$_2$, 1 mM DTT, and 350 mM NaCl. The column is eluted with this buffer at 10 ml/hr. Under these conditions the C-protein tetramers remain associated with the RNA while most other proteins do not. The C protein–RNA complexes elute in the void volume and are easily detected by the absorbance at 260 nm (Fig. 5).

[10] S. F. Barnett, W. M. LeStourgeon, and D. L. Friedman, *J. Biochem. Biophys. Methods* **16,** 87 (1988).

2. The C protein–RNA complexes (i.e., the exclusion peak) from above are adjusted to a final concentration of 10% glycerol, 1 mM CaCl$_2$, and 50 U/ml micrococcal nuclease (Boehringer-Mannheim), and the RNA is digested for 1.5 hr on ice. The low temperature and glycerol help prevent protein aggregation. A low protein concentration is also helpful. Following digestion of the RNA, the material is centrifuged to remove any precipitated material. The supernatant is then loaded onto a Mono-Q HR5/5 anion-exchange column (Pharmacia Fine Chemicals) equilibrated in 20 mM Tris-HCl (pH 8.0), 1 mM MgCl$_2$,1 mM DTT, and 280 mM NaCl. Preparative volumes (10–30 ml) can be loaded using a Pharmacia Superloop. The column is washed with 3 ml of the equilibration buffer and eluted with a 280 to 600 mM NaCl gradient applied over a period of 40 min at 0.5 ml/min. If the absorbance of the eluting material at 214 nm is monitored, the C protein is easily identified as the major peak eluting in approximately 450 mM NaCl (Fig. 6).

The 200–300 μg of native C-protein tetramer obtained by this procedure is approximately 95% pure as judged by SDS–PAGE and Coomassie

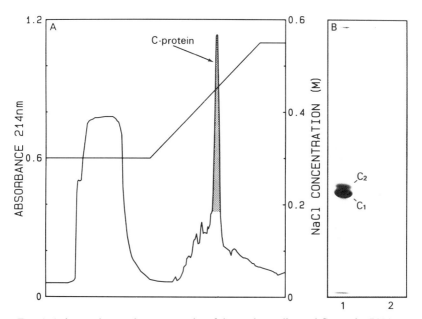

FIG. 6. Anion-exchange chromatography of the nuclease-digested C protein–RNA complex (A) obtained as in Fig. 5. (B) Purified C-protein tetramers in SDS–PAGE. Lane 2 of (B) contains one-tenth the protein load shown in lane 1.

blue staining. The C-protein tetramer can be purified to homogeneity as judged by SDS–PAGE and silver staining by sedimenting through a 5–20% sucrose gradient [10 mM Tris-HCl, 1 mM MgCl$_2$, 1 mM DTT, 100 mM NaCl (pH 8.0)] at 190,000 g and 4° for 24 hr. The tetramers sediment approximately midway in the gradient.

We have recently determined that digestion with ribonuclease A prior to gel filtration (i.e., before Step 1 above) and use of a BioGel A-1.5 column give a better first step purification. This is because the C-protein tetramers now elute just behind the void volume under these conditions and are therefore not contaminated by trace levels of denatured and aggregated protein.

In Vitro Assembly of hnRNP Particles on Defined Substrates

Many questions regarding the composition, structure, and function of hnRNP complexes have been difficult to answer experimentally because the RNA component of isolated monoparticles is as heterogeneous as the entire complement of nuclear pre-mRNA and because the core particle proteins do not protect the packaged length of RNA from endogenous nuclease action. Several advances in the past few years, however, now make it possible to study more defined experimental material, namely, the ability to synthesize large quantities of RNA *in vitro* with defined length and sequence,[11] the ability to monitor the biochemical events of RNA processing *in vitro*,[12] and the ability assemble hnRNP complexes *in vitro* on synthetic RNA.[1]

At this writing three procedures have been used to dissociate and reconstitute hnRNP complexes *in vitro*. In order to assemble hnRNP complexes on RNA substrates of defined length and sequence, it is first necessary to dissociate the particles and to remove the endogenous RNA under conditions that do not lead to protein denaturation or loss. Although isolated 40 S hnRNP particles will dissociate in 0.7 M NaCl and spontaneously reassemble on back-dialysis into low salt,[2,3] this procedure is complicated by the necessity of removing the endogenous RNA fragments under high-salt conditions to prevent particle reassembly. In the reconstitution studies of Pullman and Martin,[13] isolated hnRNP particles were dissociated with the ionic detergent sodium deoxycholate, and the endogenous RNA was then digested with micrococcal nuclease. Following enzyme inactivation with EDTA and the addition of exogenous RNA, reconstitution was achieved by sequestering the deoxycholate with Triton

[11] M. M. Konarska, P. A. Padgett, and P. Sharp, *Cell* **38**, (1984).
[12] A. R. Krainer, T. Maniatis, B. Ruskin, and M. R. Green, *Cell* **36**, 993 (1984).
[13] J. M. Pullman and T. E. Martin, *J. Cell Biol.* **97**, 99 (1983).

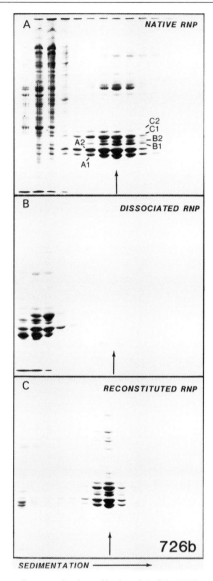

FIG. 7. Dissociation and reconstitution of isolated 40 S hnRNPs on a defined-length RNA substrate. (A) Protein composition and sedimentation of intact HeLa 40 S hnRNP particles. (B) Material collected from the 40 S peak fractions in (A) after digestion with micrococcal nuclease. Note that all of the core particle proteins remain at the top of the gradient. (C) Sedimentation and protein composition of monoparticles which assemble on a 726-nt transcript of the T4 phage genome. Note the symmetrical distribution of individual core particle proteins above and below the peak fraction and note that the protein stoichiometry is the same as for intact 40 S monomers shown in (A) and also in Fig. 1.

X-100. The procedure described below is similar to that of Wilk *et al.*[14] in that isolated hnRNP particles are directly dissociated with micrococcal nuclease, the enzyme is inactivated with EDTA, and exogenous RNA is added as the assembly substrate. This procedure for particle dissociation and reconstitution takes advantage of the fact that RNA "activates" the protein–protein binding domains required for particle assembly. The utility of micrococcal nuclease is derived from its requirement for Ca^{2+} and from the fact that divalent cation is not required for hnRNP assembly or stability in solution. Figure 7 shows the protein composition of monoparticles assembled *in vitro* on a 726-nt length of mouse β-globin pre-mRNA. Figure 8 shows electron micrographs of monoparticles, dimers, and oligomeric hnRNP complexes.

1. Combine the peak OD_{260} fractions from sucrose gradients containing the native 40 S hnRNP particles and dialyze overnight against 1000 ml of pH 8.0 STM to remove the sucrose. The dialysis tubing should be boiled and well rinsed with distilled water and clamped or tied tight above the sample to prevent a volume increase during dialysis.

2. Adjust the dialyzed material to an OD_{260} of 1.0 with pH 8.0 STM and measure the volume of the sample. Add 1/100 volume of 0.1 M DTT and 1/100 volume of 1.0 M $CaCl_2$ and mix well.

3. Add 1/32 volume of a solution containing 10,000 U/ml of micrococcal nuclease (final concentration 300 U/ml). Incubate the preparation 40 min at 25°. It is very important to assay the nuclease for protease contamination because of the incubation time. On three consecutive occasions we obtained nuclease from one vendor which digested most of the protein as well as nucleic acid. We have had good success recently with Boehringer-Mannheim (nuclease S7, *Staphylococcus aureus*, EC 3.1.31.1). We have not examined the effect of protease inhibitors on particle reconstitution.

4. Return the sample to ice and add 1/24 volume of 0.25 M EDTA (pH 8.0) to a final concentration of 10 mM. Add the synthetic RNA substrate solubilized in a small volume of TE [10 mM Tris, 1 mM EDTA (pH 8.0)]. Let the preparation stand in ice for 5–60 min, then layer the sample on cold 15–30% sucrose density gradients made to contain 10 mM EDTA and spin as described above for isolating 40 S monoparticles. As described in detail elsewhere,[1] the following conditions are assumed for the assembly of 40 S monoparticle possessing all characteristics of native hnRNPs.

[14] H.-E. Wilk, G. Angeli, and K. P. Schafer, *Biochemistry* **22**, 4592 (1983).

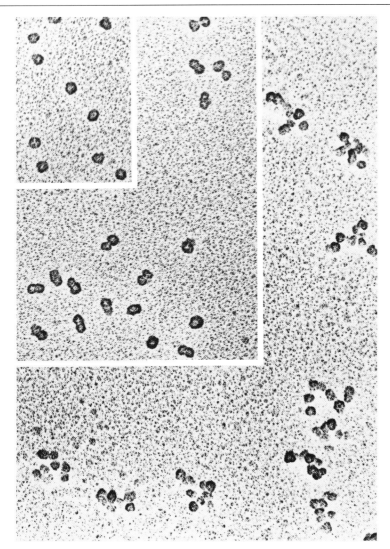

FIG. 8. Electron micrographs of 40 S monoparticles, dimers, and oligomeric hnRNP complexes that assemble *in vitro* on transcripts of defined length and sequence. At upper left are shown 40 S monoparticles that assemble on a 736-nt transcript of the mouse β-globin gene. In the middle are shown dimers that assemble on a 1509-nt length transcript of the same gene. At lower right are shown the polyparticle complexes that assemble on the φX174 single-stranded DNA genome. For each preparation the particles were absorbed to polyly-sine-coated grids, stained with 5% uranyl acetate, and platinum shadowed. The particles have a diameter of 26 ± 2 nm.

a. For the *in vitro* assembly of 40 S monoparticles, the added RNA
   or single-stranded DNA should be between 685 and 726 nt in
   length or integral multiples of 700 nt for the assembly of dimers
   and oligomeric complexes. Other RNA lengths will support the
   assembly of nonstoichiometric complexes which sediment
   through the gradients. The core proteins of 40 S hnRNP do not
   assemble in a sequence-dependent manner *in vitro*, and nucleotide triphosphates are not required as an energy source.
b. For correct particle reassembly, the exogenous RNA should be
   added at 0.7 times the measured OD of the pooled and dialyzed
   sample. In other words, add 0.7 OD units (28 μg) of exogenous
   RNA per OD unit of isolated hnRNP. Correct particle assembly
   will occur over the range 10–20 : 1 for protein to RNA. Reactant
   concentration has no significant affect on particle assembly.

[25] Physical Methods for Characterization of Heterogeneous Nuclear Ribonucleoprotein Complexes

By JERZY T. SCHÖNEICH and JOHN O. THOMAS

The major components of heterogeneous nuclear ribonucleoprotein
(hnRNP) complexes are transcripts of RNA polymerase II, a rather complex set of proteins, and small nuclear (sn) RNAs.[1,2] The sizes of the
complexes and their shapes as judged from electron micrographs depend
to a large degree on the method used for their preparation. The largest
complexes observed sediment at about 200 S. Complexes of this size can
be obtained by sonicating nuclei in the presence of RNase inhibitors.[3] In
the presence of an RNase activity, either endogenous or an added nuclease, these large complexes are degraded, yielding smaller complexes that
sediment in a rather broad band at about 40 S. These complexes contain
RNA fragments that range in size from about 100 to about 1000 nucleotides (nt).[4] When viewed by electron microscopy, these particles have an

[1] S. Y. Chung and J. Wooley, *Proteins: Struct. Funct. Genet.* **1**, 195 (1986).
[2] G. Dreyfuss, *Annu. Rev. Cell Biol.* **2**, 459 (1986).
[3] R. Sperling, P. Spann, D. Offen, and J. Sperling, *Proc. Natl. Acad. Sci. U.S.A.* **83**, 6721 (1986).
[4] Y. D. Choi and G. Dreyfuss, *Proc. Natl. Acad. Sci. U.S.A.* **81**, 7471 (1984).

average diameter of about 20–25 nm, although they are rather heterogeneous in size and shape.[5]

Because of the heterogeneous nature of hnRNP complexes, it is difficult to apply directly the physical methods traditionally used for elucidating the structures of macromolecules. One alternative is to study complexes reconstituted with nucleic acids of defined sizes and to examine the properties of the reconstituted complexes as a function of the size of nucleic acid that is in the complex. Several approaches for doing this are described here.

Principles of Analysis

The principle of this approach is to examine the structures formed between proteins and nucleic acids of defined lengths. Changes in the size and shape of the complexes as the lengths of the nucleic acids increase can be related to a stepwise buildup of the complexes and will reflect their structure and the way in which they are assembled. Depending on the size of nucleic acid that is chosen and the method that is used for analysis, several questions can be addressed. By studying the binding to a series of small oligonucleotides of increasing sizes, one can determine binding site sizes and binding constants and also measure cooperative interactions between neighboring proteins.[6] Also, by examining the effects of solvents on these parameters, some information about the nature of the binding forces (such as the number of ionic interactions) that are involved can be obtained. By using longer nucleic acids as substrates, higher order structures that involve several or many proteins can be examined. In particular, the size and composition of cooperative units can be determined by sedimentation and electrophoresis, and the molecular organization of the complexes can be observed by electron microscopy.

Preparation of Defined-Length Nucleic Acids

RNAs of defined lengths and sequences can be easily obtained by *in vitro* transcription. A number of procedures using various promoters have been described,[7,8] and several kits are commercially available. For some analyses a heterogeneous mixture of RNAs of various sizes is desirable. This can be produced either by limited alkaline hydrolysis of a transcript (see below) or by synthesizing the RNA in the presence of a limiting

[5] J. Wooley, S. Y. Chung, J. Wall, and W. LeStourgeon, *Biophys. J.* **49,** 17 (1986).

[6] J. D. McGhee and P. H. Von Hippel, *J. Mol. Biol.* **86,** 469 (1974).

[7] D. A. Melton, P. A. Krieg, M. R. Rebagliati, T. Maniatis, K. Zinn, and M. Green, *Nucleic Acids Res.* **12,** 7035 (1984).

[8] E. T. Schenborn and R. C. Mierendorf, Jr., *Nucleic Acids Res.* **13,** 6223 (1985).

amount of nucleotide, which results in premature terminations and a distribution of RNA sizes.

Although the hnRNP proteins are found in association with hnRNA, they also appear to be capable of binding single-stranded DNA.[9-14] Large quantities of defined sizes of single-stranded DNAs can be obtained by denaturing DNA fragments generated by restriction enzymes. These can be end-labeled by the Klenow fragment of DNA polymerase[15] prior to denaturation, or they can be labeled with T4 polynucleotide kinase after removal of the 5'-phosphate by phosphatase.[15] The DNA can be rendered single stranded by dissolving it in 1 mM ethylenediaminetetraacetic acid (EDTA), pH 8.0, heating it in a boiling water bath, and quickly cooling it on ice.

Homopolymers are useful for reconstitution because they provide simplified model systems that are homogeneous in composition and secondary structure. Homopolymers, which are typically several hundred nucleotides long, can be obtained from a number of commercial sources. Commercial poly(A) is usually longer than poly(U) or poly(C). Shorter homopolymers can be prepared by limited alkaline hydrolysis[16] (for deoxyhomopolymers, a limited digestion with micrococcal nuclease can be used), which leaves a free 5'-hydroxyl that can be conveniently labeled using T4 polynucleotide kinase.

To obtain defined sizes of homopolymers, a partially hydrolyzed mixture can be separated by electrophoresing through a polyacrylamide gel, cutting the gel, and eluting the fragments of the desired sizes. To elute the fragments, the end of a plastic micropipet tip is plugged with a small amount of siliconized glass wool, and the end is sealed. The gel piece is placed in the pipet tip and crushed with a glass rod. After crushing, 0.5 ml of 0.5 M ammonium acetate, 1 mM EDTA is added, and the RNA is allowed to elute overnight. The end of the tip is then cut, the solution is drained, and the RNA is precipitated with 2.5 volumes of ethanol. For larger fragments, electroelution also works well.

To produce a mixture of RNA fragments by alkaline hydrolysis,[16] the RNA or homopolymer in 50 mM sodium carbonate buffer, pH 9.0, is

[9] J. M. Pullman and T. E. Martin, *J. Cell Biol.* **97**, 99 (1983).
[10] H.-E. Wilk, G. Angeli, and K. P. Schäfer, *Biochemistry* **22**, 4592 (1983).
[11] L. Nowak, D. K. Marvil, J. O. Thomas, M. Boublik, and W. Szer, *J. Biol. Chem.* **255**, 6473 (1980).
[12] Raziuddin, J. O. Thomas, and W. Szer, *Nucleic Acids Res.* **10**, 7777 (1982).
[13] A. Kumar, K. R. Williams, and W. Szer, *J. Biol. Chem.* **261**, 11266 (1986).
[14] A. Kumar, H. Sierakowska, and W. Szer, *J. Biol. Chem.* **262**, 17126 (1987).
[15] T. Maniatis, E. F. Fritsch, and J. Sambrook, "Molecular Cloning: A Laboratory Manual." Cold Spring Harbor Laboratory, Cold Spring Harbor, New York, 1982.
[16] H. Donis-Keller, A. M. Maxam, and W. Gilbert, *Nucleic Acids Res.* **4**, 2527 (1977).

heated to 65° in a microcentrifuge tube. After 3, 10, and 25 min, samples are placed on ice and neutralized with 1/10 volume of 2.5 M sodium acetate, 10 mM MgCl$_2$, pH 5.2. Because the rate of degradation is very sensitive to the exact conditions used, particularly the pH, it is necessary to check the extent of hydrolysis of each sample. This is done by 5'-end-labeling a small portion and analyzing it by polyacrylamide gel electrophoresis and autoradiography.

Reconstitution with Total hnRNP Proteins

Methods for forming reconstituted hnRNP complexes using exogenously added nucleic acids have been described.[9,10] All of these methods suffer from the lack of a suitable structural or functional assay for "native" hnRNP complexes, so it is never certain how faithfully the reconstituted particles reflect the structure of the complexes that are formed *in vivo*. Nonetheless, *in vitro* reconstitution studies do provide a handle for investigating the RNA–protein and protein–protein interactions that can occur during the formation of the structures. This is a useful and perhaps essential framework for examining these rather multifarious particles.

The simplest method for reconstitution is to digest the endogenous RNA and then add the desired nucleic acid in its place. The proteins associate spontaneously with the added nucleic acid. The hnRNP complexes [1 mg/ml protein in 10 mM tris(hydroxymethyl)aminomethane hydrochloride (Tris-HCl), 50 mM NaCl, 1 mM CaCl$_2$, 1 mM dithiothreitol (pH 7.6)] are digested with 2.5 units/ml micrococcal nuclease. After 10 min at 37°, the nuclease is inactivated by adding EDTA to 5 mM. The exogenous RNA is then added to give a protein to nucleotide molar ratio between 1:6 and 1:12. The composition of the buffer is not particularly critical, although elevated salt concentrations tend to favor the dissociation of most nucleic acid–protein complexes and very low concentrations of salt may lead to the precipitation of hnRNP proteins. Magnesium does not appear to have much influence on the formation of the complexes. Dithiothreitol (DTT) is routinely included since the free proteins tend to precipitate in its absence.[13,14] For the reconstitution of complexes with poly(A), the endogenous RNA can be digested with RNase A. Since poly(A) is not degraded by RNase A, the complete inhibition of the enzyme is not a consideration as it is with micrococcal nuclease.

Reconstitution with Individual Proteins

A few hnRNP proteins have been purified and partially characterized.[11–14,17] They appear to share the ability to bind to RNA (and single-

[17] F. Cobianchi, R. L. Karpel, K. R. Williams, V. Notario, and S. H. Wilson, *J. Biol. Chem.* **263**, 1063 (1988).

stranded DNA) and form condensed nucleic acid–protein complexes.[12,18,19] The best characterized complexes are those formed with HD40, the major hnRNP protein of *Artemia salina*. HD40 condenses nucleic acids into helices that are only marginally stable; on average, the helix formed by the nucleoprotein complex is only about $2\frac{3}{4}$ turns.[18]

The structure of the hnRNP complexes, and the way in which the nucleic acids are condensed, can be examined by observing the particles formed when the proteins associate with nucleic acids of increasing lengths. The effect of varying the protein to nucleotide ratio and the salt conditions used for reconstitution can also be examined to give additional information about the structure of the complexes and the forces involved in their formation. Homopolymers of defined sizes are good substrates for these studies since complications arising from sequence specificities of the proteins and differences in secondary structures of the nucleic acids can be minimized. Although the complexes can be observed with almost any physical technique, electron microscopy, electrophoresis, and hydrodynamic methods are particularly informative. Typically, the nucleic acid–protein complexes are formed simply by mixing the protein and nucleic acid. It is important to keep the protein concentration as high as possible to assure that the complex is fully saturated (the complexes that have been examined have intrinsic association constants of about 10^5–10^6 M^{-1}).[11,17] Following the formation of the complex, it can be analyzed as described below.

Analysis of Reconstituted Complexes

Sedimentation. As mentioned above, sedimentation has been widely used for the isolation and characterization of hnRNP complexes. Sedimentation can also be used in conjunction with electrophoresis for the analysis of hnRNP complexes reconstituted with either defined-length RNAs or with a heterogeneous mixture of RNAs of various sizes. Following reconstitution (see above), the mixture is layered on top of a 10–30% (w/w) sucrose gradient made with a buffer containing 10 mM Tris-HCl, 25 mM NaCl, 5 mM EDTA, 1 mM DTT, pH 7.6, and centrifuged. For sucrose gradients made on a percent (w/w) basis, a good approximation of the sedimentation coefficient as a function of the position in the gradient can be easily calculated,[20] although a more exact determination should be made by comparing the rate of sedimentation with ribosomal markers. For analytical purposes, centrifugation in a Beckman SW 50.1 rotor is

[18] J. O. Thomas, S. K. Glowacka, and W. Szer, *J. Mol. Biol.* **171**, 439 (1983).
[19] J. T. Schöneich and J. O. Thomas, unpublished observations (1988).
[20] C. R. McEwen *Anal. Biochem.* **20**, 114 (1967).

convenient. After a 100-min centrifugation at 5° at 49,000 rpm, 60 S material should be located near the middle of the gradient.

Following fractionation, the gradients can be analyzed by several methods. The simplest is to plot the distribution of radioactivity through the gradient. More information can be obtained, particularly if heterogeneous RNA is used, if the RNA is isolated from each fraction by phenol extraction and ethanol precipitation and then analyzed by electrophoresis. Separation of the RNA on 3.5% polyacrylamide–8 M urea gels is convenient for resolving RNAs between 0.1 and 2 kilobases (kb) in size. For smaller RNAs, the concentration of acrylamide can be increased up to 20%. For larger RNAs, agarose–formaldehyde gels[21] should be used. After separating the RNAs on gels, the size distribution of the RNAs in each fraction can be quantitated by densitometry of autoradiograms. From one experiment in which a broad size range of RNAs is used, the sedimentation behavior of complexes incorporating this population of RNAs can be determined. This enables a direct comparison to be made between complexes formed with RNAs of many different sizes under the exact same conditions.

In addition to an analysis of the RNAs in each fraction of the gradient, the protein composition of each fraction can be determined by sodium dodecyl sulfate (SDS)–polyacrylamide gel electrophoresis.[22] For detection, the proteins can be stained either with Coomassie brilliant blue or with silver.[23] The silver stain is much more sensitive, but it should be noted that with silver the major hnRNP proteins are less intensely stained relative to the other protein components of the hnRNP complexes than they are with Coomassie brilliant blue. A serious problem with examining the proteins present in complexes separated by sedimentation through sucrose gradients is that the hnRNP proteins readily aggregate in the absence of nucleic acid, particularly in low-salt buffers in the absence of DTT.[13,14] This makes it difficult to distinguish between nucleic acid–protein complexes and aggregates of free proteins that might cosediment with the complexes.

Analytical ultracentrifugation is useful for examining the interactions between purified hnRNP proteins and nucleic acids. From sedimentation equilibrium studies, the molecular weights of complexes formed with different sizes of nucleic acids can be determined, and from the change in molecular weight as a function of nucleic acid size, the excluded binding site size can be calculated.[18] The molecular weights can also be used in conjunction with sedimentation coefficients, determined by sedimenta-

[21] H. Lehrach, D. Diamond, J. M. Wozney, and H. Boedtker, *Biochemistry* 16, 4743 (1977).
[22] U. K. Laemmli, *Nature* (*London*) 227, 680 (1970).
[23] C. R. Merril, D. Goldman, S. A. Sedman, and M. H. Ebert, *Science* 211, 1437 (1981).

tion velocity experiments using the same samples, to estimate frictional coefficients and axial ratios of the complexes.[18]

Electrophoresis on Nondenaturing Gels. Electrophoresis through nondenaturing agarose gels provides an alternative to sedimentation for the separation of reconstituted hnRNP complexes. A major advantage of electrophoresis over sedimentation is that there is a clear separation of the RNA–protein complexes from free proteins and from unbound RNAs. This enables one to characterize both the RNA and protein components of the complexes that are formed.

Reconstituted complexes [in 20 mM N-2-hydroxyethylpiperazine-N'-2-ethanesulfonic acid (HEPES), 50 mM NaCl, 5% (v/v) glycerol, 1 mM DTT, 0.001% bromphenol blue (pH 7.6)] are loaded onto a 1% agarose gel [in 20 mM HEPES, 25 mM NaCl, 5% (v/v) glycerol, 1 mM DTT (pH 7.6)]. Electrophoresis is conducted in the cold at 3.5 V/cm until the dye front has migrated about 80% of the length of the gel. The running buffer [50 mM Tris-HCl, 25 mM NaCl, 1 mM DTT (pH 7.6)] should be continuously recirculated. In nondenaturing gels, the free hnRNP proteins remain near the origin. Free nucleic acids and nucleic acid–protein complexes move toward the positive electrode, with the free nucleic acids migrating much faster than nucleic acid–protein complexes.[19] Separation of the complexes is based on a combination of size, shape, and net charge. The distribution of the RNA and protein components throughout the nondenaturing gel can be analyzed on second dimension gels: a polyacrylamide gel containing 8 M urea for the analysis of RNA or an SDS–polyacrylamide gel for the analysis of proteins.

A disadvantage of using electrophoresis for the resolution of the complexes is the fact that one cannot a priori determine the relationship between sedimentation coefficients and mobilities on gels. Rather, it is necessary to correlate empirically the electrophoretic mobilities of the complexes with their sedimentation coefficients.

Electron Microscopy. Electron microscopy is potentially an extremely powerful tool for examining the structure of macromolecular complexes. However, since the hnRNP complexes are rather heterogeneous, the value of electron microscopy has been limited. The problem of sample heterogeneity can be reduced by examining complexes reconstituted with defined sizes of oligonucleotides. This approach can be particularly powerful if the morphology of the complex is examined as a function of the size of the nucleic acid used for reconstitution.

For electron microscopy, complexes are reconstituted with the desired nucleic acid as described above. Complexes containing nucleic acids longer than 100 nt can be easily visualized, but structures containing fewer than 50 nt are difficult to distinguish reliably from background debris. Complexes reconstituted from crude hnRNP proteins are also difficult to analyze because of a large background of proteins and protein aggregates.

In order to preserve the structure of the complexes during the subsequent processing steps, it may be desirable to fix them by adding 1/20 volume of 8% glutaraldehyde to the reconstituted complexes.[18] On the other hand, the presence of the fixative might alter the structure of the complexes and may also lead to the formation of protein aggregates if fixation is excessive. Because of these problems, it is best to examine and compare both fixed and unfixed complexes. In our experience, there is usually little difference.

Specimens can be mounted either by spraying them onto a mica sheet or carbon film or by direct adsorption onto a carbon film. Adsorption is much easier. For adsorption, thin carbon films on 400-mesh copper electron microscope grids[24] are used. The carbon-coated grids are generally hydrophobic and do not interact with the nucleic acid–protein complexes. There are several methods for making the grids hydrophilic. The easiest and most satisfactory method for nucleic acids coated with proteins is to activate the grids by placing them in a glow discharge.[25] We generate a glow discharge by passing a 55 mA current through a 80–120 mTorr vacuum. The grids should be kept in the discharge for about 8 min. Insufficient activation will not render the grids hydrophilic, and overactivation will result in increased background contamination. Consequently, it is best to determine empirically the optimal conditions for each particular apparatus. The activated grids should be used the same day.

The nucleoprotein complexes are mounted on the activated grids simply by placing a 5- to 10-μl drop of sample on the grids and allowing it to adsorb for about 30 sec. The amount of sample that is adsorbed onto the grid is dependent on the adsorption time, so dilute samples may require longer times. For negative staining, the droplet is washed off of the grid with several drops of 10 mM Tris-HCl, pH 7.5, applied rapidly with a Pasteur pipet. This is followed by several drops of stain. The stain is then immediately removed by touching a piece of paper towel or filter paper to the edge of the grid. When done properly, a thin film of stain will remain on the grid. This will quickly dry in the air, and the grids can be observed immediately. A number of stains are commonly used for the negative contrasting of macromolecules.[24,26] In our experience, the stain which

[24] R. H. Haschemeyer and R. J. Myers, *in* "Principles and Techniques of Electron Microscopy: Biological Applications, Volume 2" (M. A. Hayat, ed.), p. 99. Van Nostrand-Reinhold, Princeton, New Jersey, 1972.
[25] J. D. Griffith and G. Christiansen, *Annu. Rev. Biophys. Bioeng.* **7,** 19 (1978).
[26] G. W. Seegan, C. A. Smith, and V. N. Schumaker, *Proc. Natl. Acad. Sci. U.S.A.* **76,** 907 (1979).

gives the best contrast is a 1% aqueous unbuffered solution of uranyl acetate.

The samples can also be contrasted by shadowing with platinum.[25] This is the method that we prefer because of the fine grain size of the evaporated platinum and the relatively high degree of contrast that can be obtained. Following adsorption of the sample, the grid is washed and dehydrated by briefly dipping the grid in each of a series of solutions: 10 mM Tris-HCl, pH 7.5, followed by 10, 50, 80, and 100% ethanol solutions. The grids are kept under ethanol until they are ready to be shadowed. For shadowing, they are quickly placed on a rotary table in the shadowing chamber, and without allowing the ethanol to dry (this would result in rehydration of the complexes) the chamber is evacuated. The samples are then rotary shadowed with platinum at an angle of 10°. The platinum is evaporated by passing a current through a tungsten wire wrapped with 1.5 cm of 0.2-mm diameter platinum wire.

By shadowing the samples from a single direction rather than rotary shadowing, it is possible to estimate the height of the complexes by examining the length of the shadow that is cast.[27] In calculating the height of the particle from the length of the shadow by trigonometry, it is necessary to know the angle of the shadowing. Because the grids are seldom perfectly flat, internal standards such as gold particles or latex spheres should be used to calculate the angle of the shadow.[18]

Although adsorption is much more convenient, samples can also be observed by mounting them on mica and preparing a platinum replica.[28] This procedure usually results in somewhat better preservation of the structure of the complex and, because the mica is flat, somewhat better definition. The complexes to be observed are mixed with glycerol to give a final glycerol concentration of 20–30% (v/v). The presence of the glycerol reduces drying artifacts. The samples should contain about 0.5 mg/ml protein, but several 2-fold serial dilutions should be made to determine the optimal concentration. The sample is then sprayed onto freshly cleaved mica cut into pieces that are about 7 × 15 mm. For spraying, we use a Paasche air brush modified by replacing the nozzle with a plastic micropipet tip. About 10 μl of sample is sprayed onto the mica, which is placed 10–20 cm away. Several different samples can be sprayed in succession on separate mica sheets and shadowed together. For shadowing, the samples are dried for 1–2 hr under a vacuum of 10^{-6} Torr. After drying, the samples are contrasted with platinum as described above. This is followed by evaporating a layer of carbon to provide support. The

---

[27] P. R. Smith and I. E. Ivanov, *J. Ultrastruct. Res.* **71**, 25 (1980).
[28] J. M. Tyler and D. Branton, *J. Ultrastruct. Res.* **71**, 95 (1980).

replicas are then floated off of the mica onto the surface of distilled water, and picked up with a 400-mesh copper grid.

When the samples are observed by electron microscopy, changes in the size and shape of the complexes as the lengths of the nucleic acids increase can be seen. These changes can be related to a stepwise buildup of the complex, reflecting the way in which it is assembled. Therefore, an understanding of the structure of the complexes formed should emerge from an analysis of these data. Further information about the structure of the particles can be acquired by observing complexes formed at subsaturating ratios of protein to nucleic acid and complexes formed under different ionic conditions.

Oligomer Binding. A considerable amount of information about how hnRNP proteins bind to nucleic acids can be obtained by examining the binding constants as a function of oligonucleotide size for oligonucleotides ranging in size from monomers to about 50-mers.[29] From these data, one can determine the binding site size, the intrinsic association constant, the amount of cooperativity, and the size of the cooperative unit. The relative affinities for different nucleotides and the effect of ionic strength on the binding can also be determined.

The protein of interest is combined with a mixture of 5'-end-labeled oligonucleotides obtained by alkaline hydrolysis of a homopolymer (see above). The protein-bound oligonucleotides are separated from unbound ones by passing the mixture through a 3-mm nitrocellulose filter (which can be cut from a sheet with a small hole punch) and washed briefly with the sample buffer. The amount of each oligonucleotide that is bound depends on the protein concentration and the binding constant for the protein and the particular oligonucleotide. The nucleotides bound to the filter are eluted by boiling the filter in a small amount (10 μl) of 1% SDS, and the eluted oligonucleotides are resolved by electrophoresis through a 20% polyacrylamide gel. The oligonucleotides are detected by autoradiography, and their amounts are quantitated by densitometry.

To generate titration curves, the binding experiments are done at several protein concentrations, with the oligomers from each experiment being separated on adjacent lanes of the gel. The protein concentrations used should border the intrinsic binding constant, and the total molar nucleotide concentrations must be less than the molar protein concentration. The resulting data can be combined into a series of titration curves, one for each nucleotide that can be resolved by the gel. By analyzing these titration curves, values for the binding site size and intrinsic binding constant can be calculated. The magnitude of protein–protein interactions

[29] J. T. Schöneich and J. O. Thomas, in preparation.

(cooperativity) can also be evaluated, since these interactions will result in an increase in the affinity for oligomers that are long enough to accommodate two or more proteins.[6] Since the strength of ionic interactions between proteins and nucleic acids is dependent on the salt concentration, the ionic character of the binding can be assessed by determining the intrinsic binding constant (and degree of cooperativity) as a function of ionic strength.[30]

Other Physical Methods

In this chapter, we have discussed several techniques for examining the interactions between hnRNP proteins and nucleic acids, and the structures of the complexes that they form. We have emphasized an approach in which the structures are examined as a function of the size of the nucleic acid in the complex. In addition, some other physical techniques have been widely used, and will probably continue to play an important role. Nitrocellulose binding assays are useful for determining nucleotide preferences of hnRNP proteins.[11–14,31] UV absorption spectroscopy and circular dichroism are employed for examining the helix-destabilizing properties of proteins,[11–14] and fluorescence quenching is of value for measuring the thermodynamic parameters of the interactions.[17]

[30] M. T. Record, T. M. Lohman, and P. de Haseth, *J. Mol. Biol.* **107**, 145 (1976).
[31] A. B. Sachs and R. D. Kornberg, *Mol. Cell. Biol.* **5**, 1993 (1985).

[26] Immunological Methods for Purification and Characterization of Heterogeneous Nuclear Ribonucleoprotein Particles

By SERAFÍN PIÑOL-ROMA, YANG DO CHOI, and GIDEON DREYFUSS

Heterogeneous nuclear RNAs (hnRNAs) are associated in the cell with specific proteins to form hnRNA–ribonucleoprotein (hnRNP) complexes, also referred to as hnRNP particles (reviewed by Dreyfuss[1]). hnRNP particles are one of the major components of the nucleus. The proteins of the hnRNP particles are as abundant as histones in the nucleus of growing cells, and they comprise approximately 80% of the mass of hnRNP particles. hnRNP complexes can be isolated from vertebrate nu-

[1] G. Dreyfuss, *Annu. Rev. Cell Biol.* **2**, 459 (1986).

METHODS IN ENZYMOLOGY, VOL. 181

cleoplasm by a rapid, specific, and efficient immunopurification procedure using monoclonal antibodies to hnRNP proteins.[2,3] By two-dimensional gel electrophoresis, immunopurified hnRNP particles of human HeLa cells contain about 20 polypeptides in the range of 34,000–120,000 daltons.[3] The abundant 30,000–40,000 dalton proteins A, B, and C described previously[4-6] are a subset of these polypeptides.

The immunopurification procedure has a number of advantages over methods traditionally used for studying hnRNP complexes such as sucrose gradient sedimentation of 30 S particles[4,5,7,8] and photochemical cross-linking of proteins to RNA.[9] Immunopurification circumvents the need to radioactively label the hnRNA in order to identify the hnRNP-containing fraction; it does not depend on RNase cleavage of the RNA and thus can yield hnRNP particles with intact transcripts; it is rapid and gentle, minimizing proteolysis and disruption of the complex and thereby loss of genuine components; and, most important, it is highly specific by the very nature of the monoclonal antibody reagents used. The results obtained are highly reproducible in terms of protein composition and stoichiometry, and virtually identical results have been obtained with monoclonal antibodies against at least seven different hnRNP proteins.

Unlike the UV cross-linking procedure, the immunopurification procedure preserves the proteins in native form and apparently normal relative abundance because the different efficiency of cross-linking of the various proteins is not a factor. The purity of the immunopurified hnRNP complexes is obvious by the absence of abundant nuclear proteins such as histones. Under the conditions described here, there are no significant amounts of other RNP proteins, such as those of small nuclear RNPs (snRNPs). However, it should be emphasized that, as in any cellular fractionation procedure, one cannot completely exclude the possibility that some components associate with the complex during the fractionation but are not part of it *in vivo*. It is also possible that additional authentic proteins of the complex exist in the cell but dissociate from the complex during the fractionation procedure. It is therefore best, when

[2] Y. D. Choi and G. Dreyfuss, *Proc. Natl. Acad. Sci. U.S.A.* **2**, 7471 (1984).

[3] S. Piñol-Roma, Y. D. Choi, M. J. Matunis, and G. Dreyfuss, *Genes Dev.* **2**, 215 (1988).

[4] A. L. Beyer, M. E. Christensen, B. W. Walker, and W. M. LeStourgeon, *Cell* **11**, 127 (1977).

[5] T. E. Martin, P. B. Billings, J. M. Pullman, B. J. Stevens, and A. J. Kinniburgh, *Cold Spring Harbor Symp. Quant. Biol.* **42**, 899 (1978).

[6] H. E. Wilk, H. Werr, D. Friedrich, H. H. Kiltz, and K. P. Schäfer, *Eur. J. Biochem.* **146**, 71 (1985).

[7] O. P. Samarina, E. M. Lukanidin, J. Molman, and G. P. Georgiev, *J. Mol. Biol.* **33**, 251 (1968).

[8] J. Karn, G. Vidali, L. C. Boffa, and V. G. Allfrey, *J. Biol. Chem.* **252**, 7307 (1977).

[9] S. Piñol-Roma, S. A. Adam, Y. D. Choi, and G. Dreyfuss, this series, Vol. 180, p. 410.

possible, to use both immunopurification and UV cross-linking for the verification of the authenticity of the protein in question. So far, by means of UV cross-linking it can be seen that every protein present in the immunopurified hnRNP particles for which specific monoclonal antibodies were available (A1, C1, C2, D1, D2, L, M, and U) is indeed in contact with hnRNA *in vivo* (Ref. 10 and S. Piñol-Roma and G. Dreyfuss, unpublished observations). Immunopurification of the hnRNP complex thus enables the identification and definition of its components with greater detail and certainty than afforded by previously used methods. The immunopurified complexes so isolated are also amenable to a variety of biochemical studies and manipulations, such as assays for enzymatic activities associated with the particles.

The procedures described below are used routinely in our laboratory for the immunopurification of hnRNP complexes from HeLa cells, although they apply to most other eukaryotic cells with minimal modifications.[2] It is, of course, necessary that the specific antigen against which the antibody is directed be exposed on the surface of the complex so that it is available to the antibody for efficient antibody–antigen interaction to occur. Furthermore, the antigen must remain associated with the complex under the conditions used during the cellular fractionation and immunopurification procedures. Because of the essential role of the RNA component in maintaining the integrity of the complex, it is imperative that extreme care be exerted to minimize degradation of the RNA as well as to minimize proteolysis.

Procedures

Reagents. Protein A-agarose is from Boehringer-Mannheim (Indianapolis, IN). Alternatively, we have also used protein A-Sepharose from Pharmacia (Piscataway, NJ) without any appreciable difference in yields or composition of the immunopurified complexes. The highest quality Triton X-100 available should be used during the cellular fractionation. We have observed differences between commercial batches that cause nuclear lysis or chromatin disruption, which interfere with the isolation of nucleoplasm and retention of the integrity of the hnRNP particles. We routinely use Triton X-100 purchased from New England Nuclear (Boston, MA). Aprotinin, leupeptin, and pepstatin A are from Sigma (St. Louis, MO). Empigen BB (alkylbetaine) is from Albright & Wilson (Norwood, NJ).

Buffer Solutions

RSB-100 (isotonic reticulocyte standard buffer): 10 mM tris(hydroxymethyl)aminomethane hydrochloride (Tris-HCl) (pH 7.4), 100 mM NaCl, 2.5 mM MgCl$_2$

Immunopurification buffer: RSB-100 containing 0.5% (v/v) Triton X-100, 0.5% (v/v) aprotinin, and 2 μg/ml each of leupeptin and pepstatin A

TEL buffer: 200 mM Tris-HCl (pH 7.4), 25 mM ethylenediaminetetraacetic acid (EDTA), 100 mM LiCl; sodium dodecyl sulfate (SDS) is added to 1% (w/v) for extraction of RNA from immunopurified hnRNP particles

Preparation of Antisera and of Monoclonal Antibodies against hnRNP Proteins

Various methods have been used for the preparation of antibodies for the immunopurification of hnRNP complexes. One of the most important aspects is the isolation of the appropriate antigen for the immunization steps. We have used hnRNP complexes UV cross-linked *in vivo* as the antigen for production of monoclonal antibodies against the hnRNP proteins C1, C2, and the 120-kDa U protein.[10] More recently we have used hnRNP proteins partially purified by single-stranded DNA-agarose affinity chromatography[11] for the production of monoclonal antibodies against other hnRNP proteins.[3] hnRNP monoparticles of 30 S have also been used as the source of antigen for the preparation of anti-hnRNP monoclonal antibodies.[12] We routinely test the monoclonal antibodies obtained for their recognition of proteins that cross-link to RNA *in vivo* upon exposure of intact cells to UV light.[9] Either ascites fluid or hybridoma culture supernatant is suitable for the immunopurification procedure.

One of the major difficulties in the preparation of antibodies against hnRNP proteins is the poor immunogenicity of most of these proteins, which is most probably due to their high degree of conservation. We have in some instances at least partially circumvented this problem by immunization of the immunocompromised New Zealand Black mice, rather than BALB/c mice, for the production of monoclonal antibodies against the hnRNP A1 and D proteins.[3] Anti-hnRNP protein antibodies have also become available from the recognition that at least some of the previously identified mammalian single-stranded DNA-binding proteins are indeed proteolytic fragments of genuine hnRNP proteins. Antisera against one such protein, UP1, cross-react with the A and B group hnRNP proteins.[13]

[10] G. Dreyfuss, Y. D. Choi, and S. A. Adam, *Mol. Cell. Biol.* **4,** 1104 (1984).
[11] S. Piñol-Roma, M. S. Swanson, M. J. Matunis, and G. Dreyfuss, this volume [27].
[12] G. P. Leser, J. Escara-Wilke, and T. E. Martin, *J. Biol. Chem.* **259,** 1827 (1984).
[13] O. Valentini, G. Biamonti, M. Pandolfo, C. Morandi, and S. Riva, *Nucleic Acids Res.* **13,** 337 (1985).

Cell Culture and Labeling

HeLa cells were used for most of the immunopurifications. HeLa cells grown on plates to subconfluent densities are labeled in Dulbecco's minimal essential medium containing one-tenth the normal amount of methionine and supplemented with 5% calf or fetal calf serum and 1% penicillin and streptomycin. In our hands, labeling for 20 hr with [^{35}S]methionine at 20 μCi/ml suffices for detection of hnRNP complex proteins immunopurified from less than one-half a plate with an overnight exposure following fluorography. Alternatively, for hnRNA analysis, the cells can be labeled with [^{3}H]uridine for 15 min at 50 μCi/ml. Actinomycin D is added to 40 ng/ml 15 min prior to labeling in order to inhibit ribosomal RNA synthesis.

Cellular Fractionation

The initial step in the hnRNP immunopurification procedure involves separation of the nuclear and cytoplasmic fractions, and the subsequent removal of nucleoli and insoluble chromatin from the nuclear fraction in order to generate what is operationally defined as the nucleoplasm. Procedures for the preparation of nucleoplasm are essentially according to Pederson.[14] All the steps are carried out at 4° as rapidly as possible to minimize proteolysis and RNA degradation.

HeLa cells grown on 10-cm tissue culture plates are rinsed twice with cold phosphate-buffered saline (PBS), the PBS is drained from the plate, and 1 ml of cold immunopurification buffer is added to the plate. The cells are scraped with a rubber policeman and lysed by four passages through a 25-gauge needle. The nuclei are then pelleted by brief centrifugation at 3,000 *g* (the centrifuge is stopped as soon as it reaches the speed). The supernatant, containing the cytoplasmic fraction, is discarded.

The nuclear pellet is resuspended in 500 μl of immunopurification buffer, and the nuclei are mechanically disrupted by sonicating twice for 5 sec each, using a microtip sonicator (Model w-220F, Heat System/ Ultrasonics, Plainview, NY) set at scale 2, on ice. The sonicate is layered over a 30% sucrose cushion [30% sucrose (w/v) in RSB-100] and centrifuged at 4,000 *g* for 15 min, at 4°. This step removes nucleoli and insoluble chromatin, as well as other insoluble nuclear structures. The material overlaying the cushion at the end of this centrifugation step is the nucleoplasm.

---

[14] T. Pederson, *J. Mol. Biol.* **83,** 163 (1974).

Antibody Specificity

As in any immunoprecipitation procedure, it is imperative to ascertain the specificity of the antibody used. This can be done by immunoprecipitation and by immunoblotting, preferably both. The immunoprecipitation needs to be done under conditions that completely dissociate the hnRNP complexes, so that the antibody will isolate only the protein with which it specifically reacts. In the case of hnRNP complexes, the strength of the interactions between the proteins and the RNA is such that it is difficult to immunoprecipitate only the antigen directly recognized by the antibody, free of other hnRNP proteins. We find that the zwitterionic detergent Empigen BB is particularly effective for this; it causes virtually complete disruption of hnRNP complexes and allows efficient antibody–antigen interaction.[15]

The cells are lysed, after rinsing with PBS, with 1 ml of 1% Empigen BB prepared in PBS, containing 1 mM EDTA and 0.1 mM dithiothreitol (DTT), and the protease inhibitor aprotinin at 0.5%. The cells are scraped, transferred to a tube, and sonicated 3 times for 5 sec each on ice. The sonicate is then centrifuged at 13,000 rpm for 5 min in a microcentrifuge to pellet any insoluble material. The supernatant can then be used directly for the immunoprecipitations. The washes after the 10-min incubation are also carried out in the presence of 1% Empigen.

Immunopurification of hnRNP Complexes

The antibody of interest is prebound to 25 μl (packed volume) of *Staphylococcus aureus* protein A-agarose or protein A-Sepharose for 1 hr at 4° with rocking. The ascites fluid or antiserum to be used is diluted in about 500 μl of immunopurification buffer. Alternatively, 500 μl of antibody-containing hybridoma culture fluid can be added to the beads, and Triton X-100 added to a final concentration of 0.5%. In the case of the anti-C proteins monoclonal antibody 4F4,[15] we routinely use 1 μl of the ascites fluid per 10-cm plate and per 25-μl packed volume of protein A-agarose beads. It is also essential to always include a control immunopurification with preimmune serum or immunoglobulins from the parent myeloma cell line used for the production of the hybridomas. Amounts of immunoglobulins similar to those of the specific antibody should be used for the control immunopurifications. These can be estimated from Coomassie blue staining of the antibodies that bound to the protein A beads after SDS–polyacrylamide gel electrophoresis (PAGE).

The beads are washed 3 times at the end of the incubation period with

[15] Y. D. Choi and G. Dreyfuss, *J. Cell Biol.* **99**, 1997 (1984).

immunopurification buffer. This is done by brief centrifugation to pellet the beads (in a microcentrifuge at ~13,000 rpm) and repeated replacement of the supernatant with new immunopurification buffer. We find that prebinding of the antibody to the protein A-agarose beads is much preferable to prior binding of the antibody to the antigen for several reasons: first, the preincubation of the antibody makes the whole procedure considerably faster; second, and related to the first reason, it drastically reduces the length of time during which the hnRNP complexes are exposed to agents (e.g., RNases) that may disrupt their integrity; third, we find that ascites fluids especially contain high RNase activity that degrades the complexes. Such activity is greatly reduced on prebinding of the antibodies and washing of the antibody-coated beads.

After the last wash, the nucleoplasmic fraction is added to the drained beads and incubated for 10 min at 4° with rocking. The beads are then washed 6 times with immunopurification buffer. This washing step is extremely important to ensure the purity of the isolated complexes. After the last wash, the complexes can be eluted by boiling the beads with SDS–PAGE sample buffer (for identification and analysis of the proteins) or with TEL buffer (for analysis of the RNA component) containing 1% (w/v) SDS. The RNA is then extracted twice with phenol–chloroform–isoamyl alcohol after digestion of the proteins with proteinase K (350 μg/ml) for 15 min at 37° and precipitated with 3 volumes of cold ethanol at $-20°$ overnight. It is also possible to elute the complexes from the drained beads with the urea-containing isoelectric focusing or nonequilibrium pH gradient polyacrylamide gel electrophoresis (NEPHGE) sample buffer for analysis by two-dimensional gel electrophoresis. Especially in this case it is important to drain the beads completely in order to avoid diluting the sample buffer. This can be done using a drawn-out Pasteur pipet or a 27-gauge syringe needle. The amount of RNA recovered by this procedure is typically of the order of 25% of the acid-precipitatable nucleoplasmic RNA, as determined from the amount of [³H]uridine-labeled material that is precipitated with 4F4. Similar yields of hnRNP proteins are obtained.

Characterization of hnRNP Particles with Monoclonal Antibodies

Insights into the structure of hnRNP complexes can be gained from analysis of the composition of particles immunopurified under different salt concentrations and varying degrees of nuclease digestion. These conditions can be adjusted in the nucleoplasm prior to the incubation with the corresponding antibody. Furthermore, the sensitivity of the interaction of the various proteins with the RNA to different competitors such as heparin can be readily studied[3] by immunopurification with different monoclonal antibodies. In a similar manner, one can easily study the accessibility of hnRNP proteins in these complexes to proteases and to chemical modifications.

Specific enzymatic or RNA-binding assays can be performed on immunopurified complexes in order to detect the association of such activities with hnRNP complexes. Similar studies can be carried out on individual hnRNP proteins by directly inhibiting specific activities such as splicing via antibody addition or by immunodepletion of the given protein from a mixture of proteins.[16]

Monoclonal antibodies against different hnRNP proteins can be used to detect the presence of these proteins in fractions from sucrose gradients and from column chromatography fractionation procedures. Finally, monoclonal antibodies against hnRNP proteins can be used for the immunolocalization of hnRNP complexes and of the individual proteins by immunofluorescence microscopy and/or by immunoelectron microscopy.[10,12]

Gel Electrophoresis

We have developed a modified discontinuous gel electrophoresis system in order to optimize separation and resolution of proteins over a wide range of molecular weights without the need for use of acrylamide gradients.[17] This gel system, besides the resolution it affords for hnRNP proteins, has the advantage of a much improved mechanical strength of the gels for further handling because of the modified cross-linking ratio, and the gels do not crack during drying.

The separating gel is prepared from a stock of 33.5% acrylamide and 0.3% N,N'-bisacrylamide, usually to a final concentration of 12.5% acrylamide. At this concentration proteins of molecular weight of 5,000–200,000 are well resolved. The separating gel buffer contains 0.38 M Tris-HCl (pH 9.1). The stacking gel is prepared from a stock of 30% acrylamide and 0.44% N,N'-bisacrylamide to a final acrylamide concentration of 4% in 0.125 M Tris-HCl (pH 6.8). Both gels contain 0.1% SDS and are polymerized with ammonium persulfate and N,N,N',N'-tetramethylenediamine (TEMED). The electrode tank buffer is 25 mM Tris–192 mM glycine containing 0.1% (w/v) SDS. Samples are prepared by boiling for 3 min in a 0.125 M Tris-HCl (pH 6.8) buffer containing 1% (w/v) SDS, 5% mercaptoethanol, 10% glycerol, and bromphenol blue. The final acrylamide concentration can be easily adjusted to higher (e.g., 15%) or lower amounts (e.g., 10%) for optimal resolution of smaller or

[16] Y. D. Choi, P. J. Grabowski, P. A. Sharp, and G. Dreyfuss, *Science* **231**, 1534 (1986).
[17] G. Dreyfuss, S. A. Adam, and Y. D. Choi, *Mol. Cell. Biol.* **4**, 415 (1984).

larger polypeptides. After electrophoresis of ^{35}S-labeled material, the gels are stained with Coomassie blue and impregnated with PPO (2,5-diphenyloxazole), and fluorography is performed with X-ray film.[18]

General Comments

It is essential in this immunopurification procedure to include as a control immunopurifications with nonimmune serum or with myeloma immunoglobulins. The complexes isolated with antibodies specific for hnRNP proteins contain a set of proteins very similar to those that cross-link to hnRNA *in vivo*[2], which underscores the specificity of the immunopurification of hnRNP complexes. The procedure is highly reproducible, in terms of the protein composition of the immunopurified complexes, and complexes immunopurified with monoclonal antibodies against different hnRNP proteins have similar compositions.[2,3]

It is also important to emphasize the sensitivity of hnRNP particles to RNases, proteases, and elevated salt concentration. Digestion of the RNA to average lengths below about 125 nucleotides will result in disruption of the complexes and gradual loss of protein components, the A and B group proteins being released most readily. The association of the different proteins of the complex also exhibits varying degrees of sensitivity to salt concentration, the A and B group proteins again being the most sensitive (their dissociation from the complex is already apparent at 150 mM NaCl, resulting from their dissociation from the RNA). At 600 mM NaCl, virtually all of the proteins (except for the C proteins) dissociate from the RNA and are released from the complex. These considerations underscore the need to carry out the immunopurification at low temperatures and as rapidly as possible in order to minimize loss of hnRNP proteins, owing especially to digestion of the RNA. Of course, with short incubation times the yield of hnRNP particles isolated is lower, but the essential features of this procedure are the purity and intactness of the particles.

Affinity chromatography and other chromatographic methods are more suitable for the purification of large amounts of specific hnRNP proteins of interest.[11] The conditions of digestion with endogenous or exogenous RNases and variations in the salt concentration can be adjusted prior to the incubation of the nucleoplasm with the antibody-bearing beads, in order to study the effect of these parameters on hnRNP complexes. Such experiments can provide much information on the arrangement of the proteins and the hnRNA in hnRNP particles.

---

[18] R. A. Laskey and A. D. Mills, *Eur. J. Biochem.* **56**, 335 (1975).

[27] Purification and Characterization of Proteins of Heterogeneous Nuclear Ribonucleoprotein Complexes by Affinity Chromatography

By SERAFÍN PIÑOL-ROMA, MAURICE S. SWANSON,
MICHAEL J. MATUNIS, and GIDEON DREYFUSS

Large-scale purification of proteins of heterogeneous nuclear ribonu-cleoproteins (hnRNPs) in native form is desirable for their biochemical characterization, for studies of their function, and for preparation of anti-bodies against them. Affinity chromatographic methods for purification of proteins rely on the specificity and relatively high affinity of a protein for an immobilized substrate. Where it can be applied, affinity chromatogra-phy is often an extremely efficient, usually the decisive, step in the purifi-cation. Most of the hnRNP polypeptides bind the hnRNA and are there-fore RNA-binding proteins. Work on the single-stranded DNA (ssDNA)-binding protein UP1[1] showed that this protein was, in fact, a fragment of the hnRNP protein A1.[2–3] Another hnRNP protein, the E protein, also turned out to be a previously described ssDNA-binding pro-tein, UP2.[1]

These observations raised the possibility that other hnRNP proteins may also bind efficiently to ssDNA and that this binding property could form the basis for an affinity chromatographic approach for their purifica-tion. Indeed, recent studies have shown that the majority of hnRNP pro-teins from human HeLa cells bind ssDNA columns in a heparin- and salt-resistant manner.[4,5] Fractionation of nucleoplasm on ssDNA-agarose results in an almost complete one-step purification of most of the hnRNP proteins. Furthermore, the differences in the affinities of individual hnRNP polypeptides for the column also provide an excellent means of fractionating them from one another.[4] Other non-hnRNP RNA-binding proteins that are present in the starting material may, obviously, also bind to the column and there may also be proteins that are not related to RNA

[1] G. Herrick and B. Alberts, J. Biol. Chem. 251, 2124 (1976).
[2] O. Valentini, G. Biamonti, M. Pandolfo, C. Morandi, and S. Riva, Nucleic Acids Res. 13, 337 (1985).
[3] S. Riva, C. Morandi, P. Tsoulfas, M. Pandolfo, G. Biamonti, B. Merrill, K. R. Williams, G. Multhaup, K. Beyreuther, H. Werr, B. Henrich, and K. P. Schaefer, EMBO J. 5, 2267 (1986).
[4] S. Piñol-Roma, Y. D. Choi, M. J. Matunis, and G. Dreyfuss, Genes Dev. 2, 215 (1988).
[5] M. Pandolfo, O. Valentini, G. Biamonti, P. Rossi, and S. Riva, Eur. J. Biochem. 162, 213 (1987).

metabolism which could also bind to the column (e.g., some dehydro-genases, see below). Therefore, binding of proteins to a ssDNA column, although a characteristic of many hnRNP proteins and an efficient means for their purification, is not a diagnostic tool for their identification. In combination with other chromatographic methods, such as ion-exchange chromatography, ssDNA affinity chromatography results in highly puri-fied hnRNP proteins. In addition, several hnRNP proteins have been shown to exhibit preference for specific ribonucleotide homopolymers, and this provides an additional, more specific, affinity chromatographic step for their rapid identification and purification.[6]

The procedures described here are suitable for the separation and large-scale purification of hnRNP proteins from HeLa cells and from a wide range of tissues and organisms. Among the advantages of the large-scale purification of hnRNP proteins by affinity chromatography on ssDNA are its relative rapidity and ease, its independence of the intact-ness of hnRNP particles during sample preparation, the lack of a need for radioactive labeling of the hnRNA for detection of the hnRNP-containing fractions, and the stability of the ssDNA column. These considerations are discussed further elsewhere.[4]

Procedures

Reagents. ssDNA-agarose is from Bethesda Research Laboratories (Gaithersburg, MD). Micrococcal nuclease is from Pharmacia (Piscata-way, NJ). ssDNA agarose is preferred to ssDNA-cellulose because of the superior flow properties of the former, especially since the tendency of hnRNPs to aggregate in the absence of RNA makes fast flow rates desir-able in the chromatographic procedure. Furthermore, ssDNA columns are preferred to RNA columns because of the greater stability of the former. The ribohomopolymer agarose beads (type 6) are from Pharma-cia. Triton X-100 is purchased from New England Nuclear (Boston, MA). Aprotinin, leupeptin, and pepstatin A are from Sigma (St. Louis, MO).

Buffer Solutions

RSB-100 (isotonic reticulocyte standard buffer): 10 mM tris(hydroxy-methyl)aminomethane hydrochloride (Tris-HCl) (pH 7.4), 100 mM NaCl, 2.5 mM MgCl$_2$

All elutions from the ssDNA column are done in 50 mM sodium phos-phate buffer (pH 7.4). Solid NaCl is added to the buffer in order to obtain the desired final concentration.

[6] M. S. Swanson and G. Dreyfuss *Mol. Cell. Biol.* **5**, 2237 (1988).

Cellular Fractionation

The procedure described here is for chromatography on ssDNA-agarose of hnRNP polypeptides from cultured HeLa cells, although it can easily be adapted to other cell types or tissues.

Preparation of nucleoplasm is done as described for the immunopurification of hnRNP complexes from HeLa cells.[7] This is done in the presence of protease inhibitors throughout the entire procedure, at 4°, and as rapidly as possible. Procedures for the preparation of nucleoplasm are essentially according to Pederson.[8] Briefly, HeLa cells grown in tissue culture plates are rinsed twice with cold PBS, the PBS is drained from the plate, and 1 ml of cold RSB-100 containing 0.5% Triton X-100, 0.5% aprotinin, and 1 μg/ml each of leupeptin and pepstatin A is added to the plate. The cells are scraped with a rubber policeman and lysed by four passages through a 25-gauge needle. The nuclei are then pelleted by brief centrifugation at 3,000 *g* (the centrifuge is stopped as soon as it reaches the speed). The supernatant, containing the cytoplasmic fraction, is discarded, and the nuclear pellet is resuspended in 500 μl of RSB-100 containing the protease inhibitors. The nuclei are then mechanically disrupted by mild sonication twice for 5 sec each using a microtip sonicator (Model w-220 F, Heat System/Ultrasonics, Plainview, NY) set at scale 2 on ice.

The sonicate is layered over a 30% sucrose cushion [30% sucrose (w/v) in RSB-100] and centrifuged at 4,000 *g* for 15 min, at 4°. This step removes nucleoli and insoluble chromatin, as well as other insoluble nuclear structures. The material overlaying the cushion at the end of this centrifugation step is the nucleoplasm.

The preparation of nucleoplasm can be easily modified for large-scale preparations from HeLa cells grown in spinner culture. The cells in this case can be pelleted by centrifugation at 600 *g*. The cell pellet is then washed with PBS and resuspended in at least 5 volumes of the Triton X-100-containing lysis buffer. The cells are then lysed with five strokes in a Dounce homogenizer using the B pestle. The remainder of the procedure is as described for cells grown in tissue culture plates. The nuclear pellet is resuspended in at least 4 volumes of RSB-100 containing the protease inhibitors prior to the sonication step. We usually increase the sonication time to 5 sec each, 5 times, on ice, with no apparent adverse effects on the yield from the column.

The nucleoplasm is digested with micrococcal nuclease at 100 units/ml for 15 min at 30° in the presence of 1 mM CaCl₂. The reaction is then

---

[7] S. Piñol-Roma, Y. D. Choi, and G. Dreyfuss, this volume [26].

[8] T. Pederson, *J. Mol. Biol.* **83**, 163 (1974).

stopped by addition of ethylene glycol bis (β-aminoethyl ether)-N,N,N',N'-tetraacetic acid (EGTA) to a final concentration of 5 mM, on ice. Digestion of the nucleoplasm is done in order to break down endogenous RNA and thus disrupt hnRNP complexes and free single-stranded nucleic acid-binding sites. It is absolutely necessary to add EGTA to inactivate the nuclease, since it otherwise will destroy the ssDNA-agarose column. The digested nucleoplasm should be clarified by brief centrifugation prior to loading on the column.

Chromatography on ssDNA-Agarose

The digested, clarified nucleoplasm is loaded onto the ssDNA-agarose column which has previously been equilibrated with 100 mM NaCl. Flow rates of the order of 60 ml/min are desirable to minimize protein aggregation. We routinely rerun the flow-through fraction through the column, although this has variable effects on the recovery of hnRNP proteins. After thorough washing of the column with binding buffer, the proteins can be eluted stepwise with increasing salt concentrations or with a salt gradient to 2.0 M NaCl. Such a relatively high salt concentration is needed for complete elution of the C proteins. The A and B group proteins elute between 0.2 and 0.5 M NaCl. Other proteins such as D, E, F, L, and M elute from the column at salt concentrations ranging from 0.4 to 1.0 M NaCl. The C proteins are the last ones to elute from the column with a salt gradient, with complete elution requiring up to 2 M salt.

Addition of a heparin wash (1.0 mg/ml in 100 mM NaCl) after the binding results in greater purity of the fractions, most probably owing to the competition for those proteins that bind to the column predominantly via ionic interactions with the DNA phosphate backbone. Some of the hnRNP proteins may also bind heparin tightly, as appears to be the case for the U proteins. Thus, the heparin wash results in loss of some genuine proteins such as U. For total hnRNP protein purification, the proteins can be eluted from the column with 2.0 M NaCl following the heparin wash. In this case, all of the major proteins that elute from the column at 2.0 M NaCl, and most of the minor ones, correspond to genuine hnRNP polypeptides. The fractions can then be analyzed by sodium dodecylsulfate–polyacrylamide gel electrophoresis (SDS–PAGE) following trichloroacetic acid (TCA) precipitation. In most instances, analysis of the fractions of interest by two-dimensional gel electrophoresis [nonequilibrium pH gradient polyacrylamide gel electrophoresis (NEPHGE) in the first dimension, SDS–PAGE in the second dimension] becomes absolutely essential because of the comigration of many of the hnRNPs on one-dimensional gels. Alternatively, or in addition to analysis by two-dimensional gel electrophoresis, use of polyclonal or monoclonal antibodies against specific proteins will allow their unambiguous detection throughout the elution from the column. The combination of ssDNA-agarose affinity chromatography with other column chromatography procedures, such as with DEAE resins, is an extremely powerful and straightforward tool for the purification of individual hnRNP proteins or groups of proteins.

Isolation and Purification of hnRNP Proteins by Affinity Chromatography on Ribohomopolymer-Agarose Columns

The affinity chromatography isolation of hnRNP proteins based on their differential affinities for ribohomopolymers has been recently described.[6] The starting material for the procedure is also micrococcal nuclease-digested nucleoplasm prepared as described above. The salt concentration is then adjusted to 2.0 M by addition of solid NaCl, and the nucleoplasm is incubated with the appropriate ribohomopolymer-agarose beads at 4° with rocking for 10 min. The beads are pelleted by brief centrifugation and washed at least 5 times with RSB containing 2.0 M NaCl. We also include heparin at 2 mg/ml in the first wash, both as a nonspecific polyanionic competitor of the RNA as well as an RNase inhibitor.

The bound proteins can be eluted from the beads with SDS-containing sample buffer for SDS–PAGE, or with 4.0 M guanidine-HCl. Such extreme conditions are necessary because of the strong interactions between some of the hnRNP polypeptides and specific ribohomopolymers. By two-dimensional gel electrophoretic analysis, as well as by immunoblotting with hnRNP-specific monoclonal antibodies, we have determined by this procedure that the hnRNP C proteins are poly(U)-binding proteins, although some M proteins also bind poly(U) at 2.0 M NaCl. The K and J proteins bind specifically to poly(C) under these conditions. The P protein(s) bind poly(A), and F, H, E, M, and a subset of the P proteins bind specifically to poly(G).

Adjustment of the nucleoplasm salt content to lower concentrations results in increased binding of a larger number of hnRNPs to the ribohomopolymers. This procedure can again be easily modified to be carried out in a column mode for the large-scale purification of hnRNPs based on their ribohomopolymer-binding specificities.

General Comments

We have successfully used the ssDNA-agarose chromatography procedure in combination with DEAE column chromatography for purification of the C proteins and of the D proteins to near homogeneity. We have also partially purified by similar methods the A and B group proteins for immunization of mice for the production of monoclonal antibodies. The ribohomopolymer-agarose affinity chromatography procedure has also been an invaluable analytical tool for unveiling RNA-binding specificities among hnRNP proteins and for the isolation of ribohomopolymer-specific binding (hnRNP) polypeptides from a variety of organisms including human, *Xenopus laevis,* the fruit fly *Drosophila melanogaster,* and the yeast *Saccharomyces cerevisiae.*

One should be cautioned that not all of the ssDNA-binding proteins obtained by this procedure are necessarily hnRNP proteins. There are reports of enzymes of intermediary metabolism such as lactate dehydrogenase and glyceraldehyde-phosphate dehydrogenase[9] that bind single-stranded DNA in the presence of dextran sulfate and in a high-salt-resistant manner, raising questions about the physiological relevance of the binding of some proteins to ssDNA columns. ssDNA affinity chromatography has also been used extensively with yeast to identify those proteins that may be involved in DNA replication.[10,11] The isolation by ssDNA of hnRNP (and mRNP) polypeptides does not therefore constitute proof that the proteins thus isolated are indeed authentic hnRNP (or mRNP) proteins. It is therefore necessary to use additional means to show that the proteins in question are in contact with RNA *in vivo* by means of UV cross-linking[12] and, in the case of hnRNPs, to ascertain, by means of immunofluorescence, their nuclear localization.

[9] F. Grosse, H.-P. Nasheuer, S. Scholtissek, and U. Schomburg, *Eur. J. Biochem.* **160,** 459 (1986).
[10] S. G. LaBonne and L. B. Dumas, *Biochemistry* **22,** 3214 (1983).
[11] A. Y. S. Jong, R. Aebersold, and J. L. Campbell, *J. Biol. Chem.* **260,** 16367 (1985).
[12] S. Piñol-Roma, S. A. Adam, Y. D. Choi, and G. Dreyfuss, this series, Vol. 180, p. 410.

[28] Purification and Characterization of Polyadenylate-Binding Protein

By Alan B. Sachs and Roger D. Kornberg

Introduction

Polyadenylate-binding protein (PAB) is specifically bound to the poly(A) tail of mRNA in eukaryotic cells.[1-6] PAB is similar in aspects of its primary structure to a large family of eukaryotic RNA-binding proteins.[7] A stretch of as few as 37 amino acid residues of PAB may possess the full poly(A)-binding activity of the protein and represent a widely conserved RNA-binding motif.[8] The availability of mutants in yeast PAB[8] may give some insight into the physiologic role of the protein which, like that of poly(A), remains poorly defined.

The primary structures of yeast and human PAB have been deduced from gene sequences.[3,6,9] Extending from near the amino terminus are four direct repeats of a 90-residue domain, which is highly conserved between the yeast and human proteins. A homologous domain has been identified in many other RNA-binding proteins, where it is often repeated as well.[7] Deletion analysis of yeast PAB revealed that only the first half of a single domain is required for cell growth and division.[8] Evidently, the entire 173-residue carboxy-terminal region of the protein is dispensable. Consistent with this, the yeast and human proteins are divergent in the carboxy-terminal region.

Both nuclear and cytoplasmic forms of PAB have been identified and purified to homogeneity.[2-4] The nuclear protein is derived from the cytoplasmic one by proteolytic cleavage into 53- and 17-kDa polypeptides, which remain associated during isolation.[3] The poly(A)-binding properties

[1] G. Blobel, *Proc. Natl. Acad. Sci. U.S.A.* **70**, 924 (1973).

[2] B. Baer and R. D. Kornberg, *J. Cell Biol.* **96**, 717 (1983).

[3] A. B. Sachs, M. W. Bond, and R. D. Kornberg, *Cell* **45**, 827 (1986).

[4] A. B. Sachs and R. D. Kornberg, *Mol. Cell. Biol.* **5**, 1993 (1985).

[5] B. Setyono and J. R. Greenberg, *Cell* **24**, 775 (1981).

[6] S. Adam, T. Nakagawa, M. Swanson, T. Woodruff, and G. Dreyfuss, *Mol. Cell. Biol.* **6**, 2932 (1986).

[7] B. M. Merrill, K. L. Stone, F. Cobianchi, S. Wilson, and K. R. Williams, *J. Biol. Chem.* **263**, 3307.

[8] A. B. Sachs, R. W. Davis, and R. D. Kornberg, *Mol. Cell. Biol.* **7**, 3268 (1987).

[9] T. Grange, deSa. Martins, J. Oddos, and R. Pictet, *Nucleic Acids Res.* **15**, 4771 (1987).

of the two proteins are identical.[8] They contain multiple binding sites, one of which exhibits a high affinity for poly(A) and covers 12 adenylate residues. The maximal packing density along a poly(A) strand is one protein molecule for every 25 adenylate residues.

PAB is capable of interacting with two different poly(A) strands.[8] Such interaction requires multiple 90-residue domains and creates a mechanism by which PAB can move between strands. This process increases the rate of binding equilibration but does not affect the equilibrium binding constant.[10] Because multiple domains are not required for PAB function *in vivo*, it may be imagined that rapid equilibration confers a selective advantage on a multidomain protein.

The techniques presented here for the identification, purification, and characterization of PAB should be applicable to other RNA-binding proteins. The availability of large quantities of PAB for biochemical studies, together with the ease of genetic analysis in yeast, make PAB an excellent choice for further investigation of a large and growing family of RNA-binding proteins.

Identification of PAB

Nitrocellulose Filter Binding Assay[4]

The sequence specificity of PAB [more than a 5000-fold higher affinity for poly(A) than for poly(C)] allows its rapid identification in crude extracts. Nonspecific RNA-binding proteins in an extract may be adsorbed to unlabeled poly(C), while PAB remains available to bind radioactively labeled poly(A). After capture of the PAB–poly(A) complex on a nitrocellulose filter, extensive washing can be performed to further remove nonspecific complexes (Fig. 1). The protein concentration in the assay mixture must be low enough to ensure complete adsorption of the nonspecific binding proteins by poly(C). This is easily determined by finding the protein concentration where filter retention of [^{32}P]poly(A) is unaffected by further addition of poly(C) to the assay mixture but is essentially abolished by addition of a 20- to 100-fold weight excess of unlabeled over labeled poly(A). To ensure linearity of the assay, the concentration of labeled poly(A) is adjusted so that less than 20% of the label is retained on the filter.

Assay mixtures contain 2 ng of [^{32}P]poly(A) [poly(A) 180–220 (Miles Scientific, Naperville, IL) labeled as described below], 100 μg of poly(C) (P-L Biochemicals, Madison, WI), and protein [0.05 to 0.2 unit of specific

---

[10] O. G. Berg, R. B. Winter, and P. H. von Hippel, *Biochemistry* **20,** 6929 (1981).

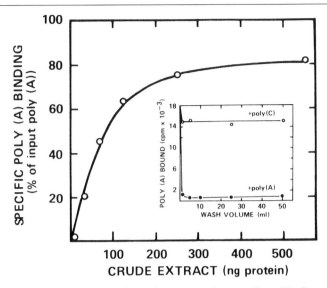

FIG. 1. Poly(A)-binding activity in nuclear extract from rat liver. The inset shows primary data for assay mixtures containing 30 ng of protein and excess unlabeled poly(C) or poly(A) as described in the text. (Reprinted from Sachs and Kornberg[4] with permission of the American Society for Microbiology.)

poly(A)-binding activity] in 100 μl of buffer A [0.1 M NaCl, 30 mM tris(hydroxymethyl)aminomethane hydrochloride (Tris-HCl) (pH 8.0), 15 mM 2-mercaptoethanol, 1 mM ethylenediaminetetraacetic acid (EDTA), 1 mM phenylmethylsulfonyl fluoride]. Assay mixtures are kept for 10 min at 25°, diluted to 1 ml with ice-cold buffer B [60 mM KCl, 10 mM Tris-HCl (pH 8.0), 1 mM EDTA], and passed through 12-mm nitrocellulose filters (HAWP, Millipore Corp., Bedford, MA) at 5 ml/min. The filters are washed with 5 ml of ice-cold buffer B, dried, and counted. Background is determined by substituting unlabeled poly(A) for poly(C). Specific poly(A)-binding refers to [^{32}P]poly(A) bound in the presence of poly(C) minus background. A unit of specific poly(A)-binding activity retains 1 ng of poly(A) on the filter. Under these conditions, 1 unit of activity corresponds to approximately 1 ng of PAB.

Poly(A) (200 ng) is 5′-end-labeled with T4 polynucleotide kinase (4 units) and [γ-^{32}P]ATP (150 μCi) in 25 μl of kinase buffer [50 mM Tris-HCl (pH 8.0), 10 mM MgCl$_2$, 5 mM dithiothreitol (DTT), 100 μM spermidine, 100 μM EDTA] at 37° for 30 min. The specific activity of the product is about 10^8 cpm/μg. The reaction is terminated by dilution to 100 μl with buffer A and extraction with an equal volume of phenol–chloroform–

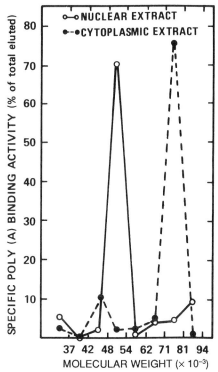

Fig. 2. Distribution of poly(A)-binding activity in SDS–polyacrylamide gels of nuclear and cytoplasmic extracts from rat liver. The divisions on the abscissa correspond to gel slices. (Reprinted from Sachs and Kornberg[4] with permission of the American Society for Microbiology.)

isoamyl alcohol (25 : 24 : 1) and chloroform–isoamyl alcohol (24 : 1). No further purification is required prior to use in the filter binding assay.

Elution from SDS Gels and Renaturation[4]

The similar binding properties of nuclear and cytoplasmic PAB make it difficult to determine the relative abundance of each species in a crude extract. Fortunately, both forms of PAB can be fractionated in sodium dodecyl sulfate (SDS)–polyacrylamide gels and then eluted and renatured with almost complete recovery of poly(A)-binding activity (Fig. 2). This procedure allows the rapid separation and quantitation of nuclear and cytoplasmic PABs without extensive biochemical fractionation.

Crude extracts (100 μg of protein) and more purified protein fractions [1500 units of specific poly(A)-binding activity] are analyzed in 1.5- and 0.5-mm-thick SDS–10% polyacrylamide gels,[11] respectively, calibrated with SDS–polyacrylamide gel electrophoresis low molecular weight standards (Bio-Rad, Richmond, CA). After Coomassie blue staining and destaining of the gel in 10% acetic acid, the gel is sliced at approximately 6- to 8-kDa intervals. Slices are soaked twice for 15 min in 10 ml of 1 mM DTT. After transfer to a silanized glass tube or a 1.5-ml microcentrifuge tube, each slice is homogenized in 1 ml of 100 mM Tris-HCl (pH 8.0), 100 mM NaCl, 100 $\mu$$M$ EDTA, 5 mM DTT, 0.1% SDS, 100 μg/ml bovine serum albumin (BSA) with a Teflon pestle. The homogenate is agitated for 8–12 hr at room temperature and then cleared by centrifugation at over 5000 g for 2 min. The supernatant is transferred to a silanized 15-ml glass centrifuge tube and combined with two 0.5-ml washes of the acrylamide pellet. Four volumes of acetone are added to the eluate, which is kept at −70° for 8–12 hr and then centrifuged at greater than 8000 g for 30 min at 4°. The supernatant is carefully removed, and the pellet is air dried for 5 min. Pellets are resuspended by vigorous pipetting in 20 μl of 6 M guanidine-HCl, 30 mM Tris (pH 8.0), diluted to 1 ml with buffer A, and kept for 1 hr at 25° to allow renaturation before assays of poly(A)-binding activity. Approximately 75% of the activity in crude extracts is recovered from gels in this way.

UV Cross-Linking

UV cross-linking of PAB to poly(A) is a powerful approach for identifying or confirming the binding of PAB to poly(A). For example, this technique provided an early indication that the nuclear and cytoplasmic PABs had different molecular weights,[5] and it helped to prove that yeast cytoplasmic PAB was associated with the poly(A) tail *in vivo*.[6]

Purification of PAB

Eukaryotic Sources

The purification of nuclear and cytoplasmic PAB from eukaryotic cells can be accomplished with two steps of column chromatography (Table I, Figs. 3 and 4). Affi-Gel Blue resin has a high affinity and capacity for PAB. The binding of PAB to the resin results in the dissociation of poly(A) from PAB. It is therefore possible to adsorb PAB from crude

---

[11] U. K. Laemmli, *Nature (London)* **227**, 680 (1970).

TABLE I
FRACTIONATION OF NUCLEAR POLY(A)-BINDING ACTIVITY[a]

Step	Volume (ml)	Total units	Protein (μg/ml)	Yield (%)	Specific activity (units/μg)
Crude extract	6.7	50,000	1,250	100	6
Oligo(dT)-cellulose	7.4	12,300	10	27	166
Affi-Gel Blue	4.1	10,000	6.5	20	375
Oligo(dT)-cellulose	1.7	2,500	2	5	735

[a] Reprinted from Sachs and Kornberg[4] with permission of the American Society for Microbiology.

extracts with the simultaneous removal of contaminating poly(A). PAB is eluted from the resin in 1.3 M guanidine-HCl, but the proteins in this eluate aggregate after removal of the denaturant. Extracts prepared by high-salt washes of nuclei are particularly prone to aggregation and are therefore a poor starting material. This difficulty is overcome by the addition of poly(A) prior to removal of the denaturant. The poly(A)–PAB complexes that are formed remain soluble during the rest of the purification procedure. In order to limit the amount of nonspecific protein binding to poly(A), only enough poly(A) is added to saturate all of the high-affinity PAB-binding sites. In general, this can be estimated by measuring the PAB activity in the eluate and then adding 10 ng of poly(A) per unit of PAB activity. Selectivity of this step is further enhanced by adding large amounts (1 mg/ml) of poly(C) to the eluate.

The second step in the purification is chromatography on oligo(dT)-cellulose, which selectively adsorbs poly(A)–PAB complexes. PAB cannot be eluted from this column with linear or step NaCl gradients without contamination by nonspecific RNA-binding proteins. Contaminating proteins can, however, be removed by extensive washing with poly(C) prior to elution of PAB. The removal of these proteins requires that they dissociate from the poly(A) before they bind to the competing poly(C). Because many of these proteins dissociate very slowly, the procedure is most effective with extended washing times (e.g., 15–20 hr).

PAB can be recovered from the column in purer form by elution as a complex with poly(A) at low ionic strength, as opposed to elution as free protein at high ionic strength. Some proteins nonspecifically bound to the column are evidently eluted at high ionic strength. If greater purification of PAB is required, both the Affi-Gel Blue and the oligo(dT)-cellulose column chromatography steps can be repeated. Note that for rat liver nuclear extracts, the first Affi-Gel Blue step is omitted, because of the

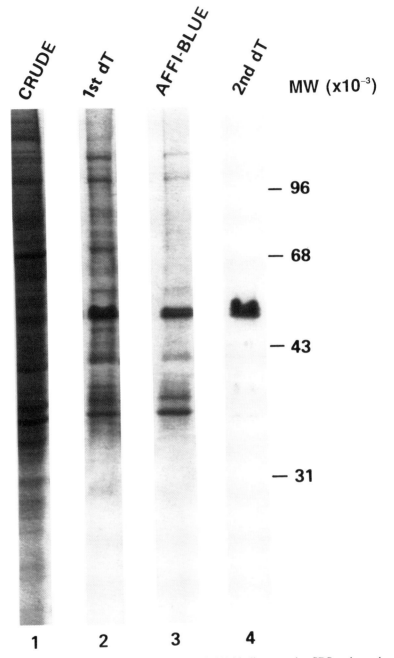

CRUDE   1st dT   AFFI-BLUE   2nd dT   MW (x10^{-3})

— 96

— 68

— 43

— 31

1   2   3   4

FIG. 3. Purification of rat liver nuclear poly(A)-binding protein. SDS–polyacrylamide gels of nuclear extract from rat liver (crude) and of eluates from oligo(dT)-cellulose and Affi-Gel Blue columns. A photograph of the Coomassie blue-stained gel is shown. (Reprinted from Sachs and Kornberg[4] with permission of the American Society for Microbiology.)

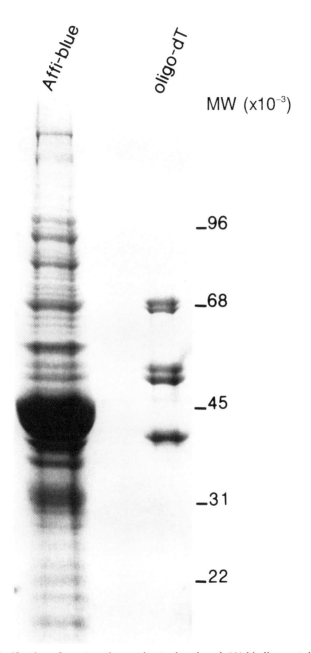

FIG. 4. Purification of yeast nuclear and cytoplasmic poly(A)-binding proteins. A yeast nuclear extract was fractionated on Affi-Gel Blue and then oligo(dT)-cellulose as described. Approximately equal amounts of poly(A)-binding activity (3000 units) from peak fractions were analyzed in a 10% polyacrylamide gel in the presence of SDS. A photograph of the Coomassie blue-stained gel is shown. (Reprinted from Sachs et al.[3] with permission of Cell Press.)

high specific activity of the protein in the crude extract. Thus, the purification of rat liver nuclear PAB begins with an oligo(dT)-cellulose column, followed by an Affi-Gel Blue and another oligo(dT)-cellulose column.

There are three major advantages of adsorbing preformed poly(A)–PAB complexes onto oligo(dT)-cellulose rather than binding PAB to poly(A)-Sepharose for purification. Because the amount of poly(A) added to the Affi-Gel Blue eluate can be carefully titrated, there is minimal nonspecific protein binding to poly(A). Likewise, the small amount of poly(A) added to the oligo(dT)-cellulose column maximizes the ratio of poly(C) to poly(A) during the washing step, which results in increased washing efficiency. Finally, the elution of an intact poly(A)–PAB complex gives a purer product, as explained above, and yields a species which is soluble at physiologic ionic strengths. For these reasons, nuclear PAB can only be purified by oligo(dT)-cellulose and not by poly(A)-Sepharose chromatography.

Fractionation of Nuclear Poly(A)-Binding Activity from Rat Liver.[4] Nuclei from 60 g of rat liver[12] are suspended in 38 ml buffer C, pH 7.0 (0.1 M NaCl, 30 mM Tris-HCl, 15 mM 2-mercaptoethanol, 1 mM CaCl$_2$, 1 mM phenylmethylsulfonyl fluoride), 2 ml of 20% Nonidet P-40 (NP-40) is added, and the mixture is kept for 10 min in ice and centrifuged at 5000 g. The pellet is washed 3 times with 15 ml of buffer C, pH 7.0, resuspended in 7 ml of buffer C, pH 8.3 (adjusted to this pH by the addition of 1 N NaOH), turned end-over-end in a 15-ml tube for 4 hr,[13] and centrifuged in a Beckman Ti50 rotor at 48,000 rpm for 45 min.

The supernatant (nuclear extract) is made 5 mM in EDTA, pH 8.0, and 80 μg/ml in poly(A), kept for 10 min at 4°, and applied to a column (3.25 cm height, 1.75 cm diameter) of oligo(dT)-cellulose (Type 3, Collaborative Research, Lexington, MA) in buffer D (buffer C, pH 8.0, containing 5 mM EDTA). Note that the large quantity of nonspecific binding protein present in this crude extract cannot be effectively saturated with poly(C), which is therefore omitted. The column is washed with 40 ml of buffer D containing 1 mg/ml poly(C), 40 ml of buffer D containing 1 mg/ml poly(C) and 0.3 M NaCl, and 4 ml of buffer D. Poly(A)-binding activity is eluted with 12 ml of 5 mM NaCl–5 mM Tris (pH 8.0) at 45°,[14] adjusted to the composition of buffer D, centrifuged at 10,000 g for 15 min to remove insoluble material, and applied to a column (0.75 cm height, 2.2 cm diameter) of Affi-Gel Blue (Bio-Rad) in buffer D. The column is washed with

[12] D. R. Hewish and L. A. Burgoyne, *Biochem. Biophys. Res. Commun.* **52**, 504 (1973).
[13] T. J. Quinlan, P. B. Billings, and T. E. Martin, *Proc. Natl. Acad. Sci. U.S.A.* **71**, 2632 (1974).
[14] S. K. Jain, M. G. Pluskal, and S. Sarker, *FEBS Lett.* **97**, 84 (1979).

12 ml of buffer D containing 4 M NaCl, 4 ml of buffer D (required to prevent PAB from eluting at the 4 M NaCl–0.5 M guanidine-HCl front), and 12 ml of buffer D containing 0.5 M guanidine-HCl. Poly(A)-binding activity is eluted with 4 ml of buffer D containing 1.3 M guanidine-HCl, supplemented with 100 μg of poly(A) and 4 mg of poly(C), dialyzed against 1 liter of buffer D for 3.5 hr, and fractionated on an oligo(dT)-cellulose column (0.75 cm height, 2.2 cm diameter) as described above except with 10 ml of each of the wash buffers and 2.2 ml of the elution buffer. Columns are run at 6 ml/hr, and procedures are performed at 4° unless otherwise indicated.

Cytoplasmic PAB from rat liver can be purified to homogeneity as in Baer and Kornberg[2] or as described here for the purification of PAB from yeast whole cell extracts. The starting material for rat cytoplasmic PAB can be as in Baer and Kornberg[2] up to the nuclease treatment step.

Isolation of Poly(A)-Binding Activity from Saccharomyces cerevisiae.[3] Yeast nuclear and whole cells extracts are prepared from *Saccharomyces cerevisiae* strain BJ926 (α/**a** trp1/+ +/his1 prc1-126/prc1-126 pep4-3/pep4-3 prb1-1122/prb1-1122 can1/can1). For nuclear extracts, a stationary culture in 2 liters of YPD (2% glucose, 2% bactopeptone, 1% yeast extract) is washed twice with deionized water and once with 500 ml of buffer E [1.1 M sorbitol, 50 mM Tris-HCl (pH 8.0), 1 mM MgCl$_2$, 1 mM phenylmethylsulfonyl fluoride, 1.4 μg/ml pepstatin, 400 ng/ml leupeptin] plus 30 mM DTT. The cell paste is resuspended in 80 ml of buffer E plus 15 mM 2-mercaptoethanol, warmed to 30°, stirred slowly for 5 min, and then made 400 μg/ml in zymolyase 100,000 (Miles Scientific). Greater than 95% of the yeast cells are converted to spheroplasts in 25 min under these conditions. Spheroplasts are washed twice in 200 ml of buffer E plus 15 mM 2-mercaptoethanol, resuspended by vortexing in 200 ml of buffer F [10 mM Tris-HCl (pH 8.0), 1.5 mM MgCl$_2$, 15 mM KCl, 0.1 mM EDTA, 15 mM 2-mercaptoethanol, 1 mM phenylmethylsulfonyl fluoride, 1.4 μg/ml pepstatin, 400 ng/ml leupeptin], and centrifuged at 6000 rpm in a Sorvall SS-34 rotor. The pellet is resuspended in 120 ml of buffer F, homogenized with 20 strokes of a glass A pestle, and centrifuged at 6000 rpm for 10 minutes in an SS34 rotor. The pellet (crude nuclei) is washed once with 160 ml of buffer G [100 mM NaCl, 30 mM Tris-HCl (pH 8.0), 1 mM CaCl$_2$, 1 mM MgCl$_2$, 15 mM 2-mercaptoethanol, 1 mM phenylmethylsulfonyl fluoride, 1.4 μg/ml pepstatin, 400 ng/ml leupeptin], resuspended in 80 ml of this buffer, adjusted to pH 8.2 by the addition of 1 N NaOH, stirred for 2 hr, and centrifuged in a Beckman Ti60 rotor at 48,000 rpm for 45 min.

The supernatant (nuclear extract), containing approximately 20 mg of protein, is made 5 mM in EDTA and applied to an 18-ml column (1.5 cm diameter) of Affi-Gel Blue in buffer H [buffer G (pH 8.0) containing 5 mM EDTA] at 50 ml/hr. The column is washed at this flow rate with 70 ml of buffer H, 400 ml of buffer H containing 4 M NaCl, 50 ml of buffer H, and 300 ml of buffer H containing 0.5 M guanidine-HCl. Poly(A)-binding proteins are eluted with 50 ml of buffer H containing 2 M guanidine-HCl, and peak protein fractions are pooled, diluted to 30 ml with buffer H, made 1 mg/ml in poly(C) and 16 μg/ml in poly(A), dialyzed against 1 liter of buffer H for 3.5 hr, and centrifuged at 10,000 g for 15 min. The supernatant is applied to a 10-ml column (2.2 cm diameter) of oligo(dT)-cellulose in buffer H at 5 ml/hr. The column is washed at this flow rate with 75 ml of buffer H containing 1 mg/ml of poly(C) and 0.23 M NaCl and then with 10 ml of buffer H. Poly(A)-binding activity is eluted with 20 ml of 5 mM NaCl, 5 mM Tris-HCl (pH 8.0) at 45°.[14] A contaminating 40-kDa protein identified as SSB-1 copurifies with PAB,[8] probably because of its high but nonspecific affinity for poly(A).

For whole cell extracts, a stationary culture in 6 liters of YPD is washed 3 times with deionized water and once with buffer H. The cell paste is resuspended in 200 ml of buffer H, combined with 150 ml of acid-washed glass beads (0.45 mm diameter) in a 350-ml bead beater cup (Biospec Products, Bartlesville, OK), and blended 8 times for 30 sec with cooling between times in an ice water bath for 1.5 min. The homogenate and a 200-ml rinse of the glass beads with buffer H are combined and centrifuged at 13,500 rpm for 30 in in a Sorvall SS34 rotor. The supernatant, containing approximately 5 g of protein, is adjusted to pH 8.0 by the addition of 1 N NaOH, loaded onto an 18-ml Affi-Gel Blue column, and further processed as described above.

Prokaryotic Sources[8]

The overproduction of eukaryotic PAB in *Escherichia coli* greatly simplifies the purification procedure. The increased abundance of the protein in crude extracts allows purification using Affi-Gel Blue and an ion-exchange resin. Interestingly, PABs purified from *E. coli* are soluble (>10 mg/ml) and do not aggregate in the absence of poly(A). The increased solubility could be due to the lack of unidentified posttranslational modifications or to the absence of seven amino acids from the amino terminus.

A plasmid for expression of wild-type PAB (p68) in *E. coli* contains nucleotides 885–3435 of the PAB gene[3] inserted at the *Bam*HI site of pAS1[15] (unrelated to yeast strain AS1 described below). Plasmids for

[15] M. Rosenberg, Y.-S. Ho, and A. Shatzman, this series, Vol. 101, p. 123.

Protein   Name

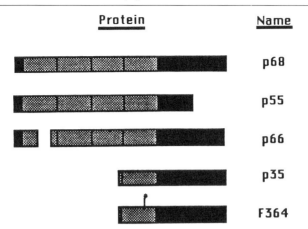

FIG. 5. Schematic representation of various PABs purified from *E. coli*. Stippled areas correspond to amino-terminal domains, dark areas to other coding regions, and blank areas to deleted regions. (Reprinted from Sachs *et al.*[8] with permission of the American Society for Microbiology.)

expression of p66, p55, and p35 (and F364) are identical except for deletion of nucleotides 1028–1131, 2274–2678, and 885–1702, respectively (Fig. 5).

Escherichia coli strain AR120 transformed with the pAS1 derivatives (24 liters) is grown to midlogarithmic phase and induced for 4 hr with 40 mg/liter of nalidixic acid[15] (Sigma, St. Louis, MO). Cells are harvested, washed once with 6 liters of STE [150 mM NaCl, 30 mM Tris-HCl (pH 8.0), 5 mM EDTA, 15 mM 2-mercaptoethanol, 1 mM phenylmethylsulfonyl fluoride, 1.4 μg/ml pepstatin, 400 ng/ml leupeptin], resuspended in 300 ml of STE, cooled to 4°, stirred with 400 μg/ml lysozyme (Sigma), frozen in liquid nitrogen, quickly thawed, sonicated, and centrifuged for 30 min at 12,000 rpm in a Beckman JA14 rotor.

The supernatant (containing ~6–7 g of protein) is loaded at 300 ml/hr on a 350-ml Affi-Gel Blue column (5.5 × 15 cm) equilibrated in STE. The column is washed at this rate with 900 ml of STE, 3500 ml of 2 M NaCl in STE, and 300 ml of STE. Protein is eluted with 1.3 M guanidine-HCl in STE, and peak protein fractions (containing ~600 mg of protein) are pooled, dialyzed 3 times against 4 liters of LS [50 mM NaCl, 5 mM Tris-HCl (pH 8.0), 0.5 mM EDTA, 15 mM 2-mercaptoethanol, 1 mM phenylmethylsulfonyl fluoride], and centrifuged at 11,000 rpm in a Sorvall SS34 rotor. The supernatant is loaded at 60 ml/hr on either a 60-ml DEAE-Sephacryl (Pharmacia) column (2.5 × 15 cm) for p55, p66, and p68, or a 60-ml carboxymethylcellulose column (Whatman, Maidstone, Kent) for p35 and F364. The columns are washed with 180 ml of LS buffer at this rate, and protein is eluted with a 500-ml linear gradient of 0–333 mM NaCl in LS. Peak protein fractions (eluting between 70 and 120 mM NaCl) are pooled and either precipitated with ammonium sulfate (45, 70, and 55% of saturation for p55, p66, and p68, respectively), dialyzed against LS, and stored at 10 mg/ml at −70°, or simply stored at −70° (p35 and F364). All steps are carried out at 4°.

Final yields of protein are approximately 20 mg for p35 and F364 and 100 mg for p55, p66, and p68. Purity is estimated to be between 90 and 95% from Coomassie blue-stained SDS–10% or 12.5% polyacrylamide gels.[11] Protein concentrations are determined from the absorbance at 257 nm, using the calculated molar extinction coefficients (M^{-1} cm^{-1}) for tryptophan and tyrosine residues (p68, 36,420; p66, 33,740; p55, 29,720; p35 and F364, 17,420).

Binding Measurements by Protein Fluorescence[8]

Binding of PAB to poly(A) diminishes the intrinsic tyrosine fluorescence of the protein (Fig. 6). The availability of large quantities of PAB allows this property to be exploited for the rapid and quantitative determination of binding parameters, as defined by von Hippel et al.[16,17] The affinity of PAB for various lengths of oligo(A) may be used to calculate a binding site size of 12 nucleotides (Fig. 7). The titration of PAB with poly(A) or of poly(A) with PAB allows the determination of a packing density of one molecule of wild-type protein for every 25 adenylate residues. Unfortunately, this method is not sensitive enough to determine accurately the affinity of PAB for poly(A) or to assess if PAB binds poly(A) cooperatively.

Poly(A) and poly(C) (P-L Biochemicals) are dialyzed for 48 hr against water. Poly(A) with a degree of polymerization greater than 800, obtained from the void volume of a Sepharose 4B (Pharmacia, Piscataway, NJ) column,[18] is used in all experiments. Oligo(A) 8–12 residues in length may be purchased from P-L Biochemicals. Oligo(A) of 13–23 residues is prepared by heating poly(A) (1.5 mg/ml) in 0.2 N NaOH for 50 min at 37°, followed by neutralization, dialysis against water, lyophilization, and electrophoresis in a 1.5-mm-thick 15% polyacrylamide gel containing 7.5 M urea and half-strength TBE.[19] For poly(A) of 180–220 residues,

---

[16] R. C. Kelly, D. E. Jensen, and P. H. von Hippel, J. Biol. Chem. **251,** 7240 (1976).

[17] S. C. Kowalczykowski, L. S. Paul, N. Lonberg, J. W. Newport, J. A. McSwiggen, and P. H. von Hippel, Biochemistry **25,** 1226 (1986).

[18] T. Lohman, Biochemistry **23,** 4656 (1984).

[19] T. Maniatis, E. F. Fritsch, and J. Sambrook, in "Molecular Cloning: A Laboratory Manual." Cold Spring Harbor Laboratory, Cold Spring Harbor, New York, 1982.

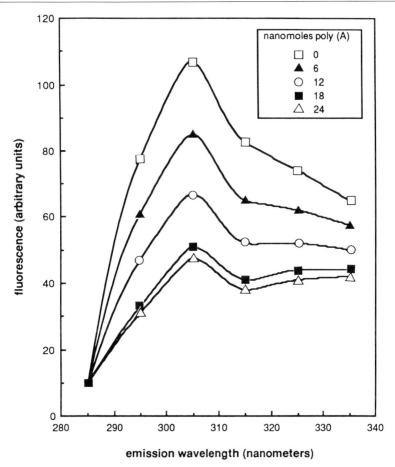

FIG. 6. Poly(A) binding can be monitored by quenching of PAB tyrosine fluorescence. The effects of the indicated amounts of poly(A) on the fluorescence emission spectrum of 0.62 nmol of wild-type PAB are shown. The difference in intensity between the spectra obtained with 18 and 24 nmol of poly(A) represents nonspecific quenching by the polymer. (Reprinted from Sachs *et al.*[8] with permission of the American Society for Microbiology.)

shorter heating times are required. Gel slices are crushed and eluted with 3 ml of 10 mM Tris-HCl (pH 8.0), 0.1 mM EDTA for 20 hr. The eluates are centrifuged, desalted by filtration through a Sephadex G-25 (Pharmacia) column (1.5 × 6 cm), lyophilized, and resuspended in water at a final concentration of 100 μM.

Fluorescence measurements are made in duplicate or triplicate at 23° with the use of an SLM 8000 fluorimeter (SLM Instruments, Urbana, IL). Protein fluorescence is monitored at excitation and emission wavelengths of 277 and 306 nm.

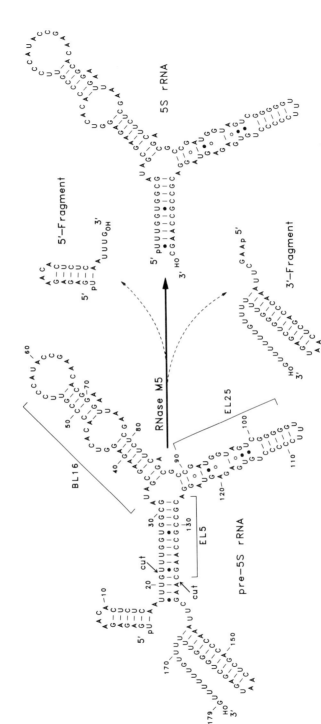

FIG. 1. *Bacillus subtilis* RNase M5 reaction. As described in the text, RNase M5 removes 5'- and 3'-terminal precursor segments in producing mature 5 S rRNA. The order of scission (3'-precursor segment removed before the 5' cleavage?) is indicated, as are the approximate binding regions for the ribosomal proteins discussed in the text.

Substrates and Assays for Terminal Maturases

The simplest assay for terminal processing nucleases is achieved by incubating an isotopically labeled precursor rRNA or RNP with appropriate cell extracts and scoring the production of mature-size products by polyacrylamide gel electrophoresis. Early work with RNase M5 utilized as a substrate uniformly ^{32}P-labeled pre-5 S rRNAs that accumulate in *B. subtilis* in the presence of chloramphenicol (an inhibitor of protein synthesis). The purification of pre-5 S rRNAs by phenol extraction and polyacrylamide gel electrophoresis as well as the gel assay for RNase M5 have been detailed previously.[6] Polyacrylamide gel assays are unwieldy for large numbers of samples, however. When it was discovered that the precursor-specific segments had no role in the recognition of pre-5 S rRNA by RNase M5,[12] it was possible to fabricate a more convenient assay substrate (Fig. 2). As detailed,[14] RNA ligase was used to add a synthetic oligonucleotide, usually UUUG (the *in vivo* sequence), to the 5' end of mature 5 S rRNA. The new 5' end then was labeled using [γ-^{32}P]ATP and polynucleotide kinase, and release of the synthetic "precursor" segment by RNase M5 was scored by loss of acid-insoluble radioactivity or by analysis using polyethyleneimine thin-layer chromatography. In the latter assay, with the chromatography solvents employed, the uncleaved substrate (120 nucleotides) remains at the origin of the chromatogram, and the released, isotopically labeled oligonucleotide (4 nucleotides) migrates at a characteristic rate (Fig. 2, inset). Quantitative results are obtained by scissoring out or scraping radioactive spots that are located by autoradiography and counting them in a scintillation spectrometer.

The enzymatic activities that produce the mature termini of 16 S rRNA in *E. coli* have been assayed *in vitro* using as substrates pulse-labeled preribosome particles[2,3] or ribosomes derived from a mutant defective in the maturation of the 5' end of 16 S rRNA.[4] Naked pre-16 S rRNA is not a substrate for these maturation reactions. There have been no reports of attempts to reconstitute substrates for maturases using purified ribosomal proteins and pre-rRNAs isolated from cells or synthesized *in vitro* with transcription vectors; however, this approach seems viable. The pre-rRNA-containing RNPs that accumulate in cells in the absence of protein synthesis are not likely to be effective substrates, since they probably do not contain a normal complement of ribosomal proteins.

---

[14] B. Meyhack, B. Pace, O. C. Uhlenbeck, and N. R. Pace, *Proc. Natl. Acad. Sci. U.S.A.* **75,** 3045 (1978).

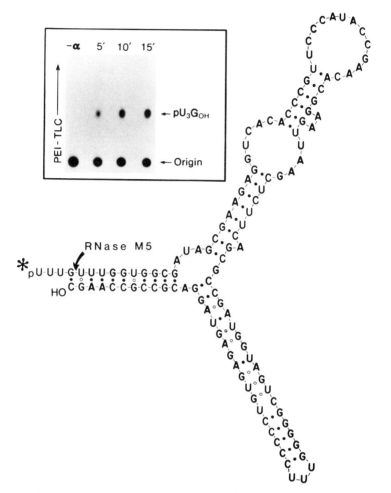

FIG. 2. Assay of RNase M5 using a partially synthetic substrate. As described in the text, a synthetic oligonucleotide (UUUG in the example shown) is appended to the 5' end of mature 5 S rRNA using RNA ligase. Then the artificial substrate is labeled at its 5' end using [γ-³²P]ATP and polynucleotide kinase. Incubation with RNase M5 releases the synthetic precursor segment, which migrates at a characteristic rate during thin-layer chromatography on polyethyleneimine-impregnated cellulose (inset).

Nonionic Detergents and Rare Proteins

The terminal rRNA maturases are rare: each *B. subtilis* cell probably contains no more than about 100 molecules of RNase M5. The extensive purification of rare enzymes such as RNase M5 inevitably results in low solution concentrations of protein, a condition that is considered to predispose enzymes to denaturation and loss of activity. (Hence, bovine serum albumin commonly is added to pure enzymes.) During early studies RNase M5, too, seemed "unstable" at low concentrations of protein.[6] However, this proved to be due to the tenacious adherence of the enzyme to glass and plastic surfaces. Losses in enzyme activity during purification and storage could be countered by the inclusion of a nonionic detergent in all buffers that are used for dilute solutions of proteins.[9] Two detergents, Nonidet P-40 (NP-40) and Brij 58, at concentrations of 0.01–0.1% were interchangeably employed in studies with RNase M5. NP-40 has satisfactory solubility properties in all buffers utilized, but it absorbs light strongly in the ultraviolet. Brij 58 has no significant UV absorbance, but it forms micelles at high ionic strengths and in phosphate-containing buffers in the cold. The inclusion of a detergent in buffers is to be recommended generally for the purification of proteins or other macromolecules. We believe that many anecdotes regarding enzyme denaturation at low solution concentrations of protein can be accounted for by adsorption to surfaces: storage tube walls, flow tubes, chromatographic supports, etc.

Experimental Procedures

Assay of RNase M5. Ten-microliter reactions contain 10 mM tris(hydroxymethyl)aminomethane hydrochloride (Tris-HCl) (pH 7.4), 5 mM MgCl$_2$, 30% (w/v) glycerol, 2 mM dithiothreitol, 0.05% Brij 58, 1–10 ng ^{32}P-labeled pre-5 S rRNA (at a specific radioactivity of $\sim$10^6 cpm/μg RNA), and test amounts of RNase M5 and BL16 proteins. Reactions are mixed on ice, then incubated for appropriate times at 37°. In the thin-layer chromatographic assay,[13] reactions are terminated by chilling in an ice bath and adding 1 μl of 0.5% formic acid, and aliquots of the reactions are spotted onto plastic-backed, polyethyleneimine-impregnated thin-layer plates (Brinkman No. 801063). Glycerol and salts are washed off the origin by ascending chromatography with water to a few centimeters past the origin, then, without drying, ascending chromatography is continued with 1 M formic acid adjusted to pH 4.3 with pyridine. Reaction products are visualized by autoradiography. In the use of the polyacrylamide gel electrophoresis assay, reactions are terminated by adding 1 μl of 10% sodium dodecyl sulfate (SDS) 50% (w/v) sucrose, and products are resolved by electrophoresis in 8% polyacrylamide gels as detailed previously.[6]

Protein Gel Electrophoresis. Protein purifications are monitored by silver staining[15] on 15% polyacrylamide gels following electrophoresis in the second dimension buffer of O'Farrell.[16] The *B. subtilis* β protein was

[15] B. R. Oakley, D. R. Kirsch, and N. R. Morris, *Anal. Biochem.* **105,** 361 (1980).

[16] P. H. O'Farrell, *J. Biol. Chem.* **250,** 4007 (1975).

identified as BL16 ribosomal protein by Dr. Eiko Otaka (Research Institute for Nuclear Medicine and Biology, Hiroshima University, Japan), using the two-dimensional polyacrylamide gel electrophoresis system of Kaltschmidt and Wittmann.[17]

Purification of RNase M5. Frozen *B. subtilis* 168 cell paste (175 g) is thawed in 170 ml of PA Buffer [50 mM Tris-HCl (pH 7.4), 20 mM MgCl$_2$, 5% (w/v) glycerol, 60 mM NH$_4$Cl] adjusted to 200 mM NH$_4$Cl and 1 μg/ml DNase I (Sigma Type II). All enzyme solutions are manipulated and stored at 4° or lower. Cells are broken by passage through an Aminco French pressure cell at 20,000 psi. Cellular debris and ribosomes are removed by centrifugation, first at 18,000 rpm in a Sorvall SS34 rotor for 40 min and then at 45,000 rpm for 6 hr in a Beckman 50.2 Ti rotor. The supernatant is dialyzed overnight against 2 changes of PA buffer. After a fine precipitate is removed by centrifugation at 10,000 rpm for 20 min, the dialyzate is loaded onto a 500-ml DEAE-cellulose column (Whatman DE-52) equilibrated in PA buffer, then washed with the same buffer. Flow-through and wash fractions are assayed for RNase M5 activity (supplementing assay reactions with ribosomal protein BL16), and the relevant fractions are pooled. A nonionic detergent, NP-40 (Sigma, St. Louis, MO) or Brij 58 (Sigma), is then added to 0.1%, and is included in all subsequent buffers in order to prevent adsorption to surfaces at low concentrations of protein.

After dialysis against SB buffer [50 mM Tris-HCl (pH 8.0), 30 mM NH$_4$Cl, 4 mM 2-mercaptoethanol, 0.1 mM ethylenediaminetetraaceticacid (EDTA), 0.1% NP-40], the enzyme preparation is loaded onto a 150-ml phosphocellulose column (Whatman P-11). SB buffer is used for column equilibration, for washing after loading, and for the linear 30–600 mM NH$_4$Cl gradient. Column fractions are assayed, and the RNase M5 peak, at about 300 mM NH$_4$Cl, is pooled and dialyzed against SB buffer. It is then loaded onto a 5-ml Affi-Gel Blue (Bio-Rad, Richmond, CA), column, which has been washed extensively with SB buffer containing 600 mM NH$_4$Cl in order to strip UV-absorbing materials from the matrix and then equilibrated to SB buffer before use. The peak RNase M5 activity elutes at 150 mM in a 30–300 mM NH$_4$Cl gradient in SB buffer.

Most of the reported studies of RNase M5 used enzyme at this level of purity. Preparations are greatly enriched in RNase M5 activity and free of other nucleases. A spectrophotometric scan of a silver-stained, polyacrylamide gel electrophoresis track of such an RNase M5 preparation indicated that the enzyme constitutes about 3% of the silver-staining material, presuming equivalent staining of all bands present. This type of estimate is used to calculate nominal enzyme concentrations in RNase M5

[17] E. Kaltschmidt and H. G. Wittmann, *Anal. Biochem.* **36**, 401 (1970).

reactions. It is not known what fraction of the protein in the RNase M5 band is in fact catalytically active.

Further purification can be achieved by sequential chromatography, at pH 6.5 and 4.0, on 2-ml phosphocellulose columns, using 0–300 mM NaCl gradients in 50 mM sodium phosphate buffer, 4 mM 2-mercaptoethanol, 0.1% NP-40, 6 M urea. A 1-ml carboxymethylcellulose (Whatman CM-52) column with a 0–300 mM NaCl gradient in 50 mM sodium acetate (pH 4.0), 4 mM 2-mercaptoethanol, 0.1% NP-40, 6 M urea also can afford some purification. Final purification of the RNase M5 activity to homogeneity is achieved by denaturing (SDS) polyacrylamide gel electrophoresis, recovering the protein according to the protocol of Hager and Burgess.[18] A single gel band of M_r 24,000 is highly active against the BL16–pre-rRNA RNP substrate; however, the number of peptide chains required to form the active enzyme is not known.

Purification of Ribosomal Proteins BL16 and EL18. Ribosomal proteins BL16 and EL18 originally were purified as activities (β) that are required for the RNase M5 reaction. When it was discovered that the β activities are ribosomal proteins, the following protocol was devised. It is similar to methods commonly used to purify ribosomal proteins.[11]

Frozen *E. coli* or *B. subtilis* cell paste (50 g) is thawed in 50 ml of buffer B [20 mM Tris-HCl (pH 7.8), 30 mM NH$_4$Cl, 10 mM magnesium acetate, 10 mM 2-mercaptoethanol] containing 1 μg/ml DNase I (Sigma Type II) and broken by passage through an Aminco French pressure cell at 20,000 psi. All steps in the purification schedule are carried out at or below 4°. The lysate is diluted with 50 ml of buffer B and centrifuged for 20 min at 18,000 rpm in a Sorvall SS34 rotor. The supernatant is centrifuged for 5 hr at 45,000 rpm in a Beckman 50.2 Ti rotor, and the ribosome pellet resuspended in buffer B. The NH$_4$Cl concentration is adjusted to 1 M and the volume to 120 ml by the addition of buffer B, and the ribosomes are again pelleted. The salt-washed ribosomes are resuspended in 75 ml of buffer B. After dialysis overnight against 20 mM Tris-HCl (pH 7.8), 10 mM NH$_4$Cl, 2 mM magnesium acetate, 10 mM 2-mercaptoethanol, 0.05% Brij 58 (or NP-40), ribosomes are dispersed by addition of urea crystals to 8.5 M. After 24 hr on ice, the urea concentration is reduced to 6.5 M by addition of 50 mM sodium phosphate (pH 6.5), and a fine precipitate is removed by centrifugation for 5 min at 18,000 rpm in a Sorvall SS34 rotor.

Approximately 25,000 A_{260} units of supernatant is loaded onto a 500-ml phosphocellulose column equilibrated with 50 mM sodium phosphate (pH 6.5), 6 M urea, 2 mM methylamine, 0.05% NP-40 and washed with several hundred milliliters of the same buffer. Proteins are eluted with a 0–700

[18] D. A. Hager and R. R. Burgess, *Anal. Biochem.* **109,** 76 (1980).

mM NH$_4$Cl gradient in the same buffer; the activities that complement the RNase M5 reaction elute at about 300 mM. Peak activity-containing fractions are pooled, dialyzed against 50 mM acetic acid (pH 3), 0.1 mM EDTA, 0.1% NP-40, 30 mM NH$_4$Cl, loaded onto a 15-ml phosphocellulose column in the same buffer, and eluted with a 0.03–1.2 M gradient of NH$_4$Cl in that buffer. Both BL16 and EL18 elute late, at approximately 800 mM NH$_4$Cl, and are nearly pure at this stage. A third column chromatography is carried out on a 2-ml phosphocellulose column in 40 mM Tris-HCl (pH 7.4), 0.1 mM EDTA, 0.1% NP-40, 30 mM NH$_4$Cl, eluting with a 30–700 mM NH$_4$Cl gradient; the ribosomal proteins elute at approximately 400 mM. A fourth column, required only for rigorous purification of the EL18, is the same as the third except that all buffers contain 6 M urea. The EL18 elutes at approximately 150 mM in the 20–400 mM NH$_4$Cl gradient. Pooled fractions are dialyzed against buffer B containing 0.1% NP-40 and stored as aliquots frozen at $-70°$.

Acknowledgments

The authors' research is supported by National Institutes of Health Grant GM34527 to N.R.P. We thank David Smith for artwork.

Section III

Processing of Transfer RNAs

A. Genetics
Articles 31 through 33

B. Enzymology of Processing
Articles 34 through 42

[31] Yeast Suppressor Mutations and Transfer RNA Processing

By MARK NICHOLS, IAN WILLIS, and DIETER SÖLL

Introduction

Transfer RNA molecules are essential for protein synthesis in all organisms. The biosynthesis of these molecules in eukaryotes involves many steps including the following: tRNA gene transcription to yield a precursor RNA molecule; 5'- and 3'-end-maturation; removal of an intron (tRNA splicing) if present; 3'-terminal nucleotide (-CCA) addition; and a significant number of nucleotide modifications. Once mature-sized tRNAs are produced, further interactions with enzymes or structural proteins are necessary for their use in protein biosynthesis: transport from the nucleus to the cytoplasm, aminoacylation by the cognate aminoacyl-tRNA synthetase, binding of initiation or elongation factors, and ribosomal interactions during translation and proofreading. Because of the many steps leading to the expression of tRNA function, many potential control points exist for regulating the synthesis of these molecules. Many of these steps are also dependent on previous ones to provide appropriate substrates. How and to what degree conserved nucleotides and structures play roles in this pathway can be determined by analyzing mutations that cause the loss of tRNA function. With the use of suppressor genetics,[1] the products of a single tRNA gene within a multigene family can be assayed as a result of its dominant phenotype. This phenotype detects only the *final*, mature products of one tRNA gene, and thus it can potentially be used to find mutations which inhibit any or all processes that are required for expression of tRNA function. In addition, tRNAs have small genes which can be easily sequenced to determine the cis-acting mutations responsible for loss of function *in vivo*. The length of the primary transcripts is usually less than 100 nucleotides.[2]

We have used two general approaches to find defects in the expression of a nonsense suppressor tRNA. One method detects the loss of suppressor tRNA function *in vivo* by making use of yeast strains containing nonsense alleles of several structural genes whose activity can be screened. Marker genes exist which also allow visual determination of the

[1] G. Eggertsson and D. Söll, *Microbiol. Rev.* **52**, 354 (1988).

[2] S. Sharp, J. Schaack, L. Cooley, D. J. Burke, and D. Söll, *CRC Crit. Rev. Biochem.* **19**, 107 (1985).

limiting adenine, add 3 mg/liter adenine and use 30 ml of a 0.1 mg/ml solution).

Reagents

Restriction enzymes: Used with conditions suggested by the supplier (New England Biolabs)

rNTPs, dNTPs: Resuspended to 10 mM nucleoside triphosphate, at pH 7.0; store in aliquots at $-20°$

Radiochemicals ([α-^{32}P]GTP, [α-^{32}P]dATP, [γ-^{32}P]ATP) (Amersham Inc.)

PMSF (phenylmethylsulfonyl fluoride): Make a 150 mM solution in 100% acetone; store at $-20°$ (this is a toxic protease inhibitor!)

DTT (dithiothreitol): Prepare a 1 M solution; store aliquots at $-20°$

20× SSC: 3 M sodium chloride, 300 mM sodium citrate; adjust to pH 7.0 with HCl

50× Denhardt's solution: 1% poly(vinylpyrrolidone) (PVP), 1% bovine serum albumin (BSA), 1% Ficoll; store at $-20°$

Generation of Suppressor tRNA Mutants

Mutant tRNA genes can be generated in numerous ways both *in vitro* and *in vivo*. A number of articles and methods applicable to this kind of analysis have been published.[7-11] Some of these methods may be more appropriate than others in different circumstances. The general principles of the specific methods that we have used for *in vitro* and *in vivo* mutagenesis are summarized below.

Genetic Screen for Loss of Suppressor tRNA Function in Vivo

In vivo screening of suppressor tRNA function allows rapid identification of suppressor alleles which exhibit loss of the suppression phenotype. Mutagenesis of strains containing a suppressor tRNA gene can be used to increase the rate of mutagenic changes or, alternatively, a genetic screen for suppression can be used to detect spontaneous mutations which cause the loss of phenotype, as has been done recently for a plas-

[7] F. Hofer, H. Hollenstein, F. Janner, M. Minet, P. Thuriaux, and U. Leupold, *Curr. Genet.* **1**, 45 (1979).

[8] M. S. Ciampi, D. A. Melton, and R. Cortese, *Proc. Natl. Acad. Sci. U.S.A.* **79**, 1388 (1982).

[9] W. R. Folk and H. Hofstetter, *Cell* **33**, 585 (1983).

[10] J. A. Tobian, L. Drinkard, and M. Zasloff, *Cell* **43**, 415 (1985).

[11] C. N. Giroux, J. R. A. Mis, M. K. Pierce, S. E. Kohalmi, and B. A. Kunz, *Mol. Cell. Biol.* **8**, 978 (1988).

mid-borne suppressor tRNA gene.[11] To have a visual assay of suppressor activity in *S. cerevisiae*, the appropriate nonsense allele of either the *ADE1* or *ADE2* gene is ideal. Each of these genes encodes a different enzyme involved in the adenine biosynthetic pathway, which has red-pigmented substrates. If the enzyme is not active (i.e., there is no suppression), the accumulation of these precursors leads to a red colony phenotype, when grown on plates containing limited adenine (MMA plates with 3 mg/liter of added adenine or standard YEPD plates; see Fig. 6). The amount of added adenine is critical to the color formation and can be varied empirically to give both reasonable growth and red color in Ade⁻ colonies. Excess supplemented adenine in the plates will result in white Ade⁻ strains. Suppression of the nonsense allele (either *ade1⁻* or *ade2⁻*) by the appropriate tRNA suppressor restores enough enzymatic activity to give white Ade⁺ colonies. Intermediate levels of suppression will give intermediate, pink-colored colonies. No suppression of the nonsense allele will result in red Ade⁻ colonies.

Use of Chromosomal Suppressor tRNA Gene in Vivo. If the suppressor tRNA locus has not been cloned, the loss of suppressor function in red or pink colonies must be mapped to the locus of the tRNA gene by further analysis. This can be achieved by mating the mutant strain with appropriate tester strains, such as a strain containing the suppressor-inactive, wild-type anticodon for the tRNA in question. With a predictably low frequency after sporulation (about 10^{-6}), a crossing-over recombination event should occur within the tRNA gene between the inactivating mutation and the suppressor anticodon to restore the suppression phenotype. This crossover is relatively rare because it must occur within the few nucleotides between the anticodon and the inactivating tRNA mutation. A gene, unlinked to the tRNA gene, which causes the red colony formation (e.g., a nonsuppressible second mutation in *ade1⁻*) should give 2.5 × 10^{-1} or 25% (i.e., 1 : 3 in tetrad spore analyses) restoration of suppression in spores. This high frequency results from the fact that there are many kilobases of DNA for a cross-over event between the tRNA gene and the unlinked gene, or the two genes may simply be on different chromosomes. Further analysis of the chromosomal mutant tRNA alleles requires that the active suppressor locus is cloned first by using the suppressor phenotype as an assay in yeast clone bank transformations. Cloning of an appropriate-sized genomic fragment containing the mutant tRNA genes can then proceed by screening *E. coli* transformants in colony hybridizations,[12,13] followed by DNA sequencing to determine the muta-

[12] I. Willis, H. Hottinger, D. Pearson, V. Chisholm, U. Leupold, and D. Söll, *EMBO J.* **3,** 1573 (1984).

[13] J. Kurjan and B. D. Hall, *Mol. Cell. Biol.* **2,** 1501 (1982).

inhibition of transcription of a second gene.[26] A quantitation of the transcripts from the second tRNA gene then gives a measure of the affinity of the first gene for the sequestered transcription factors. A 0% competitor gene is one that allows as much second tRNA gene transcription as a plasmid vector which contains no tRNA gene, because no factor is specifically sequestered. A 100% competition level describes a tRNA gene (or mutant gene) which reduces the amount of second tRNA gene transcription as much as the wild-type, first tRNA gene does (owing to the unavailability of sequestered factors). Experimentally, the competition experiments are done as in the transcription activity experiments, with slight changes. The first tRNA gene (10 μg/ml) is preincubated for 5 min under transcription conditions (in 50-μl reactions). Then an equal amount of the second, reference tRNA gene is added, and the reaction is continued for 40 min at 15°. After purification of the RNA (as described above) and autoradiography, the primary transcript band of the second tRNA is excised, and the transcription inhibition is quantitated, after Cerenkov counting.

In Vivo Analysis of tRNA Mutants

As the S. pombe sup3-e and sup9-e suppressor tRNA genes are active in both S. pombe and S. cerevisiae, in vivo studies of the mutant alleles can be performed in both systems. The S. pombe tRNA gene products can be directly identified by Northern analysis of RNA isolated from transformed S. cerevisiae strains, because there is low background hybridization of the S. pombe gene probe to S. cerevisiae tRNAs.[12] Normally the multigenic nature of isoaccepting tRNA genes masks in vivo analysis of a single mutant allele in a homologous system. However, for the S. pombe suppressors, confirmation of the results obtained for the various mutants from the in vitro transcription and processing system can be corroborated in vivo in S. cerevisiae.

Northern Hybridization Analysis

Reclones in standard yeast plasmid vectors of the various mutant sup 3-e and sup 9-e alleles are transformed into an appropriately marked S. cerevisiae strain. Colonies are grown with selection for the plasmid marker to an A_{600} of 0.6 in 50 ml of appropriately supplemented minimal medium, and low molecular weight RNAs are prepared as follows.[27] Two milliliters of buffer [20 mM Tris-HCl (pH 7.5), 5 mM EDTA] and 2 ml of

[26] J. Schaack and D. Söll, Nucleic Acids Res. 13, 2803 (1985).
[27] G. Rubin, Methods Cell Biol. 12, 45 (1975).

buffer-saturated phenol are added to the cell pellet and mixed gently at 37° for 45 min. The resulting aqueous phase is phenol extracted twice and ethanol precipitated. The concentration of RNA is determined by reading the A_{260}.

The RNAs are resolved on 10% polyacrylamide gels containing 8.3 M urea (0.7 mm thick). Each lane contains 40 μg of RNA. After electrophoresis, the gels are stained in 0.5 μg/ml ethidium bromide solution for 10 min and photographed under UV light to verify approximately equal amounts of undegraded tRNA for each sample. Following equilibration of the gels in TAE buffer [10 mM Tris–acetate (pH 7.8), 5 mM sodium acetate, 0.5 mM EDTA], the RNA is electrophoretically transferred in TAE buffer to zeta-probe membranes (70 mA for 4 hr) using a Bio-Rad Trans-Blot cell, following the manufacturer's instructions (Bio-Rad, Richmond, CA). Filters are hybridized in a buffer (100 μl/cm^2 filter) containing 50% deionized formamide, 5× SSC, 50 mM sodium phosphate (pH 6.5), 2× Denhardt's solution, and 0.1 mg/ml boiled calf thymus DNA for 20 hrs at 42°. The hybridization probes (10^6 cpm/ml) consist of DNA fragments containing the *sup3-e* or *sup9-e* gene labeled to a specific activity of about 1 × 10^8 cpm/μg using T4 DNA polymerase and [α-^{32}P]dATP (3000 Ci/mmol).[28] The final posthybridization wash is at 65° in 0.1× SSC and 0.1% SDS for 30 min. A sample Northern hybridization for several mutant alleles of *sup9-e* is shown in Fig. 5. The mutant A_{19} is an example of a transcription mutant, and $A_{37:13}$ shows greatly reduced splicing, as the serine plus IVS species has accumulated.

Suppression Assays

The *in vivo* suppressor activity is assayed in *S. cerevisiae* strain 3A84 (*MATa ura3-52 ade1*$_{op}$ *leu2-2*$_{op}$ *his 4-260*$_{op}$). Parental and mutant *sup3-e* genes, cloned on the single-copy YCp50 vector, are transformed by the lithium acetate whole-cell transformation method[29] into strain 3A84 with selection for the *URA3* gene on the vector. Opal suppressor activity is screened by monitoring the degree of red color formation which develops in the yeast, grown on limiting adenine plates (3 mg/liter), if the *ade1*$_{op}$ mutation is not suppressed (see Fig. 6). Suppression is quantitated in some cases by spotting logarithmic dilutions of saturated cultures on minimal plates containing uracil and selecting for suppression of all three genes with opal mutations. The resulting growth indicates subtle differ-

[28] T. Maniatis, E. F. Fritsch, and J. Sambrook, "Molecular Cloning: A Laboratory Manual." Cold Spring Harbor Laboratory, Cold Spring Harbor, New York, 1982.
[29] H. Ito, Y. Fukuda, K. Murata, and A. Kimura, *J. Bacteriol.* **153**, 163 (1983).

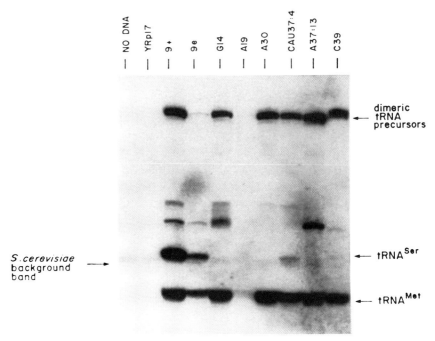

FIG. 5. Northern hybridization analysis of RNA from *S. cerevisiae* transformants. Transformants of *S. cerevisiae* strain YH-D5 (*MATα trp1-1*$_{am}$ *leu2-2*$_{op}$ *his4-260*$_{op}$), each containing a mutant *S. pombe sup9-e* tRNA gene on plasmid YRp17, were grown at 30° prior to preparation of RNA. Each lane contained equal amounts (50 μg) of RNA. The wild-type *sup9*$^{+}$ allele is designated 9+ , and the suppressor-active *sup9-e* allele is designated 9e. After transfer of the RNA to the filter, hybridization was carried out using the *sup9-e* tRNA^{Ser}– tRNA$^{Met}_{i}$ dimeric gene as a probe.

ences in the level of suppression by the various mutant tRNA alleles. Both assays give similar, relative measures of suppression.

In some cases, it may be interesting to increase the copy number of the mutant tRNA allele to see if this increases suppression. For example,

FIG. 6. Color assay for suppression. Various mutant alleles of *sup3-e* and *sup9-e* were transformed into *S. cerevisiae* strain 3A84 (*MATa ura3-52 ade1*$_{op}$ *leu2-2*$_{op}$ *his4-260*$_{op}$) on single-copy plasmids. They were plated on a minimal plate, supplemented with histidine, leucine, and 3 mg/liter (limited) adenine. The phenotypes range from suppressor active (Ade^{+}, white colonies) to suppressor inactive (Ade^{-}, red colonies). Intermediate pink colonies result from intermediate levels of suppression of the *ade1*$_{op}$ allele. Each of the transformed *sup3-e* alleles have three independent isolates grown next to each other. From top to bottom: Left column, C_2G_{71}, G_{71}, A_1T_{72}, T_{72}, G_8C_{14}, $T_{18}A_{55}$, YRp17, *sup9*$^{+}$, *sup9-e*, $G_{37:10}$-9e, A_{53}-9e; central column, A_{16}, T_{18}, A_{19}, C_{21}, Ai_{26}, A_{30}, C_{33}, A_{47}, $T_{47:6}$, C_{49}, A_{55}, T_{58}, Ai_{61}; right column, *sup3*$^{+}$, *sup3-i*, *sup3-e*, A_1, C_2, G_6, G_8, T_9, T_{11}, A_{12}, T_{13}, G_{14}, C_{14}, C_{15}.

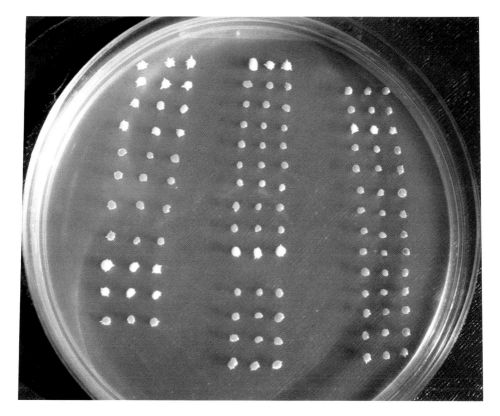

Fig. 6.

a greater number of primary transcripts may result in a sufficiently greater number of mature tRNA molecules to detect an improved suppression pheotype. Suppressor activity at higher gene dosages can be examined in *S. cerevisiae* strains using replicative or episomal plasmid vectors.[12,30]

Biosynthetic Processes Affected

In eukaryotes, tRNA genes are transcribed by RNA polymerase III. With very few exceptions, they are transcribed into monomeric units, as opposed to the large, polycistronic transcripts of tRNA operons in *E. coli* and *B. subtilis*.[2] tRNA gene transcription by RNA polymerase III is also known to require two general factors, TFIIIB and TFIIIC. Various studies of tRNA transcription have revealed that two internal control regions (5′-ICR and 3′-ICR, or A-box and B-box) are involved in transcription promotion.[31,32] These studies have delimited the two regions from about base pairs 8–20 and 48–61 in the gene, corresponding to the D-stem and loop and the T-stem and loop in the tRNA, respectively. Both regions are highly conserved in tRNAs of eukaryotes and prokaryotes.

Transfer RNA genes have variable abilities to sequester factors leading to their transcription. *In vitro* studies, using separated and partially purified factors TFIIIB, TFIIIC, and RNA polymerase III, have shown that TFIIIC binds stably to a gene through multiple rounds of transcription[33] (for review, see Ref. 34). Footprinting and binding experiments show that TFIIIC binds tightly and primarily to the 3′-ICR.[35] Mutations in the 3′-ICR can alter significantly the affinity of TFIIIC for the tRNA gene, whereas mutations in the 5′-ICR have much less of an effect on TFIIIC binding.[35] For the *sup3-e* mutants in the *S. cerevisiae* S-100 extract, the 5′-ICR mutants that show reduced transcriptional efficiency (e.g., G_8, C_{14}, C_{15}, A_{16}) do not affect competitive strength to a large extent. A typical 5′-ICR mutant competes about 90% as well as *sup3-e*. However, the 3′-ICR mutants such as A_{55}, C_{57}, and T_{58} reduce the competitive ability to about 50% of the wild-type level. Others (C_{53} and G_{56}) may be as low as 5–15%. The TFIIIB factor is thought to be involved in protein–protein interactions, since it does not footprint tRNA genes by itself.[36]

[30] I. Willis, M. Nichols, V. Chisholm, D. Söll, W.-D. Heyer, P. Szankasi, H. Amstutz, P. Munz, and J. Kohli, *Proc. Natl. Acad. Sci. U.S.A.* **83**, 7860 (1986).

[31] S. Sharp, D. DeFranco, T. Dingermann, P. Farrell, and D. Söll, *Proc. Natl. Acad. Sci. U.S.A.* **78**, 6657 (1981).

[32] C. Galli, H. Hofstetter, and M. L. Birnstiel, *Nature (London)* **294**, 626 (1981).

[33] J. Schaack, S. Sharp, T. Dingermann, and D. Söll, *J. Biol. Chem.* **258**, 2447 (1983).

[34] E. P. Geiduschek and G. P. Tocchini-Valentini, *Annu. Rev. Biochem.* **57**, 873 (1988).

[35] R. Baker, O. Gabrielson, and B. D. Hall, *J. Biol. Chem.* **261**, 5275 (1986).

[36] M. S. Klekamp and P. A. Weil, *J. Biol. Chem.* **261**, 2819 (1986).

Because genes have differential abilities to sequester the TFIIIC factor, different levels of *in vivo* transcription and gene expression could result if the amount of active TFIIIC becomes limiting or is altered by the growth conditions.

As transcription is the first step in tRNA production, mutations within the two ICRs can lead to a reduced number of initial transcripts for subsequent maturation enzymes. In addition, as the internal control regions correspond to highly conserved parts of tRNAs, mutations there also destabilize the tRNA structure necessary for recognition by many enzymes. As a result, the concentration of final, mature-sized product may be very low. However, there are mutants where increased gene dosage (therefore increased transcription) can help to restore suppression. For example, the *sup3-e* double mutant $A_{30}A_{67}$ is not active as a suppressor in *S. cerevisiae* when present on a single-copy plasmid but is active when present on a high-copy plasmid,[30] presumably because of an increase in the number of tRNA primary transcripts.

Once tRNA primary transcripts are formed, nucleotides at both ends of the molecule must be removed. RNase P is the ribonucleoprotein enzyme which must be able to recognize and remove the extra 5'-flanking nucleotides from all cellular tRNA precursors. Consequently, it must recognize common features of all tRNAs. Accordingly, mutations throughout the tRNA which alter tertiary structure also inhibit the removal of the 5'-flank. Since this processing step occurs early in the maturation of tRNA, usually the step after transcription, it is needed to generate substrates for most of the further processing events. Mutants which inhibit the action of RNase P on the tRNA precursor are found in the anticodon-intron stem ($A_{37:13}$), the anticodon stem (A_{30}), the D- and T-loops and stems (G_{10}, C_{14}, C_{15}, T_{18}, C_{49}, A_{55}), and the acceptor stem, where the most severe effects are seen (A_1, C_2, A_{68}, A_{70}). In addition, nucleotides in the transcript next to the cutting site of RNase P are important in the reaction for efficient cleavage.[15]

After 5'- and 3'-end-trimming have occurred, intron-containing tRNA precursors are substrates for the tRNA splicing apparatus in *S. cerevisiae*.[37] This splicing requires a general structure which is common to the precursors of *S. cerevisiae* tRNAs with introns.[38–40] It appears that the nucleotides next to the 5'- and 3'-splice sites do not determine the sites of endonucleolytic cutting,[41] as in pre-mRNA.[42] Instead a proper, base-

[37] C. L. Greer, D. Söll, and I. Willis, *Mol. Cell. Biol.* **7,** 76 (1987).
[38] M.-C. Lee and G. Knapp, *J. Biol. Chem.* **260,** 3108 (1985).
[39] R. C. Ogden, M.-C. Lee, and G. Knapp, *Nucleic Acids Res.* **12,** 9367 (1984).
[40] H. Swerdlow and C. Guthrie, *J. Biol. Chem.* **259,** 5197 (1984).
[41] D. Colby, P. S. Leboy, and C. Guthrie, *Proc. Natl. Acad. Sci. U.S.A.* **78,** 415 (1981).
[42] R. Breathnach and P. Chambon, *Annu. Rev. Biochem.* **50,** 349 (1981).

paired conformation of the lower half of the general tRNA L-form is required for efficient splicing. For example, the 3'-cut site occurs in a single-stranded, bulge-loop region,[38] and the anticodon region base pairs with a portion of the intron to extend the anticodon stem.[16,43] A mutation (Ai_{26}) in *sup3-e* which inserts a nucleotide and potentially adds a base pair to the anticodon stem, shifts both cut sites by one nucleotide.[37] Hence, the endonucleolytic cutting sites which initiate the splicing reaction are measured from a fixed region or structure near the corner of the tRNA L-form. Generally mutations in the lower part of the tRNA (D, anticodon, intron, and extra arm stems and loops) influence the ability of the *sup3-e* and *sup9-e* precursors to be spliced. For example, C_{14}, G_{14}, C_{15}, $A_{37:13}$, A_{39}, and C_{39} mutants all have reduced ratios of mature tRNA^{Ser} to tRNA^{Ser} plus IVS species (see Figs. 4 and 5).

Reconstituted Systems

For further study of the enzymatic processes in more detail, one can purify the specific enzymes and study particular tRNA mutants in kinetic analyses. To study RNA polymerase III transcription of tRNA genes, procedures exist to isolate both TFIIIB[36] and TFIIIC.[44] RNA polymerase III can be isolated as described by Valenzuela.[45] An *S. cerevisiae* tRNA splicing system is described elsewhere in this volume.[46] RNase P purification has been described previously by Leontis *et al.*[47] and Engelke *et al.*[48] The enzyme(s) responsible for 3'-processing of tRNAs are described (Deutscher [35] this volume).[49,50]

Conclusions

The tRNA biosynthetic pathway has been studied in prokaryotes for several of the T4 phage tRNAs,[51] as well as for some *E. coli* tRNA-containing transcripts.[50] In prokaryotes, the first RNA processing step is usually an endonucleolytic cleavage at or near the 3' end of the mature tRNA, followed by removal of the 5' flank by RNase P. In the eukaryote *S. cerevisiae,* an order of processing events is apparent from *in vitro*

[43] J. Abelson, *Annu. Rev. Biochem.* **48,** 1035 (1979).
[44] A. Ruet, S. Camier, W. Smagowicz, A. Sentenac, and P. Fromageot, *EMBO J.* **3,** 343 (1984).
[45] P. Valenzuela, G. I. Bell, F. Weinberg, and W. J. Rutter, *Methods Cell Biol.* **18,** 1 (1977).
[46] Q. Xu, E. M. Phizicky, C. L. Greer, and J. N. Abelson, this volume [38].
[47] N. Leontis, A. DaLio, M. Strobel, and D. Engelke, *Nucleic Acids Res.* **16,** 2537 (1988).
[48] D. R. Engelke, P. Gegenheimer, and J. Abelson, *J. Biol. Chem.* **260,** 1271 (1985).
[49] D. Frendewey, T. Dingermann, L. Cooley, and D. Söll, *J. Biol. Chem.* **260,** 449 (1985).
[50] M. P. Deutscher, *CRC Crit. Rev. Biochem.* **17,** 45 (1984).
[51] J. G. Seidman, B. G. Barrell, and W. H. McClain, *J. Mol. Biol.* **99,** 733 (1975).

studies in an extract active in transcription and RNA processing and from *in vivo* analyses by Northern blots of RNA isolated from transformed strains. Initially, the tRNA gene must be transcribed by RNA polymerase III. Then 5'- and 3'-processing occur in an order depending on the particular tRNA substrate, after which the intron, if present, is removed.[50] Presumably, nucleotide modifications by specific enzymes are occurring during these processes *in vivo*.[52] Transport of tRNAs out of the nucleus appears to require that the tRNA is mature sized, since tRNA precursors are not found in the cytoplasm.[53,54] Aminoacylation by a specific aminoacyl-tRNA synthetase is followed by elongation (or initiation) factor binding. This precedes ribosomal binding and proofreading during translation. The loss of suppression is often the result of progressive diminution in the amounts of successive pathway intermediates by a reduction in their stability, a specific block, or a combination of these effects. Generally, the amount of a mature tRNA that will be produced is inversely proportional the time required for a tRNA precursor to be processed to the mature form.

Because the shape and function of tRNA molecules are dependent on every nucleotide, owing to the small size of tRNAs, almost any nucleotide change can inactivate them. In almost all cases, changes in the stem regions which remove a Watson–Crick base pair also inactivate the tRNA. Mutations which can retain G-U pairing in the stems (e.g., G_7, T_{40}, $T_{47:6}$) generally retain suppressor activity and seem to cause little detriment at the tRNA level, except for increased temperature sensitivity. This would seem to be a plausible mechanism for tRNAs or any other structured RNA to allow evolutionary sequence change without necessarily requiring a multigene family. No *sup3-e* tRNA mutants have been confirmed in misaminoacylation since a proper screening assay has not been devised. Therefore, nucleotide changes in the tRNA that permit sufficient charging by the seryl-tRNA synthetase or some other aminoacyl-tRNA synthetase will be detected as suppressor active. Future studies will focus on the structures, expression, and mechanisms of action of specific enzymes and factors responsible for tRNA expression.

Acknowledgments

We would like to thank Urs Leupold and Jürg Kohli for experimental support and encouragement. This research was supported by a grant from the National Institutes of Health.

[52] A. K. Hopper, *in* "Processing of RNA" (D. Apirion, ed.), p. 91. CRC Press, Boca Raton, Florida, 1984.
[53] A. K. Hopper, F. Banks, and V. Evagelidis, *Cell* **14,** 211 (1978).
[54] A. K. Hopper, L. D. Schultz, and R. A. Shapiro, *Cell* **19,** 741 (1980).

[32] Genetics of Transfer RNA Processing in *Escherichia coli*

By YOSHIRO SHIMURA and HIDEAKI SHIRAISHI

For the study of tRNA processing in *Escherichia coli*, the genetic approach has been shown to be useful, as has been the case with other biosynthetic pathways.[1-4] The major strategy in the genetic approach is, of course, to isolate mutants that have defects in tRNA processing. Analyses of such mutants should provide vital information about tRNA processing. Several approaches have been used to isolate mutants defective in the biosynthesis of tRNA molecules.[5-7] Since mutational blocks in tRNA biosynthesis might well be lethal to cells, we must look for conditional lethal mutants in which the defect in tRNA synthesis occurs only under restrictive conditions (e.g., at high temperature). A problem in the isolation of such mutants is that mature tRNA molecules, synthesized before the temperature shift, are still active in protein synthesis even at the restrictive temperature. Thus, preexisting tRNA molecules would interfere with the identification of mutants defective in tRNA maturation at the restrictive temperature. The procedures described here[6] were devised to overcome this impasse by introducing a transducing phage carrying a suppressor tRNA gene after the temperature shift, followed by screening those cells incapable of synthesizing the suppressor tRNA at high temperature. Mutants in tRNA synthesis were enriched under conditions where cells expressing suppressor activity were killed by two virulent phages.

Principle of Mutant Isolation

The parental strain *E. coli* 4273 carries mutations in the *lac* gene and in the genes specifying the receptors for two virulent phages, T6 and BF23. Phage λcI857 was lysogenized to *E. coli* 4273, and the resulting *E. coli* 4273 (λcI857) was mutagenized with *N*-methyl-*N'*-nitro-*N*-nitrosoguani-

[1] J. Abeleson, *Annu. Rev. Biochem.* **48**, 1035 (1979).

[2] J. Smith, *Progr. Nucleic Acid Res. Mol. Biol.* **16**, 25 (1976).

[3] Y. Shimura and H. Sakano, *in* "Nucleic Acid–Protein Recognition" (H. J. Vogel, ed.), p. 293. Academic Press, New York, 1977.

[4] Y. Shimura and H. Ozeki, *Adv. Biophys.* **4**, 191 (1973).

[5] P. Shedl and P. Primakoff, *Proc. Natl. Acad. Sci. U.S.A.* **70**, 2091 (1973).

[6] H. Sakano, S. Yamada, T. Ikemura, Y. Shimura, and H. Ozeki, *Nucleic Acids Res.* **1**, 355 (1974).

[7] M. P. Deutscher, C. W. Marlor, and R. Zaniewski, *Proc. Natl. Acad. Sci. U.S.A.* **82**, 6427 (1985).

dine (NG). The NG-treated cells were infected with phage λpsu^+3, which carries a single copy of the $tRNA^{Tyr}$ gene, and subsequently incubated with phages T6 and BF23 at 42°. Those cells that could make functional su^+3 $tRNA^{Tyr}$ were thus eliminated via infection by the two virulent phages. The heat-inducible $\lambda cI857$ lysogen was employed to exclude cells that escaped λpsu^+3 infection. The introduction of the cI^+ gene into the cells along with λpsu^+3 infection is essential to rescue them from thermal induction of $\lambda cI857$. In this way, uninfected cells that would behave like the desired mutants are eliminated together with those cells capable of synthesizing the suppressor $tRNA$, thereby enriching the mutants.

Materials

Bacteria and Phages. *Escherichia coli* strain 4273 (su^-, lac_{am}, $T6^r_{am}$, $BF23^r_{am}$) is used as the parental strain.[6] λpsu^+3, which carries the su^+3 gene, is constructed from $\phi 80psu^+3$ (Cambridge strain[8]) by replacing both sides of the su^+3 transducing segment with the λ genome.[6] $\phi 80psu^+3$ used in other experiments is a Kyoto strain described by Andoh and Ozeki.[9]

Media and Buffers. The λ broth used for bacterial growth and phage propagation contains, in 1 liter, 10 g Bacto-tryptone (Difco) and 2.5 g NaCl adjusted to pH 7.2 with HCl. The selection medium used is λ broth supplemented with 40 $\mu g/ml$ L-tryptophan, 0.1 M NaCl, and 1 mM CaCl$_2$. M9–casaminoglycerol contains, in 1 liter, 6 g Na$_2$HPO$_4$, 3 g KH$_2$PO$_4$, 0.5 g NaCl, 1 g NH$_4$Cl, 2 mM MgSO$_4$, 0.1 mM CaCl$_2$, 2 g casamino acid (Difco), and 0.2% glycerol. Media are solidified with 1.2% agar powder. The low-phosphate medium used for ^{32}P-labeling contains, in 1 liter, 1.5 g KCl, 5 g NaCl, 1 g NH$_4$Cl, and 12.1 g Trizma base [tris(hydroxymethyl)-aminomethane; Sigma, St. Louis, MO] adjusted to pH 7.4 with HCl. Before use, KH$_2$PO$_4$ (0.1 mM), MgSO$_4$ (1 mM), and 0.02 volume of low-phosphate Bacto-peptone are added. Low-phosphate Bacto-peptone is prepared as follows[10]: 20 g of Bacto-peptone (Difco) is dissolved in 150 ml of water and adjusted to pH 10 with concentrated NH$_4$OH. Then 1 ml of 1 M MgCl$_2$ is added, and the solution is chilled on ice for 60 min. The solution is then centrifuged at 0°, after which the upper two-thirds of the supernatant is recovered and adjusted to pH 7.4 with HCl. Adsorption of

[8] R. L. Russell, J. N. Abelson, A. Landy, M. L. Gefter, S. Brenner, and J. D. Smith, *J. Mol. Biol.* **47**, 1 (1970).

[9] T. Andoh and H. Ozeki, *Proc. Natl. Acad. Sci. U.S.A.* **59**, 792 (1968).

[10] G. Pieczenik, B. G. Barrell, and M. L. Gefter, *Arch. Biochem. Biophys.* **152**, 152 (1972).

$\phi80$ was carried out in $\phi80$ adsorption buffer, which contains 10 mM Tris-Cl (pH 7.3), 10 mM MgSO$_4$, 0.5 mM CaCl$_2$, and 0.001% gelatin.

Methods

Mutagenesis. Escherichia coli strain 4273(λcI857) is grown to 2 × 10^8 cells/ml in λ broth, washed twice with Tris–malate buffer (pH 6.0), and treated with 200 μg/ml of NG in the same buffer at 30° for 30 min. The NG-treated cells are washed once with λ broth and subsequently grown overnight at 30°.

Screening. The NG-treated *E. coli* 4273(λcI857) cells are grown to 2 × 10^8 cells/ml, washed once with 10 mM MgSO$_4$, and suspended in the same solution to give a density of 2 × 10^9 cells/ml. The cells are infected with λp*su$^+$3* at a multiplicity of 10 at 42°. Ten minutes after infection, the cells are diluted with selection medium (prewarmed at 45°C) to 2 × 10^8 cells and incubated at 42° with shaking. Thirty minutes after dilution, T6 and BF23 are added at a multiplicity of 20 for each phage and subsequently incubated for 3 hr at 42° with shaking. The number of viable cells after these treatments is about 1–10% of the original number. The culture is spread on λ plates after adequate dilutions.

Among the survivors of phage selection, those cells showing temperature-sensitive growth are selected by replica plating, because mutants unable to synthesize tRNA at the high temperature are expected to be, in all likelihood, temperature sensitive for growth. Since most of the survivors are still lysogenic for λcI857, the growth test is performed after superinfection with λcI$^+$ to antagonize thermal induction of prophage λcI857. About 25% of the survivors are unable to grow at 42°.

In order to screen the temperature-sensitive mutants of tRNA function, the amount of β-galactosidase synthesized by these mutants at 42° after $\phi80$p*su$^+$3* infection is measured. Since the parental strain carries the *lac*$_{am}$ mutation, the synthesis of β-galactosidase in the cells after $\phi80$p*su$^+$3* infection depends on the synthesis of *su$^+$* tRNA at 42°. On the other hand, when the cells are infected with $\phi80$p*lac*, β-galactosidase might be synthesized even after the temperature shift, since tRNA molecules made at 30° may still function. Accordingly, the mutants in tRNA function should be those that synthesize β-galactosidase at 42° after $\phi80$p*lac* infection but not after $\phi80$p*su$^+$3* infection. Typically, of 2000 mutants showing temperature-sensitive growth, about 10% prove to satisfy this criterion.

Phage infection is performed as follows. Cells grown to 2 × 10^8 cells/ml at 30° in M9–casaminoglycerol are concentrated in the $\phi80$ adsorption

TABLE I
SYNTHESIS OF β-GALACTOSIDASE AFTER INFECTION OF
TRANSDUCING PHAGES[a]

	E. coli ts241		E. coli 4273	
Cells	30°	42°	30°	42°
ϕ80psu^+3 infected	40	1	53	51
ϕ80plac infected	167	53	220	113
ϕ80psu^+3 lysogen	34	15	56	47

[a] β-Galactosidase activity was calculated by the following formula:

Enzyme activity =
$$(A_{420} - 1.65A_{550}) \times 10^3/\text{reaction time (min)}$$

buffer and infected with appropriate phage at a multiplicity of 10. The infected cells are diluted with the same medium to 2×10^8 cells/ml and incubated at 30 or 42°. For both the infected cells and the lysogen, isopropyl-β-D-thiogalactoside (0.5 mM) is added 5 min after the shift of temperature. Enzyme activity is measured at 30 min after induction. The enzyme is assayed by the method of Pardee et al.[11]

In Table I, the amounts of β-galactosidase synthesized in one of the mutants (ts241) after infection with ϕ80psu^+3 or ϕ80plac are compared with the amounts produced by parental strain (4273).[12] When this mutant is previously lysogenized with ϕ80psu^+3, thus allowing the synthesis of su^+3 tRNA, β-galactosidase is synthesized even at 42°. Therefore, the temperature-sensitive synthesis of β-galactosidase after ϕ80psu^+3 infection is not due to a temperature-sensitive mutation in the lac gene of the cell. Essentially the same kinetic pattern of β-galactosidase synthesis is observed with other mutants. These results strongly suggest that the synthesis of su^+3 tRNA is temperature sensitive in the mutants. For further experiments, cured strains missing λcI857 are selected from each of the mutants by examining the survivors after a heat-pulse treatment.

Analysis of Mutants

To analyze tRNA synthesis in the isolated mutants, RNAs accumulated in them are labeled with ^{32}P. Each strain, grown to 2×10^8 cells/ml

[11] A. B. Pardee, F. Jacob, and J. Monod, J. Mol. Biol. 1, 165 (1959).
[12] H. Ozeki, H. Sakano, S. Yamada, T. Ikemura, and Y. Shimura, Brookhaven Symp. Biol. 26, 89 (1975).

in 4 ml of low-phosphate medium, is divided into two tubes, and one is transferred to 43° while the other is kept at 30° (time 0). At 10 min, [^{32}P]orthophosphate is added to give a final concentration of 0.2 mCi/ml. At 60 min, labeling is terminated by the addition of an equal volume of warmed phenol (45°) saturated with 0.1 M Tris-Cl (pH 8.5). The mixture is shaken vigorously for 30 min at room temperature, then centrifuged for 30 min at 5000 g. Nucleic acids are recovered from the aqueous phase by ethanol precipitation, and small RNAs are fractionated by electrophoresis in a 10% polyacrylamide gel. The ratio of acrylamide to bisacrylamide in the gel is 29:1 (w/w). Electrophoresis is carried out in 0.1 M Tris–acetate buffer (pH 8.3) at 4°.

Many RNA species which had never been detected in wild-type cells were accumulated in some of the mutants incubated at 43°.[6] These RNAs were shown to be tRNA precursors specific to RNase P mutants, as judged by RNA fingerprint analysis and by *in vitro* cleaving reactions.[13,14] Genetic analyses of the mutants have shown that the RNase P activity of *E. coli* is specified by at least two loci, *rnpA* and *rnpB*.[15] These genes code for the protein component and the RNA component of the enzyme, respectively.

A complication in the genetic analysis of the RNase P mutants is that the *rnpB* gene carried by relaxed (multicopy) plasmids complements not only *rnpB* mutants but also *rnpA* mutants (e.g., strains A49[16] and ts241[17]), although the *rnpB* gene on the F′ factor never complements *rnpA* mutants.[15] To avoid the dosage effect, we employed the F replicon to introduce the *rnpB* gene into cells when we carried out a complementation test between isolated genes and an RNase P mutant, ts709.[18] This is achieved by introducing a *polA*⁻ mutation into ts709 by P1 transduction. A DNA fragment which carries the gene to be examined is inserted into a composite plasmid of pBR322 and mini-F, and the resulting plasmids are introduced into ts709*polA*. This plasmid replicates using the replication origin of pBR322 in wild-type (*polA*⁺) cells. In the ts709*polA* mutant, however, it replicates as a stringent plasmid using the replication origin of F factor since the pBR322 replicon is inactive in a *polA*⁻ background. Using this system, the complementation test is unambiguously carried out. A close

[13] T. Ikemura, Y. Shimura, H. Sakano, and H. Ozeki, *J. Mol. Biol.* **96,** 69 (1975).
[14] H. Sakano and Y. Shimura, *J. Mol. Biol.* **123,** 287 (1978).
[15] Y. Shimura, H. Sakano, S. Kubokawa, F. Nagawa, and H. Ozeki, *in* "Transfer RNA: Biological Aspects" (D. Söll, J. N. Abelson, and P. R. Schimmel, eds.), p. 43. Cold Spring Harbor Laboratory, Cold Spring Harbor, New York, 1980.
[16] S. K. Jain, M. Gurevitz, and D. Apirion, *J. Mol. Biol.* **162,** 515 (1982).
[17] H. Shiraishi, unpublished observation.
[18] H. Shiraishi and Y. Shimura, *EMBO J* **5,** 3673 (1986).

examination of the structure–function relationship of the RNA component of RNase P is also carried out using the *rnpB* mutants, including those isolated by the method described here.[18-20] Among the mutants isolated by this method, only RNase P mutants have been extensively characterized. If there are any unidentified processing enzymes that are essential for tRNA synthesis, they should be isolated by the procedure described above.

[19] H. Sakamoto, N. Kimura, F. Nagawa, and Y. Shimura, *Nucleic Acids Res.* **11**, 8237 (1983).
[20] H. Shiraishi and Y. Shimura, *EMBO J.* **7**, 3817 (1988).

[33] Genetic Methods for Study of Trans-Acting Genes Involved in Processing of Precursors to Yeast Cytoplasmic Transfer RNAs

By ANITA K. HOPPER

Introduction

Efficient processing of precursors to transfer RNAs (pre-tRNAs) has two major requirements: appropriate substrates and particular trans-acting components. The trans-acting components include catalytic activities and hypothetical ancillary factors possibly involved in facilitating proper pre-tRNA conformation, efficient enzyme–substrate interactions, or appropriate subcellular localization of substrates or catalytic activities. It has been possible to use a genetic approach to probe the features of the substrate that are necessary for efficient processing. Such an approach is described by Nichols *et al.*[1] This chapter addresses the genetic methods to identify trans-acting products involved in pre-tRNA processing and/or modification and is restricted to those genes involved in the processing of nuclear-encoded tRNAs. Many different strategies for mutant hunts have been employed to identify lesions of yeast genes that cause defects in pre-tRNA processing. These have been organized into four categories: (1) screens for conditional lethal yeast defective in pre-tRNA processing; (2) screens assessing expression of nonsense suppressors; (3) correction of processing-defective substrates; and (4) reverse genetics. Each of these

[1] M. Nichols, I. Willis, and D. Söll, this volume [31].

strategies has proved successful as summarized in the Appendix. Several lesions that affect removal of intervening sequences (IVS) from pre-tRNA, tRNA modification, and nucleotidyltransferase have been described. However, to date no lesions that affect 5' or 3' nucleolytic cleavages have been identified. Furthermore with the exceptions of tRNA ligase, nucleotidyl transferase, and various tRNA modification activities, the exact biochemical roles of many of the identified genes have remained elusive.

Strategies to Identify Mutations in Genes Important to tRNA Processing

Screening Conditional Lethal Yeast for Defects in Pre-tRNA Processing

There have been a number of efforts to screen collections of conditional lethal yeast for isolates that are unable to process pre-tRNA at the nonpermissive temperature. These screens are based on the premise that mutations causing defective processing of pre-tRNA will have lethal consequences. The *rna1-1*,[2] *sen1-1*,[3] *sen2-1*,[3] and *sen2-2* (M. Winey and M. Culbertson, personal communication) alleles have been identified from screens of the collection generated by Hartwell,[4] and the *sen2-3 cca1-1*, and *pta1-1* alleles have been found in independent collections generated by Vijayraghavan *et al.*[4a] and J. O'Connor and C. Peebles (personal communication).

Assay of in Vivo Incorporation of Radioisotopes into Precursor Species. Although tedious, assay of *in vivo* incorporation of radioisotopes into precursor species allows for inspection of only those RNAs generated at nonpermissive conditions as well as assessment of the stability of particular RNA species. Furthermore, the *in vivo* labeled RNAs can be isolated and utilized as substrates for *in vitro* processing reactions.[5,6] This approach originally demonstrated that the *rna1-1* mutant accumulates pre-tRNAs.[2]

The screening procedure involves labeling exponentially growing mutant yeast at nonpermissive conditions, isolating RNA, resolving the RNA species by electrophoresis, and monitoring precursor species by

---

[2] A. K. Hopper, F. Banks, and V. Evangelidis, *Cell* **14**, 211 (1978).

[3] M. Winey and M. R. Culbertson, *Genetics* **118**, 609 (1988).

[4] L. H. Hartwell, *J. Bacteriol.* **93**, 1662 (1967).

[4a] U. Vijayraghavan, M. Company, and J. Abelson, *Genes Dev.* **3**, 1206 (1989).

[5] G. Knapp, J. Beckmann, P. Johnson, S. Fuhrman, and J. Abelson, *Cell,* **14**, 221 (1978).

[6] P. Z. O'Farrell, B. Cordell, P. Valenzuela, W. J. Rutter, and H. M. Goodman, *Nature, (London)* **274**, 438 (1978).

radiography or fluorography. Many variations on this theme have been used successfully. Both [^{3}H]uracil[2] and ortho[^{32}P]phosphate[5,6] have been used to label pre-tRNA species. For [^{3}H]uracil labeling, approximately 10 ml of yeast cells are grown in complete synthetic medium to early log phase ($\sim1 \times 10^7$ cells/ml). Ninety minutes prior to labeling, the cells are collected onto 0.45-μm Millipore filters and resuspended in synthetic medium containing 0.25 μg/ml uracil. The cells are shifted to the nonpermissive temperature (34–37°), and 5–30 min later 5–25 μCi/ml [5,6-^{3}H]uracil is added to the culture and incubated for 5 (pulse label) to 60 min before harvesting the cells. For ^{32}P labeling, the procedure is similar except that cultures are grown in rich medium depleted for phosphate by Mg^{2+} precipitation.[7] Mutants defective in pre-tRNA processing at the nonpermissive temperature should accumulate radiolabeled precursor species under these conditions.

After labeling, the cells are rapidly chilled to 4° and collected by centrifugation at 4°, and low molecular weight RNAs are isolated by phenol extraction of whole cells.[7] Cell pellets are resuspended in 3 ml of ice-cold TSE [10 mM tris(hydroxymethyl)aminomethane (Tris), 10 mM ethylenediaminetetraacetic acid (EDTA), 0.1 M NaCl (pH 7.5], and 3 ml of TSE–saturated phenol is added. The mixture is shaken at 35° for about 1 hr and incubated at 4° for 2 hr. The phases are separated by centrifugation and the aqueous phase reextracted with phenol. RNA is collected by precipitation with ethanol at −20°, reprecipitated one time, and then redissolved in approximately 50 μl of distilled water, and the specific activity is determined. Either equal quantities (~10 μg) or equal amounts of radioactivity ($\sim2 \times 10^5$ cpm) of each sample are resolved by electrophoresis on 28-cm-long, 0.75-mm-thick 10% polyacrylamide gels containing 7 M urea, 90 mM Tris–borate, 2.5 mM EDTA (pH 8.3) at 200 V for 18 hr. For ^{32}P-labeled RNAs, polyphosphates are removed prior to electrophoresis.[5]

[^{3}H]Uracil-labeled RNA species are detected by fluorography. Following electrophoresis, the gel is embedded with 2,5-diphenyloxazole (PPO)[8] and dried. The dried gel is placed on X-ray film and exposed at −70°. ^{32}P-Labeled RNAs are detected by autoradiography.

RNA Blot Analysis. RNA blot analysis is another approach to screen conditional lethal yeast for defects in tRNA processing. One advantage of this approach is the relative ease of the procedure. A potential disadvantage is that both steady-state RNAs and those species generated under nonpermissive conditions are detected together, and there are considerable amounts of initial tRNA transcripts and processing intermediates in

[7] G. M. Rubin, *Methods Cell Biol.* **12,** 45 (1975).
[8] W. M. Bonner and R. A. Laskey, *Eur. J. Biochem.* **46,** 83 (1974).

logarithmically grown wild-type cells.[9] If, however, the mutation in question causes stable accumulation of pre-tRNA species, the relative increase in quantity of a particular pre-tRNA species can be readily detected. In fact, the pre-tRNA species accumulating in *rna1-1* cells have been detected by RNA blot analysis,[10] and *sen2-3*[10a] and *pta1-1* lesions have been identified using this approach (J. O'Connor and C. Peebles, personal communication, 1989). Interestingly, a mutation of the *CCA1* locus that encodes nucleotidyltransferase was first discovered as a lesion that extends mRNA half-life; presumably, a deficiency in nuceloti-dyltransferase slows down protein synthesis which, in turn, slows down mRNA turnover.[10b]

The screens are variations of the Northern or dot-blot procedures.[11] For Northern analysis, RNAs isolated from cells incubated at the nonpermissive temperature as described above are resolved by polyacrylamide gel electrophoresis. The RNAs are transferred electrophoretically from the gel to a Gene Screen Plus (Du Pont Co.) or alternative membrane. Membranes are prehybridized in $4\times$ SSC ($1\times$ SSC is 0.15 M NaCl plus 15 mM sodium citrate), 20 mM EDTA, 0.5% sodium dodecyl sulfate (SDS), and 100–500 μg/ml heat-denatured fragmented salmon sperm DNA. The blots are hybridized in the same buffer with oligonucleotide probes that are end-labeled with [γ-^{32}P]ATP using T4 polynucleotide kinase[12] at hybridization temperatures below the melting point (T_m) of each DNA–RNA hybrid.[13] Alternatively, the RNA species can be detected by hybridization to plasmid DNA containing tRNA genes radiolabeled with [^{32}P]dCTP or [^{32}P]dTTP by nick translation.[12] Blots are exposed to XAR-5 X-ray film (Eastman Kodak Co., Rochester, NY) at $-70°$. Before rehybridization with a second probe, the radioactive probes are stripped from the membrane by boiling in 0.1$\times$ SSC and 1% SDS for 5–30 min.

J. O'Connor and C. Peebles (personal communication, 1988) have circumvented the need to resolve RNA species on polyacrylamide gels by directly applying RNA samples onto filters (dot blots) and probing the blots with synthetic oligonucleotides corresponding to intervening sequences. The obvious advantage of this is that the screening procedure is

[9] A. K. Hopper and J. Kurjan, *Nucleic Acids Res.* **9**, 1019 (1981).

[10] S. S. Wang and A. K. Hopper, *Mol. Cell. Biol.* **8**, 5140 (1988).

[10a] C. K. Ho, R. Rauhut, U. Vijayraghavan, and J. Abelson, submitted for publication (1990).

[10b] M. Aebi, G. Kirchner, J.-Y. Chen, U. Vijayraghavan, A. Jacobson, N. Martin, and J. Abelson, in preparation.

[11] P. S. Thomas, this series, Vol. 100, p. 255.

[12] T. Maniatis, E. F. Fritsch, and J. Sambrook, "Molecular Cloning: A Laboratory Manual," p. 122. Cold Spring Harbor Laboratory, Cold Spring Harbor, New York, 1982.

[13] G. A. Beltz, K. A. Jacobs, T. H. Eickbush, P. T. Cherbas, and F. C. Kafatos, this series, Vol. 100, p. 266.

streamlined. The disadvantage is that the actual step at which processing is blocked is not revealed, and only quantitative and not qualitative changes in intron-containing sequences can be monitored.

Assessment of Cellular Extracts for Particular tRNA Processing Activities. Assessment of cellular extracts for RNA processing activities is a third approach for screening conditional lethal yeast for defects in pre-tRNA processing. One potential problem with this approach is the difficulty of the screen. That is, one must prepare cellular extracts from many mutant candidates, use the extracts in *in vitro* assays, and monitor activity by inspection of products resolved by polyacrylamide electrophoresis. The advantage of this approach, however, is that it allows one to focus in on the particular processing step of interest. For example, Winey and Culbertson (personal communication, 1988),[3] by assaying for tRNA splicing endonuclease activity, identified the *sen1-1* and the *sen2-1* and *sen2-2* alleles. The following procedure has been utilized by Winey and Culbertson[3] to identify mutants defective in tRNA splicing endonuclease:

Preparation of Extracts. Approximately 1 ml of log-phase cells are collected by centrifugation in a 1.5-ml microcentrifuge tube. The cells are resuspended and incubated for 15 min at room temperature in 35 μl of buffer [1.0 M sorbitol, 50 mM Tris (pH 7.9), 10 mM MgCl$_2$, 10 mM dithiothreitol]. The cells are again pelleted and resuspended in 35 μl of the same buffer containing 0.1 mM dithiothreitol and 0.15 mg/ml Zymolyase 60,000 or 100,000 (Kirin). Following incubation for 1 hr at room temperature with gentle agitation, the spheroplasts are collected by centrifugation at 4°, and 100 μl ice cold lysing buffer [15 mM KCl, 10 $\mu$$M$ N-2-hydroxy-ethylpiperazine-N'-2-ethanesulfonic acid (HEPES) (pH 7.9), 5 mM MgCl$_2$, 0.1 mM EDTA, 3 mM dithiothreitol] is added to the pellet. The resuspended spheroplasts are lysed by grinding with a pestle (Kontes, Chicago, IL) at 4°, and the lysate is spun in a microcentrifuge for 10 sec. The supernatant is discarded, and the pellet containing nuclei is resuspended in 20 μl ice-cold tRNA endonuclease buffer [25 mM Tris (pH 8.0), 0.5 mM EDTA, 100 mM (NH$_4$)$_2$SO$_4$, 10 mM spermidine, 1% v/v Triton X-100, 5.0 mM 2-mercaptoethanol, 0.5 mM phenylmethylsulfonyl fluoride, 20% v/v glycerol] and used for enzyme assays.

Substrates. The end-matured pre-tRNA substrates containing IVS can be obtained by four different methods: (1) by *in vivo* labeled pre-tRNA species that accumulate in *rna1-1* cells at nonpermissive temperature (see above); (2) by labeling steady-state pre-tRNA *in vitro* utilizing nucleotidyltransferase;[14,15] (3) by *in vitro* transcription and partial processing of

[14] M.-C. Lee and G. Knapp, *J. Biol. Chem.* **260**, 3108 (1985).
[15] H. Sternbach, F. von der Haar, E. Schlimme, E. Gaentner, and F. Cramer, *Eur. J. Biochem.* **22**, 166 (1971).

pre-tRNAs, using an RNA polymerase III transcription system,[16] an IVS-containing tRNA gene, and [^{32}P]UTP (Amersham, Arlington Heights, IL, 400 Ci/mmol); or (4) by T7-directed transcription of a synthetic construct containing a T7 promotor adjacent to a pre-tRNA gene sequence.[16a] Labeled pre-tRNAs are eluted from polyacrylamide gels and ethanol-precipitated prior to use in the endonuclease reaction.

Reactions. Thirty-five-microliter reactions containing 7 μl extract, 40 mM Tris-HCl (pH 8.0), 0.4 mM EDTA, 20 mM (NH$_4$)$_2$SO$_4$, 6 mM spermidine, 1 mM dithiothreitol, 0.2% (v/v) Triton X-100, 10% (v/v) glycerol, and 3.5 × 10^{-5} mM pre-tRNAs were assembled on ice. Aliquots of 10 μl are incubated at the desired temperature, 25, 30, or 34°, for 15 min. The reaction is monitored by resolving the labeled RNA species on polyacrylamide gels followed by autoradiography. Radiolabeled precursor species should be processed to 5' halves, 3' halves, and IVS species unless the extract lacks splicing endonuclease.

Screens Utilizing Inability to Express Suppressor tRNAs

As indicated above, screening conditional lethal yeast for defects in pre-tRNA processing has proved effective for the identification of yeast genes important to this process. However, since conditional lethal yeast can contain lesions in any essential biosynthetic or metabolic process, other types of screens that might start with collections enriched for members defective in tRNA processing have also been designed. Furthermore, such procedures have served to identify genes that code for products unessential for generating mature species. Such genes are thought to encode tRNA modification activities and/or activities that perform ancillary roles in pre-tRNA processing. Screening for the ability of yeast to express suppressor tRNAs is one such alternative approach to identify mutants defective in pre-tRNA processing. The *los1, mod5-1, sin1, sin3,* and *tpd5-47* alleles have been found using variations of this screen. The screen for loss of ability to suppress nonsense suppressors is based on the fact that nonsense suppressors of yeast are altered tRNAs that insert amino acids at translational termination codons. This led to the expectation that lesions which affect nonsense suppression could be encoded by lesions of the suppressing tRNA gene or by mutant genes involved in the production of functional suppressor tRNA.

There are a large number of *Saccharomyces cerevisiae* and *Schizosaccharomyces pombe* suppressors that can be utilized for such screens (re-

[16] D. R. Engelke, P. Gegenheimer, and J. Abelson, *J. Biol. Chem.* **260,** 1271 (1985).
[16a] V. M. Reyes and J. Abelson, *Anal. Biochem.* **166,** 90 (1987).

viewed by Hawthorne and Leupold[17]). For *S. cerevisiae* the following nomenclature is adopted. An uppercase three-letter code followed by superscript plus (i.e., $SUPX^+$) indicates the wild-type locus coding for an aminoacylating tRNA species, an uppercase code without a superscript (i.e., $SUPX$) indicates a nonsense suppressor, and an uppercase code followed by a superscript minus (i.e., $SUPX^-$) indicates an altered non-sense suppressor that is not active or only partially active. The *S. cerevisiae* amino acid-inserting nonsense suppressors have been placed into categories based on efficiencies of suppression. Such efficiency is presumably related to the expression of the tRNAs. For example, the tyrosine-inserting UAA suppressors ($SUP2$ to $SUP8$ and $SUP11$) are very efficient, as is the serine-inserting UAG SUP-$RL1$ suppressor; serine-inserting suppressors $SUP16$, $SUP17$, and $SUP19$ show moderate to low efficiency; leucine-inserting UAG suppressors $SUP52$ to $SUP56$ show low efficiency; and the leucine-inserting UAA suppressors $SUP26$ to $SUP29$, $SUP32$, and $SUP33$ show very low efficiency.[18] The choice of a suppressor will be dependent on not only its efficiency and spectrum of nonsense alleles available (see below) but also whether the locus contains an IVS. The $SUP2^+$ to $SUP8^+$, $SUP11^+$, $SUP53^+$ (presumably also $SUP52^+$ and $SUP54^+$ to $SUP56^+$), and SUP-$RL1^+$ loci contain IVS sequences, whereas the $SUP16^+$, $SUP17^+$, and $SUP19^+$ loci do not.

To avoid isolation of mutations at the loci encoding the nonsense mutations and to assess consequences of the mutation in question on nonsense suppression, generally two or more nonsense alleles are scored at the same time. For *S. cerevisiae* a large number of nonsense alleles have been characterized. Different alleles can be suppressed to different extents; the efficiencies of suppression for various alleles with ochre mutations is summarized in Table I. Presumably, the extent of suppression is related to the ability of a given amino acid replacement to restore full enzyme activity as well as to the quantity of a gene product necessary for particular reactions. The choice of which nonsense alleles to score depends on both the ease of assay and the efficiency of suppression. Two of the most convenient ochre nonsense alleles, *ade2-1º* and *can1-100º* (the superscript o indicates an ochre mutation; see below for details), are also two of the most difficult alleles to suppress (see Table I). The difficulty of suppression can be used to an advantage since mutations that cause as little as 50% reduction in $SUP2º$ to $SUP8º$ expression affect suppression

[17] D. C. Hawthorne and U. Leupold, *Curr. Top. Microbiol. Immunol.* **64,** 1 (1974).
[18] F. Sherman, *in* "The Molecular Biology of the Yeast *Saccharomyces:* Metabolism and Gene Expression" (J. N. Strathern, E. W. Jones, and J. R. Broach, eds.), p. 463. Cold Spring Harbor Laboratory, Cold Spring Harbor, New York, 1982.

TABLE I

SUPPRESSION OF OCHRE (UAA) ALLELES BY COMMONLY UTILIZED *Saccharomyces cerevisiae* NONSENSE SUPPRESSORS[a]

| Allele | Tyrosine inserting | | Serine inserting | | | | | | Leucine inserting SUP27° |
| | SUP2°-SUP8° | SUP11° | SUP16° | | SUP17° | | SUP19° | | |
			ψ^{+b}	ψ^-	ψ^+	ψ^-	ψ^+	ψ^-	
trp5-2°	+[c]	±	−	−			−	−	
ade2-1°	+	±	+	−	+	−	±	−	−
can1-100°	+	+	±	−	−	−	−	−	−
lys2-52°	+	+	−	−	−	−	−	−	−
tyr1-1°	±	±							
lys1-1°	+	+	+	−	+	−	+	−	−
lys2-1°	+	+							
his5-2°	+	+	+	±	+	−	+		−
met4-1°	+	+	+	+	+	+			+
his4-1°	+	+	+	+	+	+			+
arg4-17°	+	+	+	+	+	+			+
leu2-1°	+	+	+	+	+	+	+	+	+
trp5-48°	+	+	+	+					
ilv1-2°	+	+	+		+		+		+
ura4-1°	+	+	−	−	−	−	−	−	+
his4-1176°	−	−	+	+	+	+	+	+	−
lys2-255°	−	−	+	+	+	+	+	+	+
cyc1-72°		60%[d]	30%	<3%	20%	<3%	20%	<3%	
cyc1-9°		<3%	<3%	<3%	<3%	<3%			

[a] Information derived from Hawthorne and Leupold,[17] Sherman,[18] and B.-I. Ono, N. Willis, J. W. Stewart, R. F. Gesteland, and F. Sherman, *J. Mol. Biol.* **150**, 361 (1981).
[b] ψ^+ is a cytoplasmically inherited element that enhances the efficiency of suppression.
[c] +, Good suppression; ±, poor suppression; −, no suppression.
[d] Numbers indicate approximate percentage of normal levels of 1-isocytochrome *c*.

of *ade2-1°* and *can1-100°*. Thus, mutations can be categorized as strong if they affect both "easy and difficult to suppress" nonsense alleles or weak if they affect only "difficult to suppress" alleles. Strains containing various nonsense suppressors and suppressible alleles are available from the Yeast Genetics Stock Center (Department of Biophysics and Medical Physics, University of California, Berkeley, CA 94720) and workers in the field.

"*ade–can*" *Screening Procedure.* Wild-type yeast cells are sensitive to canavanine, an analog of arginine. Mutations in the *CAN1* locus, the gene encoding arginine permease, prevent uptake of arginine and canavanine and thereby allow cells to grow in the presence of canavanine.

Suppression of a nonsense mutation at the *can1* locus (*can1-100*°) results in drug sensitivity; on the other hand, loss of suppressor activity results in drug resistance. The *ade2-1*° allele leads to the formation of red colonies owing to defects in adenine biosynthesis.[19] Nonsense suppression of *ade2-1*° leads to white colonies, and loss of suppression results in red colonies again. The *can1-100*° and *ade2-1*° alleles, then, provide for a double screen for the loss of suppressor activity[20] (Table II).

The following adaptation of this screen has been utilized to identify mutants defective in pre-tRNA processing.[21] A yeast strain (relevant genotype *SUP4*° *ade2-1*° *can1-100*° *lys1-1*° *his5-2*° *trp5-48*°) was subjected to ethylmethane sulfonate mutagenesis[22] and plated for single colonies at a permissive temperature (23°) on rich medium [YEPD; 2% w/v glucose, 2% w/v Bacto-peptone (Difco), 1.5% w/v agar]. The colonies on this master plate were replicated to two plates containing all requirements except adenine [complete minus adenine or −*ade* media contains 2% w/v glucose, 0.67% w/v yeast nitrogen base minus amino acids (Difco), 40 μg/ml uracil, 10 μg/ml histidine, 60 μg/ml leucine, 40 μg/ml lysine, 10 μg/ml methionine, 10 μg/ml tryptophan, 60 μg/ml isoleucine, 50 μg/ml threonine, 50 μg/ml tyrosine, 60 μg/ml phenylalanine, 10 μg/ml arginine, 1.5% w/v agar], two plates containing canavanine (*can* plates; as above except containing 40 μg/ml adenine, 60 μg/ml canavanine, and lacking arginine), and one YEPD plate. The YEPD replica and one of each of the −*ade* and *can* replicas were incubated at a nonpermissive temperature (35°). The remaining plates were incubated at a permissive temperature (23°). One day later the YEPD plate incubated at the nonpermissive temperature was replicated to −*ade* and *can* media and these incubated at 35°. After 48–72 hr the colonies were screened for color on −*ade* medium and for growth on *can* medium. Colonies that failed to suppress at 23° and failed to grow at 35° or colonies that fail to suppress after a period of growth at the nonpermissive temperature (to dilute out functional suppressor tRNA) were considered candidates for mutants altered in tRNA biosynthesis. Variations of this procedure using the red colony color or canavanine resistance assay alone and other suppressor tRNAs can be used.

Lesions at the tRNA locus can be distinguished from lesions in trans-acting loci by scoring the phenotypes of diploids formed by mating mutant candidates with a *SUP4 ade2-1*° *can1-100*° strain and a strain lacking a

[19] R. K. Mortimer and D. C. Hawthorne, *in* "The Yeasts" (A. H. Rose and J. S. Harrison, eds.), Vol. 1, p. 385. Academic Press, New York, 1969.
[20] F. Rasse-Messenguy and G. R. Fink, *Genetics* **75**, 459 (1973).
[21] A. K. Hopper, L. D. Schultz, and R. A. Shapiro, *Cell* **19**, 741 (1980).
[22] G. R. Fink, this series, Vol. 17A, p. 59.

TABLE II
GENETIC SCREENS TO ASSESS EXPRESSION OF NONSENSE SUPPRESSION

		ade–can screen	
Genotype	Color on rich medium	Growth and color on medium lacking adenine	Growth on canavanine-containing medium
ADE2 CAN1	White	+, White	−
ade2-1° can 1-100°	Red	−, Red[a]	+
ade2-1° can1-100° SUP4°	White	+, White	−
ade2-1° can1-100° SUP4° losx[b]	Red	−, Red	+

		ade–gal screen	
	Color on rich glucose medium	Growth and color on glucose-containing medium lacking adenine	Growth and color on galactose-indicator medium
ADE2 GAL1 GAL7	White	+, White	+, White, yellow halo
ade2-1° gal1° gal7	Red	−, Red	−, Gray-white[a]
ade2-1° gal1° gal7 SUP4°	White	+, White	−, Yellow
ade2-1° gal1° gal7 SUP4° losx	Red	−, Red	−, Gray-white

	Conditional expression of suppressor tRNAs			
	23°		35°	
	Color on X-Gal	Invertase activity	Color on X-Gal	Invertase activity
hmlα::SUP4° hmra::SUC2 leu2::GAL10– lacZ°	White	−	Blue	+
hmlα::SUP4° hmra::SUC2 leu2::GAL10– lacZ° mutx[b]	White	−	White/light blue	+

[a] Although replicas of colonies do not grow significantly on these media, the color changes can be readily detected.
[b] *losx* and *mutx* refer to mutations that affect the expression of the suppressor.

nonsense suppressor (*SUP4⁺ ade2-1° can1-100°*). If the mutation is recessive and at the *SUP4°* locus, only the diploid formed by mating the suppressor-containing strain with the mutant should be capable of suppression at the nonpermissive temperature. If the mutation is recessive and not at the *SUP4°* locus, then both diploids would be capable of suppression. If the mutation is dominant, neither diploid should be capable of suppression.

It should be noted that since growth on canavanine media is required in order to detect loss of suppression, only mutations in genes that are unessential for growth or "leaky" mutations in essential genes will be detected by this procedure. In fact, this and other variations of the *ade–can* screen or the *ade* screen alone have failed to identify lesions of essential genes. Moreover, in "reconstruction" experiments it was shown that the *rna1-1* lesion, a conditional lethal mutation affecting the removal of IVS from pre-tRNA, did not affect nonsense suppression at nonpermissive or semipermissive temperatures of growth.[23] Nevertheless, screens based on canavanine resistance and/or red colony color have proved effective in identifying lesions of genes encoding the suppressing tRNA[24–26] and unessential genes that affect the removal of IVS from pre-tRNA (*LOS1*[21,27]) and various nucleoside modifications such as isopentenyladenosine (*MOD5*,[28] *SIN1*[29]) and 5-(methoxycarbonylmethyl)-2-thiouridine (*SIN3*[30]).

"ade–gal" Screening Procedure. The *ade–gal* screen provides a double-color assay for suppressor function but does not require that cells be able to grow more than a few or several generations at nonpermissive conditions in order to monitor suppression.[23] Utilization of galactose is dependent on three regulatory loci and four inducible structural genes.[19] Mutations in any of the structural genes (*GAL1, GAL7*, and *GAL10*) prevent galactose utilization. On rich medium containing bromthymol blue as a pH indicator and galactose as the carbon source (1% Bacto-peptone, 1% w/v yeast extract, 1% w/v galactose, 1.5% w/v agar, 0.004% w/v uracil, and 10 ml/liter of a 3% v/v stock of bromthymol blue brought into solution by titrating with NaOH), replicas of *gal1* colonies are gray-white whereas replicas of *gal7* mutant colonies are yellow. Since *GAL1* codes for the first enzyme and *GAL7* for the second enzyme in the pathway metabolizing galactose to glucose 1-phosphate, *gal1 gal7* double mutants have a "*gal1*" phenotype. Strains with a *gal1* ochre (*gal1°*) mutation and a *gal7* missense mutation produce colonies which are gray-white in

[23] A. K. Hopper, S. L. Nolan, J. Kurjan, and A. Hama-Furukawa, *in* "Molecular Genetics in Yeast" (D. von Wettstein, J. Friis, M. Kielland-Brandt, and A. Stenderup, eds.), p. 302 (Alfred Benzon Symposium 16). Munksgaard, Copenhagen, 1981.

[24] J. Kurjan, B. D. Hall, S. Gillam, and M. Smith, *Cell* **20**, 701 (1980).

[25] D. Colby, P. Leboy, and C. Guthrie, *Proc. Natl. Acad. Sci. U.S.A.* **78**, 415 (1981).

[26] F. Hofer, H. Hollenstein, F. Janner, M. Minet, P. Thuriaux, and U. Leupold, *Curr. Genet.* **1**, 45 (1979).

[27] D. J. Hurt, S. S. Wang, Y.-H. Lin, and A. K. Hopper, *Mol. Cell. Biol.* **7**, 1208 (1987).

[28] H. Laten, J. Gorman, and R. M. Bock, *Nucleic Acids Res.* **5**, 4329 (1978).

[29] F. Janner, G. Vogeli, and R. Fluri, *J. Mol. Biol.* **139**, 207 (1980).

[30] W.-D. Heyer, P. Thuriaux, J. Kohli, P. Ebert, H. Kersten, C. Gehrke, K. C. Kuo, and P. F. Agris, *J. Biol. Chem.* **259**, 2856 (1984).

the absence of an ochre suppressor and yellow in the presence of the functional suppressor. The gray-white/yellow and the red *ade* colors can be scored even if replicas cannot generate colonies. Strains containing *gal1°, gal7, ade2-1°,* and *SUP4°* alleles therefore provide a double-color assay for loss of suppression (Table II). Two alleles of the *LOS1* locus were isolated utilizing this procedure.[23]

Conditional Expression of Suppressor tRNAs. A major problem with the two above screening procedures is that tRNAs are stable molecules. Therefore, the supply of preexisting suppressor tRNAs must be depleted before loss of suppression can be scored. If, as one would predict, lesions affecting pre-tRNA processing are lethal, then cells possessing such lesions may not be able to grow enough to dilute out preexisting tRNAs. That is, conditional lethal lesions are unlikely to be detected using the *ade–can* and *ade–gal* screening procedures.

To circumvent this problem, van Zyl *et al.*[31] devised a system to allow for conditional expression of a tyrosine-inserting ochre suppressor (*SUP4°*). The procedure was fashioned after selections devised for *Escherichia coli*[31a] that utilized phage-encoded suppressor tRNAs. This yeast screen is based upon the observations of Schnell and Rine[32] that a tRNA gene placed in the silent mating-type cassettes (*HML* or *HMR*) is repressed by the products of the *SIR* loci. The *SUP4°* gene was cloned into the *HMLα* locus and introduced into a strain containing a thermosensitive *sir4* lesion. The same strain contained two reporter genes: *GAL10–lacZ°,* encoding galactose-regulated β-galactosidase, and *hmra::SUC2,* encoding SIR-regulated invertase.

The screen was performed as follows. Strains possessing the *hmlα:: SUP4°, hrma::SUC2,* and *leu2::GAL10–lacZ°* alleles were subjected to mutagenesis and plated on −uracil +glucose synthetic medium and incubated at 23° until colonies developed. The colonies were replica plated to −uracil +galactose synthetic medium (pH 7) with 70 μg/ml 5-bromo-4-chloro-3-indolyl-*b*-D-galactoside (X-Gal) and incubated at 23° until the colonies developed again. The plates were shifted to 35° for 2 days. In wild-type cells *SUP4°* expression would be induced at 35° and functional suppressor tRNA would be produced, resulting in suppression of the ochre mutation in the *lacZ°* mRNA and production of blue colonies. In contrast, for a mutant clone deficient in transcription or processing of tRNA, no functional suppressor tRNA would be produced, no β-galactosidase would be synthesized, and the colony would remain white. The

[31] E. van Zyl, N. Wills, and J. R. Broach, *Genetics* **121,** 55 (1989).
[31a] Y. Shimura and H. Shiraishi, this volume [32].
[32] R. Schnell and J. Rine, *Mol. Cell. Biol.* **6,** 494 (1986).

expression of invertase was used to assess whether mutant cells were defective for protein synthesis, in general, or specifically in tRNA biosynthesis. Desired mutants would be invertase positive[33] and β-galactosidase negative at the nonpermissive temperature (Table II). The plates were thus screened for white or light blue colonies among the majority of blue colonies after the incubation at 35°. Among five mutants identified by this screen, one conditional lethal mutation, *tpd5-47*, causes accumulation of end-matured, IVS-containing pre-tRNA species as assessed by Northern analysis and *in vivo* labeling.

Correction of Processing-Defective Substrates

Correction of processing-defective substrates is based on the premise that mutations within a tRNA locus that result in reduced or abolished processing of pre-tRNA species might define positions or structures on the precursor that are recognized by components of the processing machinery. Such processing defects might be expected to be corrected by increased quantities of rate-limiting components of the processing machinery or by processing enzymes with altered specificity (i.e., second-site suppressors mapping to genes encoding components of the processing activities).

Scores of point mutations of *S. cerevisiae* suppressors and the *S. pombe sup3-e* and *sup9-e* genes (for details and references, see Nichols *et al.*[1]) have been characterized. Several of these mutant loci have been shown to produce transcripts and/or processing intermediates that are inefficient or totally defective in conversion to mature tRNAs as assessed by *in vitro* processing systems[34] or by biosynthesis in surrogate[35] or endogenous hosts.[25] These mutations of suppressor tRNA loci have served as a starting point for the identification of genes able to correct the processing defects.

Correction by Increased Quantities of Trans-Acting Gene Products. One gene involved in pre-tRNA processing was identified by the ability to correct the suppression and processing defects of a mutation of the *SUP4°* locus. The *SUP4⁻°(G37)* gene possesses an A → G mutation at position 37, the 5' splice junction of *SUP4°* pre-tRNA^{Tyr}.[24,25] This lesion results in a decreased level of *SUP4°*-mediated nonsense suppression. Cells containing the *SUP4⁻°(G37)* lesion are able to suppress the *met4-1°* mutation but are unable to suppress the *lys2-1°*, *trp5-2°*, *ade2-1°*, and *can1-100°*

[33] A. Goldstein and J. O. Lampen, this series, Vol. 42, p. 504.
[34] D. Pearson, I. Willis, H. Hottinger, J. Bell, A. Kumar, U. Leupold, and D. Söll, *Mol. Cell. Biol.* **5,** 808 (1985).
[35] K. Nishikura, J. Kurjan, B. D. Hall, and E. M. DeRobertis, *EMBO J.* **2,** 263 (1982).

ochre alleles. *In vivo* studies have shown that end-matured, IVS-containing pre-tRNA^{Tyr} suppressor species is inefficiently but correctly processed.[25]

Assuming that $SUP4^{-o}(G37)$ pre-tRNA was a poor substrate for splicing and that processing might be limited *in vivo* by components of the splicing machinery, Wang and Hopper[10] designed the following strategy to search for a gene that when amplified causes increased efficiency of $SUP4^{-o}(G37)$ suppression. A $SUP4^{-o}(G37)$ strain was transformed[36] with a yeast library inserted in the multicopy vector YEp24.[37] Approximately 2000 resulting transformants were screened for increased efficiency in the ability to suppress nonsense alleles by scoring for the ability to grow on media lacking lysine and by the *ade2-1°* colony color assay (see above). Lysine prototrophic, pink candidates were checked for suppression of other nonsense alleles. Confirmation that the phenotype of candidates is due to plasmid-encoded sequences was determined by streaking for single colonies on nonselective media and assessment of whether colonies that have lost the plasmid marker (i.e., *URA3*) have also lost the ability to suppress *lys2-1* and *ade2-1* (and vice versa).

Since such plasmid sequences could encode nonsense suppressors or products that generally aid suppression ("allosuppressors") as well as products involved in pre-tRNA processing, the plasmids were tested for the ability to suppress nonsense mutations in the absence of chromosomal *SUP* loci and the ability to aid suppression of *SUP* loci encoded by genes that lack an IVS (*SUP16°* in this case). Candidate plasmids were therefore transformed into *SUP4⁺* and *SUP16°* strains and the transformants assessed for the ability to suppress nonsense alleles. Those plasmids that aided the defective *SUP* lesion (or lesions that have related biochemical defects) but not *SUP4⁺* or *SUP16°* were analyzed biochemically. The *STP1* gene was identified in this manner.

The effect of multiple copies of *STP1* on the processing of $SUP4^{-o}(G37)$ precursor species was assessed by Northern analysis employing oligonucleotides that preferentially hybridize to mutant pre-tRNA and not to wild-type precursor species. When *STP1* is present in multiple copies in yeast cells possessing the $SUP4^{-o}(G37)$ lesion, the cells produce greater quantities of the $SUP4^{-o}(G37)$ mature species than do cells harboring vector alone. Conversely, cells possessing disruptions of this locus, although viable, show the reduced suppressor activity of a normal *SUP4°* tRNA and accumulate one form of intron-containing precursors to tRNA^{Tyr}, tRNA^{Lys}, tRNA^{Phe}, tRNA^{Pro}, and tRNA$^{Ser}_{UCG}$ species, but not pre-tRNA$_3^{Leu}$ or pre-tRNA^{Ile} species.[10]

[36] H. Ito, Y. Fukuda, K. Murata, and A. Kimura, *J. Bacteriol.* **153,** 163 (1983).
[37] M. Carlson and D. Botstein, *Cell* **28,** 145 (1982).

Correction by Second-Site Suppressors. Starting with a processing-defective suppressor tRNA, one can also select or screen cells with the processing defect for mutants that restore nonsense suppression. The use of such second-site suppressors is based on the pioneering work of Jarvik and Botstein,[38,39] who showed that second-site suppressors can identify genes whose products interact with the defective product of the first gene.

In practice one begins with haploid or diploid cells carrying the *SUPX⁻* lesion and several nonsense alleles that cannot be suppressed by the *SUPX⁻* locus. One selects or screens for mutants able to suppress some or all of these alleles. Such variants could map to the defective *SUPX⁻* locus, to another gene encoding a tRNA, to a locus that generally affects translation, or to the desired type of gene that codes for a product involved in pre-tRNA processing. One way to begin to distinguish the later from the former possibilities is to generate a diploid strain heterozygous for the *SUPX⁻* locus (i.e., *SUPX⁻/SUPX⁺*) and homozygous for various nonsense mutations. (This may be the starting strain in a diploid selection.) If the mutant locus encodes a functional suppressor tRNA, on sporulation the diploid will segregate 2 : 2 for ability to suppress to inability to suppress. On the other hand, if the mutant locus encodes a product that aids suppression or corrects the processing defect, asci with the following segregation pattern will be obtained: 1 (2 : 2, suppression capable to suppression incapable) to 4 (1 : 3, suppression capable to suppression incapable) to 1 (0 : 4 suppression capable to suppression incapable).

Mutations that generally aid nonsense suppression (allosuppressors) can be distinguished from those that specifically correct the processing defect either by biochemical procedures to assess pre-tRNA processing of the *SUPX⁻* tRNA[10] or by additional genetic analysis. For example, if the second-site suppressor corrects a splicing defect, then it should have no effect on the expression of a suppressor encoded by a gene that lacks an IVS (e.g., *SUP16°*).

Although this strategy of selection of second-site suppressors has proved successful for identifying genes involved in many aspects of yeast biochemistry and cell biology, to this author's knowledge it has identified only one promising candidate for a mutation of a gene involved in tRNA processing. Selecting among diploid *S. cerevisiae* for variants able to express the *S. pombe, sup3-e-A39* lesion, Kolman and Soll[39a] found a recessive lethal mutation that in the heterozygous state corrected the splicing defect of this mutant substrate. Part of the difficulty in identifying

[38] J. Jarvik and D. Botstein, *Proc. Natl. Acad. Sci. U.S.A.* **70,** 2046 (1973).
[39] J. Jarvik and D. Botstein, *Proc. Natl. Acad. Sci. U.S.A.* **72,** 2738 (1975).
[39a] C. Kolman and D. Söll, in preparation (1990).

mutations by this approach is that allosuppressors and other nonsense suppressors are frequently obtained.

Reverse Genetics

In one case a gene important to tRNA processing has been identified by the so-called reverse genetics approach. Phizicky *et al.*[40] purified tRNA ligase, the enzyme possessing the activity required to join the two cognate half-molecules together following removal of the IVS, and determined the sequence of amino acids 2 to 15 of the approximately 90 kDa protein. Sets of oligonucleotides able to code for N-terminal peptides were utilized to identify candidate recombinant DNAs coding for this activity. Substantial data including DNA sequence analysis, amplified activity in plasmid-containing yeast cells, and tRNA ligase expression in *E. coli* demonstrated that the structural gene, *LIG1*, for tRNA ligase was obtained. This sequence has been used to study the consequences of mutations of *LIG1* on pre-tRNA processing. These studies employ sophisticated molecular technologies that are only outlined here.

Deletions generated *in vitro* followed by gene replacement[41] of the endogenous *LIG1* gene by the altered *lig1* sequences have shown that *LIG1* is an essential gene. Two approaches have been taken to study further the role of tRNA ligase in yeast. First a centromere-containing plasmid was constructed in which *LIG1* gene was placed under the control of the yeast inducible *GAL* system.[40] Cells possessing a genomic disruption of *LIG1* and containing the *GAL*-regulatable *LIG1* sequences were utilized to determine the consequences of ligase depletion as a response to removal of galactose (E. Phizicky, personal communication, 1989). Second, thermosensitive conditional alleles of *LIG1* were generated by treating the *LIG1* gene *in vitro* with formic acid or hydroxylamine. When these alleles replace the endogenous *LIG1* sequence by gene replacement, tRNA ligase activity is thermosensitive, and the yeast cells accumulate tRNA half-molecules and small quantities of end-matured, IVS-containing pre-tRNAs (E. Phizicky, personal communication, 1989).

Characterization of Mutants

Assessment of Biochemical Consequences of Mutations

Those mutants identified on the basis of accumulation of a pre-tRNA species (see above) by definition possess lesions affecting pre-tRNA pro-

[40] E. M. Phizicky, R. C. Schwartz, and J. Abelson, *J. Biol. Chem.* **261**, 2978 (1986).
[41] R. J. Rothstein, this series, Vol. 101, p. 202.

cessing. Mutants identified by screening for *in vitro* defects of particular enzyme activities may not have pre-tRNA processing defects *in vivo*. For example, the *sen2-1* allele causes a temperature-dependent decrease of tRNA splicing endonuclease activity *in vitro* but does not result in accumulation of pre-tRNA species *in vivo* (M. Winey and M. Culbertson, personal communication, 1988). Yet there is another mutation of this locus, *sen2-3*, identified by the accumulation of pre-tRNA processing intermediates that is clearly defective in pre-tRNA processing.[10a] More recent studies suggest that *SEN2* encodes one of the three peptides of the tRNA splicing endonuclease.[10a] Those mutants identified on the basis of reduced suppressor activity or correction of a defective suppressor need not code for products related to pre-tRNA processing. For example, of the scores of mutants identified by the *ade–can* and *ade–gal* assays only those possessing the *los1* alleles showed defects in pre-tRNA processing when analyzed by *in vivo* labeling procedures.[21] And only a few other loci affecting tRNA modifications were identified when the mutants were screened for nucleoside modifications by thin-layer chromatography[28,29] or high-performance liquid chromatography.[30] Even those mutants generated by reverse genetics need not have defective pre-tRNA processing. For example, a deletion of the *CHS1* gene encoding chitin synthase I is not lethal because yeast cells possess a second nonhomologous gene also encoding a chitin synthase activity that has different divalent cation requirements than does chitin synthase I.[42] Thus, it is important to investigate the biochemical consequences of the lesions in question.

Initially, the effects of mutations on pre-tRNA accumulation can be assessed by the *in vivo* labeling or Northern procedures described above. The exact nature of the accumulated pre-tRNA species can be determined by oligonucleotide mapping procedures.[5,6] The possibility that mutations affect particular nucleoside modifications is most readily assessed by high-performance liquid chromatography of nucleosides derived from enzymatic digestions of the tRNA species,[30,43]

Even when the molecular defect for pre-tRNA processing and/or modification is understood, it may not be obvious what activity is encoded by the wild-type counterpart of the gene in question. For example, the *rna1-1* and *los1* alleles cause accumulation of end-matured, IVS-containing pre-tRNA species, but the *RNA1* and *LOS1* genes do not encode tRNA splic-

[42] C. E. Bulawa, M. Slater, E. Cabib, J. Au-Young, A. Sburlati, W. L. Adair, Jr., and P. W. Robbins, *Cell* **46**, 213 (1986).
[43] A. K. Hopper, A. H. Furukawa, H. D. Pham, and N. C. Martin, *Cell* **28**, 543 (1982).

ing endonuclease or tRNA ligase activities. Although a mutation of *SEN1* affects tRNA splicing endonuclease activity *in vitro*, *SEN1* may not encode endonuclease activity; *SEN1* could code for activities necessary to activate the endonuclease, for example. One of the most exciting outcomes of the efforts to understand the functions of the genes in question is that the mutants may serve to identify processes or products that were unanticipated from current understandings of the RNA processing pathways.

Genetic Analysis of the Mutations

It is extremely important that the biochemical defect in pre-tRNA processing and the original phenotype used to identify the mutant be shown to cosegregate. There are many examples where this has not been the case. That is, although the *sen2-1, sen2-2*, and *sen2-3* alleles were identified among members of collections of conditionally lethal yeast, none of these alleles cause heat-dependent lethality[3,10a] (M. Winey and M. Culbertson, personal communication, 1988). The *trm2* mutation was isolated in a search for processing defects among a collection of yeast with reduced nonsense suppression. Yet the *trm2* mutation that affects m⁵U in tRNA does not cause defects in nonsense suppression.[43] The reason for the lack of cosegregation is that mutagenized yeast generally possess multiple lesions, each resulting in different phenotypes.

Cosegregation of processing defects and conditional lethality is easy to follow. The mutant yeast are mated to haploid wild-type yeast of opposite mating type. Diploids are induced to sporulate by nitrogen and glucose starvation, and the resulting asci are separated into the four progeny. Each of the progeny is tested for both phenotypes. If the processing defects and conditional lethality are due to the same lesion all progeny will be either wild-type or mutant for both defects. Alternatively, one can screen for coreversion of the two phenotypes. Approximately 10^8 yeast can be plated on rich medium at the nonpermissive temperature. One would expect revertants at a frequency of 10^{-6}–10^{-8}. Colonies able to grow at this temperature can be tested for restoration of pre-tRNA processing.

Cosegregation of reduced efficiency of suppression and processing defects can be tested by mating the mutants to a strain carrying the same nonsense suppressor and suppressible alleles as the parent. Cosegregation is assessed by monitoring the suppression phenotype and *in vivo* processing of pre-tRNAs of the meiotic segregants on sporulation and dissection. To assess cosegregation of a lesion that corrects a pre-tRNA processing defect and restoration of suppression, the mutant $SUPX^-$ strain carrying the mutation in question is mated to a $SUPX^-$ strain of opposite mating type, and the meiotic segregants are analyzed.

Complementation assays are utilized as a first test to determine whether two independently derived mutations are lesions of the same or different genes. The two mutant strains in question are mated, and the diploids are isolated and scored for the phenotype in question. If a wild-type phenotype does not result, then the two mutations fail to complement. Failure to complement may be taken as an indication that the two mutations are lesions of the same gene. However, there are now examples of noncomplementing mutations that *do not* map to the same locus.[44,45] Thus, complementation studies need be confirmed by segregation analysis. If the two mutations are indeed alleles of the same gene, then the vast majority of asci will segregate 4:0 for mutant to wild-type phenotype. If the two mutations are lesions of unlinked genes, then ratios of 1:4:1 of asci segregating 4:0, 3:1, and 2:2 mutant to wild-type phenotypes will be obtained. The ratios will be skewed toward 4:0 and 3:1 segregations if the genes are linked. If the gene is not cloned, the chromosomal location of mutations can be determined by tetrad analysis.[19] If the gene is cloned, then it can be mapped by a combination of readily available techniques including hybridization to electrophoretically separated chromosomes (OFAGE[46] or FIGE[47] techniques) and loss of genetic information by $2\mu m$ recombination[48] in combination with tetrad analysis.[19]

Summary

All of the strategies to identify mutants defective in pre-tRNA processing have proved to be useful (see Appendix). However, in some of these strategies (such as inability to derepress a nonsense suppressor, correction of processing-defective substrates, and reverse genetics) only a single mutant has been isolated. This is not because the methods have proved to be inefficient, but rather because the field is still in infancy. Further work is needed to saturate the types of mutants found by each procedure. Undoubtedly, new strategies will be designed.

[44] K. D. Atkinson, *Genetics* **111,** 1 (1985).
[45] J. M. Song and S. W. Liebman, *Genetics* **115,** 451 (1987).
[46] G. F. Carle and M. V. Olson, *Proc. Natl. Acad. Sci. U.S.A.* **82,** 3756 (1985).
[47] G. F. Carle, M. Frank, and M. V. Olson, *Science* **232,** 65 (1986).
[48] S. Falco and D. Botstein, *Genetics* **105,** 857 (1983).

Appendix: Yeast Genes Involved in Pre-tRNA Processing

Gene	Method of identification	Description/biochemical lesion
RNA1 (<u>RNA</u> synthesis)	Conditional lethal lesion causing defects in production of cytoplasm-localized RNAs;[a] screen of conditional lethals for pre-tRNA accumulation[2]	Essential gene; defective in removal of IVS from pre-tRNA,[5,6] in pre-rRNA processing,[2] and in production of mRNA; biochemical function unknown
PTA1 (<u>p</u>re-<u>t</u>RNA <u>a</u>ccumulation)	Screen of conditional lethal yeast for pre-tRNA accumulation[b]	Essential gene; inefficient removal of IVS[b]
SEN2 (allele *sen2-3*) (<u>s</u>plicing <u>en</u>donuclease)	Screen of mutagenized yeast for pre-tRNA accumulation[10a]	Accumulation 5′ half-molecules and IVS (two-thirds molecules);[10a] defect in splicing endonuclease activity *in vitro*,[c] likely subunit of splicing endonuclease[10a]
SEN2 (alleles *sen2-1, sen2-2*) (<u>s</u>plicing <u>en</u>donuclease)	Screen of mutagenized yeast for defect in tRNA splicing endonuclease[3]	Defect in splicing endonuclease activity *in vitro*[3]
SEN1 (<u>s</u>plicing <u>en</u>donuclease)	Screen of mutagenized yeast for defect in tRNA splicing endonuclease[c]	Essential gene; defective splicing activity *in vitro*; accumulation of end-matured, IVS-containing pre-tRNAs *in vivo*[3,c]
CCA1 (<u>CCA</u> addition)	Screen of conditional lethal yeast for alterations in mRNA half-life[10b]	Defect in nucleotidyltransferase[10b]
LOS1 (<u>l</u>oss <u>o</u>f <u>s</u>uppression)	Reduced efficiency of suppression;[21] reverse genetics[27]	Unessential gene;[27] affects removal of IVS from end-matured pre-tRNAs at elevated temperatures;[21,27] biochemical function unknown
TPD5 (<u>t</u>RNA <u>p</u>rocessing <u>d</u>efect)	Inability to express derepressed nonsense suppressor[31]	Essential gene; accumulation of end-matured IVS containing pre-tRNAs at nonpermissive temperatures[31]
STP1 (species-<u>s</u>pecific <u>t</u>RNA <u>p</u>rocessing)	Correction of splicing-defective suppressor tRNA[10]	Unessential gene; accumulation of end-matured, IVS-containing precursors to tRNA^{Tyr}, tRNA^{Lys}, tRNA^{Phe}, tRNA^{Pro}, and tRNA$^{Ser}_{UCG}$; no effect on processing

(*continued*)

Appendix: (*continued*)

Gene	Method of identification	Description/biochemical lesion
		of pre-tRNA$_3^{Leu}$ or pre-tRNA^{Ile10}; biochemical function unknown
MOD5 (modification)	Reduced efficiency of suppression[28]	Unessential gene;[d] encodes Δ^2-isopentenylpyrophosphate : tRNA 2-isopentenyltransferase, the enzyme necessary to modify position A-37 to isopentenyladenosine (i^{6}A)[28,e]
SIN1 (suppressor interacting)	Reduced efficiency of suppression[29]	S. pombe gene affects modification of i^{6}A[29]
SIN3 (suppressor interacting)	Reduced efficiency of suppression[30]	S. pombe gene affects modification of 5-(methoxycarbonylmethyl)-2-thiouridine[30]
TRM1 (tRNA methylation)	Assay for mutagenized cells to produce methyl-deficient tRNAs[f]	Unessential gene; encodes N^2,N^2-dimethylguanosine-specific tRNA methyltransferase,[g] enzyme necessary to modify G to m$_2^2$G
LIG1 (ligase)	Reverse genetics[40]	Essential gene; [h]encodes tRNA ligase[40]
MIA1 (multiple isoaccepting species)	Altered column chromatography for several tRNAs; serendipity[i]	No phenotype; tRNAs are deficient in dihydrouracil[i]
TRM2 (tRNA methylation)	Altered migration of tRNAs on polyacrylamide gels; serendipity[43]	No phenotype; tRNAs lack 5-methyluracil (rT)[43]

[a] H. T. Hutchinson, L. H. Hartwell, and C. S. McLaughlin, *J. Bacteriol.* **99**, 807 (1969).

[b] J. O'Connor and C. Peebles, personal communication, 1989.

[c] M. Winey and M. Culbertson, personal communication.

[d] L. Slusher, E. Gillman, and A. Hopper, unpublished.

[e] M. E. Dihanich, D. Najarian, R. Clark, E. C. Gillman, N. C. Martin, and A. K. Hopper, *Mol. Cell. Biol.* **7**, 177 (1987).

[f] J. Phillips and K. Kjellin-Straby, *J. Mol. Biol.* **26**, 509 (1967).

[g] S. R. Ellis, M. J. Morales, J.-M. Li, A. K. Hopper, and N. C. Martin, *J. Biol. Chem.* **261**, 9703 (1986).

[h] E. Phizicky, personal communication, 1989.

[i] R. Y. C. Lo, J. B. Bell, and K. L. Roy, *Nucleic Acids Res.* **10**, 889 (1982).

Acknowledgments

I thank J. O'Connor and C. Peebles; E. Phizicky; W. H. van Zyl and J. Broach; C. Ho and J. Abelson; C. Kolman and D. Söll; M. Winey and M. Culbertson; and N. Martin for unpublished information. M. Culbertson, E. Gillman, R. Keil, C. Peebles, H. Traglia, S. Wang, and M. Winey provided valuable comments on the manuscripts.

[34] Ribonucleases Active at 3' Terminus of Transfer RNA

By Murray P. Deutscher

Introduction

Although considerable information about the processing of tRNA precursors has accumulated in recent years, our understanding of processing reactions at the 3' termini of these molecules is still extremely limited.[1] Both endonucleolytic and exonucleolytic 3' processing reactions have been observed that vary with the system under study. In *Escherichia coli* it is thought that 3' processing is initiated by an endonucleolytic cleavage downstream from the mature 3' end followed by one or two exonucleolytic trimming reactions that stop at the encoded -CCA sequence. However, the enzyme(s) involved in these processing reactions has not been conclusively established.

Several *E. coli* exoribonucleases able to act at the 3' terminus of tRNA and tRNA precursors have been identified and characterized to varying degrees.[2] These include RNase II,[3] RNase D,[4] RNase BN,[5] RNase T,[6] polynucleotide phosphorylase,[7] and RNase PH.[8] RNase II and polynucleotide phosphorylase, which have been known for many years, are relatively nonspecific for RNA substrates. The other nucleases display a higher degree of specificity for tRNA-like substrates, but it is not yet clear if they function *in vivo* in the processing of *E. coli* tRNA precursors. From

---

[1] M. P. Deutscher, *Crit. Rev. Biochem.* **17**, 45 (1984).
[2] M. P. Deutscher, *Cell* **40**, 731 (1985).
[3] R. S. Gupta, T. Kasai, and D. Schlessinger, *J. Biol. Chem.* **252**, 8945 (1977).
[4] H. Cudny, R. Zaniewski, and M. P. Deutscher, *J. Biol. Chem.* **256**, 5627 (1981).
[5] P. K. Asha, R. T. Blouin, R. Zaniewski, and M. P. Deutscher, *Proc. Natl. Acad. Sci. U.S.A.* **80**, 3301 (1983).
[6] M. P. Deutscher and C. W. Marlor, *J. Biol. Chem.* **260**, 7067 (1985).
[7] U. Z. Littauer and H. Soreq, *in* "The Enzymes" (P. D. Boyer, ed.), 3rd Ed., Vol. 15, Part B, p. 517. Academic Press, New York, 1982.
[8] M. P. Deutscher, G. T. Marshall, and H. Cudny, *Proc. Natl. Acad. Sci. U.S.A.* **85**, 4210 (1988).

studies of mutant strains, it has been shown that RNase BN plays a role in the 3' processing of certain phage T4 tRNA precursors[9] and that RNase T is involved in the end turnover of tRNA.[10] However, a multiple RNase-deficient strain (RNase D⁻, II⁻, BN⁻, T⁻, I⁻) still retains viability and processes tRNA precursors, leaving open the possibility that these enzymes do not function in tRNA processing or, alternatively, that an additional nuclease can substitute when these are absent. One candidate for an additional nuclease is the recently discovered RNase PH which trims tRNA precursors *in vitro* in a phosphorolytic manner.[8,11]

RNase II, polynucleotide phosphorylase, RNase D, and RNase T have been purified to homogeneity and studied in some detail.[2] In addition, the genes for the first three of these enzymes, *rnb*,[12] *rnd*,[13] and *pnp*,[14] have been cloned, and the *rnd*[15] and *pnp*[16] genes have also been sequenced. A partial purification scheme has been developed for RNase BN.[17] This chapter focuses on RNase D, RNase T, and RNase BN, three RNases showing specificity for tRNA-type substrates.

RNase D

Assay Method

Principle. The assay for RNase D measures the conversion of a specific radioactive tRNA substrate to an acid-soluble form.[18,19] Two different substrates have been used for this purpose, tRNA-CCA-[¹⁴C]C₃ and [³²P]phosphodiesterase-treated tRNA (dtRNA) as follows:

$$\text{tRNA-CCA-}[^{14}\text{C}]\text{C}_3 \rightarrow \text{tRNA-CCA} + [^{14}\text{C}]\text{CMP} \qquad (1)$$
$$[^{32}\text{P}]\text{dtRNA} \rightarrow [^{32}\text{P}]\text{RNA} + [^{32}\text{P}]\text{mononucleotides} \qquad (2)$$

The first substrate is an analog of a tRNA precursor, the presumed substrate for RNase D *in vivo*.[18] The second substrate is a mixture of tRNA molecules lacking two or three of the 3'-terminal residues.[19] Removal of terminal residues from some tRNA species alters their conformation and renders them susceptible to RNase D action.[19] A similar effect can be

[9] J. G. Seidman, F. J. Schmidt, K. Foss, and W. H. McClain, *Cell* **5**, 389 (1975).
[10] M. P. Deutscher, C. W. Marlor, and R. Zaniewski, *Proc. Natl. Acad. Sci. U.S.A.* **82**, 6427 (1985).
[11] H. Cudny and M. P. Deutscher, *J. Biol. Chem.* **263**, 1518 (1988).
[12] W. P. Donovan and S. R. Kushner, *Nucleic Acids Res.* **11**, 265 (1983).
[13] J. Zhang and M. P. Deutscher, *J. Bacteriol.* **170**, 522 (1988).
[14] C. Portier, *Mol. Gen. Genet.* **178**, 343 (1980).
[15] J. Zhang and M. P. Deutscher, *Nucleic Acids Res.* **16**, 6265 (1988).
[16] P. Regnier, M. Greenberg-Manago, and C. Portier, *J. Biol. Chem.* **262**, 63 (1987).
[17] D. Neri-Cortes and M. P. Deutscher, unpublished results (1986).
[18] M. P. Deutscher and R. K. Ghosh, *Nucleic Acids Res.* **5**, 3821 (1978).
[19] R. K. Ghosh and M. P. Deutscher, *J. Biol. Chem.* **253**, 997 (1978).

obtained by heating and quick-cooling intact tRNA.[19] tRNA-CCA-[^{14}C]C$_3$ is a somewhat more specific substrate for RNase D than [^{32}P]dtRNA; approximately 80% of the activity of a crude *E. coli* extract against this substrate under standard assay conditions is due to RNase D, whereas for [^{32}P]dtRNA the value is about 50%.[20] The latter substrate, however, is more readily available (see below).

Preparation of Substrates. tRNA-CCA-[^{14}C]C$_3$ is prepared by addition of [^{14}C]CMP residues to *E. coli* tRNA using [^{14}C]CTP (~10^4 cpm/nmol) and rabbit liver tRNA nucleotidyltransferase as described.[18] The incorporation of CMP into intact tRNA occurs at a slow rate and is due to an anomalous reaction carried out by the rabbit liver enzyme.[21] The reaction is carried out in the presence of MnCl$_2$, rather than MgCl$_2$, since Mn^{2+} stimulates the anomalous reaction. A ratio of 2–3 units of enzyme per milligram of tRNA is used, and the reaction is carried out for about 3 hr at 37°. Under these conditions an average of about three CMP residues are added to the tRNA chains. Rabbit liver tRNA nucleotidyltransferase is prepared as described.[22] *Escherichia coli* tRNA nucleotidyltransferase has not been effective for preparations of this substrate.[23]

[^{32}P]dtRNA is prepared by snake venom phosphodiesterase (Worthington, Freehold, NJ) treatment of *E. coli* [^{32}P]tRNA as described.[20] Phosphodiesterase treatment is carried out until about 2.5% of the tRNA radioactivity is made acid soluble.

Procedure. The standard assay mixture for RNase D activity contains the following in 0.1 ml: 20 m*M* glycine–NaOH (pH 8.9), 5 m*M* MgCl$_2$, 20 μg radioactive tRNA substrate, and enzyme fraction.[4] After incubation at 37°, 0.3 ml of carrier yeast RNA (10 mg/ml) and 0.4 ml of cold 20% trichloroacetic acid are added. The samples are left in ice for 10 min and then centrifuged for 10 min at 10,000 *g*. A portion (0.4 ml) of the supernatant fluid is removed, added to 0.6 ml of 1 *M* tris(hydroxymethyl)aminomethane (Tris) in a scintillation vial, and radioactivity determined with 10 ml of Liquiscint (National Diagnostics, Manville, NJ) scintillation fluid. One unit of activity with either substrate is the amount of enzyme that will solubilize 1 μmol of nucleotide per hour.[4]

Enzyme Source

The *E. coli rnd* gene encoding RNase D at about 40 min on the *E. coli* chromosome[24] has been cloned and sequenced.[15] A strain that overex-

[20] R. Zaniewski, E. Petkaitis, and M. P. Deutscher, *J. Biol. Chem.* **259,** 11651 (1984).
[21] M. P. Deutscher, this series, Vol. 29, p. 706.
[22] M. P. Deutscher and P. Masiakowski, *Nucleic Acids Res.* **5,** 1949 (1978).
[23] M. P. Deutscher, this volume [35].
[24] R. Zaniewski and M. P. Deutscher, *Mol. Gen. Genet.* **185,** 142 (1982).

presses RNase D was constructed by transforming *E. coli* 20-12E/18-11 (RNase II⁻, D⁻, BN⁻, T⁻, I⁻) with plasmid pDB14.[10,13] This plasmid contains a 1.4-kb DNA fragment encoding the complete *rnd* gene inserted into the *Bam*HI–*Sma*I site of plasmid pUC18. This *E. coli* strain is mutant for five RNases, and overexpresses RNase D from the plasmid to a level about 20 times that of the wild type. Based on maxicell analysis, N-terminal sequencing, and amino acid composition, the overexpressed RNase D is the normal *E. coli* protein.[13,15]

Overexpressing cells are prepared by growth in YT medium (8 g tryptone, 5 g yeast extract, 5 g NaCl per liter) containing ampicillin at 12.5 mg/liter. Cells are grown to an A_{550} of 0.8 and stored frozen at −20° until use. In order to achieve the highest specific activity of RNase D, cells are grown to a relatively low density because the culture tends to be overgrown by faster growing cells with a lower plasmid copy number and less RNase D. At these levels of overexpression, RNase D is deleterious to *E. coli* and leads to slowed growth.[13]

Purification Procedure

All operations are carried out in a cold room at 4°. Buffer A contains 20 m*M* Tris-Cl (pH 7.5), 5 m*M* MgCl₂, 0.1 m*M* dithiothreitol, and 0.1 m*M* phenylmethylsulfonyl fluoride.

Steps 1 and 2: Preparation of Supernatant Fraction. Frozen cells (40 g) are partially thawed and suspended in 2 volumes of buffer A containing 150 m*M* KCl. The cells are disrupted by two passages through an Aminco French press at about 10,000 psi. The crude extract is treated with DNase I (10 μg/ml) for 45 min in ice and centrifuged for 40 min at 0–5° at 17,000 *g* to remove cell debris. The supernatant fraction (I) is then centrifuged for 2 hr at 45,000 rpm in a Beckman Ti 50.2 rotor. The resulting high-speed supernatant fraction II is used for further purification. This step, carried out in the presence of 150 m*M* KCl, leaves less than 10% of the RNase D bound to ribosomes.

Step 3: DEAE-Sephadex Chromatography. Supernatant fraction II is dialyzed against buffer A containing 40 m*M* KCl to lower the KCl concentration and applied to a DEAE-Sephadex column (5 × 18 cm) equilibrated with buffer A containing 60 m*M* KCl. The column is washed with this buffer until the A_{280} is below 0.2, and RNase D is eluted with a 7-liter linear gradient from 60 to 300 m*M* KCl in buffer A. This step leads to a severalfold purification of RNase D and also to a large increase in the number of units of activity because of the removal of a fraction of the inhibitory nucleic acid present in supernatant II ($A_{280}/A_{260} = 0.9$).

Step 4: Chromatography on Affi-Gel Blue. Fractions containing RNase D activity are combined, dialyzed against buffer A to remove KCl, and applied to a column of Affi-Gel Blue (2.5 × 10 cm) equilibrated with

TABLE I
PURIFICATION OF OVEREXPRESSED *Escherichia coli* RNase D[a]

Step	Total activity (units)	Specific activity (units/mg)	Relative purification (-fold)	Yield (%)
1. Supernatant fraction I	2,200	1.80	1.0	100
2. Supernatant fraction II	2,100	2.54	1.4	95
3. DEAE-Sephadex	3,700	14.1	7.8	168
4. Affi-Gel Blue	5,200	96.6	54	236
5. Hydroxyapatite	4,300	271	150	195
6. Ultrogel AcA 44	2,100	326	180	95

[a] Adapted from Ref. 15.

buffer A. RNase D is eluted with a 4-liter linear gradient from 0 to 500 mM KCl in buffer A. This step results in a further increase in the total number of units of RNase D owing to removal of the remaining nucleic acid present (A_{280}/A_{260} = 1.7). Active fractions are combined, concentrated to 10 ml with an Amicon YM10 membrane in a filtration apparatus, and dialyzed against buffer A.

Step 5: Hydroxyapatite Chromatography. The dialyzed sample is added to a hydroxyapatite column (1.5 × 5 cm) equilibrated with buffer A. RNase D is eluted with a 150-ml linear gradient from 10 to 200 mM potassium phosphate (pH 7.5) in buffer A. Active fractions are combined and concentrated, as above, to 3 ml.

Step 6: Chromatography on Ultrogel AcA 44. Separate 1.5-ml portions of the concentrated sample are applied to a column of Ultrogel AcA 44 (1.5 × 80 cm) equilibrated with buffer A containing 200 mM KCl. RNase D is eluted with the same buffer. Active fractions are combined and frozen at −20° until use.

The purification scheme is summarized in Table I. This procedure leads to a homogeneous preparation of RNase D with an apparent yield of 95%. However, if one assumes that the number of units of enzyme present after Step 4 more accurately reflects the amount of RNase D present, the actual recovery is a maximum of 40%, and most likely is lower, since some enzyme undoubtedly is lost in Steps 3 and 4. Thus, the apparent overall purification of 180-fold required to attain homogeneity from extracts of the overexpressing strain actually is closer to 50-fold.

Properties of Purified RNase D

Purity. RNase D prepared by the new procedure described here has a specific activity (300 units/mg) identical to that of the homogeneous enzyme prepared earlier by a more cumbersome purification scheme.[4] The purified RNase D is homogeneous by electrophoresis under denaturing conditions or by isoelectric focusing. In addition, the amino acid composition of the purified protein agrees very closely with that expected based on the nucleotide sequence of the *rnd* gene, and N-terminal sequence analysis of the purified protein shows only a single amino acid residue at each position.[15]

Structural Properties. Based on sodium dodecyl sulfate–polyacrylamide gel electrophoresis the molecular weight of *E. coli* RNase D is 41,000. This agrees with a value of 42,679 determined from the nucleotide sequence of the coding region of the *rnd* gene.[15] The native molecular weight of RNase D based on its elution position from Ultrogel AcA 44 is 40,000, indicating that the protein is a monomer. The isoelectric point of RNase D based on isoelectric focusing is 6.2. This value also is in agreement with that derived from the gene sequence. RNase D is a relatively hydrophobic protein; over 43% of the residues are the hydrophobic amino acids Ala, Ile, Leu, Met, Phe, Trp, or Val.[15] RNase D also is relatively insensitive to sulfhydryl reagents although at least one of the five half-cysteines present in the protein should be in the sulfhydryl form.

Catalytic Properties. RNase D is a 3′-exoribonuclease with a high degree of specificity for tRNA-like molecules.[25] It can remove extra residues from the 3′ terminus of a tRNA precursor in a random fashion and regenerate amino acid acceptor activity.[26] Its rate of hydrolysis of intact tRNA is less than 5% of its rate on tRNA precursors. However, once the -CCA sequence is removed from tRNA, a number of these species unfold to some degree and again become substrates for the enzyme.[19] RNase D requires a free 3′-hydroxyl group for activity, but the identity of the terminal nucleotide residue does not appear to have any effect. Likewise, the resistance of intact tRNA is not due to the -CCA sequence per se since a second -CCA sequence following the first is removed rapidly. Other RNA molecules such as synthetic polyribonucleotides and rRNA are poor substrates.

RNase D requires a divalent cation for activity. This requirement can be satisfied by Mg^{2+}, Mn^{2+}, or Co^{2+} with similar effectiveness. The pH optimum for RNase D is about 9.0–9.5, but this may partly reflect an effect of loosening the structure of the RNA substrate. RNase D is sensitive to elevated ionic strength. Fifty percent inhibition is obtained at 150 mM KCl.

Function. The *in vitro* specificity of RNase D strongly suggested that it played a role in the 3′ processing of tRNA precursors.[25] However, a

[25] H. Cudny, R. Zaniewski, and M. P. Deutscher, *J. Biol. Chem.* **256**, 5633 (1981).
[26] H. Cudny and M. P. Deutscher, *Proc. Natl. Acad. Sci. U.S.A.* **77**, 837 (1980).

mutant strain containing a deletion of the *rnd* gene and totally devoid of RNase D activity grows and processes its tRNA normally.[27] Thus, the *in vivo* function of RNase D remains undetermined. The possibility still exists that RNase D does function in tRNA processing but that a backup enzyme can function in its absence. Overexpression of RNase D is deleterious to *E. coli*,[13] and recent work has shown that strains lacking tRNA nucleotidyltransferase are extremely sensitive to RNase D overexpression.[28] These findings establish that RNase D can work at the 3' terminus of tRNA *in vivo*.

RNase T

Assay Method

The assay for RNase T measures the release of [¹⁴C]AMP from radioactive tRNA-CC[¹⁴C]A into an acid-soluble form as follows:

$$\text{tRNA-CC[}^{14}\text{C]A} \rightarrow \text{tRNA-CC} + \text{[}^{14}\text{C]AMP} \tag{3}$$

The radioactive tRNA substrate is prepared from tRNA-CC and [¹⁴C]ATP (~10⁴ cpm/nmol) using tRNA nucleotidyltransferase as described.[18] tRNA nucleotidyltransferase from any source can be used for this purpose. tRNA-CC[¹⁴C]A is a highly specific substrate for RNase T. In crude extracts, under standard RNase T assay conditions, over 80% of the activity against this substrate is due to RNase T.

The standard assay mixture for RNase T contains the following in 0.1 ml: 50 mM glycine–NaOH (pH 8.9), 5 mM MgCl$_2$, 30 μg tRNA-CC[¹⁴C]A, and enzyme fraction. After incubation at 37° with an amount of enzyme that releases less than 30% of the radioactivity, acid-soluble material is determined exactly as described above for RNase D. RNase T is somewhat unstable during assays when present in purified form, leading to nonlinear release of [¹⁴C]AMP. However, addition of 10 μg of bovine serum albumin to the assay mixture helps to increase the linearity of the assay. One unit of RNase T activity is the amount of enzyme that will solubilize 1 μmol of AMP per hour.

Purification Procedure

In order to simplify purification of RNase T, a mutant *E. coli* strain, C3/5-7 (RNase D⁻, II⁻, I⁻), lacking several potential contaminating RNases, is used as the source of starting material.[6] Cells are grown in YT medium containing 0.4% glucose and tetracycline (10 mg/liter) to an A_{550}

[27] R. T. Blouin, R. Zaniewski, and M. P. Deutscher, *J. Biol. Chem.* **258,** 1423 (1983).
[28] J. Zhang and M. P. Deutscher, *J. Biol. Chem.* **263,** 17909 (1988).

of about 2.5. Buffer B contains 50 mM glycine–NaOH (pH 8.9), 1 mM dithiothreitol, 0.1 mM ethylenediaminetetraacetic acid (EDTA), and 0.1 mM phenylmethylsulfonyl fluoride, and buffer C contains 10 mM Tris-Cl (pH 7.5), 1 mM dithiothreitol, 0.1 mM EDTA, 0.1 mM phenylmethylsulfonyl fluoride, and 10% glycerol. All operations are carried out in a cold room at 4°.

Step 1: Preparation of Supernatant Fraction. Frozen cells (250 g) are partially thawed, suspended in 2.5 volumes of buffer B, and ruptured by two passes through an Aminco French press at 10,000 psi. DNase I (10 μg/ml) is added to the extract, and the sample is incubated for 45 min at 0°. The treated extract is centrifuged at 27,000 g for 40 min, and the supernatant fraction obtained is centrifuged further at 45,000 rpm for 2 hr in a Beckman Ti 50.2 rotor.

Step 2: Chromatography on Affi-Gel Blue. The high-speed supernatant fraction is made 10 mM in Tris-Cl (pH 7.5) and 0.1 M in KCl and applied to a column of Affi-Gel Blue (5 × 15 cm) equilibrated with buffer C containing 0.1 M KCl. The column is washed with this buffer until the A_{280} is below 1, then washed with 1 liter of buffer C containing 0.3 M KCl. RNase T is eluted with buffer C containing 0.8 M KCl. This step is very effective for removing the nucleic acid present in the extract and as well leads to about a 20-fold purification of RNase T. The apparent purification is even greater because the removal of inhibitory nucleic acid increases the total RNase T activity about 4-fold.

Step 3: DEAE-Sephadex Chromatography. The combined eluate is dialyzed against buffer C to reduce the salt concentration and added to a DEAE-Sephadex column (5 × 13 cm) equilibrated with buffer C. The column is washed with this buffer until the A_{280} is below 0.01. RNase T is eluted with a linear 2.5-liter gradient from 0 to 1 M KCl in buffer C. This step does not give much purification, but serves to separate RNase T from most of the RNase BN present since the latter enzyme is eluted at a lower salt concentration.

Step 4: Hydroxyapatite Chromatography. The active fractions are combined and dialyzed against buffer C containing 50 mM potassium phosphate (pH 7.5) instead of Tris-Cl. The dialyzed sample is applied to a hydroxyapatite column (2.5 × 10 cm), containing 10% cellulose (to improve flow rates), equilibrated with buffer C containing 100 mM potassium phosphate (pH 7.5). RNase T is eluted with a 500-ml gradient from 100 to 500 mM phosphate in buffer C. The peak fractions, which elute at about 350 mM phosphate, are combined and concentrated to 5 ml with an Amicon YM5 Diaflo membrane.

Step 5: Chromatography on Ultrogel AcA 44. The concentrated RNase T sample was added to a column (2.5 × 92 cm) of Ultrogel AcA 44

TABLE II
PURIFICATION OF *Escherichia coli* RNase T[a]

Step	Total activity (units)	Specific activity (units/mg)	Relative purification (-fold)	Yield (%)
1. High-speed supernatant fraction	104	0.008	1.0	100
2. Affi-Gel Blue	426	0.791	99	410
3. DEAE-Sephadex	293	0.996	125	280
4. Hydroxyapatite	156	12.5	1,560	150
5. Ultrogel AcA 44	65	195	24,400	62

[a] Adapted from Ref. 6.

equilibrated with buffer C containing 1 M KCl to prevent adsorption to the resin. RNase T elutes as a single sharp peak corresponding to a peak in the absorbance profile. The active fractions are combined, concentrated by ultrafiltration, as above, and stored frozen at −20°. Purified RNase T retains 80–90% of its activity over 6 months when stored this way provided that dithiothreitol is present at 1 mM.

The purification scheme is summarized in Table II. This scheme leads to an apparent purification of RNase T of about 25,000-fold with a recovery of about 60%. However, since there is a 4-fold increase in total activity after Step 2, it is clear that RNase T activity is underestimated in crude extracts. Addition of a boiled crude fraction to purified RNase T leads to about 85% inhibition, indicating that a 6-fold correction is needed. Using this estimate, the overall purification of RNase T is about 4,000-fold with a recovery of about 10%.

Properties of Purified RNase T

Purity. Based on sodium dodecyl sulfate–polyacrylamide gel electrophoresis and silver staining, RNase T purified by this procedure is over 95% pure. On nondenaturing gel electrophoresis, RNase T activity corresponds with the major protein band. Elution and rerunning of this band on denaturing gels confirmed its identity with the major band on these gels. Several closely migrating bands are observed on isoelectric focusing gels, but the explanation for this heterogeneity is not known. It may be due to some proteolysis during purification or to an undetermined protein modification.

Structural Properties. Based on migration on denaturing gels, RNase T has a monomer molecular weight of 25,000. However, the size of native

RNase T based on gel filtration (32 Å) and glycerol gradient centrifugation (3.8 S) corresponds to a globular protein of about 50,000 daltons. Based on these data, RNase T is probably an α_2 dimer.

The A_{280}/A_{260} ratio of purified RNase T is about 1.0, suggesting that about 3% nucleotide material is still present. However, there is no evidence for an RNA involvement in activity since treatment with micrococcal nuclease has no effect on activity.

RNase T is readily inactivated by sulfhydryl reagents, suggesting that one or more -SH groups are important for maintaining structure or activity of this enzyme. This finding is consistent with the requirement for dithiothreitol to prevent inactivation by oxidation.

RNase T is also quite sensitive to temperature inactivation, losing 80% of its activity in 5 min at 37°. KCl at 1 M concentration and tRNA protect to some degree. Nevertheless, this problem, as well as sensitivity to dilution, contribute to nonlinear assays for RNase T.

Catalytic Properties. RNase T displays maximal activity at pH values between 8 and 9 in either Tris or glycine buffers. A divalent cation is required for activity, with 2–5 mM Mg^{2+} or 1 mM Mn^{2+} satisfying the requirement. Fifty percent inhibition of RNase T activity was found at 100 mM KCl.

RNase T is an exoribonuclease that removes the 3′-terminal AMP residue from intact tRNA. RNase T can work on all tRNA species in a bulk tRNA population; however, its rate of action on individual tRNAs is not known. The apparent K_m value for tRNA in the reaction is 14 μM. Aminoacyl-tRNA and tRNA-CCp are not substrates for RNase T, and tRNA-CC is hydrolyzed at less than 1% the rate of intact tRNA. Thus, RNase T is highly specific for a free 3′-terminal residue on mature tRNA. However, the tRNA precursor analog tRNA-CCA-C$_3$ is hydrolyzed at about 40% the rate of intact tRNA[29]; natural tRNA precursors prepared by SP6 RNA polymerase also are substrates, but 3′ trimming proceeds into the -CCA sequence.[30]

Function. In vivo, RNase T is responsible for the end turnover of tRNA. Normally, *E. coli* strains lacking tRNA nucleotidyltransferase contain about 15% defective tRNA molecules, and grow slowly. However, if RNase T is also deficient (the best *rnt* mutant still retains 20% activity), the level of defective tRNA decreases about 4-fold and cells grow almost normally.[10] In fact, this observation can be used to select for *rnt* mutants. The *rnt* mutation displays no phenotype in *cca*$^+$ cells. These findings demonstrate that RNase T does participate in end turnover of

[29] M. P. Deutscher, C. W. Marlor, and R. Zaniewski, *Proc. Natl. Acad. Sci. U.S.A.* **81**, 4290 (1984).

[30] H. Cudny and M. P. Deutscher, unpublished results (1987).

tRNA. However, since end turnover of tRNA is not yet known to serve any physiological function, it is not clear whether this is the only role of RNase T *in vivo*. The ability of RNase T to also act on tRNA precursors may support a role in this process as well.

RNase BN

Assay Method

The assay for RNase BN measures the release of [^{14}C]UMP from radioactive tRNA-C[^{14}C]U into an acid-soluble form as follows:

$$\text{tRNA-C}[^{14}\text{C}]\text{U} \rightarrow \text{tRNA-C} + [^{14}\text{C}]\text{UMP} \tag{4}$$

The tRNA substrate is an analog of a precursor to bacteriophage T4 tRNA^{Pro} which has been shown to require RNase BN for its maturation *in vivo*.[9] tRNA-C[^{14}C]A can also be used as a substrate with equal effectiveness.

The radioactive tRNA substrate is prepared from tRNA-C and [^{14}C]UTP ($\sim10^4$ cpm/nmol) using tRNA nucleotidyltransferase.[18] Either rabbit liver or *E. coli* enzyme can be used for this purpose, although the former is somewhat more effective.

The standard assay mixture for RNase BN contains the following in 0.1 ml: 20 mM Tris–maleate (pH 6.0), 0.2 mM CoCl$_2$, 200 mM KCl, 20 μg tRNA-C[^{14}C]U, and enzyme fraction. After incubation at 37° under conditions that release less than 30% of the radioactivity, acid-soluble material is determined as described above for the other RNases. One unit of RNase BN is the amount of enzyme that will solubilize 1 μmol of UMP per hour.

Purification Procedure

No definitive procedure for purifying RNase BN to homogeneity has yet been developed. However, several initial purification steps have been used with repeated success that lead to a partially purified enzyme fraction ($\sim$30-fold),[17] and these are described below. In addition, several other steps that result in further purification, but with which we have limited experience, are also mentioned. In one of these procedures, RNase BN was purified approximately 1,000-fold, and the material obtained by this method was used to characterize the enzyme.

In order to eliminate as many potential contaminating RNases as possible, purification of RNase BN is carried out with extracts prepared from *E. coli* strain 680#6 (RNase I$^-$, II$^-$, D$^-$, R$^-$). This strain was constructed from strain S296-680 (RNase$^-$, II$^-$, R$^-$)[5] by phage P1-mediated transduction of a deletion mutation in the *rnd* gene and selection for tetracycline resistance owing to a nearby Tn*10* transposon.[27] Cells were grown in the high-density growth medium described in another chapter of this volume,[23] but containing 10 mg/liter of tetracycline. Buffer D contains 10 mM Tris-HCl (pH 7.5), 0.1 M NH$_4$Cl, 20 mM MgCl$_2$, 1 mM dithiothreitol. All operations are carried out in a cold room at 4°.

Step 1: Purification of Supernatant Fraction. Frozen cells (50 g) are partially thawed, suspended in 2.5 volumes of buffer D, and a high-speed supernatant fraction prepared as described for RNase T.

Step 2: DEAE-Sephadex Chromatography. The supernatant fraction is diluted with buffer C to reduce the NH$_4$Cl concentration to 50 mM and applied to a column of DEAE-Sephadex (2.5 × 25 cm) equilibrated with buffer C. RNase BN is eluted with a linear 1200-ml gradient from 0 to 0.7 M KCl in buffer C. RNase BN is eluted slightly ahead of RNase T in this procedure. The DEAE-Sephadex chromatography routinely leads to a 5- to 8-fold purification and also serves to remove much of the nucleic acid present.

Step 3. Chromatography on Affi-Gel Blue. Fractions containing RNase BN activity are combined, dialyzed against buffer C to remove KCl, and applied to a column of Affi-Gel Blue (2.5 × 13 cm) equilibrated with buffer C. RNase BN is eluted with a 600-ml linear gradient from 0 to 1 M KCl in buffer C. This column leads to an additional 5-fold purification and removal of the remaining nucleic acid.

These two steps lead to a purification of 30-fold. Additional purification of RNase BN can be obtained on hydroxyapatite, phosphocellulose, Phenyl-Sepharose, and Ultrogel AcA 44, but the various steps have not yet been combined into a reproducible purification scheme. In one preparation that employed DEAE-Sephadex, hydroxyapatite, Ultrogel AcA 44, and Affi-Gel Blue, in that order, a purification of about 1000-fold was obtained with an increase in specific activity from about 0.0009 to 0.81 units/mg.[17] This preparation has been used for the characterization of RNase BN described below.

A major difficulty in the purification of RNase BN is the considerable loss of activity that accompanies dialysis and ultrafiltration, apparently owing to its tight binding to the membranes. Addition of 0.1% Triton X-100 alleviates this problem to some degree.

Properties of RNase BN

Based on gel filtration, the native molecular weight of RNase BN is about 70,000. RNase BN is an exoribonuclease, releasing [^{14}C]UMP or [^{14}C]AMP from tRNA-C[^{14}C]U and tRNA-C[^{14}C]A, respectively. These

TABLE III
COMPARISON OF RNase D, RNase T, AND RNase BN

Enzyme	Molecular weight	Subunit structure	Preferred substrate	Mode of action	pH optimum	Preferred cation
RNase D	42,000	α	tRNA-CCA-Cn	3′-exo	9.0–9.5	Mg^{2+}
RNase T	50,000	α_2	tRNA-CCA	3′-exo	8.0–9.0	Mg^{2+}
RNase BN	~70,000	Unknown	tRNA-CU	3′-exo	6.0–6.5	Co^{2+}

two tRNAs are the preferred substrates for RNase BN, and, based on studies of BN mutants, activity against these substrates is lost when RNase BN is absent. The enzyme displays no activity against synthetic polynucleotides or rRNA. A low level of activity against tRNA-CC[^{14}C]A and tRNA-CCA[^{14}C]C$_3$ is observed, but it is not certain that this is not due to a contaminating activity.

RNase BN is unusual in that the preferred divalent cation for activity is Co^{2+}, at 0.2 mM. Mg^{2+} or Mn^{2+} are less effective at all concentrations tested. Likewise, the pH optimum for RNase BN is 6.0 to 6.5, considerably lower than for other exoribonucleases. Optimal activity is obtained at a salt concentration of 200 mM.

The function of RNase BN in uninfected *E. coli* is not known. The enzyme is absent in extracts prepared from strains BN and CAN,[5,31] and these strains do not support the growth of amber-suppressor derivatives of bacteriophage T4.[9,32] Furthermore, 3′ processing of certain T4 tRNA precursors is affected by the absence of this RNase.[9] No phenotype has been observed when this enzyme is absent from uninfected cells.

Comparison of RNases D, T, and BN

A summary of the properties of RNases D, T, and BN are presented in Table III. The three RNases differ structurally, catalytically, and genetically.

[31] P. K. Asha and M. P. Deutscher, *J. Bacteriol.* **156**, 419 (1983).
[32] A. N. Maisurian and E. A. Buyanovska, *Mol. Gen. Genet.* **120**, 227 (1973).

[35] Transfer RNA Nucleotidyltransferase

By MURRAY P. DEUTSCHER

All tRNA molecules contain at the 3' terminus the identical -CCA sequence which is required for their acceptor and transfer functions.[1] An enzyme, termed tRNA nucleotidyltransferase, that incorporates nucleotide residues into this terminal sequence has been identified in a wide variety of cells and, where examined, has been found in both the cytoplasmic and mitochondrial compartments.[2] To date, in all eukaryotic tRNA genes examined, the -CCA sequence is not encoded and must be added posttranscriptionally during the processing of tRNA precursors, presumably by this enzyme.[1] In prokaryotic cells, no discrete pattern has emerged. In laboratory strains of *Escherichia coli,* all tRNA genes sequenced encode the -CCA sequence,[3] and mutant strains devoid of tRNA nucleotidyltransferase retain viability, although they grow slowly.[4] In other prokaryotic systems, some tRNA genes encode -CCA and others do not.[5]

Procedures for the purification of tRNA nucleotidyltransferase from a number of sources have been described.[2] The purification procedure for the enzyme from rabbit liver was included in an earlier volume of this series.[6] Owing to the low level of this enzyme in cells, earlier purifications led to the isolation of relatively small amounts of purified protein that were insufficient for detailed structural characterization, and, until recently, little was known about the structure of this protein from any source.[2] The *E. coli cca* gene, encoding tRNA nucleotidyltransferase, has now been cloned and sequenced,[7] and the protein has been overexpressed several hundredfold.[8] This has made available a rich source for purification of large amounts of the enzyme. The purification procedure described below affords a rapid route to partially purified tRNA nucleotidyltrans-

[1] M. P. Deutscher, *in* "Enzymes of Nucleic Acid Synthesis and Modification" (S. T. Jacob, ed.), p. 159. CRC Press, Boca Raton, Florida, 1983.
[2] M. P. Deutscher, *in* "The Enzymes" (P. D. Boyer, ed.), 3rd Ed., Vol. 15, Part B, p. 183. Academic Press, New York, 1982.
[3] M. J. Fournier and H. Ozeki, *Microbiol. Rev.* **49,** 379 (1985).
[4] L. Zhu and M. P. Deutscher, *EMBO J.* **8,** 2473 (1987).
[5] B. S. Vold, *Microbiol. Rev.* **49,** 71 (1985).
[6] M. P. Deutscher, this series, Vol. 29, p. 706.
[7] H. Cudny, J. R. Lupski, G. N. Godson, and M. P. Deutscher, *J. Biol. Chem.* **261,** 6444 (1986).
[8] H. Cudny and M. P. Deutscher, *J. Biol. Chem.* **261,** 6450 (1986).

ferase that is suitable for most enzymological studies as well as homogeneous enzyme for detailed structural analysis.

Assay Method

The most commonly used assay for tRNA nucleotidyltransferase measures the incorporation of radioactive ATP into an acid-insoluble product as follows:

$$tRNA-CC + ATP^* \rightarrow tRNA-CCA^* + PP_i \tag{1}$$

The incorporation of radioactive CTP into tRNA-N or tRNA-C may also be used, and in some extracts with active poly(A) polymerizing activities this may be preferable in the early steps of a purification.

The most convenient tRNA substrate for routine use during purification is commercial bakers' yeast tRNA. Generally, this material contains about 50% tRNA-CC and is active with tRNA nucleotidyltransferases from most sources. If more defined preparations of tRNA-CC, tRNA-C, or tRNA-N are required, they can be prepared by single or multiple cycles of periodate oxidation and alkaline phosphatase treatment of any tRNA.

The standard assay mixture contains the following in 0.1 ml: 50 mM glycine–NaOH (pH 9.4), 10 mM MgCl$_2$, 0.2 mM [^{14}C]ATP ($\sim$1,000 cpm/nmol), 200 μg yeast tRNA, and enzyme fraction. After incubation at 37° for 3–5 min (using an amount of enzyme that incorporates 0.2–1.5 nmol of AMP), the reaction is terminated by placing in ice and adding 3 ml of cold 10% w/v trichloroacetic acid containing 20 mM sodium pyrophosphate. After 10 min to allow complete precipitation of tRNA, the mixture is filtered through Whatman GF/C filters and washed with six 3-ml portions of 2.5% w/v trichloroacetic acid–20 mM sodium pyrophosphate and then one 5-ml portion of ethanol–ether (1 : 1) (v/v). The filters are placed along the side of miniscintillation vials, dried under an infrared lamp for 10 min, and counted in a scintillation counter with 3 ml of scintillation fluid. Blank values under these conditions are within 30 cpm of the machine background.

One unit of tRNA nucleotidyltransferase activity is the amount of enzyme catalyzing the incorporation of 1 μmol of AMP per hour under the standard conditions. It should be noted that the standard conditions are suboptimal for [^{14}C]ATP but are used to conserve material.

Enzyme Source

Overexpression of tRNA Nucleotidyltransferase. Several *E. coli* strains that overexpress tRNA nucleotidyltransferase have been constructed.[8] Purification of the enzyme from one of these strains,

UT481/pEC4, is described below. This strain was constructed by transforming *E. coli* UT481 [(Δ*lac-pro*), *hsdS*(r^-m^-), F'*lacI*^{q}, Δ*lac*M15] with plasmid pEC4. This plasmid contains a 1.9-kb DNA fragment encoding the *cca* gene inserted into the *Eco*RI–*Sma*I site of plasmid pUC8.[7] This strain overexpresses tRNA nucleotidyltransferase activity approximately 100-fold relative to the wild type, and expression is dependent on the *cca* gene promoter. Based on maxicell analysis and study of the purified protein, the overexpressed tRNA nucleotidyltransferase is probably the normal *E. coli* protein. Other strains in which tRNA nucleotidyltransferase expression is dependent on the *lac* promoter overproduce the enzyme to as much as 5-fold higher levels, but in these cases a fusion protein is produced.[8] The fusion protein has not been studied further.

Growth of Cells. High-density growth medium contains 32 g tryptone and 20 g yeast extract per liter, M9 salts, 0.1 mM MgSO$_4$, 1 μM FeCl$_3$, 0.2% glucose, and 0.1 mg/ml ampicillin (all added after autoclaving separately). Bacterial cells are grown at 37° in a 12-liter fermentor to an A_{600} of about 20, harvested by centrifugation, and stored frozen at $-20°$ until use. Approximately 20 g (wet weight) of cells per liter of culture can be obtained under these growth conditions.

Purification Procedure

All operations are carried out in a cold room at 4°.

Reagents

Buffer A: 20 mM tris(hydroxymethyl)aminomethane (Tris)-Cl (pH 7.6), 0.1 mM phenylmethylsulfonyl fluoride, 10% glycerol

Buffer B: 20 mM potassium phosphate (pH 7.5), 0.1 mM phenylmethylsulfonyl fluoride, 0.1 mM dithiothreitol, 10% glycerol

Buffer C: 20 mM potassium phosphate (pH 6.5), 0.1 mM dithiothreitol, 0.1 mM phenylmethylsulfonyl fluoride

Buffer D: 20 mM potassium phosphate (pH 7.0), 0.1 mM dithiothreitol, 20% (w/v) ammonium sulfate

Step 1: Preparation of Crude Extract. Frozen UT481/pEC4 cells (460 g) are partially thawed and suspended in 5 volumes of buffer A. The cells are disrupted by two passages through an Aminco French press at 6000 psi, and the resulting suspension is centrifuged for 30 min at 0–5° at 15,000 rpm to remove cell debris. The low-speed supernatant fraction is recentrifuged at 45,000 rpm for 60 min at 0–5° in a Beckman Ti 50.2 rotor to remove ribosomes. The resulting high-speed supernatant fraction is the starting material for further purification.

Step 2: DEAE-Cellulose Chromatography. The supernatant fraction is applied to a DEAE-cellulose column (8 × 30 cm) equilibrated with buffer

B and washed with this buffer until the A_{280} is below 1. The enzyme is eluted with a linear gradient of 3 liters of buffer B and 3 liters of buffer B containing 200 mM monobasic potassium phosphate (pH 4.5) instead of phosphate (pH 7.5). This step serves to remove a large amount of nucleic acid and as well as leads to a 3- to 5-fold purification of the enzyme.

Step 3: Chromatography on Controlled-Pore Glass. Active fractions from DEAE-cellulose are pooled and added to a column (2.5 × 25 cm) of controlled-pore glass (Electro-Nucleonics, mean pore size ~700 Å, 120/200 mesh) equilibrated with buffer C. The column is washed with this buffer until the A_{280} is below 0.05. Additional contaminating protein can be removed by washing with 0.1 M Tris-Cl (pH 8.8), 0.1 mM dithiothreitol. tRNA nucleotidyltransferase is eluted as a sharp peak using the latter buffer containing 1 M NaCl. The basis for enzyme binding to the glass surface is not completely understood, but it presumably involves both hydrophobic and electrostatic interactions. Nevertheless, this step affords good recovery of activity and is very effective both for purifying the enzyme and removing essentially all of the remaining nucleic acid.

Step 4: Phenyl-Sepharose Chromatography. The active fractions from controlled-pore glass are combined and applied to a Phenyl-Sepharose column (1.5 × 20 cm) equilibrated with buffer D. After loading, the column is washed with this buffer until the A_{280} is below 0.01. The column is then washed successively with buffer D lacking ammonium sulfate and buffer D containing 10% v/v ethylene glycol. tRNA nucleotidyltransferase is eluted batchwise with buffer D containing 30% v/v ethylene glycol and 10% glycerol.

Step 5: Chromatography on Affi-Gel Blue. The active eluate from Phenyl-Sepharose is placed on a column of Affi-Gel Blue (2.5 × 10 cm) equilibrated with 50 mM Tris-Cl (pH 8.8) containing 0.1 mM dithiothreitol. The column is washed with 3 volumes of the starting buffer containing 0.2 M NaCl to remove poorly bound proteins, and tRNA nucleotidyltransferase is eluted with 50 mM Tris-Cl (pH 7.0), 0.1 mM dithiothreitol, 2 M KBr. The final eluate is dialyzed against this buffer lacking KBr. Purified tRNA nucleotidyltransferase is somewhat unstable. Inclusion of 30% ethylene glycol helps in stabilizing the enzyme.

The purification procedure is summarized in Table I. This procedure leads to homogeneous enzyme in about 10% yield. However, tRNA nucleotidyltransferase of about 40% purity, sufficient to be used as a reagent or for enzymatic studies, can be prepared by the first two chromatographic steps in about 50% yield. Enzyme at this stage is free of ribonuclease activity against a variety of tRNA substrates. The final purification to homogeneity requires two additional chromatographic steps in order to remove all the minor contaminating proteins present. A variety of columns were tried for the final two steps including phosphocellulose,

TABLE I
PURIFICATION OF OVEREXPRESSED *Escherichia coli* tRNA NUCLEOTIDYLTRANSFERASE[a]

Step	Total activity (units)	Specific activity (units/mg)	Relative purification (-fold)	Yield (%)
1. High-speed supernatant	582,000	27.0	1	100
2. DEAE-cellulose	333,000	150	5	60
3. Controlled-pore glass	300,000	1200	43	54
4. Phenyl-Sepharose	65,000	1800	67	12
5. Affi-Gel Blue	55,000	3500	130	10

[a] Adapted from Ref. 8.

hydroxyapatite, Ultrogel AcA 44, Phenyl-Sepharose, and Affi-Gel Blue, and each was successful in giving some additional purification. Most combinations of two columns will provide pure enzyme, but no single column was sufficient for removing all the contaminants remaining after Step 3.

Properties of the Purified Enzyme

Purity. *Escherichia coli* tRNA nucleotidyltransferase prepared by the procedure described here is homogeneous by several criteria. Electrophoresis on denaturing gels and isoelectric focusing in acrylamide gels both showed only a single stained protein band. Antibody prepared against the purified enzyme cross-reacted against only a single protein in Western blots of crude high-speed supernatant fractions. Purified tRNA nucleotidyltransferase displays an A_{280}/A_{260} ratio of 1.6, indicating that it is devoid of nucleic acid and confirming that a template is not required for synthesis of the -CCA sequence.

Structural Properties. Based on sodium dodecyl sulfate–polyacrylamide gel electrophoresis, the molecular weight of *E. coli* tRNA nucleotidyltransferase is 47,000. This agrees with a value of 46,408 determined from the nucleotide sequence of the coding region of the *cca* gene. The amino acid composition of the purified protein also agrees very closely with that expected from the nucleotide sequence. The native molecular weight of tRNA nucleotidyltransferase based on gel filtration is about 50,000, indicating that the protein is a monomer. The isoelectric point of the enzyme based on isoelectric focusing in polyacrylamide gels is 5.9.

Catalytic Properties. Purified *E. coli* tRNA nucleotidyltransferase catalyzes the incorporation of both AMP and CMP residues into the appropriate tRNA acceptor. The enzyme requires Mg^{2+} for activity, and optimum incorporation of nucleotides occurs at pH 9–10. The apparent K_m

values for ATP and CTP are 0.3 and 0.03 mM, respectively, and for tRNA substrates, about 10–20 μM.

Under certain conditions, tRNA nucleotidyltransferases can make errors in nucleotide incorporation, and these anomalous reactions have been used successfully for the synthesis of tRNA molecules with altered 3' termini. The *E. coli* enzyme can be used to synthesize altered molecules such as tRNA-CU and tRNA-CA; however, the rate of these reactions is not as rapid as with the liver enzyme. Nevertheless, the ability to make much larger amounts of the *E. coli* enzyme should make it useful as a reagent for synthesizing altered tRNAs or for repairing the 3' end of tRNA molecules.

Function. In *Escherichia coli,* tRNA nucleotidyltransferase is not an essential enzyme.[4] At least in laboratory strains, tRNA biosynthesis is not affected by the presence or absence of this enzyme. It does participate in repair of tRNA molecules that are lacking 3'-terminal residues since mutant strains devoid of the enzyme accumulate defective tRNAs,[9] but such mutant cells can still grow slowly.[4] Loss of terminal residues occurs when tRNA molecules are uncharged[10] owing to the action of the enzyme, RNase T.[11] The significance of end turnover of tRNA *in vivo* is not known; however, the existence of enzymes to specifically carry out this process suggests that it does have physiological relevance.

[9] M. P. Deutscher and R. H. Hilderman, *J. Bacteriol.* **118,** 621 (1974).
[10] M. P. Deutscher, J. J. Lin, and J. A. Evans, *J. Mol. Biol.* **117,** 1081 (1977).
[11] M. P. Deutscher, C. W. Marlor, and R. Zaniewski, *Proc. Natl. Acad. Sci. U.S.A.* **82,** 6427 (1985).

[36] Yeast Extracts for Transfer RNA Gene Transcription and Processing

By CLAIRE F. EVANS and DAVID R. ENGELKE

Introduction

The synthesis and processing of tRNA precursors have been studied extensively in *Saccharomyces cerevisiae*. This is due in part to the relatively large number of tRNA genes that have been cloned and sequenced[1]

[1] C. Guthrie and J. Abelson, *in* "Molecular Biology of the Yeast *Saccharomyces cerevisiae: Metabolism and Gene Expression*" (J. N. Strathern, E. W. Jones, and J. R. Broach, eds.), p. 487. Cold Spring Harbor Laboratory, Cold Spring Harbor, New York, 1982.

METHODS IN ENZYMOLOGY, VOL. 181

and in part to the accessibility of the system to genetic manipulation. A collection of altered tRNA genes has also been cloned[2-8] that harbors defects in transcription and RNA processing. Analysis of these altered genes has shown that mutations leading to severe defects in transcription reside primarily in intragenic control regions (ICRs) within the sequences encoding the tRNA arms (reviewed in Ref. 9). Additional DNA sequences upstream of the coding region also appear to influence the efficiency of promoter utilization *in vivo*[10] and *in vitro*.[11]

The biochemical characterization of yeast tRNA gene expression requires the preparation of extracts that serve as sources for the purification of the individual transcription and processing components. The use of these extracts to also synthesize pre-tRNA substrates for examining RNA processing events can have a number of advantages over synthesizing the substrates from bacteriophage promoters *in vitro*.[12] The labor required to construct recombinant plasmid templates with correct transcription initiation and termination sites for the phage RNA polymerases can be prohibitive if a large number of different gene products are to be studied and large quantities of RNA are not required. Further, it is often desirable to obtain maturation intermediates rather than primary transcripts as substrates for later processing enzymes in an ordered pathway. Soluble yeast extracts have previously been reported[13-15] that accurately initiate and terminate transcription on exogenous tRNA gene templates and proceed with steps in nucleolytic processing and nucleoside modification. Here we detail the preparation and use of a reproducibly active subcellular extract from *S. cerevisiae* haploid cells that can be manipulated so as to provide several different processing intermediates.

[2] J. Kurjan, B. D. Hall, S. Gillam, and M. Smith, *Cell* **20**, 701 (1980).
[3] R. A. Koski, S. G. Clarkson, J. Kurjan, B. D. Hall, and M. Smith, *Cell* **22**, 415 (1980).
[4] G. Carrara, G. Di Segni, A. Otsuka, and G. P. Tocchini-Valentini, *Cell* **27**, 371 (1981).
[5] A. J. Newman, R. C. Ogden, and J. Abelson, *Cell* **35**, 117 (1983).
[6] V. M. Reyes, A. Newman, and J. Abelson, *Mol. Cell. Biol.* **6**, 2436 (1986).
[7] M. C. Strobel and J. Abelson, *Mol. Cell. Biol.* **6**, 2663 (1986).
[8] R. E. Baker, S. Camier, A. Sentenac, and B. D. Hall, *Proc. Natl. Acad. Sci. U.S.A.* **84**, 8768 (1987).
[9] E. P. Geiduschek and G. P. Tocchini-Valentini, *Annu. Rev. Biochem.* **57**, 873 (1988).
[10] K. J. Shaw and M. V. Olson, *Mol. Cell. Biol.* **4**, 657 (1984).
[11] J. D. Johnson and G. J. Raymond, *J. Biol. Chem.* **259**, 5990 (1984).
[12] J. F. Milligan and O. C. Uhlenbeck, this series, Vol. 180, p. 51.
[13] M. S. Klekamp and P. A. Weil, *J. Biol. Chem.* **257**, 8432 (1982).
[14] R. A. Koski, D. S. Allison, M. Worthington, and B. D. Hall, *Nucleic Acids Res.* **10**, 8127 (1982).
[15] J. M. Huibregtse, C. F. Evans, and D. R. Engelke, *Mol. Cell. Biol.* **7**, 3212 (1987).

Preparation of Cell Extracts

Protocol. A forerunner of the procedure described here was originally an unpublished method developed by Carl Parker and Joanne Topol (California Institute of Technology) and was improved by suggestions from Craig Peebles (University of Pittsburgh). We have found that considerable variability is permissible in the details of the procedure. The simplest and most reproducible of these variations is given below, with the rationale and alternatives noted. The volumes given assume the starting material is from a 12-liter yeast culture grown to an A_{600} of 6–8.

1. Grow a 50-ml saturated culture in YPD (1% yeast extract, 2% Bacto-peptone, 2% glucose). [We routinely make extracts from two protease-deficient haploid strains, BJ411 (α, *his1, prb1-1122, prc1-126;* gift of Beth Jones, in our previous publications and those of others incorrectly referred to as EJ101) and PP1002 (*pep4;* gift of P. Piper and K. Stråby), that have been tested for retention of the proteinase lesion.[16] The PP1002 strain gives slightly more reproducibly active extracts, and the data described here are derived from extracts of this strain. For some RNA processing work, however, BJ411 may be preferred. Strain PP1002 is defective in the 3′-end-processing of 5 S rRNA[17] and may be defective in a tRNA 3′-endonuclease.[18]]

2. Inoculate 12 liters of YPD with 12 ml of the saturated culture. Grow at 30° with good aeration until an A_{600} of 6.0–8.0. [Exponentially growing cultures were originally used until it was found that the extracts from freshly saturated cultures gave equal or greater transcription activity per gram of cells. Cultures have not been tested from high-density fermentors because of reports that the cells become difficult to lyse.[19]]

3. Harvest the cells at room temperature by centrifugation at 3000 rpm for 5 min in a Sorvall GS3 rotor. Resuspend cell pellets in 400 ml of sorbitol buffer [1 M sorbitol, 50 mM tris(hydroxymethyl)aminomethane (Tris) (pH 7.9), 10 mM MgCl$_2$, 30 mM dithiothreitol (DTT)]. Incubate at room temperature for 15 min. Spin at 3000 rpm for 5 min in a Sorvall GS3 rotor. Decant the supernatant.

4. Resuspend the cells in 400 ml of sorbitol buffer containing 3 mM DTT and 0.45 mg/ml zymolyase 20,000 (Seikagaku, Tokyo). Homogenize with one stroke of a loose-fitting pestle in a Dounce homogenizer to ensure an even suspension. Incubate at 30° with gentle swirling for

---

[16] E. W. Jones, *Genetics* **85**, 23 (1977).
[17] P. W. Piper, J. A. Bellatin, and A. Lockheart, *EMBO J.* **2**, 353 (1983).
[18] P. Piper and K. Stråby, personal communication, (1988).
[19] J. Jaehning, personal communication, (1984).

40–60 min. Spin at 3000 rpm for 5 min in a Sorvall GS3 rotor. Decant the supernatant.

5. The surface of the cell pellet is usually slightly mottled, and the pellet is more difficult to resuspend. Resuspend the cell pellet in 400 ml of sorbitol buffer containing 3 mM DTT. Spin at 3000 rpm for 5 min in a Sorvall GS3 rotor at 4°. Decant the supernatant.

6. [All subsequent steps are performed at 0–4° with precooled solutions and equipment.] Resuspend the pellet in 250 ml of ice-cold hypotonic buffer [15 mM KCl, 10 mM N-2-hydroxyethylpiperazine-N'-2-ethanesulfonic acid (HEPES) (pH 7.9), 5 mM MgCl$_2$, 0.1 mM ethylenediaminetetraacetic acid (EDTA), 3 mM DTT]. Homogenize with one stroke of a loose-fitting pestle. Incubate for 20 min on ice. Homogenize with 5 strokes of a tight-fitting pestle. [Treatment with a Dounce homogenizer is not absolutely required but is included to aid cell disruption.] Spin at 10,000 rpm for 20 min in a Sorvall GSA rotor. Decant the supernatant. [The supernatant of the postlysis centrifugation lacks significant levels of RNA polymerases I/III and RNase P, although the latter may be masked by inhibitors. RNA polymerase II is found in variable amounts in this fraction. Early attempts to obtain more highly purified nuclei (by differential centrifugation) as starting material for the high-salt extraction did not materially affect the specific activity of the extracts and were abandoned. However, a recent report of a nuclear extract capable of specific initiation by RNA polymerase II[20] has suggested that isolated nuclei may be superior starting material for other purposes.]

7. Resuspend the pellet in 100 ml of hypotonic buffer. Homogenize with one stroke of a loose-fitting pestle. Measure the volume to calculate the "nuclear pellet" volume (total volume minus volume of hypotonic buffer added). Add one-fifth of the total volume of 4 M (NH$_4$)$_2$SO$_4$ (pH 7.9) and mix well. Incubate on a slow tilter or on ice for 30 min. [Immediate and vigorous mixing of the 4 M ammonium sulfate and pellet resuspension is required before the increased viscosity of the mixture causes clumping. Usually we add aliquots of the resuspension to individual, screw-cap ultracentrifuge tubes and shake well as the ammonium sulfate is added to each tube.]

8. Spin at 34,000 rpm for 60 min in a Beckman Ti 35 ultracentrifuge rotor. Filter the supernatant through cheesecloth into a graduated cylinder (discard pellet). Measure the volume. Slowly add 0.25 g of (NH$_4$)$_2$SO$_4$ per milliliter of filtrate while stirring gently (takes 30–60 min to dissolve completely). Stir an additional 15–20 min. Centrifuge at 10,000 rpm for 30 min in a Sorvall HB4 rotor. Decant the supernatant.

[20] N. F. Lue and R. D. Kornberg, *Proc. Natl. Acad. Sci. U.S.A.* **84,** 8839 (1987).

9. Carefully rinse the sides of the tubes with water to remove traces of $(NH_4)_2SO_4$ and then wipe dry with Kimwipes. [Wiping and rinsing the sides of the centrifuge tube are essential only if the dissolved extract is not to be dialyzed and the salt concentration needs to be held to a minimum. Dialysis of the extracts leads to variable recovery of activity in our hands and is avoided.] Resuspend in one-half of the "nuclear pellet" volume (Step 7) of HGED buffer [10% glycerol, 20 mM HEPES (pH 7.9), 0.1 mM EDTA, 1 mM DTT]. Freeze extract at $-80°$. [Aliquots of extract are quick-frozen on crushed dry ice in microcentrifuge tubes. Slower freezing in larger volumes has sometimes led to partial loss of transcription activity, but the small aliquots generally retain activity after multiple rounds of freezing and thawing for periods of up to several years.]

Conditions for Transcription Reactions

Standard Reactions. Analytical transcription reactions are carried out with 1–4 μl extract in a total volume of 20 μl containing 20 mM HEPES buffer (pH 7.9), variable KCl and $(NH_4)_2SO_4$, 4 mM $MgCl_2$, 0.1–0.4 mM dithiothreitol, 0.5–2% glycerol, 0.2 mM each ATP, CTP, and GTP, 0.02 mM UTP (2.5 Ci/mmol); and 0.0001–10 μg DNA template. Reactions are stopped by the addition of 4 μl of 2% sodium dodecyl sulfate (SDS), 0.1 M EDTA (pH 8), 1 mg/ml proteinase K, 1 mg/ml *Escherichia coli* tRNA and incubation at $37°$ for at least 30 min. Products from small reactions are then recovered by ethanol precipitation and analyzed by size on 8% polyacrylamide sequencing gels[21] containing 8 M urea. $MgCl_2$ is optimal between 4 and 6 mM. In HEPES buffer, pH is optimal at 7.5–8.2. (In 50 mM Tris the optimum is 7.0–7.3 for unknown reasons.) We recommend that all parameters be rechecked with individual extracts if maximum transcription rates are desired. Under optimum conditions a peak transcription rate of about two transcripts/gene/minute is achieved.

Ionic Strength, Time, and Temperature Optima. The degree of nucleolytic processing, as well as transcription rate, can be manipulated by changing the ionic strength, the length of the reactions, or the temperature. Ionic strength is optimal between 40 and 70 mM $(NH_4)_2SO_4$ or between 100 and 160 mM KCl, with slightly better transcription in KCl. As seen in Fig. 1A,[22] higher salts [KCl or $(NH_4)_2SO_4$] favor end-matured products. On addition of extract to the reaction mixture at $30°$, tRNA synthesis undergoes a lag of 2–4 min and then active transcription for 6–10 min (Fig. 1B). After this period, transcription drops at varying rates,

[21] F. Sanger and A. R. Coulson, *FEBS Lett.* **87,** 107 (1978).
[22] D. R. Engelke, P. Gegenheimer, and J. Abelson, *J. Biol. Chem.* **260,** 1271 (1985).

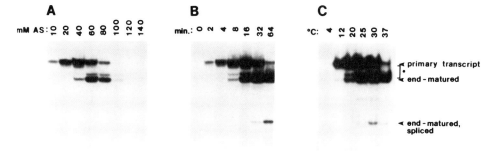

FIG. 1. Ionic strength, time, and temperature dependency of transcription reactions. Transcription reactions and polyacrylamide gel analyses of products were performed as described in the text. Reactions contained yeast extract from strain PP1002 and 1 μg of plasmid template containing the *S. cerevisiae* tRNA[SUP53] gene.[5] The tRNA[SUP53] primary transcript and major processing products[7,22] are indicated at the right of (C). RNAs indicated by the starred bracket are partially end-matured intermediates. RNAs 2–4 nucleotides shorter than the primary transcript are 3'–5' exonuclease products. The intermediate slightly larger than the end-matured form does not appear when extracts of strain BJ411 are used and may represent either incomplete 3'-end-removal or CCA addition (not determined). (A) Reactions contained 1 μl of extract and the indicated concentration of ammonium sulfate (AS) and were incubated at 30° for 15 min. (B) Reactions contained 1 μl of extract and 140 mM potassium chloride and were incubated at 30° for the indicated lengths of time. (C) Reactions contained 3 μl of extract and 140 mM potassium chloride and were preequilibrated and incubated at the indicated temperatures for 15 min.

depending on the particular extract preparation, and processed RNAs accumulate to the exclusion of primary transcripts. Nucleolytic processing can be slowed by performing the reactions at reduced temperatures (Fig. 1C), thus favoring the primary transcripts and early intermediates.

Template Effects. A variety of cloned DNA templates containing tRNA genes or tRNA-like promoters are efficiently transcribed in the yeast extracts. We have tested the genes for eight different yeast tRNAs, *Bombyx mori* tRNA[Ala] (gift of K. Sprague), adenovirus VA RNA (gift of T. Shenk), and *Xenopus laevis* Satellite I DNA (gift of D. Carroll). The *S. cerevisiae* tRNA[SUP53] (*leu3*) gene has been used for the experiments shown here because its transcription and processing products in this extract have been thoroughly characterized.[7] The optimum template concentration varies slightly between extracts and should also be tested for each type of template.

Template titrations are shown in Fig. 2 for 1 μl (Fig. 2A) or 3 μl (Fig. 2B) of extract. (Using more than 4 μl of extract does not result in linearly increased transcription.) For 1 μl, a peak of tRNA transcripts is observed between 0.01 and 0.1 μg plasmid template (~5–50 fmol). With

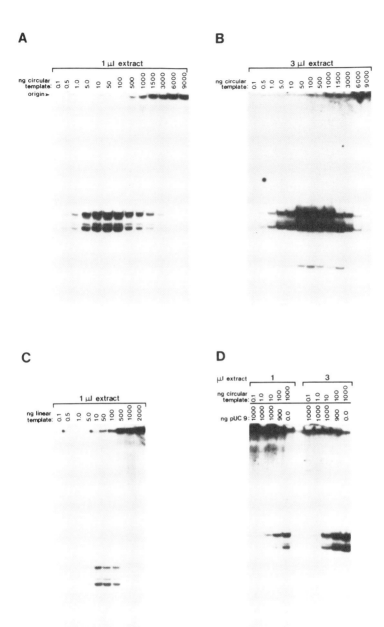

FIG. 2. Template effects. Transcription template titrations were carried out in reactions containing 140 mM potassium chloride for 15 min at 30°. The tRNA gene template was plasmid pSUP53, a 3400-bp pUC9 derivative containing one copy of the *S. cerevisiae* tRNA^{SUP53} gene.[5,7] (A) and (B) Transcription reactions contained 1 μl (A) or 3 μl (B) of yeast extract and the indicated amount of circular pSUP53 template (no carrier). (C) Reactions contained 1 μl of yeast extract and the indicated amount of linearized pSUP53 template (no carrier). The plasmid was linearized at a *Hin*dIII site 430 bp downstream of the tRNA gene. (D) Reactions contained either 1 or 3 μl of yeast extract and the indicated amounts of circular pSUP53 template and pUC9 carrier DNA. The total amount of DNA was held constant at 1 μg.

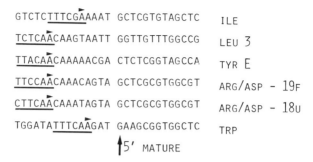

FIG. 3. Transcription initiation sites. The sequences of several yeast tRNA genes have been aligned with respect to the mature 5' ends of the tRNA products (arrow). Transcription initiation sites are indicated with arrowheads above the initiating nucleotide, and a consensus sequence at the initiation site is underlined.

3 μl of extract a proportionately higher level of transcripts is obtained with an optimum of 0.05–0.5 μg template. Quantitation of the synthesized RNA shows that the transcription rate (transcripts per gene) is highest at low template concentrations and decreases with increased template. The total tRNA synthesized increases, however, until the maximum for transcripts/reaction is reached and then falls sharply. This could be due to the binding of RNA polymerase III and transcription factors to separate template molecules at increased template concentrations, resulting in the increase in nonspecific transcripts and decrease in tRNA transcripts. Consistent with this, there appears in Fig. 2A,B to be an increase in labeled high molecular weight material (not characterized) as tRNA synthesis diminishes.

To test if the loss of specific transcription was a function of tRNA promoter sequences in the excess DNA, the total amount of DNA in the reactions was held constant at 1.0 μg with plasmid vector (pUC9) while the concentration of the gene was varied (Fig. 3D). Overall, the synthesis of tRNA transcripts was much less than observed with no carrier, reaching a maximum when the DNA was entirely the tRNA gene plasmid. We conclude that inhibition of specific transcription is primarily due to nonspecific DNA–DNA competition, but the precise mechanism of this interference has not been determined. Plasmids linearized at a unique restriction endonuclease site were also tested for transcription (Fig. 3C). Maximum transcription occurs at the same template concentration as with circular DNA but is much less efficient at all template concentrations. This might be due to a preferred topology adopted by circular templates (not tested), but it could also be due to binding of limiting transcription components to free ends on the DNA molecules. Inclusion of as little as 0.1 μg of sheared pUC9 carrier DNA (200–300 bp fragments) completely inhibits transcription by the extract. In summary, transcription is maximized by the use of circular template with a high promoter to DNA ratio.

Initiation and Termination of Transcription. The sites of transcription initiation have been determined for several yeast tRNA gene templates by incorporating [γ-^{32}P]ATP as the initiating nucleotide and subsequently sequencing the transcripts.[22] The start sites are shown in Fig. 3 for a tRNA^{Ile} gene,[23] the tRNA^{SUP53} gene,[24] the tRNA^{SUP6} gene,[25] two Arg/Asp dimeric tRNA transcription units,[26] and a tRNA^{Trp} gene.[27] No tRNA gene has yet been found that initiates with G, C, or U, and initiation may be blocked entirely by the lack of an appropriately positioned A residue.[6] On all genes examined, transcripts initiated at the last position of the sequence (Py)$_3$CPuA, which is often found as part of the longer consensus (Py)$_3$CAACAA. The influence of additional upstream sequences observed by others[10,11] has not been tested in this system, and the initiating nucleotide appears to be offset at least one position from a related system.[27a] Transcripts terminate with either five or six Us,[22,28] rather than four Us as in higher eukaryotic systems (reviewed in Ref. 9).

Synthesis of Pre-tRNAs as Substrates for RNA Processing

Synthesis with Whole Extract. The simplest method for obtaining intermediates in nucleolytic processing is to vary time and temperature as described above, separate the products by polyacrylamide gel electrophoresis, and excise gel slices containing the desired products. RNAs are separated using 6–8% polyacrylamide sequencing gels and eluted from gel slices by agitating in elution buffer (10 mM sodium acetate, 1 mM EDTA, 0.1 M KCl, 0.4% SDS) for 1–2 days at 30°. RNA is recovered by precipitation with 2.5 volumes of ethanol and resuspended in water.

The order in which the 5'-leader and 3'-trailing sequences are removed from the primary transcripts varies depending on the tRNA gene

[23] R. Ogden, personal communication, (1982).
[24] A. Andreadis, Y.-P. Hsu, G. B. Kohlhaw, and P. Schimmel, *Cell* **31**, 319 (1982).
[25] R. B. Wallace, P. F. Johnson, S. Tanaka, M. Schold, K. Itakura, and J. Abelson, *Science* **209**, 1396 (1980).
[26] O. Schmidt, J. Mao, R. Ogden, J. Beckmann, H. Sakano, J. Abelson, and D. Soll, *Nature* (*London*) **287**, 750 (1980).
[27] H. S. Kang, R. C. Ogden, and J. Abelson, *in* "Mobilization and Reassembly of Genetic Information," p. 317. Academic Press, New York, 1980.
[27a] G. A. Kassavetis, D. L. Riggs, R. Negri, L. H. Nguyen, and P. E. Geiduschek, *Mol. Cell. Biol.* **9**, 2551 (1989).
[28] D. Engelke, unpublished observations, (1983).

and is not strongly dependent on ionic strength, divalent cation concentration, or temperature. For pre-tRNA^{SUP53} (Figs. 1 and 2), the first three positions of the 3' terminus are subject to attack by a 3' to 5' exonuclease (Fig. 1, upper bands in starred bracket), but the 3' end is resistant to further processing until the RNase P cleavage has been made to yield the mature 5' terminus. Afterward, the mature 3' end (minus CCA) is rapidly obtained. These observations have been confirmed by cleaving with the separated nuclease activities, and sequence analysis of similarly ordered intermediates has been reported.[22]

The exonuclease resistance of the 3' end might be explained by an extended aminoacyl stem that can be formed by Watson–Crick base pairing between the A/G-rich 5' leaders and U-rich 3'-trailing sequences of the primary transcripts. In cases where the 3' end has been shown to be processed first,[22,28] this base pairing leads to a shorter extended stem than in pre-tRNA^{SUP53}. It is not clear whether these extracts contain an endonuclease capable of making a specific cleavage at the mature 3' terminus. Processing of the 3' end by both the whole extract and chromatographic fractions gives intermediates characteristic of exonuclease cleavage, but the presence of an endonuclease cannot be ruled out. This is true regardless of which strain is used (BJ411 or PP1002) and has also been observed in an extract from rho° petite strain. It has not been established whether one type of 3' activity is used exclusively *in vivo*.

Splicing of the precursors occurs primarily after terminal maturation in this extract. This may not merely reflect the low splicing activity of this preparation method relative to others,[29,30] since tRNA^{SUP53} primary transcripts are spliced at a reduced rate relative to the end-matured molecules by the separated components.[31,32] Removal of the termini prior to splicing is consistent with the observed accumulation of end-matured intermediates *in vivo* when splicing is blocked.[33,34]

Fractionation of Activities. If necessary, particular processing intermediates can be obtained by transcribing the tRNA genes with partially purified transcription factors and treating the transcripts with individual processing enzymes before purification through polyacrylamide gels. Several fractionation schemes have been tested to segregate the extract components required for tRNA gene transcription from the nucleolytic pro-

[29] C. L. Peebles, R. C. Ogden, G. Knapp, and J. Abelson, *Cell* **18,** 27 (1979).
[30] C. L. Greer, C. L. Peebles, P. Gegenheimer, and J. Abelson, *Cell* **32,** 537 (1983).
[31] C. L. Greer, D. Soll, and I. Willis, *Mol. Cell. Biol.* **7,** 76 (1987).
[32] C. L. Greer, personal communication, (1982).
[33] A. K. Hopper, F. Banks, and V. Evangelidis, *Cell* **14,** 211 (1978).
[34] G. Knapp, J. S. Beckmann, P. F. Johnson, S. A. Fuhrman, and J. Abelson, *Cell* **14,** 221 (1978).

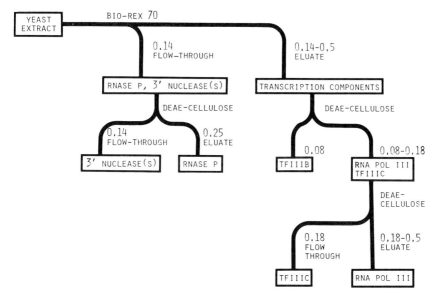

FIG. 4. Yeast extract fractionation. Extract from 70 g of cells is applied to a 30-ml column of Bio-Rex 70 (Bio-Rad) cation exchanger to separate transcription components from nucleolytic processing enzymes. The volumes of the subsequent DEAE-cellulose and DEAE-Sephadex columns used for transcription components are 15 and 4 ml, respectively. The DEAE column used for the processing enzymes is 20 ml. All chromatography buffers are HGED, with the molarity of ammonium sulfate given at which the relevant activities elute. Rationale and additional details can be found in the text.

cessing activities by varying established protocols.[22,35,36] A flow chart is shown in Fig. 4 for the scheme currently in use. The fractionation of the other tRNA processing enzymes, including the splicing activities, has not been investigated for this procedure.

Three components are required for tRNA gene transcription: RNA polymerase III and two transcription factors (TFIIIB and TFIIIC) that have been partially characterized (reviewed in Ref. 9). These components bind to cation exchangers (e.g., Bio-Rex 70, Bio-Rad, Richmond, CA) at higher ionic strengths than do the nucleolytic enzymes and are eluted in a single step to minimize column size and provide a fraction capable of synthesizing full-length primary transcripts.[22] The Bio-Rex 70 high-salt eluate can be either precipitated with ammonium sulfate or dialyzed against HGED buffer containing 70 mM ammonium sulfate. Precipitation

[35] J. Segall, T. Matsui, and R. G. Roeder, *J. Biol. Chem.* **255**, 11986 (1980).
[36] M. Taylor and J. Segall, *J. Biol. Chem.* **260**, 4531 (1985).

involves the slow addition of 0.45 g of ammonium sulfate (ultrapure) per milliliter of eluate over a period of 1 hr with gentle stirring. The precipitate is collected by centrifugation at 10,000 rpm for 30 min in a Sorvall HB4 rotor and resuspended in HGED buffer to an ammonium sulfate concentration of 80 mM. If dialysis is used, an aggregate that forms is removed by centrifugation without loss of activity. Dialysis to lower salt concentrations sometimes results in activity loss with the aggregate. The three transcription components are separated[15] by passage through DEAE-cellulose at 80 mM ammonium sulfate to split off TFIIIB in the flow-through and elution of TFIIIC and RNA polymerase III in a single step at 180 mM ammonium sulfate. RNA polymerase III is separated from TFIIIC by passage through DEAE-Sephadex, from which it elutes at an anomalously high ionic strength. TFIIIB tends to be contaminated with TFIIIC, which can be removed by rechromatography on DEAE-cellulose with a 20–80 mM ammonium sulfate gradient. Further purification procedures can be found elsewhere for each of the components.[37-39]

The terminal processing enzymes are recovered from the first Bio-Rex 70 flow-through by ammonium sulfate precipitation as described above. We find that this provides faster and more consistent recoveries than dialysis to reduce the ionic strength for subsequent steps. RNase P is separated from the 3′ processing nuclease(s) by rechromatography on Bio-Rex 70 followed by DEAE-cellulose and has subsequently been purified by velocity gradient centrifugation.[40]

Acknowledgments

We would like to thank Carl Parker and Joanne Topol for communicating unpublished results and Jon Huibregtse, Alexandra Krikos, and Dennis Thiele for helpful comments on the manuscript. This work was supported by Grant DMB-8603115 from the National Science Foundation.

[37] G. L. Hager, M. J. Holland, and W. J. Rutter, *Biochemistry* **16,** 1 (1977).
[38] M. S. Klekamp and P. A. Weil, *J. Biol. Chem.* **261,** 2819 (1986).
[39] O. S. Gabrielsen, N. Marzouki, A. Ruet, A. Sentenac, and P. Fromageot, *J. Biol. Chem.* **264,** 7505 (1989).
[40] J.-Y. Lee and D. Engelke, *Mol. Cell. Biol.* **9,** 2536 (1989).

[37] Enzymatic Addition of Guanylate to Histidine Transfer RNA

By JACINTA B. WILLIAMS, LYNN COOLEY, and DIETER SÖLL

Introduction

All histidine tRNAs with known sequences derived from many organisms are unique in that they possess an additional nucleotide at their 5′ end. The extra nucleotide is guanosine in all species except bacteriophage T5, in which case the histidine tRNA possesses an additional uridine residue at its 5′ terminus.[1]

In *Escherichia coli,* as well as other prokaryotic organisms, the guanosine residue is encoded in the histidine tRNA gene at position −1 and is transcribed by RNA polymerase III as part of the 5′-leader sequence of the precursor tRNA. During maturation of these precursor histidine tRNAs, the additional guanosine residue is retained in the mature products as a result of an unusual cleavage reaction catalyzed by RNase P.[2] Conversely, examination of eukaryotic histidine tRNA gene sequences indicates that the guanosine residue is not encoded.[1] Rather, the additional guanylate is added in an unusual posttranscriptional modification reaction. The steps of histidine tRNA biosynthesis in eukaryotes is schematically presented in Fig. 1. The primary tRNA gene transcript is first cleaved by RNase P, and the 3′ end is formed by the action of an endo- or exonuclease and of the tRNA nucleotidyltransferase (CCA–enzyme). Then the guanylate moiety is added by a specific enzyme, the tRNA guanylyltransferase.[3] In the following we describe our current knowledge of this reaction. Attempts to purify the enzyme from *Saccharomyces cerevisiae* and *Drosophila melanogaster* have been made. They are described below.

Assay

Two principal assays are available for the tRNA guanylyltransferase. The addition assay is based on the *de novo* addition of radioactive guanylate (using [α-32P]GTP) to the 5′ end of *in vitro* synthesized tRNA^His

---

[1] M. Sprinzl, T. Hartmann, F. Meissner, J. Moll, and T. Vorderwülbecke, *Nucleic Acids Res.* **15**, r53 (1987).

[2] O. Orellana, L. Cooley, and D. Söll, *Mol. Cell. Biol.* **6**, 525 (1986).

[3] L. Cooley, B. Appel, and D. Söll, *Proc. Natl. Acad. Sci. U.S.A.* **79**, 6475 (1982).

METHODS IN ENZYMOLOGY, VOL. 181

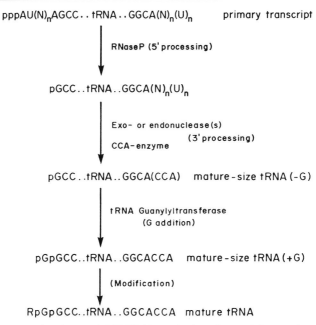

FIG. 1. Proposed pathway of tRNA[His] biosynthesis: scheme of the reactions involved in the maturation of the 5' end of eukaryotic histidine tRNAs. The nucleotide sequences depicted are common to the *Drosophila* and *S. cerevisiae* tRNA[His] genes.

molecules. These substrate molecules are nonradioactive (or are of very low specific radioactivity) and lack nucleotide -1. The reaction products can be separated by gel electrophoresis. This reaction could also be adapted to a filter disk assay once the enzyme is partially purified.

The exchange assay measures the exchange of GMP (using $[\alpha\text{-}^{32}P]GTP$) for the preexisting nonlabeled guanylate residue at the 5' end of mature tRNA. The substrate for this assay is mature histidine tRNA (contained in readily available unfractionated tRNA). The reaction products need to be separated by gel electrophoresis.

Addition Assay

The reaction mixture (usually 40 μl) contains 30 mM N-hydroxyethylpiperazine-N'-2-ethanesulfonic acid (HEPES)–KOH (pH 8.0), 3 mM dithiothreitol (DTT), 8 mM creatine phosphate, 100 mM KCl, 3 mM MgCl$_2$, 0.5 mM ATP, 1–20 μM $[\alpha\text{-}^{32}P]GTP$ (20–40 Ci/mmol), precursor RNA (synthesized in a *Drosophila* transcription extract[4]) and 20 μl

[4] T. Dingermann, S. Sharp, B. Appel, D. DeFranco, S. Mount, H. Reinhard, O. Pongs, and D. Söll, *Nucleic Acids Res.* **9,** 3907 (1981).

($\sim$200 μg protein) of *Drosophila* Kc cell extract[4] or aliquots of purified enzyme fractions.[5] After incubation for 90 min at 24° the reaction products are extracted and separated on polyacrylamide gel electrophoresis.

Exchange Assay

The reaction mixture (usually 50 μl) contains 8 mM HEPES–NaOH (pH 7.9), 0.4 mM DTT, 120 mM NaCl, 10 mM MgCl$_2$, 8% (v/v) glycerol, 3 mM ATP, 5 μM [α-^{32}P]GTP (8–10 Ci/mmol), and 0.25 A_{260} units of bakers' yeast tRNA. The reaction is initiated by the addition of 2 μl of *S. cerevisiae* S100 or enzyme fraction. After 45 min at 25° the reactions are terminated by the addition of 5μl of 0.25 M ethylenediaminetetraacetic acid (EDTA) (pH 8.0), and the mixture is phenol extracted. If necessary, subsequent extractions with an equal volume of 2 : 1 phenol–chloroform followed by chloroform are performed. In order to remove unreacted radiolabeled nucleotide triphosphate, the tRNA is then precipitated twice in the presence of an equal volume of 4 M ammonium acetate and 2 volumes of 100% ethanol. Following the final precipitation, the RNA pellet is resuspended in 5 μl of a solution consisting of 80% formamide, 1 mM EDTA, 1% xylene cyanol FF, and 4% bromphenol blue. The RNA products in the sample are resolved on an 8 M urea–8% polyacrylamide (20 : 1) gel. Following autoradiography of the wet gel on Kodak XAR-5 X-ray film at -70° for 15 hr, the activity of the enzyme is visualized by the appearance of a radiosignal coincident with histidine tRNA-sized molecules. Partial nucleotide sequence analysis of the product has ascertained that the band does represent the expected reaction product.

Partial Purification of tRNA Guanylyltransferase

From Yeast

Preparation of S100. Saccharomyces cerevisiae strain 20B-12 (α, *trpl*, *pep4-3*) is grown in YPD medium (1% yeast extract, 1% peptone, 2% dextrose) at 30° to a density of 6 A_{600} units. The cells are harvested by centrifugation and quick frozen at -70°. All subsequent operations are performed at 4° unless otherwise indicated. Approximately 250 g (wet weight) of cells is suspended in solubilization buffer [200 mM tris(hydroxymethyl)aminomethane (Tris) (pH 8.1), 10% (v/v) glycerol, 10 mM MgCl$_2$, 10 mM 2-mercaptoethanol] at 2 ml/g of cells. Cell lysis is achieved by passage through a Dynomill homogenizer in the presence of acid-

---

[5] L. Cooley, Ph.D. thesis, University of Texas, Austin, Texas, 1984.

washed glass beads (600 μm). The S100 extract is prepared from the resultant homogenate according to the protocol of Klekamp and Weil.[6] The extract is dialyzed against 100 volumes of buffer A [20 mM HEPES (pH 7.9), 0.2 mM EDTA, 1 mM DTT, 20% (v/v) glycerol] plus 100 mM (NH$_4$)$_2$SO$_4$ and used immediately for purification of the enzyme.

DEAE-Cellulose. In order to prepare enzyme extracts that are devoid of endogenous histidine tRNA and are consequently dependent on the addition of added histidine tRNA for activity, the S100 extract is passed over a DEAE-cellulose (DE-52, Whatman) column (10 mg protein/ml column bed volume) which has been equilibrated in buffer A plus 100 mM (NH$_4$)$_2$SO$_4$. The column is washed with 2 volumes of the same buffer. The eluant and washes are pooled and assayed for activity in both the presence and absence of added tRNA. As Fig. 2 shows, the DE-52 enzyme fractions are now totally dependent on added tRNA.

Phosphocellulose. The DE-52 fractions are pooled and loaded directly on a phosphocellulose (P-11, Whatman) column (10 mg protein/ml column bed volume) which is equilibrated in buffer A plus 100 mM (NH$_4$)$_2$SO$_4$. The column is washed with 2 column volumes of the same buffer, and the bound protein is eluted with a linear gradient (3 column volumes) of 100 mM–1 M (NH$_4$)$_2$SO$_4$ in buffer A. The tRNA guanylyltransferase activity elutes at approximately 500 mM (NH$_4$)$_2$SO$_4$. The peak fractions of enzyme activity are 10- to 15-fold purified (relative to the DE-52 fraction).

Heparin-Agarose. As is true in general for nucleotide-binding proteins, heparin-agarose (Affi-Gel Heparin, Bio-Rad) has proved to be an efficacious resin for binding tRNA guanylyltransferase activity. The peak fractions of activity in the eluant of the previous column are dialyzed against buffer A plus 100 mM NaCl. The enzyme fractions are passed over a heparin-agarose column which is equilibrated in the dialysis buffer. The tRNA guanylyltransferase is eluted from the column with a linear gradient (5 column volumes) of 100 mM–1 M NaCl in buffer A. The enzyme activity elutes very late from this column. The peak fractions of enzyme activity are at least 1000-fold purified (relative to the DE-52 fraction).

From Drosophila

Preparation of Drosophila Kc Cell Extract. The preparation of *Drosophila* Kc cell extract has been described previously.[4] All steps are carried out at 4°; the enzyme activity is assayed in the exchange assay

---

[6] M. S. Klekamp and P. A. Weil, *J. Biol. Chem.* **257**, 8432 (1982).

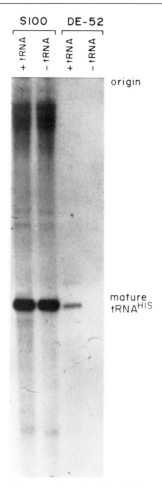

FIG. 2. Polyacrylamide gel electrophoresis of G-addition reaction products in the exchange assay using the yeast enzyme. The S100 and DE-52 fractions are used with and without the presence of exogenous tRNA.

using 5 μg unlabeled unfractionated *Drosophila* tRNA as substrate. The extract is subjected to ammonium sulfate fractionation to yield two pools (0–50% and 50%-saturation). The pellets are resuspended and dialyzed against buffer B [20 mM HEPES–KOH (pH 7.9), 1 mM DTT, 0.1 mM EDTA, 20% glycerol, 50 mM KCl].

Phosphocellulose. The 50%-saturation pool of protein is applied to a phosphocellulose (P-11, Whatman) column which is equilibrated in buffer

B (pH 7.9). The column is washed with buffer B and the flow-through fractions are collected and pooled. This chromatographic step does not result in a significant purification; however, it does remove or destroy an uncharacterized enzymatic activity in the extract which incorporates label from [α-^{32}P]GTP into 5 S-sized RNA.

DEAE-Cellulose. The phosphocellulose column eluant is applied to a DEAE-cellulose (DE-52, Whatman) column which is equilibrated in buffer B (pH 7.9). The column is washed with buffer B, followed by buffer B plus 100 mM KCl, and then developed with a linear gradient of 150–400 mM KCl in buffer B. The tRNA guanylyltransferase elutes at approximately 220 mM KCl.

Heparin-Agarose. The active fractions from the DEAE-cellulose column are pooled, dialyzed against buffer B (pH 7.9), and applied to a heparin-agarose (Sigma, St. Louis, MO) column equilibrated in buffer B. The column is washed with buffer B, followed by buffer B plus 300 mM KCl, and then developed with a linear gradient of 300 mM–1 M KCl in buffer B. The tRNA guanylyltransferase elutes at approximately 750 mM KCl. The fraction containing the highest specific enzyme activity contains highly purified (>50%) enzyme as judged by sodium dodecyl sulfate (SDS)–polyacrylamide gel electrophoresis.[5]

General Properties of tRNA Guanylyltransferase

Physical

From Yeast. The molecular weight of the enzyme is determined by gel filtration chromatography on Ultrogel AcA 34 (LKB). Based on the elution position of proteins serving as molecular weight markers, the enzyme migrates at a position which corresponds to a molecular weight of 48,000. It is not yet known whether this figure is indicative of a monomeric or oligomeric subunit structure.

In order to explore the possibility that the guanylyltransferase is a ribonucleoprotein, the enzyme fractions are subjected to extensive digestion with *Staphylococcus aureus* micrococcal nuclease. After inactivation of the nuclease by ethylene glycol bis(β-aminoethyl ether)-N,N,N',N'-tetraacetic acid (EGTA) chelation of Ca^{2+}, full activity of the tRNA guanylyltransferase is measured. While it is certainly not possible to exclude the possibility that the RNA moiety of such a complex may be sufficiently protected to escape nuclease digestion, it seems unlikely that this is the case.

From Drosophila. Electrophoretic analysis of the heparin-agarose fractions indicates that one major protein band is present after silver staining. The apparent molecular weight (as judged by SDS–polyacryl-amide gel electrophoresis) is approximately 38,000.[5] As described above for the yeast enzyme it is very likely that the *Drosophila* enzyme does not contain RNA, as micrococcal nuclease treatment does not inactivate the enzyme.[5]

Characteristics of the Reaction

Mechanism and Reaction Intermediates

The mechanism of guanylate addition was studied with the purified *Drosophila* enzyme. If the guanylyltransferase addition assay is per-formed with unlabeled, nonguanylated tRNA[His] in the presence of [α-^{32}P]ATP and in the absence of a guanosine compound, a tRNA-sized intermediate is generated. After digestion of the labeled tRNA with RNase P1, a radiolabeled dinucleotide which comigrates with A5'pp5'G is detected (Fig. 3, lane 1-a). Digestion of the A(5')pp(5')G molecule with snake venom phosphodiesterase liberates labeled 5'-AMP (Fig. 3, lane 1-b). The formation of A(5')pp(5')G at the 5' end of tRNA[His] is confirmed in the following way. Nonguanylated tRNA[His] labeled with [α-^{32}P]GTP during *in vitro* transcription is incubated with ATP in the presence of guanylyltransferase. The tRNA[His] is reisolated and digested with nuclease P1 to liberate the 5'-terminal oligonucleotide A(5')pp(5')G (Fig. 3, lane 2-a). Digestion of the A(5')pp(5')G with snake venom phosphodiesterase liberates the expected labeled 5'-GMP (Fig. 3, lane 2-b). Therefore, it was proposed[5] that the 5'-terminal phosphate in the tRNA-sized precursor [mature ($-$G)tRNA[His] in Fig. 1] is activated by pyrophosphate formation with ATP. The activated pG in the resulting A5'pp5'GtRNA[His] is then attacked by the 3'-hydroxyl of GTP with concomitant release of AMP. In this fashion the new 3'→5'-phosphodiester bond at the terminus of mature histidine tRNA [mature ($+$G)tRNA[His]] is formed.

The activation of RNA substrates through adenylylation by ATP is part of the mechanism of RNA ligases from T4-infected *E. coli* cells[7] and RNA ligases in tRNA splicing in yeast and wheat germ.[8] In the splicing reactions, the 5'-phosphate of the tRNA 3' half-molecule is adenylylated by transfer of AMP from the adenylylated ligase forming a 5'-end struc-ture of A(5')pp(5')N. AMP is displaced by the joining of the activated end to the 3'-hydroxyl of the 5' half-molecule.

[7] O. C. Uhlenbeck and R. I. Gumport, *in* "The Enzymes" (P. Boyer, ed.), 3rd Ed., Vol. 15, p. 31. Academic Press, New York, 1982.
[8] R. C. Schwartz, C. L. Greer, P. Gegenheimer, and J. Abelson, *J. Biol. Chem.* **258,** 8374 (1983).

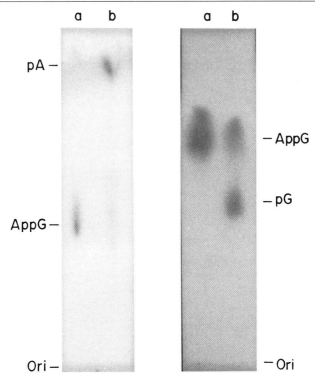

FIG. 3. Identification of adenylylated tRNA^{His}. Adenylation reactions were carried out by incubation of either unlabeled or labeled (with [α-^{32}P]GTP) nonguanylated tRNA^{His} (formed by *in vitro* transcription of a *Drosophila* tRNA^{His} gene) and purified *Drosophila* tRNA guanylyltransferase. Addition assay conditions were used except that GTP was omitted and 20 μM [α-^{32}P]ATP (2–4 Ci/mmol) or 1 mM ATP, respectively, were included. The products of these reactions were resolved by polyacrylamide gel electrophoresis and eluted. The 5′-terminal oligonucleotide (obtained after digestion with RNases T2 and A) was isolated by thin-layer chromatography and then further digested. (a) RNase P1 digestion products. (b) RNase P1 followed by snake venom phophodiesterase treatment. (Left) Analysis of 5′-terminal oligonucleotides from unlabeled tRNA^{His} substrate by chromatography on polyethyleneimine (PEI)-cellulose using 1 M ammonium formate (pH 3.5) as the solvent. (Right) Analysis of 5′-terminal oligonucleotides from labeled substrate by chromatography on PEI-cellulose with 1 M LiCl as the solvent. The positions of authentic markers are indicated.

The presence of an activated tRNA^{His} intermediate with the 5′-end structure of A(5′)pp(5′)G suggests that the mechanism for guanylate addition resembles the RNA ligase mechanism. This implies that an adenylylated enzyme intermediate also might exist during guanylation. We have been unable to identify such an adenylylated protein intermediate *in vitro*. This may be because the intermediate is not stable enough to survive the assay conditions employed or because the intermediate is too short-lived for detection.

The adenylylated tRNA[His] intermediate formed in the presence of [α-^{32}P]ATP is not observed when GTP or dGTP is present in the reaction mixture. We assume that the absence of a labeled tRNA in these experiments is the result of the release of AMP from an activated intermediate in the presence of GTP or dGTP. This implies that pyrophosphate is released from the GTP during the addition:

$$pppG + AppG\text{-}tRNA \rightarrow AMP + pGpG\text{-}tRNA + PP_i$$

The presence of more than one phosphate at the 5' end of mature tRNA[His] either from transcription reactions or from processing reactions has not been observed.

In approximately 70% of the mature histidine tRNAs, the added residue undergoes modification of the 5'-terminal phosphate[3] by an enzyme which is found in the crude *Drosophila* extract, but not in the purified *Drosophila* tRNA guanylyltransferase. The precise nature of the guanylate modification is not clear. The R group (see Fig. 1) covalently attached to the 5'-phosphate is linked through a bond that can be cleaved by snake venom phosphodiesterase. Enzymatic analysis of pGp and of RpGp (isolated as the 5' terminus of guanylated tRNA[His] made in crude *Drosophila* extract[3]) is shown in Fig. 4. A compound resulting from snake venom phosphodiesterase digestion of RpGp comigrates with pGp in two chromatography systems. This suggests that the guanosine molecule itself is not modified.

Cofactors

An absolute requirement for ATP has been determined for the enzyme from yeast and from *Drosophila*. Titration experiments with ATP show that the optimal concentration is 1.5 mM for the yeast enzyme.

In addition to the requirement for ATP, the guanylyltransferase requires Mg^{2+}. For the yeast and *Drosophila* enzymes the optimal Mg^{2+} concentration is 10 mM. Further studies with the yeast enzyme indicate that replacement of Mg^{2+} with other cations leads to reduced transferase activity. Mn^{2+} and Zn^{2+} are accepted by the enzyme, but there is no detectable activity when Ca^{2+} and Co^{2+} are supplied (Fig. 5).

Guanosine Substrate Requirements

In order to determine which guanosine derivatives may serve as donors for the guanylation reaction, the *in vitro* formation of *Drosophila*

A   B

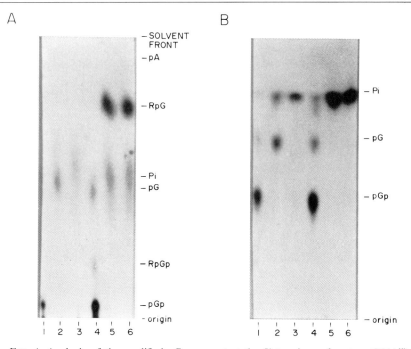

FIG. 4. Analysis of the modified pGp present at the 5' terminus of mature tRNA[His]. Unmodified pGp and RpGp [isolated chromatographically from a digest (RNases T2 and A) of guanylated tRNA[His] made in crude *Drosophila* extract[3]] were digested with the nucleases indicated below. Both phosphates in these compounds have been made radioactive by transcription and guanylation using [α-^{32}P]GTP. The analysis was performed by thin-layer chromatography with either 1 *M* ammonium formate (pH 3.5) (A) or 0.75 *M* potassium phosphate (pH 4.0) (B). Lanes 1–3, pGp; lanes 4–6, RpGp. Lanes 1 and 4, snake venom phosphodiesterase; lanes 2 and 5, RNase P1; lanes 3 and 6, alkaline phosphatase.

tRNA[His] in nucleotide-depleted *Drosophila* extract has been studied in detail with a number of different guanosine derivatives.[3,5] Utilizing this assay, it has been determined that guanosine 5'-triphosphate, 2-deoxyguanosine 5'-triphosphate, and, to a lesser extent, guanosine 5'-diphosphate are utilized as substrates whereas guanosine 5'-monophosphate, guanosine 3'-monophosphate, cyclic guanosine monophosphate, dideoxyguanosine 5'-triphosphate, and cytidine 5'-triphosphate do not participate in the reaction.

Inhibitors

As is evidenced by the requirement for Mg[2+], the reaction is inhibited by high concentrations of EDTA, i.e., in excess of 0.8 m*M*. In studies

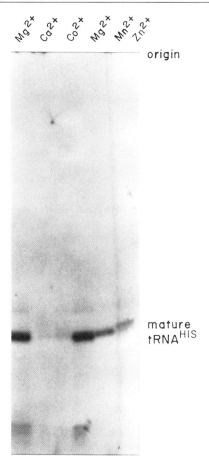

FIG. 5. Polyacrylamide gel electrophoresis of G-addition reaction products. G-addition by yeast tRNA guanylyltransferase (DE-52 fraction) supplemented with the metal ions indicated is illustrated.

with the yeast enzyme, the presence of greater than 10 mM inorganic phosphate greatly reduces the activity of the enzyme, as do high ionic strength buffers (salt concentrations in excess of 200 mM). The activity is stable over a pH range of 6.0–8.5, with the optimum determined to be pH 7.9. The optimal temperature is 25°.

tRNA Requirements

The tRNA specificity of the yeast enzyme has been tested by addition of tRNA from various representative organisms. The tRNAs were present at 0.25 A_{260} units of unfractioned tRNA in the standard reaction using a tRNA-depleted extract. Total tRNA from wheat germ, calf liver, *Drosophila melanogaster*, and *Schizosaccharomyces pombe* were able to support the reaction. Unfractionated *E. coli* tRNA as well as a *E. coli* histidine tRNA precursor (made by *in vitro* transcription with SP6 polymerase and *in vitro* processing at the 3′ end) were not acceptable substrates of the yeast enzyme. The structural parameters underlying this substrate specificity have not been determined.

Role of tRNA Guanylyltransferase

Why is the biosynthesis of histidine tRNA more complex than that of other tRNAs? What roles does histidine tRNA play which other tRNAs do not? Is there any rationale for this unusual reaction? These are the questions which are not yet answered. It seems reasonable to postulate that tRNA^His, with the unusual structure at the 5′ end (and the metabolic investment the cell has made to preserve that structure) serves a specific purpose, possibly as a control element in cellular metabolism. The additional modification of the 5′-terminal G of the tRNA[3] may be involved in this reaction.

As noted above, the prokaryotic (*E. coli*) tRNA does not appear to meet substrate requirements for the *S. cerevisiae* tRNA guanylyltransferase. Although such substrate specificity is not unusual (e.g., some eukaryotic aminoacyl-tRNA synthetases fail to recognize prokaryotic tRNAs), it is interesting to point out that no guanylyltransferase activity has been detected in *E. coli*. It is intriguing then, to speculate that, from an evolutionary standpoint, the tRNA guanylyltransferase has evolved because of an absolute requirement for its activity in response to mutations in the primary nucleotide sequence of the eukaryotic histidine tRNA genes. This would imply an essential function of the gene encoding the guanylyltransferase. If essential, the enzyme may exert an important regulatory role over the production of mature histidine tRNA, as both +G and −G histidine tRNAs have been determined to be generated as transcription products of the *S. cerevisiae* and *Drosophila melanogaster* histidine tRNA genes *in vitro*.

[38] Purification of Yeast Transfer RNA Ligase

By Qi Xu, Eric M. Phizicky, Chris L. Greer, and
John N. Abelson

Introduction

The splicing of yeast tRNA precursors requires the action of at least two proteins, an endonuclease which cleaves the two splice junctions to release the linear intron and a ligase which joins the resulting tRNA half-molecules together.[1,2] The mechanism of the reaction is depicted in Fig. 1. The tRNA ligase has three independent activities, each of which is required to complete the ligation. A phosphodiesterase activity opens the 2',3'-cyclic phosphate at the 3' terminus of the 5' half-molecule. A polynucleotide kinase activity transfers the γ-phosphate of ATP to the 5'-hydroxyl of the 3' half-molecule. Finally, the ligase activates the 5'-phosphate by joining AMP in a 5',5'-phosphoanhydride bond and then catalyzes the ligation with the consequent release of AMP and the formation of a 2'-phosphomonoester-3',5'-phosphodiester linkage.[3]

We have previously purified the ligase to homogeneity from yeast extracts. It is a single 95-kDa protein. In the presence of ATP it is adenylylated at lysine-114 (Q. Xu, unpublished results). The gene for this protein has been cloned[4] and sequenced.[5] The protein can be overexpressed in *Escherichia coli* so that this is now the convenient source of the protein.[4] In this chapter we describe the purification of tRNA ligase from the overproducing strain.

Assays

Each of the activities of tRNA ligase can be assayed independent of the overall reaction. We have routinely assayed the enzyme activity during purification by the ligation of tRNA half-molecules produced by the action of tRNA endonuclease. The pre-tRNA[Phe] substrate can now conveniently be synthesized *in vitro* by T7 RNA polymerase transcription of a synthetic gene.[6,7] The specific activity of these transcripts is approxi-

---

[1] C. L. Peebles, R. C. Ogden, G. Knapp, and J. Abelson, *Cell* **18**, 27 (1979).
[2] C. L. Peebles, P. Gegenheimer, and J. Abelson, *Cell* **32**, 525 (1983).
[3] C. L. Greer, C. L. Peebles, P. Gegenheimer, and J. Abelson, *Cell* **32**, 537 (1983).
[4] E. M. Phizicky, R. C. Schwartz, and J. Abelson, *J. Biol. Chem.* **261**, 2978 (1986).
[5] S. K. Westaway, E. M. Phizicky, and J. Abelson, *J. Biol. Chem.* **263**, 3171 (1988).
[6] V. M. Reyes and J. Abelson, *Anal. Biochem.* **166**, 90 (1987).
[7] V. M. Reyes and J. N. Abelson, this series, Vol. 180, p. 63.

METHODS IN ENZYMOLOGY, VOL. 181

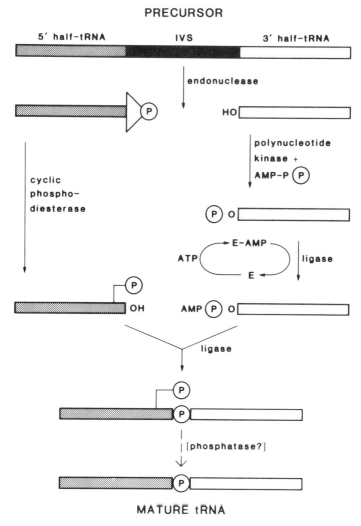

FIG. 1. Mechanism of yeast tRNA splicing.

mately 10^5 dpm/pmol. A 10-μl reaction mixture contains 20 mM N-2-hydroxyethylpiperazine-N'-2-ethanesulfonic acid (HEPES) (pH 7.5), 5 mM MgCl$_2$, 2.5 mM spermidine-HCl (pH 7.5), 0.1 mM dithiothreitol, 0.4% Triton X-100, 10% glycerol, 2 mM ATP, 0.1 unit of partially purified yeast endonuclease,[8] tRNA precursor ($\sim$1000 cpm) and tRNA ligase. The

[8] P. R. Green and J. N. Abelson, this volume [39].

reactions are carried out for 10 min at 30° and stopped by addition of 0.2% sodium dodecyl sulfate (SDS), 10 mM ethylenediaminetetraacetic acid (EDTA), and 0.2 mg/ml proteinase K. After a further incubation for 20 min at 50°, the substrate and products are separated by electrophoresis in a 10% polyacrylamide gel containing 4 M urea. The product is quantified by measuring the Cerenkov radiation in gel slices. One unit of tRNA ligase can form 1 pmol of mature tRNA in 10 min at 30°.

The polynucleotide kinase assays are performed as described by Pick and Hurwitz.[9] tRNA ligase can transfer the γ-phosphate of ATP to a variety of polynucleotides. Reaction mixtures (50 μl) contain 20 mM tris(hydroxymethyl)aminomethane hydrochloride (Tris-HCl) (pH 8.0), 6 mM MgCl$_2$, 2 mM dithrothreitol, 50 pmol of poly(A) (average chain length 600), 160 μCi of [γ-^{32}P]ATP ($\sim$7000 Ci/mmol), and tRNA ligase. After a 15-min incubation at 37° the mixture is diluted to 0.2 ml with distilled water and extracted with an equal volume of phenol–chloroform (1 : 1, v/v). Bovine serum albumin (50 μg) and 0.5 ml of cold 10% trichloroacetic acid are added to the aqueous phase. After 7 min at 0°, the insoluble material is collected by centrifugation and resuspended in 0.2 ml of 0.1 M Tris-HCl (pH 8.0). Cold 5% trichloroacetic acid (5 ml) and 0.1 ml of 0.1 M sodium pyrophosphate are added, and acid-insoluble material is collected on a glass fiber filter. The filters are washed with cold 1% trichloroacetic acid, ethanol dried, and counted.

The cyclic phosphodiesterase assay is performed as described by Greer et al.[3] and Phizicky et al.[4] The scheme for generating the substrate for this assay, U$_n$G > P labeled with ^{32}P in the 3′-terminal cyclic phosphate, is diagrammed in Fig. 2. Poly(U$_n$,G) (U : G = 10 : 1) (100 μg) is digested with RNase T1 (40 U) for 10 min at 24° in 50 mM HEPES (pH 7.5) followed by extraction 3 times with phenol and then precipitation with ethanol. The oligonucleotide mixture is labeled by reaction with T4 polynucleotide kinase (4 U) and [γ-^{32}P]ATP (200 μCi) for 1 hr at 30° in buffer containing 50 mM Tris-HCl (pH 8.0) and 10 mM MgCl$_2$. The RNA is extracted with phenol and then ethanol precipitated. The oligonucleotide is circularized by incubation with T4 RNA ligase (7.5 U) for 1 hr at 37° in 50 mM HEPES (pH 7.5), 10 mM MgCl$_2$, 3 mM dithiothreitol, and 1 mM ATP, followed by phenol extraction and ethanol precipitation. The circles are opened to linear molecules with 5′-hydroxyl and 2′,3′-cyclic phosphate termini by digestion with RNase N1 [6 × 10^{-4} U/μg oligo(U,G)] for 10 min at 24°. The reaction is stopped by heating in a boiling water bath for 2 min, followed by phenol extraction and ethanol precipitation.

[9] L. Pick and J. Hurwitz, *J. Biol. Chem.* **261,** 6684 (1986).

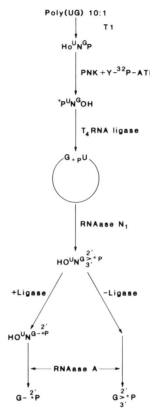

FIG. 2. Scheme for the synthesis of oligo(U_n,G) >P in which the terminal cyclic phosphate is labeled with ^{32}P. This oligonucleotide is a substrate for the phosphodiesterase activity of tRNA, ligase which opens the cyclic phosphate to give terminal ribose 2'-phosphate. The terminal nucleotide is released by digestion with RNaseA, and 2'-GMP can be separated from 2',3'-cyclic GMP by thin-layer chromatography.

Cyclic opening assays (5 μl) contain 50 mM HEPES (pH 7.5) and 1 μg oligo(U_n,G) >^{32}P ($\sim$10,000 cpm/μg) and tRNA ligase. Reactions are incubated at 30° for 20 min, stopped by boiling for 1 min, and treated with RNase A (0.5 μg) at 37° for 30 min. One microliter of each reaction mixture is applied to a cellulose plate which is developed in solvent containing saturated ammonium sulfate, 1 M sodium acetate, 2-propanol (40 : 9 : 1) and then analyzed by autoradiography. The amounts of cyclic GMP and 2'-GMP are quantitated by cutting the spots from the thin-layer plate and counting them in a scintillation counter.

The covalent addition of AMP to tRNA ligase can be demonstrated by incubating the protein with $[\alpha\text{-}^{32}\text{P}]\text{ATP}$ (10 μCi) in a reaction mixture containing 100 mM HEPES (pH 7.5), 10 mM MgCl_2, 3 mM dithiothreitol, and tRNA ligase. After incubation for 10 min at 30°, an equal volume of 125 mM Tris-HCl (pH 6.8), 1.5% SDS, 135 mM 2-mercaptoethanol, 20% glycerol, and 0.1% bromphenol blue is added. The mixture is then boiled for 2 min and analyzed by electrophoresis in an SDS–polyacrylamide gel. The adenylylated protein is reasonably stable at neutral pH with little loss of AMP after incubation for 1 hr at 30° or overnight at 4°.

Enzymes and Materials

T4 polynucleotide kinase, T4 RNA ligase, T7 RNA polymerase, and poly(A) are obtained from Pharmacia (Piscataway, NJ). Poly (U,G) (10 : 1) is obtained from P-L Biochemicals, Inc. (Madison, WI). RNase N1 is obtained from Seikagaku Kogyo (Tokyo). RNase T1 is obtained from Boehringer-Mannheim Biochemicals (Indianapolis, IN). RNase A is obtained from Worthington Biochemical Corporation (Freehold, NJ). Alkaline phosphatase from calf intestine is obtained from Boehringer-Mannheim. $[\alpha\text{-}^{32}\text{P}]\text{ATP}$ and $[\alpha\text{-}^{32}\text{P}]\text{UTP}$ are obtained from Amersham Corporation (Arlington Heights, IL), and $[\gamma\text{-}^{32}\text{P}]\text{ATP}$ is from ICN Radiochemicals (Lisle, IL). Heparin is coupled to Sepharose CL-4B by the method of Davison et al.[10] Blue Trisacryl M is obtained from IBF Biotechnics (France). Bio-Gel HTP (hydroxyapatite) is obtained from Bio-Rad (Richmond, CA) Polymin P is from Miles Laboratories (Elkhart, IN). Pepstatin A can be obtained from Boehringer-Mannheim. The protease inhibitors aprotinin, bacitracin, phenylmethylsulfonyl fluoride (PMSF), and ethylene glycol bis(β-aminoethyl ether)-N,N,N',N'-tetraacetic acid (EGTA) are from Sigma (St. Louis, MO).

Purification Procedure

The construction of the expression vector for tRNA ligase has previously been described.[4] The sequence at the N-terminal methionine codon was modified and linked to the p-tac promoter in the expression vector pKK223-3 to give the plasmid pKK223-3-RLGX. This system provides good overexpression of the 95-kDa ligase protein, but we observed some instability of the protein when it was expressed in a normal E. coli strain.[4] Consequently, we transferred the plasmid to the protease-deficient strain

---

[10] B. L. Davison, T. L. Leighton, and J. C. Rabinowitz, J. Biol. Chem. **254**, 9220 (1979).

SW1064 (Δlon, $htpR$, $lacZ_{am}$, trp_{am}, pho_{am}, Sup C^{ts} mal $rpsL$ phe rel [F' $laci^Q$ $lacZ::$Tn5]). This strain is designated SW1068. The protein is considerably more stable in this strain, and fewer lower molecular weight forms of tRNA ligase can be detected by rabbit anti-tRNA ligase IgG in Western blots. We now have good evidence that tRNA ligase is organized into three domains containing, respectively, the ligase, the kinase, and the phosphodiesterase activities. The boundaries between these domains are apparently very sensitive to proteases, and consequently care must be taken to obtain the intact 95-kDa proteins.

Growth of Cells

Escherichia coli SW1068 is grown at 37° in LB broth containing 25 μg/ml ampicillin. The cells are grown to an A_{600} of 0.5, and expression of the construct is induced by addition of 0.2 mM β-isopropyl-D-thiogalacto-pyranoside (IPTG) for 4 hr and harvested at an A_{600} of 2. The cells are collected by centrifugation and frozen immediately in liquid nitrogen. Alternatively, more recent experiments have shown that induction at an A_{600} of 1.0 and harvesting 45 min later may result in extracts with higher specific activity.

Preparation of the Extract

Two hundred grams of cell paste is resuspended in 400 ml of extraction buffer [20 mM Tris-HCl (pH 7.5), 1 mM EDTA, 10% glycerol, 0.2 M NaCl]. To prevent protease degradation a protease inhibitor mix is added to this and all subsequent buffers. The final concentrations of the inhibitors in chromatography buffers is 0.1 mM EGTA, 0.1 mM benzamidine, 0.2 μg/ml aprotinin, 10 μg/ml bacitracin, 0.2 μg/ml pepstatin A, and 0.02 mM PMSF. In extraction buffer the inhibitors are present at twice this concentration. PMSF is freshly prepared as a 200 mM solution in 2-propanol. The cell slurry is sonicated for 1.5 min in 20-sec bursts and centrifuged for 40 min at 25,000 g to obtain the crude extract (700 ml).

Polymin P Precipitation

The cell slurry is made 0.15% in Polymin P by addition of a 10% solution of polymin P. After stirring slowly for 30 min, the solution is centrifuged at 25,000 g for 40 min and the supernatant is collected. This step removes most nucleic acids from the extract (620 ml).

Heparin-Agarose Chromatography 1

Heparin-agarose paste, previously equilibrated with buffer A [20 mM Tris-HCl (pH 7.5), 1 mM EDTA (pH 8.0), 10% glycerol, 1 mM 2-mercaptoethanol] containing 0.2 M NaCl is added to the Polymin P supernatant in a 2-liter beaker. After the slurry is stirred slowly for 40 min it is poured into a glass column and washed with buffer A containing 0.2 M NaCl until no more protein is eluted. The column is then eluted with a 2000-ml linear gradient of 0.2–1 M NaCl in buffer A. The peak fractions elute between 0.4 and 0.5 M NaCl and are pooled (360 ml).

Heparin-Agarose Chromatography 2

The peak fractions are dialyzed against 2 liters of 0.2 M NaCl in buffer A and loaded on a smaller heparin-agarose column (2.2 × 30 cm, 110 ml). The column is washed with 500 ml of 0.2 M NaCl in buffer A and eluted with a 600-ml linear gradient of 0.2–2 M NaCl in buffer A. This step is optional and is usually used in larger scale purifications.

Blue Trisacryl M Chromatography

The peak fractions of the smaller heparin-agarose column are dialyzed against 2 liters of 0.2 M NaCl in buffer A and loaded onto a Blue Trisacryl M column (2.2 × 9.5 cm, 36 ml) previously equilibrated with 0.2 M NaCl in buffer A. The column is washed with 200 ml of 0.2 M NaCl in buffer A and eluted with a 600-ml linear gradient of 0.2–2 M NaCl in buffer A. The peak fractions, eluting between 0.9 and 1.1 M NaCl, are pooled (170 ml).

Hydroxyapatite Chromatography

The peak fractions from the Blue Trisacryl M column are loaded onto a 10-ml BioGel HTP column (1.6 × 5 cm) previously equilibrated with 0.2 M NaCl in buffer A. The column is washed successively with 40 ml of 0.2 M NaCl in buffer A, 40 ml of buffer B [50 mM potassium phosphate buffer (pH 7.5), 10% glycerol, 1 mM 2-mercaptoethanol] and eluted with a 300-ml linear gradient of 0.05–0.5 M potassium phosphate (pH 7.5). The active fractions eluting at approximately 250 mM potassium phosphate are pooled and dialyzed for 20 hr against 1 liter of buffer A plus 0.2 M NaCl (70 ml).

Sephadex G-150 Gel Filtration

The hydroxyapatite peak fractions are concentrated using a stirred ultrafiltration cell (Amicon with a YM10 membrane) to 2–3 ml and loaded onto a Sephadex G-150 column (2.2 × 100 cm, 380 ml) previously equilibrated with 0.2 M NaCl in buffer A. The flow rate is controlled at 10 ml/hr for the column. The peak fractions, eluting between 75 and 95 ml, are pooled (100 ml).

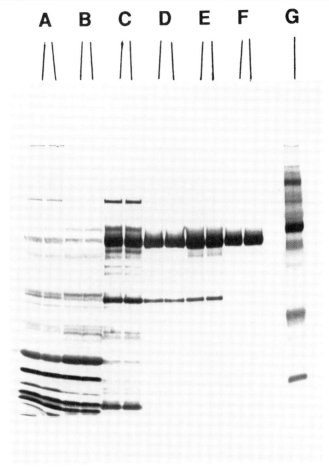

Fig. 3. SDS–polyacrylamide gel electrophoresis of fractions from the purification of tRNA ligase. Peak fractions from the purification (loaded in duplicate) were resolved on an SDS–polyacrylamide gel and visualized by silver staining. Lane A, Crude extract (70 μg); lane B, Polymin P supernatant (60 μg); lane C, heparin-agarose peak fraction (8.5 μg); lane D, Blue Trisacryl M peak fraction (3.7 μg); lane E, hydroxyapatite peak fractions (6 μg); lane F, Sephadex G-150 peak fractions (2.5 μg); lane G, molecular weight standards.

Figure 3 shows the results of an electrophoretic separation of proteins at each stage of the purification. The major band which contaminates the 95-kDa tRNA ligase in the hydroxyapatite and Blue Trisacryl M fractions is a fragment of tRNA ligase and is not removed until the final gel filtration step.

TABLE I
PURIFICATION OF YEAST tRNA LIGASE FROM *Escherichia coli* OVERPRODUCING STRAIN

Step	Total volume (ml)	Total protein	Total activity (kU)	Specific activity (U/mg)	Purification (-fold)	Recovery (%)
Crude extract	700	18 g	190	10.5	1	100
Polymin P supernatant	620	12.4 g	170	13.7	1.3	89.4
Heparin-agarose	360	305 mg	110	360	34.3	58
Blue Trisacryl M	170	64 mg	70	1093	104	37
Hydroxyapatite	70	42 mg	60	1428	136	32
Sephadex G-150	100	25 mg	58	2320	220	30

Conclusions

This procedure provides 25 mg of pure tRNA ligase with a 30% recovery of the activity (Table I). The peak fractions from the final gel filtration step can be concentrated by ultrafiltration and are stored in buffer A containing 0.2 M NaCl and 50% glycerol at $-20°$. The main problem that had to be solved in this purification was the sensitivity of the protein to protease. Three measures were taken to prevent degradation: the gene was expressed in a protease-deficient strain, the first three steps of the protocol were done as rapidly as possible, and a mixture of protease inhibitors are included in all of the chromatography steps. The tRNA ligase, as purified, possesses all three of the constituent activities.

[39] Highly Purified Transfer RNA Splicing Endonuclease from *Saccharomyces cerevisiae*

By PHILLIP R. GREEN and JOHN N. ABELSON

Intervening sequences are found in and spliced from various messenger, ribosomal, and transfer RNAs. The first splicing reaction demonstrated *in vitro* was for tRNA precursors in crude extracts from the yeast *Saccharomyces cerevisiae*.[1] Isolated labeled tRNA precursors[1,2] are

[1] G. Knapp, J. J. Beckmann, P. F. Johnson, S. A. Fuhrman, and J. Abelson, *Cell* **14**, 221 (1978).

[2] A. K. Hopper, F. Banks, and V. Evangelidis, *Cell* **14**, 211 (1978).

spliced via a two-step mechanism.[3] First, an endonuclease cleaves the precursor at the 5' and 3' splice sites, releasing the linear intron and tRNA half-molecules. The endonuclease cleavage products contain 5'-hydroxyls and 2',3'-cyclic phosphates.[4] In the second step, the tRNA half-molecules are ligated in a complex, ATP-dependent reaction catalyzed by the splicing ligase.[5]

Our goal is to study biochemically and genetically tRNA splicing in yeast to determine its possible role in gene expression. One approach is to purify the protein and use sequence data to isolate the gene. This approach has been successfully used for the tRNA splicing ligase.[6] The ligase, a single polypeptide of 90,000 Da, catalyzes all of the reactions involved in the ligation of the cut precursors. The endonuclease has, to this time, been refractory to complete purification. Like the ligase, it is present in small amounts. It also acts as an integral membrane-bound protein. In this chapter, a method used to extensively purify small amounts of the endonuclease is presented. The data suggest that the enzyme may be multimeric.

Assay of Activity[3,4]

A ^{32}P-labeled synthetic precursor tRNA^{Phe} is used as the substrate.[7] A modified, intron-containing tRNA^{Phe} gene is synthesized *de novo*. The bacteriophage T7 RNA polymerase promoter is inserted immediately 3' of the gene, and a *Bst*NI restriction endonuclease site is included at the 5' end. The gene is cloned into pUC13, transformed into *Escherichia coli* strain JM101, and plasmid DNA isolated. The plasmids are cut with *Bst*NI and used to make T7 RNA polymerase runoff transcripts in the presence of [α-^{32}P]UTP. Precursor of about 3×10^6 dpm/pmol is obtained.

The reaction is performed in 10 μl of buffer containing 40 mM tris(hydroxymethyl)aminomethane hydrochloride (Tris-HCl) (pH 8.0), 0.5 mM cthylcncdiaminctctraacetic acid, disodium salt (Na$_2$ EDTA), 4.0 mM spermidine-HCl, 0.2% Triton X-100 (w/v), 10% glycerol, and 1×10^{-15} mol (1,000–2,000 cpm) of tRNA precursor. To this buffer, 1 μl of appropriately diluted [into 10% glycerol (w/v), 0.5% Triton X-100, 25 mM Tris-HCl (pH 8.0), and 5 mM 2-mercaptoethanol] endonuclease is added at 0°, mixed, and incubated for 10 min at 30°. The reaction is stopped by adding 1 μl of a solution containing 2% sodium dodecyl sulfate (SDS), 100 mM

[3] C. L. Peebles, R. C. Ogden, G. Knapp, and J. Abelson, *Cell* **18**, 27 (1979).

[4] C. L. Peebles, P. Gegenheimer, and J. Abelson, *Cell* **32**, 525 (1983).

[5] C. L. Greer, C. L. Peebles, P. Gegenheimer, and J. Abelson, *Cell* **32**, 537 (1983).

[6] E. M. Phizicky, R. C. Schwartz, and J. Abelson, *J. Biol. Chem.* **261**, 2978 (1986).

[7] V. Reyes and J. Abelson, *Anal. Biochem.* **166**, 90 (1987).

EDTA, and 2.0 mg/ml proteinase K and incubated at 50° for 15 min. The products are mixed with 5 μl of 8 *M* urea, 20% sucrose, bromphenol blue, and xylene cyanol, heated to 65° for 5 min, and separated by electrophoresis through a 10% polyacrylamide (acrylamide–bisacrylamide, 30 : 1) gel containing 90 m*M* Tris–borate (pH 8.3), 2.5 m*M* EDTA, and 4 *M* urea.[3] The ^{32}P-labeled half-molecules are located by autoradiography and cut out of the gels, and the Cerenkov radiation is measured assuming a counting efficiency of 30%. A unit of activity is defined as the amount of enzyme needed to catalyze the cleavage of 1×10^{-12} mol of tRNA precursor.

Protein Determination

Samples are precipitated with 5% trichloroacetic acid,[8,9] and the protein concentration is estimated by the method of Lowry *et al.*,[10] using bovine serum albumin as the standard. Solutions contain 1% SDS to prevent interference by Triton X-100.

Electrophoresis

Protein samples are precipitated in 1 ml of 5% trichloroacetic acid. After a 5-min centrifugation in a microcentrifuge, Tris base is added to the solution to a concentration of 0.3 *M*. Triton X-100 redissolves upon mixing, and precipitated protein is isolated by centrifugation for 10 min. The pellet is resuspended in 15 μl of sample buffer plus 2 μl of 2 *M* Tris base, heated for 3 min at 100°, and separated on a 5–15% polyacrylamide gradient gel according to Laemmli.[11] Separated proteins are visualized by silver staining.[12]

Growth of Yeast

The protease-deficient *S. cerevisiae* strain 20B-12-1 (α-*pep4-3, prc1, prb1, his*)[5] is grown to an A_{600} of 4–6 at 30° in 350 liters of 1% yeast extract, 2% peptone, and 2% glucose. The yeast are stored at −70° until needed.

Purification Scheme

Pellets and solutions are stored at −20°, and all steps are performed at 4°.

---

[8] A. Bensadoun and D. Weinstein, *Anal. Biochem.* **70**, 241 (1976).

[9] P. R. Green, A. H. Merrill, Jr., and R. M. Bell, *J. Biol. Chem.* **256**, 11151 (1981).

[10] O. H. Lowry, N. J. Rosebrough, A. L. Farr, and R. J. Randall, *J. Biol. Chem.* **193**, 265 (1951).

[11] U. K. Laemmli, *Nature (London)* **227**, 680 (1970).

[12] W. Wray, T. Boulikas, V. P. Wray, and R. Hancock, *Anal. Biochem.* **118**, 197 (1981).

TABLE I
PURIFICATION OF YEAST tRNA SPLICING ENDONUCLEASE[a]

Purification step	Total activity (units)	Total protein (mg)	Specific activity (units/mg)	Purification (-fold)	Yield (%)
Triton X-100 extract	570	5,400	0.11	1	100
Heparin-agarose 1	767	439	1.8	16	135
Heparin-agarose 2	637	135	4.7	43	112
Hydroxyapatite flow-through	590	50	11.8	107	104
Hydroxyapatite	513	11.6	44.2	402	90
Affi-Gel Blue	95	0.89	107	970	17
tRNA elution	87	0.40[b]	218	1,980	15

[a] The endonuclease was purified from 3 kg of yeast. Details of the steps employed are described in the text.

[b] Protein was estimated by comparing volumes of the fraction loaded onto the SDS–polyacrylamide gel and the intensity of the stained protein bands between the Affi-Gel Blue and tRNA elution steps (see Fig. 2, lanes 7 and 8).

Preparation of Detergent Extract. Frozen cells are thawed in 2 volumes of 100 mM Tris-HCl (pH 8.0), 20 mM EDTA, 10% glycerol, 5 mM spermidine-HCl, 3 mM dithiothreitol, 1.0 M ammonium sulfate, and 1 mM phenylmethylsulfonyl fluoride (PMSF). The cells are lysed with glass beads (0.20–0.30 mm diameter; Glen Mills, Inc., Maywood, NJ) by passing them twice through a Dynomill chilled at $-10°$. The extract is centrifuged at 4,000 g for 10 min and the supernatant at 150,000 g for 2 hr. The membranes are homogenized into 2 volumes of lysis buffer using a Potter–Elvehjem homogenizer and centrifuged as above. The membranes are washed twice further by homogenizing into 20% glycerol, 5 mM 2-mercaptoethanol, 25 mM Tris-HCl (pH 8.0), 0.2 mM PMSF (buffer E), adding 0.02% Triton X-100, and stirring for 1 hr. The washed membranes are homogenized into 2 volumes of buffer E (20–40 mg/ml protein), Triton X-100 is added to 0.7%, and the mixture is stirred for 1 hr and centrifuged. This extract contains 5–50% of the extractable activity. The remaining activity is extracted as above using 0.9% Triton X-100 plus 0.1 M ammonium sulfate. Since activity is difficult to quantitate while membrane bound, the detergent extracts are defined to be 1-fold purified (Table I, Fig. 2, lane 2). The extracts may be stored indefinitely at $-20°$.

Because of the low yield of endonuclease activity, we find it necessary to process kilogram quantities of yeast. To obtain higher yields of membranes, it is critical to centrifuge a full 2 hr. Still, much activity is found in the milky supernatant, perhaps in very small vesicles.

Heparin-Agarose Chromatography. The extracts are made 2.5% in Triton X-100 by adding a 20% stock solution. The extract is applied to a 4 × 12 cm column of heparin-agarose that has been equilibrated with buffer C [25 mM Tris-HCl (pH 8.0), 10% glycerol, 5 mM 2-mercaptoethanol, and 0.5% Triton X-100] containing 0.1 M ammonium sulfate. Bound protein is eluted at 150 ml/hr with a 2-liter gradient of 0.1–0.6 M ammonium sulfate in buffer C (Fig. 1A and Fig. 2, lane 3). Active fractions are combined, diluted 3-fold with buffer C, and applied to a 2 × 38 cm heparin-agarose column. Activity is eluted at 70 ml/hr with a 1-liter gradient of 0.1–0.6 M ammonium sulfate (Fig. 1B and Fig. 2, lane 4).

Hydroxyapatite Chromatography. The active heparin-agarose fractions are combined but cannot be dialyzed without extensive precipitation and loss of activity. They are applied to a 2 × 35 cm column of hydroxyapatite (BioGel HTP, Bio-Rad) equilibrated with buffer C containing 0.35 M ammonium sulfate. Most of the activity flows through the column (Fig. 2, lane 5). The flow-through is collected, diluted 3-fold with buffer C, and reapplied to the hydroxyapatite column equilibrated with buffer C. Activity is eluted at 70 ml/hr with a 1,100-ml gradient of 10% glycerol, 5 mM 2-mercaptoethanol, 0.5% Triton X-100, and 0–0.45 M potassium phosphate (pH 7.0) (Fig. 1C and Fig. 2, lane 6).

Affi-Gel Blue Chromatography. The active hydroxyapatite fractions are combined and dialyzed against buffer C until the potassium phosphate concentration is less than 75 mM, then applied to a 1 × 16 cm column of Affi-Gel Blue (Bio-Rad). Activity is eluted at 30 ml/hr using a 1,100-ml gradient of buffer C containing 0–2.0 M NaCl (Fig. 1D and Fig. 2, lane 7). The yield of activity is consistently low (10–20%). It can be increased by using a smaller column (~7 ml); however, this greatly decreases the efficiency in removing certain contaminants.

tRNA Affinity Elution. Active fractions from the Affi-Gel Blue step are combined, dialyzed against buffer C, and applied to a 1 × 8 cm column of CM-Sepharose. The column is washed with 25 ml of buffer C containing 75 mM NaCl, and activity is eluted at 40 ml/hr with 100 ml of the same buffer plus 100 mg/ml yeast tRNA (Type X-S, Sigma, treated with proteinase K, phenol extracted, and ethanol precipitated) (Fig. 2, lane 8).

Glycerol Gradients. The affinity eluant is concentrated by applying it to a 1-ml column of hydroxyapatite and step eluting with 10 ml of buffer C containing 0.4 M potassium phosphate (pH 7.0). A portion (300 μl) is layered on an 11-ml, 10–16% glycerol gradient in buffer C plus 0.4 M potassium phosphate (pH 7.0). The gradients are centrifuged at 38,000 rpm (180,000 g) for 48 hr in an SW41 rotor.

The purification is summarized in Table I and Fig. 2. The overall purification through the tRNA affinity column is about 2,000-fold over

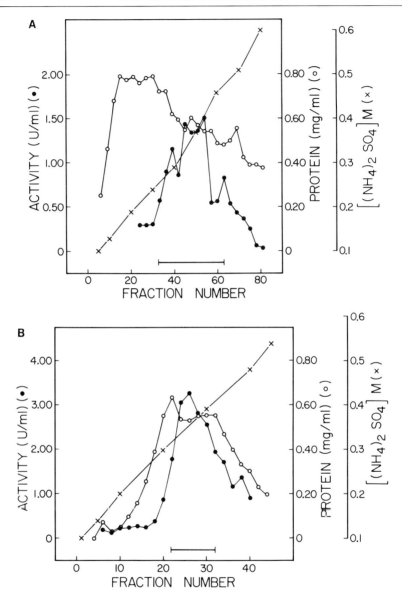

FIG. 1. Elution profiles. (A) Chromatography of Triton X-100 extract over heparin-agarose. (B) Second chromatography over heparin-agarose. (C) Chromatography over hydroxyapatite. (D) Chromatography over Affi-Gel Blue. Active fractions combined for the next step are indicated by a bar in the figures.

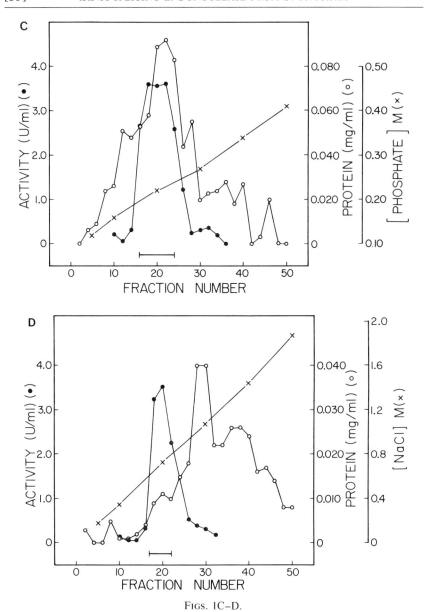

FIGS. 1C–D.

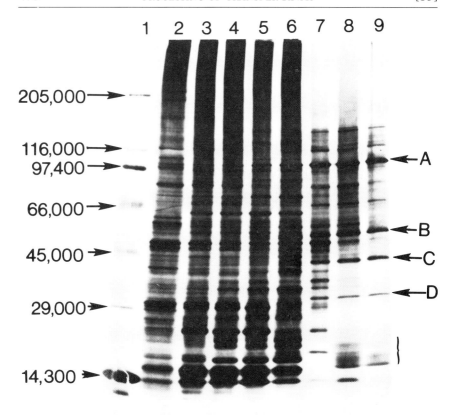

FIG. 2. SDS–polyacrylamide gel electrophoresis analysis of the purification. Lane 1, standards: rabbit muscle myosin (205,000); *E. coli* β-galactosidase (116,000); rabbit muscle phosphorylase *b* (97,400); bovine serum albumin (66,000); ovalbumin (45,000); bovine erythrocyte carbonate dehydratase (29,000); egg white lysozyme (14,300). Lane 2, Triton X-100 extract (9.0 μg protein, 0.001 units of activity). Lane 3, first heparin-agarose (9.0 μg protein, 0.02 units). Lane 4, second heparin agarose (7.8 μg protein, 0.04 units). Lane 5, hydroxyapatite flow-through, (5.8 μg protein, 0.07 units). Lane 6, hydroxyapatite chromatography (6.5 μg protein, 0.29 units). Lane 7, Affi-Gel Blue (2.7 μg protein, 0.29 units). Lane 8, tRNA affinity elution (2.7 μg protein, 0.58 units). Lane 9, glycerol gradient sedimentation (0.66 units). Arrows denote the major, highly purified polypeptides of M_r 100,000 (A), 52,500 (B), 40,700 (C), and 29,500 (D). The parentheses include the region of stained tRNA.

Triton X-100 extract. This is an estimate since an accurate measurement of the total protein is prevented by the presence of large amounts of tRNA. After the glycerol gradient step, the active fraction consists of tRNA plus four major and four minor proteins (Fig. 2, lane 9). Thus, severalfold more purification is needed to obtain a homogeneous preparation.

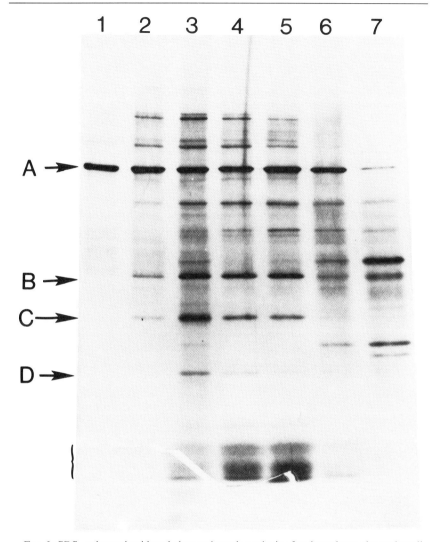

FIG. 3. SDS–polyacrylamide gel electrophoresis analysis of endonuclease glycerol gradient sedimentation. Gradient fractions (1 ml) were trichloroacetic acid precipitated and electrophoresed. Lanes 1–7 represent denser toward lighter fractions, respectively. Activity loaded into each lane is as follows: Lane 1, 0 units; lane 2, 0.2 units; lane 3, 0.8 units; lane 4, 0.8 units; lane 5, 0.4 units; lane 6, 0 units; lane 7, 0 units. Arrows and parentheses denote the polypeptides and tRNA as described in Fig. 2.

Polypeptide Components of the Endonuclease

The most highly purified fraction contains major polypeptides of M_r 100,000, 52,500, 40,700, and 29,500 (Fig. 2, lane 9, bands A, B, C, and D) and four minor polypeptides of M_r 140,000, 120,000, 90,000, and 70,000. The relative amounts of the major polypeptides all increase in conjunction with the purification, and the total amounts match the increase of units of activity analyzed on the SDS–polyacrylamide gel (Fig. 2, lanes 2–9). It was noticed that bands B, C, and D copurify with each other and with endonuclease activity when using Affi-Gel Blue chromatography (data not shown), hydroxyapatite chromatography (data not shown), and glycerol gradient sedimentation (Fig. 3). Polypeptide A elutes across the peak of active fractions but is also seen in inactive fractions (Fig. 3, lanes 1 and 6). Based on silver stain intensity, polypeptides B, C, and D are present in approximately equal amounts. They also cosediment 4–7 fractions into a 12-fraction glycerol gradient (Fig. 3, lanes 2–5), which is where other polypeptides of M_r 100,000–140,000 sediment. Polypeptides of M_r 25,000–55,000 sediment near the top of the gradient (Fig. 3, lanes 6 and 7). The data suggest that polypeptides B, C, and D exist as a trimeric complex and represent the tRNA splicing endonuclease.

[40] Purification and Characterization of Wheat Germ RNA Ligase and Associated Activities

By Leslie Pick, Henry M. Furneaux, and Jerard Hurwitz

Introduction

Wheat germ RNA ligase was discovered by Filipowicz and co-workers, who identified a novel enzymatic activity in crude extracts of wheat germ that catalyzed the formation of an unusual type of bond.[1,2] In the presence of ATP, RNA molecules containing 2′,3′-cyclic phosphate as well as 5′-hydroxyl or 5′-phosphate termini were circularized, and a 2′-phosphomonoester–3′,5′-phosphodiester bond was formed at the ligated junction. The phosphate in the phosphodiester bond originated from a

[1] M. Konarska, W. Filipowicz, H. Domdey, and H. J. Gross, *Nature (London)* **293**, 112 (1981).

[2] M. Konarska, W. Filipowicz, and H. J. Gross, *Proc. Natl. Acad. Sci. U.S.A.* **79**, 1474 (1982).

phosphate residue at the 5' terminus of the RNA substrate while the 2'-phosphomonoester was derived from the phosphate present in the cyclic phosphate bond. RNAs containing 5'-hydroxyl termini were substrates for ligation because of the presence of a 5'-hydroxyl polynucleotide kinase activity in wheat germ extracts. Thus, when a 5'-hydroxyl–2',3'-cyclic phosphate terminated RNA was used as the substrate, a phosphate residue from the γ position of ATP was transferred to the 5'-hydroxyl terminus of the RNA and was subsequently incorporated into the 3',5'-phosphodiester bond.

The wheat germ extract also contained a 2',3'-cyclic phosphodiesterase activity that hydrolyzed the cyclic bond to yield a 2'-phosphate terminus. This activity was thought to be responsible for the conversion of the cyclic phosphate terminus to the 2'-phosphomonoester either prior to or during the ligation reaction. Since the ligated junction contained a phosphate residue in the 2' position, this site was not cleaved by alkali as well as a variety of nucleases. This nuclease resistance allowed for the identification of the type of bond produced by wheat germ RNA ligase *in vitro* and also was thought to stabilize the newly formed bond against nucleolytic attack *in vivo*. Since the unusual 2'-phosphomonoester was stable in the wheat germ crude extract, it was suggested that this structure might serve as a signal for further processing or for transport following the splicing reaction *in vivo*. Alternatively, if an enzyme were to remove the phosphomonoester *in vivo*, this would serve to block the reversal of the splicing reaction. To date, however, no specific 2'-phosphatase activity has been identified.

On the basis of the data summarized above, a scheme for ligation in the wheat germ system was proposed.[2] Cleavage of a precursor RNA would release an intron and generate exons containing 2',3'-cyclic phosphate and 5'-hydroxyl termini. The 5'-hydroxyl terminus would be phosphorylated by a 5'-hydroxyl polynucleotide kinase activity, and ligation would then proceed, resulting in the formation of a 2'-phosphomonoester–3',5'-phosphodiester bond. Thus, a 5'-hydroxyl polynucleotide kinase, a 2',3'-cyclic phosphodiesterase, and an RNA ligase were thought to play a role in RNA splicing in wheat germ.

Wheat germ RNA ligase is active on a variety of synthetic substrates containing the appropriate termini. This property was used to develop a rapid assay that monitored RNA ligase activity during purification[3,4] (see below). Biochemical analysis of highly purified RNA ligase led to the surprising observation that all three enzymatic activities described above,

[3] H. Furneaux, L. Pick, and J. Hurwitz, *Proc. Natl. Acad. Sci. U.S.A.* **80,** 3933 (1983).
[4] L. Pick and J. Hurwitz, *J. Biol. Chem.* **261,** 6684 (1986).

5'-hydroxyl polynucleotide kinase, 2',3'-cyclic phosphodiesterase, and RNA ligase, reside in a single polypeptide chain.[4-6] Using the procedure described below, RNA ligase was purified approximately 6000-fold from crude extracts of raw wheat germ. Independent assays for both the 5'-hydroxyl polynucleotide kinase and 2',3'-cyclic phosphodiesterase activities were used to monitor these activities during the course of purification. Both activities copurified with the RNA ligase through all chromatographic steps and cosedimented with RNA ligase activity on glycerol gradient centrifugation carried out in the presence of high salt and urea. Further investigation of the physical relationship between the ligase and kinase activities showed that their heat inactivation profiles were identical. The intrinsic kinase activity, however, appeared to be more susceptible to proteolysis than the ligase activity, which may result in the preferential loss of this activity during purification. Thus, although the three activities appear to be intrinsic to a single protein species, the active sites of at least the ligase and kinase are separate. It should be noted that crude extracts of wheat germ also contain unrelated 2',3'-cyclic phosphodiesterase[7] and 5'-hydroxyl polynucleotide kinase (L. Pick and J. Hurwitz, unpublished observations, 1986) activities that are readily separated from the RNA ligase-associated activities during purification.

A similar RNA ligase activity has been purified from yeast by Abelson and co-workers.[8,9] This ligase also catalyzes an ATP-dependent formation of a 2'-phosphomonoester–3',5'-phosphodiester bond. Interestingly, however, this enzyme utilizes synthetic substrates rather poorly; tRNA half-molecules generated by the yeast tRNA endonuclease, which contain 5'-hydroxyl and 2',3'-cyclic phosphate termini, are the preferred substrates for the yeast tRNA ligase. As is the case for the wheat germ enzyme, highly purified preparations of yeast tRNA ligase contain 5'-hydroxyl polynucleotide kinase and 2',3'-cyclic phosphodiesterase activities, and all three activities cosediment on glycerol gradient centrifugation.

Some clues about a potential functional significance of the coexistence of three enzymatic activities on a single polypeptide chain come from the analysis of the mechanism of action of the wheat germ enzyme.[5,10,11] The

---

[5] L. Pick, H. Furneaux, and J. Hurwitz, *J. Biol. Chem.* **261**, 6694 (1986).

[6] L. Pick, Ph.D. thesis, Institute Albert Einstein College of Medicine, Bronx, New York (1986).

[7] K. Tyc, C. Kellenberger, and W. Filipowicz, *J. Biol. Chem.* **262**, 12994 (1987).

[8] C. L. Greer, C. L. Peebles, P. Gegenheimer, and J. Abelson, *Cell* **32**, 537 (1983).

[9] E. M. Phizicky, R. C. Schwartz, and J. Abelson, *J. Biol. Chem.* **261**, 2978 (1986).

[10] P. Gegenheimer, H.-J. Gabius, C. L. Peebles, and J. Abelson, *J. Biol. Chem.* **258**, 8365 (1983).

[11] R. C. Schwartz, C. L. Greer, P. Gegenheimer, and J. Abelson, *J. Biol. Chem.* **258**, 8374 (1983).

purified enzyme ligates substrates containing 5'-hydroxyl or 5'-phosphate and 2',3'-cyclic phosphate or 2'-phosphate termini. In all cases, ATP is absolutely required for ligation. For substrates containing 5'-phosphate termini (and therefore not requiring the activity of the ligase-associated kinase), ligation of 1 mol of substrate is accompanied by the hydrolysis of 1 mol of ATP to AMP and PP_i. Like prokaryotic RNA and DNA ligases, the wheat germ enzyme catalyzes an $ATP-PP_i$ exchange reaction and forms a covalent enzyme–adenylate (E–AMP) complex. However, the unique feature of the reactions catalyzed by the wheat germ enzyme is the marked stimulation by the addition of RNA. RNA molecules with different termini vary in their abilities to stimulate ATP hydrolysis. RNAs that are substrates for both the kinase and ligase activities support maximal stimulation; RNAs that are substrates for one or the other but not both of these activities stimulate partially; finally, RNAs that are not substrates for either activity do not stimulate ATP hydrolysis. These observations led to the proposal that the RNA substrate reacts with the ligase prior to the interaction of the enzyme with ATP, perhaps acting as a cofactor to alter the conformation of the enzyme, which facilitates either the binding or hydrolysis of ATP.

Furthermore, the initial interaction of the RNA substrate with the enzyme appears to reflect a functional relationship between the kinase and ligase activities. This relationship is further supported by the substrate preferences of both the 2',3'-cyclic phosphodiesterase and ligase activities. In both cases, RNA molecules containing 5'-hydroxyl termini are utilized more efficiently than those containing 5'-phosphate termini. Thus, RNA molecules phosphorylated by the kinase activity of the wheat germ enzyme will be preferentially ligated. Taken together, these observations suggest that the three independent activities of the wheat germ RNA ligase act in a concerted fashion to join RNA molecules containing 5'-hydroxyl and 2',3'-cyclic phosphate termini, the presumed products of an endonucleolytic or autocatalytic cleavage event.

Purification of RNA Ligase

RNA ligase is purified from raw wheat germ, which was obtained as a gift from Pillsbury Mills.

Buffers

Extraction buffer: 50 mM tris(hydroxymethyl)aminomethane hydrochloride (Tris-HCl) buffer (pH 7.9), 5 mM dithiothreitol (DTT), 0.2 M ammonium sulfate, and 10× protease inhibitor mixture

10% Polymin P: Prepared as described[12] and adjusted to pH 7.9

[12] J. J. Jendrisak and R. R. Burgess, *Biochemistry* **14**, 4639 (1975).

Buffer A: 25 mM Tris-HCl buffer (pH 7.9), 2 mM DTT, 10% glycerol, 0.01% Nonidet P-40 (NP-40)

Buffer B: 1 mM sodium phosphate buffer (pH 7.9), 2 mM DTT, 10% glycerol, 0.01% NP-40

Protease inhibitor mixture: 0.1 mM ethylenediaminetetraacetic acid (EDTA), 0.1 mM ethylene glycol bis(β-aminoethyl ether)-N,N,N',N'-tetraacetic acid (EGTA), 0.1 mM benzamidine, 0.2 μg/ml bacitracin, 0.2 μg/ml leupeptin, 0.1 μg/ml antipain, 10 μM phenylmethylsulfonyl fluoride (PMSF); PMSF is prepared as a 100 mM stock solution in 2-propanol, and all other inhibitors are prepared in distilled water and stored at $-20°$

Purification Procedure

Since proteolysis can interfere with the isolation of intact RNA ligase, all operations are carried out at 4°, and the protease inhibitor mixture is added to all buffers just prior to use. Sample column profiles for four steps of the purification procedure, outlined below, are presented in Fig. 1. Also shown is the copurification of RNA ligase and 5'-hydroxyl polynucleotide kinase activities at each step (see below).

Step 1: Crude Extract. Raw wheat germ is suspended in extraction buffer (4 ml/g) in a glass beaker. After 20 min on ice with occasional stirring, a further 10× aliquot of protease inhibitor mixture is added and the suspension is homogenized in a Waring blendor (2 times at low speed and 2 times at high speed for 10 sec each at 2-min intervals). The suspension is centrifuged for 30 min at 100,000 g. The supernatant is removed and stored on ice. The pellets are resuspended in extraction buffer (2 ml/g of starting material) and centrifuged as above. The combined supernatants are recentrifuged, and the resulting supernatant is filtered through sterile cheesecloth.

Step 2: Polymin P. Polymin P is added dropwise to the crude extract to a final concentration of 0.1%. The extract is stirred slowly for an additional 5 min and centrifuged for 10 min at 10,000 g, and the supernatant is collected. At this concentration of Polymin P, RNA ligase remains in the supernatant. RNA ligase can be precipitated using higher concentrations of Polymin P. However, ligase activity was not successfully resolubilized following precipitation (unpublished observations).

Step 3: DEAE-Cellulose. The Polymin P fraction is diluted with buffer A to a conductivity equivalent to 0.2 M NaCl and loaded onto a DE-52 column (Whatman, Clifton, NJ) previously equilibrated with 0.2 M NaCl in buffer A (5–10 mg protein/ml resin, at a rate of 0.5 column volume/hr). The column is washed with 4 column volumes of the same buffer. The

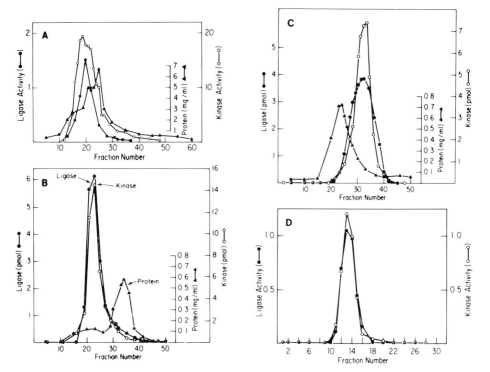

FIG. 1. Chromatography of RNA ligase and 5'-hydroxyl polynucleotide kinase activities. Chromatography on Bio-Rex (A), phosphocellulose (B), and ATP-agarose (C) columns and sedimentation in 10–30% glycerol gradients (D) were carried out as described in the text. The NaCl gradients used in these procedures are not shown. The activities of RNA ligase, measured using $[5'-{}^{32}P]p(Ap)_nA > p$ as substrate, and 5'-hydroxyl polynucleotide kinase, assayed as described in the text, are presented as pmol/μl of fraction/30 min and pmol/μl of fraction/15 min, respectively. Protein content was determined according to the method of M. Bradford, *Anal. Biochem.* **72**, 248 (1976). [Portions of this figure (B and C) are reprinted with permission from L. Pick and J. Hurwitz, *J. Biol. Chem.* **261**, 6684 (1986).]

flow-through fractions are assayed for RNA ligase activity and pooled accordingly. Yellow-colored material from the extract remains bound to the resin in this step.

Step 4: Bio-Rex. The DEAE-cellulose fraction is diluted with an equal volume of buffer A and loaded onto a Bio-Rex 70 (Bio-Rad, Richmond, CA) column, previously equilibrated with buffer A containing 0.1 M NaCl (30–50 mg protein/ml resin, at a rate of 2 column volumes/hr). The column is washed with 3 column volumes of the same buffer. RNA ligase activity is eluted using a 5-fold linear gradient of 0.1 to 0.7 M NaCl in buffer A. The peak fractions of activity elute between 0.2 and 0.275 M NaCl (Fig. 1A). It is important to adjust the pH of the Bio-Rex resin carefully prior to use. At pH 7.9, more than 60% of the total protein flows through the column. Since this resin has a very high capacity for binding, the DEAE-cellulose fraction can be concentrated approximately 30-fold in this step.

Step 5: Phosphocellulose. The Bio-Rex fraction is diluted with buffer A to a concentration of 0.1 M NaCl and loaded onto a phosphocellulose column (Whatman) previously equilibrated with the same buffer (7 mg protein/ml resin, at a rate of 2 column volumes/hr). The column is washed with 4 column volumes of buffer A containing 0.1 M NaCl and eluted with a 6-fold linear gradient of 0.1 to 1 M NaCl in buffer A. RNA ligase activity elutes between 0.275 and 0.375 M NaCl (Fig. 1B). The phosphocellulose column can be replaced by a Mono-S FPLC column (Pharmacia, Piscataway, NJ). This procedure results in a similar extent of purification in a much shorter time.

Step 6: Hydroxyapatite. The phosphocellulose fraction is diluted to a concentration of 0.2 M NaCl in buffer A and loaded onto a BioGel HTP (Bio-Rad) column, previously equilibrated with the same buffer (5 mg protein/ml resin; 0.5 column volume/hr). The column is washed with 4 column volumes of 0.2 M NaCl in buffer A and then 4 column volumes of 1 M NaCl in buffer A. RNA ligase is step eluted from the column with 4 column volumes of 1 M NaCl in buffer B. The eluted fractions are immediately dialyzed against 0.5 M NaCl in buffer A, and then 50 mM NaCl in buffer A.

Step 7: ATP-Agarose. The hydroxyapatite fraction is loaded onto an ATP-agarose column (Pharmacia) previously equilibrated with 50 mM NaCl in buffer A (1–2 mg protein/ml resin, 1 column volume/hr). The column is washed with 4 column volumes of the same buffer and eluted with a 5-fold linear gradient of 50 mM to 1 M NaCl in buffer A. RNA ligase elutes from the column in fractions containing approximately 55 mM NaCl (Fig. 1C). Peak fractions are pooled and stored in small aliquots at $-70°$. Since the ATP-agarose fraction is free of contaminating activities that interfere with the activities of RNA ligase, this fraction can be used for most purposes. To obtain highly purified enzyme, glycerol gradient centrifugation is performed.

Step 8: Glycerol Gradient. A 0.2-ml aliquot of the ATP-agarose fraction is layered onto a 5-ml linear gradient of either 10–30% or 15–35% glycerol in buffer A containing 0.5 M NaCl plus the protease inhibitor mixture. The gradient is centrifuged at 48,000 rpm for 24 hr in a Beckman SW50.1 rotor. Twelve-drop fractions are collected from the bottom of the

TABLE I
PURIFICATION OF WHEAT GERM RNA LIGASE ACTIVITY[a]

Fraction	Protein (mg)	Activity (units)	Specific activity (units/mg)
Crude extract	4900	37,000	7.6
Polymin P	N.D.	432,000	—
DEAE-cellulose	3700	491,000	133
Bio-Rex	350	220,000	629
Phosphocellulose	48	138,000	2880
Hydroxyapatite	14	82,000	5860
ATP-agarose	1.7	22,000	12,900
Glycerol gradient	0.24	11,000	45,800

[a] RNA ligase activity was assayed as described in the text using $[5'-{}^{32}P]p(Ap)_n A > P$ as substrate. One unit of RNA ligase activity renders 1 pmol of $5'-{}^{32}$P-labeled termini resistant to BALP after 30 min at 37°. N.D., Not determined.

tube. Most of the recovered RNA ligase activity is detected in three fractions (Fig. 1D).

General Comments

The procedure described above results in a 6000-fold purification of RNA ligase activity as summarized in Table I. Beginning with 25 g of raw wheat germ, approximately 0.24 mg of protein (glycerol gradient-purified material) is obtained.[4]

RNA ligase activity is stable when stored at −70°. The ATP-agarose fraction can be frozen and thawed at least 3 times without loss of activity. After storage for 2 weeks at 4°, 70% of the RNA ligase activity remained. RNA ligase is extremely unstable when stored at low ionic strength. It is recommended that the ATP-agarose or glycerol gradient fractions be maintained in buffer A containing approximately 0.5 M NaCl for long-term storage.

RNA ligase sediments with a coefficient of 6.2 S, suggesting a native molecular weight of approximately 100K. Silver staining of polyacrylamide gels reveals the presence of a doublet of approximately 110K, which is the only species that copurifies with RNA ligase activity. Furthermore, the same species is identified in the enzyme–AMP complex

(E–AMP complex) assay (see below). These results strongly suggest that RNA ligase has a native molecular weight of approximately 100K and that it behaves as a monomer in solution. RNA ligase migrates as a doublet on SDS–polyacrylamide gels, most likely because of proteolysis of the larger species. Attempts to elute active enzyme from these gels, in order to further investigate this problem, were unsuccessful (unpublished observations).

Assays of RNA Ligase and Ligase-Associated Activities

Substrate Preparations

Preparation of [5'-^{32}P]p(Ap)$_n$A > p. Poly(A) containing 5'-hydroxyl and 2',3'-cyclic phosphate termini, with an average chain length of 70 nucleotides, is prepared as follows. Reaction mixtures (1 ml) containing 4.5 mg of poly(A) and 25 mM potassium phosphate buffer (pH 7.5), are incubated with 0.1 mg of pancreatic RNase for 1 hr at 37°. Reaction mixtures are treated with 0.1 mg of proteinase K in the presence of 0.1% sodium dodecyl sulfate (SDS) for 20 min at 37°, extracted once with an equal volume of phenol–chloroform (1 : 1), and precipitated with 3 volumes of ice-cold ethanol in the presence of 0.3 M sodium acetate buffer (pH 5.3). The 5'-hydroxyl termini are labeled with [^{32}P]phosphate, using a mutant form of T4 polynucleotide kinase that lacks the intrinsic 3'-phosphatase activity (*pse T1* mutant, obtained from New England Nuclear, Boston, MA), as follows. Reaction mixtures (1 ml) containing 50 mM Tris-HCl buffer (pH 7.9), 5 mM DTT, 10 mM MgCl$_2$, 0.2 mM [γ-^{32}P]ATP (30–60 cpm/fmol), and 3 nmol of termini of 5'-hydroxyl terminated poly(A) are incubated with 15 units of kinase for 30 min at 37°. After 30 min, an additional 15 units of kinase is added and the incubation continued for 30 min. Reaction products are extracted once with an equal volume of phenol–chloroform (1 : 1) and once with an equal volume of chloroform, and unincorporated [γ-^{32}P]ATP is removed by filtration through Sephadex G-50 fine, using 50 mM ammonium acetate as the eluant.

Preparation of [2',3'-^{32}P](C)$_{11}$G > p. [2',3'-^{32}P](C)$_{11}$G > p is prepared by the addition of [5'-^{32}P]pCp to $_{OH}$(Cp)$_{11}$G$_{OH}$ (obtained as a gift from Dr. O. Uhlenbeck, University of Colorado) using T4 RNA ligase (P-L Biochemicals, Madison, WI), followed by cleavage with RNase N1 (Sigma) to yield 2',3'-cyclic phosphate termini, as follows. Reaction mixtures (50 μl) containing 50 mM N-hydroxyethylpiperazine-N'-ethanesulfonic acid (HEPES) buffer (pH 8.3), 20 mM DTT, 20 mM MgCl$_2$, 0.5 mM ATP, 20 μM [5'-^{32}P]pCp (100–300 cpm/fmol), 1 nmol of termini of $_{OH}$(Cp)$_{11}$G$_{OH}$, and 8 units of T4 RNA ligase are incubated for 12 hr at 30°. Reaction products are extracted once with an equal volume of phenol–chloroform (1 : 1) and once with an equal volume of chloroform, and the aqueous phase is adjusted to 0.5 ml with distilled water and incubated with 0.25 unit of RNase N1 for 30 min at 37°. Products are extracted once with an equal volume of phenol–chloroform (1 : 1) and once with an equal volume of chloroform. The aqueous phase is adjusted to 0.1 M Tris-HCl buffer (pH 7.9), and incubated with 0.14 unit of bacterial alkaline phosphatase (BALP, *Escherichia coli,* BAPF, Worthington Biochemical, Freehold, NJ) at 37°. After 30 min, an additional 0.14 unit of BALP is added, and the incubation is continued for 30 min. Products are extracted once with an equal volume of phenol–chloroform (1 : 1), once with an equal volume of chloroform and separated from unincorporated ^{32}P by filtration through Sephadex G-25 (fine), using 50 mM ammonium acetate as the eluant. $[2',3'-^{32}$P]p(Cp)$_{11}$G > p is prepared by treatment with T4 *pse T1* polynucleotide kinase, as described above. Nuclease P1 digestion of the (Cp)$_{11}$G > p substrates normally yields pG > p (95%) and pC (5%), while digestion with RNase T2 yields Gp (95%) and Cp (5%). These results indicate that the preparations of $_{OH}$(Cp)$_{11}$G$_{OH}$ may contain small amounts of $_{OH}$(Cp)$_n$C$_{OH}$.

2'-Phosphate-terminated $[2'-^{32}$P](Cp)$_{11}$Gp is prepared by incubation of $[2',3'-^{32}$P](Cp)$_{11}$G > p with nucleoside 2',3'-cyclic phosphate 2'-nucleotide hydrolase [Sigma, St. Louis, MO 0.1 unit in the presence of 50 mM Tris-HCl buffer (pH 7.5) for 30 min at 37°] or with purified wheat germ RNA ligase in reaction mixtures that lack ATP (8 units of ATP-agarose fraction for 30 min at 37°, using reaction conditions described for assaying 2',3'-cyclic phosphodiesterase activity). 3'-Phosphate-terminated substrates are prepared by incubation with RNase T1 [5 units in the presence of 20 mM Tris-HCl buffer (pH 7.9) and 2.5 mM EDTA for 60 min at 37°].

Preparation of Substrates Used to Assay 5'-Hydroxyl Polynucleotide Kinase Activity. Poly(A) containing 5'-hydroxyl and 2',3'-cyclic phosphate termini is prepared as described above. 5'-Hydroxyl-3'-phosphate-terminated poly(A), poly(C), and poly(I), with an average chain length of 50 nucleotides, are prepared by digestion of 0.5 mg of polynucleotide with micrococcal nuclease [P-L Biochemicals, 0.4 unit in the presence of 20 mM Tris-HCl buffer (pH 8.4) and 2 mM CaCl$_2$ for 30 min at 37°]. 5'-Hydroxyl-3'-hydroxyl-terminated poly(A) is prepared by treatment of 3'-phosphate-terminated poly(A) with BALP [0.014 unit in the presence of 50 mM Tris-HCl buffer (pH 7.9) for 60 min at 60°]. 5'-Hydroxyl-terminated salmon sperm DNA and 5'-hydroxyl-terminated poly(dA) are prepared using BALP, as described above. All reaction products are extracted once with an equal volume of phenol–chloroform (1 : 1) and once with an equal volume of chloroform and are precipitated with ethanol.

The concentration of 5'-hydroxyl termini in these preparations is determined using T4 polynucleotide kinase.

Assays of RNA Ligase Activity

Assay of RNA Ligase Using [5'-^{32}P]p(Ap)$_n$A > p as Substrate. The activity of RNA ligase can be detected using a simple assay based on the conversion of the 5'-phosphate termini of 5'-^{32}P-labeled RNA substrate to a phosphatase-resistant form. Since it is rapid, this assay is recommended for monitoring ligase activity during the course of purification.

Reaction mixtures (50 μl) containing 2.5 pmol of 5'-^{32}P-labeled termini (30–60 cpm/fmol), 20 mM Tris-HCl buffer (pH 7.9), 6 mM MgCl$_2$, 2 mM DTT, 1 mM ATP, and enzyme are incubated at 37° for 30 min. Reliable assays of crude extracts require the addition of large amounts of unlabeled poly(A) chains (~100 pmol of 5'-hydroxyl termini) to reaction mixtures. Following incubation, reaction mixtures are diluted to 0.2 ml with distilled water and extracted once with an equal volume of phenol–chloroform (1:1). For assays of the crude extract, Polymin P, and DEAE-cellulose fractions, reaction mixtures are treated with 20 μg of proteinase K in the presence of 0.1% SDS for 20 min at 37° prior to extraction with phenol–chloroform. Bovine serum albumin (50 μg) and 0.5 ml of cold 10% trichloroacetic acid are added to the aqueous phase. After 7 min on ice, the acid-insoluble material is collected by centrifugation, resuspended in 0.2 ml of 0.1 M Tris-HCl buffer (pH 7.9), and incubated in the presence of 0.014 unit of BALP for 30 min at 60°. The reaction is halted by the addition of 5 ml of cold 5% trichloroacetic acid and 0.1 ml of 0.1 M sodium pyrophosphate. After 5 min on ice, acid-insoluble material is collected on glass fiber filters by filtration under vacuum. The filters are washed with cold 1% trichloroacetic acid and with ethanol, then dried and counted using Econofluor as the scintillation fluid. One unit of RNA ligase activity renders 1 pmol of 5'-^{32}P-labeled termini resistant to BALP after 30 min at 37°.

Assay of RNA Ligase Using [2',3'-^{32}P](Cp)$_{11}$Gp as Substrate. Reaction mixtures (20 μl) containing 20 mM Tris-HCl buffer (pH 7.9), 2 mM DTT, 6 mM MgCl$_2$, 1 mM ATP, [2',3'-^{32}P](Cp)$_{11}$G > p (100–300 cpm/fmol), and enzyme are incubated for 10–30 min at 37°. Reaction mixtures are extracted once with an equal volume of phenol–chloroform (1:1), once with an equal volume of chloroform, and treated with 1 μg of pancreatic RNase for 30 min at 37°. Following extraction with phenol–chloroform (1:1) and chloroform, the reaction products are separated by polyethylenimine (PEI)-cellulose thin-layer chromatography (TLC) using 1 M LiCl as solvent. Regions corresponding to the resistant dinucleoside triphosphate (G$\stackrel{2'P}{<}$$_p$-Cp), and to the authentic marker compound G > p, located by ultraviolet illumination, are excised and quantitated by liquid scintillation spectroscopy using Econofluor.

Alternatively, conversion of linear molecules to circular molecules can be monitored by gel electrophoresis. Following phenol–chloroform extraction, products are ethanol precipitated and loaded onto 15% polyacrylamide gels (acrylamide–bisacrylamide 30 : 1), containing 50% urea in TBE buffer, pH 8.3 (0.1 M Tris base, 0.1 M boric acid, 1 mM EDTA). Circular molecules migrate faster than linear molecules under these conditions.

General Comments. In all cases, ligation should be completely dependent on the addition of enzyme, ATP, and MgCl$_2$. Control reactions in which enzyme and/or ATP are omitted should be routinely performed.

RNA ligase can be used to prepare large amounts of circular RNA with 2′-phosphate groups marking the ligated junction. Such RNA molecules may be useful, for example, for the detection of a 2′-specific phosphatase activity. For this purpose, reaction volumes can be increased and incubation can be carried out at 37° for at least several hours without any significant loss of activity (unpublished observations).

Assay of RNA Ligase-Associated Activities

As discussed above, RNA ligase appears to contain three intrinsic activities that all interact during the ligation event. Utilizing appropriate substrates and assay conditions, each of these activities can be monitored independently.

Assay of 5′-Hydroxyl Polynucleotide Kinase. Kinase activity is measured by the incorporation of ^{32}P from [γ-^{32}P]ATP into an acid-insoluble form. Reaction mixtures (50 μl) containing 20 mM Tris-HCl buffer (pH 7.9), 6 mM MgCl$_2$, 2 mM DTT, 50 pmol of 5′-hydroxyl termini of poly(A) (average chain length of 70) or other appropriate 5′-OH terminated substrate, 0.1 mM ATP (1500–4000 cpm/pmol), and enzyme are incubated at 37° for 15 min. The mixture is diluted to 0.2 ml with distilled water and extracted with an equal volume of phenol–chloroform (1 : 1). Precipitation with trichloroacetic acid is carried out as described above for the RNA ligase assay.

One unit of kinase activity renders 1 pmol of ^{32}P acid-insoluble after 15 min at 37°. Since crude extracts incorporate a substantial amount of ^{32}P in the absence of RNA, assays of crude fractions (crude extract, Polymin P, and DEAE-cellulose, see below) should include digestion with proteinase K prior to the phenol–chloroform extraction. With more highly purified

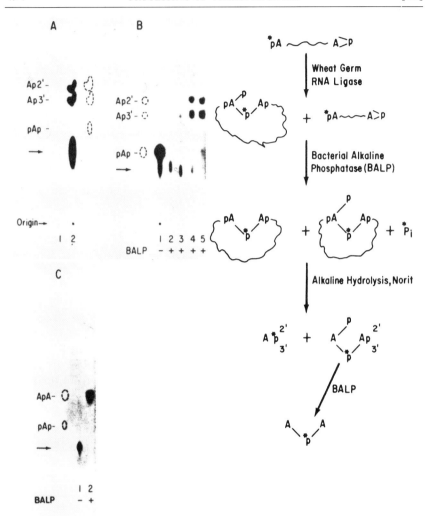

FIG. 2. Analysis of wheat germ RNA ligase reaction product. Reactions were carried out as described in the text using 5 pmol of $[5'-^{32}P]p(Ap)_nA > p$. (A) Reactions were carried out in the absence (lane 1) or presence (lane 2) of 0.2 unit of RNA ligase followed by alkaline hydrolysis and TLC as described. (B) Reactions were carried out as above except that the BALP treatment, prior to alkaline hydrolysis, was varied as follows: lane 1, no BALP added; lanes 2–5, incubation with BALP as described in the text except that the time was varied: lane 2, 10 min; lane 3, 45 min; lane 4, 2 hr; lane 5, 4 hr. (C) Alkaline-resistant material was isolated and treated in the absence (lane 1) or presence (lane 2) of BALP as described in the text. For (A) and (B) PEI-cellulose TLC was carried out using 1 M LiCl–saturated boric acid as the solvent. For (C), 0.3 M LiCl was used as solvent. Autoradiograms are shown. The positions of authentic marker compounds, located by ultraviolet illumination, are indi-

(2′,3′)Ap and decreasing amounts of the alkali-resistant dinucleoside triphosphate.

For further analysis of the nucleotide linkage, resistant $A\underset{p}{\overset{p}{\lessgtr}}Ap$ is isolated from PEI-cellulose plates and subjected to treatment with various enzymes. One example is shown in Fig. 2C. The isolated compound is incubated with 0.014 unit of BALP [in 0.1 M Tris-HCl buffer (pH 7.9) for 60 min at 60°]. Reaction products are extracted once with an equal volume of phenol–chloroform (1 : 1) and once with chloroform, and an aliquot of the aqueous phase is applied to a PEI-cellulose TLC plate that is developed in 0.3 M LiCl in acetic acid. BALP removes the terminal phosphate residues, resulting in the conversion of the dinucleoside triphosphate into a dinucleoside monophosphate, as diagrammed schematically.

Analysis of RNase T2-Resistant Ligase Product

For further analyses, the wheat germ RNA ligase product can be prepared by treatment with RNase T2 (P-L Biochemicals) rather than with alkali, yielding a resistant compound that contains exclusively a 3′-phosphate terminus, rather than the mixture of 2′- and 3′-phosphate groups generated by treatment with alkali. Reaction products, prepared as described above, are incubated with 0.1 unit of RNase T2 in the presence of 0.2 M sodium acetate buffer (pH 5.5) for 30 min at 37°. Reaction mixtures are extracted once with an equal volume of phenol–chloroform (1 : 1), and the aqueous phase is applied to a PEI-cellulose TLC plate that is developed in 0.3 M LiCl–1 N acetic acid. The RNase T2-resistant material is isolated and subjected to any of the following treatments. Products are analyzed by TLC using 1 M LiCl–saturated borate as solvent. An example of an autoradiogram is shown in Fig. 3.

Nuclease P1. Products are incubated with nuclease P1 [P-L Biochemicals, 0.1 μg in the presence of 50 mM sodium acetate buffer (pH 5.5) for 30 min at 30°]. This treatment removes 3′-terminal phosphate groups, resulting in the dinucleoside diphosphate $A\underset{p}{\overset{p}{\lessgtr}}A$, as shown schematically in Fig. 3.

β-Elimination. The product of nuclease P1 treatment contains vicinal 2′- and 3′-hydroxyl groups and is thus susceptible to periodate oxidation

---

cated by the broken lines. The arrow indicates the position of the alkali-resistant reaction product. At right, the procedure used for analysis is diagrammed schematically. [A, B, and C are reprinted from H. Furneaux, L. Pick, and J. Hurwitz, *Proc. Natl. Acad. Sci. U.S.A.* **80**, 3933 (1983).]

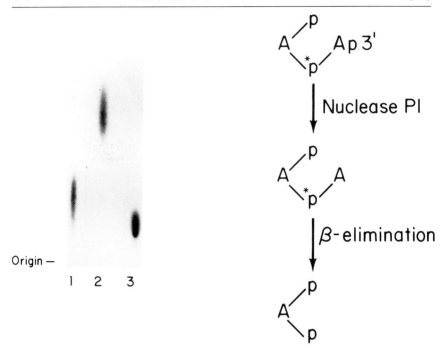

FIG. 3. Analysis of RNase T2-resistant material. Reaction products were prepared as described in the text using ATP-agarose-purified RNA ligase. Isolated T2-resistant product was treated in the absence (lane 1) or in the presence (lane 2) of nuclease P1. In lane 3, nuclease P1 treatment was followed by periodate oxidation and lysine cleavage. Products were separated by PEI-cellulose TLC using 1 M LiCl–saturated boric acid as the solvent. An autoradiogram is shown. At right, the procedure used is diagrammed schematically.

followed by amine cleavage. Following nuclease P1 treatment, products are subjected to the lysine cleavage method described by Neu and Heppel[18] in the presence of 42 nmol of unlabeled ApA, used as carrier and as an internal control. Reaction products are adsorbed to Norit and prepared for TLC as described above. This treatment produces a mononucleoside diphosphate, $A{<}^P_P$, as shown schematically.

Snake Venom Phosphodiesterase. The dinucleoside triphosphate obtained by RNase T2 treatment of the ligation products is resistant to treatment with snake venom phosphodiesterase [Worthington Biochemical, 50 μg/ml in the presence of 10 mM Tris-HCl buffer (pH 7.9) and

[18] H. C. Neu and L. A. Heppel, *J. Biol. Chem.* **239**, 2927 (1964).

5 mM MgCl$_2$ for 30 min at 37°] since it contains a phosphate group at the 3' terminus (data not shown).

Identification of Intermediates in Ligation Reaction

As is the case for other RNA and DNA ligases, wheat germ RNA ligase forms a covalent E–AMP complex as an intermediate in the ligation reaction. Formation of this intermediate can be used to identify RNA ligase, and the complex can be isolated in large quantities for further analysis.

Assay of E–AMP Complex. Reaction mixtures (20 μl) containing 20 mM Tris-HCl buffer (pH 7.9), 2 mM DTT, 6 mM MgCl$_2$, 2.4 μM [α-^{32}P]ATP (300–600 cpm/fmol), and enzyme are incubated for 10 min at 37°. Reactions are halted by the addition of 0.2 ml of cold 10% trichloroacetic acid and 30 μg of bovine serum albumin. After 10 min on ice, acid-insoluble material is collected by centrifugation and rinsed with 10% trichloroacetic acid followed by ether, and the pellets are dried under vacuum. The collected material is electrophoresed through SDS–polyacrylamide gels (10% acrylamide; acrylamide–bisacrylamide 30:0.8, w/w), according to the procedure of Laemmli.[19] Gels are dried, and ^{32}P-labeled products are visualized by autoradiography. RNA ligase–AMP complex is detected as a doublet on SDS–polyacrylamide gels, having an apparent molecular weight of approximately 110K (see Ref. 5). Formation of this complex requires the presence of a divalent cation, Mg^{2+} or Mn^{2+}, and is greatly stimulated (up to 50-fold) by the addition of a ligatable RNA substrate to reaction mixtures.

Characterization of E–AMP Complex. For analysis of the protein-nucleotide linkage, the E–AMP complex can be isolated by gel filtration through Sephadex G-50, in the presence of SDS (see Ref. 5). Boiling under acidic conditions (0.15 N HCl) or hydrolysis with hydroxylamine at acidic pH results in quantitative release of AMP from the complex, whereas the linkage is stable to boiling at neutral or basic pH. These results suggest that AMP is linked to the enzyme via a phosphamide bond. The amino acid to which AMP is attached has not yet been identified.

ATP–PP$_i$ Exchange Reaction. When RNA ligase is incubated with ^{32}PP$_i$ and ATP, an ATP–PP$_i$ exchange reaction can be detected. This exchange reaction is not supported by the other three ribonucleoside triphosphates; dATP supports a low level of exchange. As is the case for E–AMP complex formation, the ATP–PP$_i$ exchange reaction is greatly stimulated by the addition of RNA to the reaction. The magnitude of

---

[19] U. K. Laemmli, *Nature* (*London*) **277,** 680 (1968).

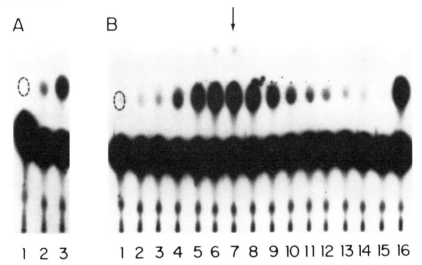

FIG. 4. Analysis of products of ATP–PP$_i$ exchange reaction. (A) ATP–PP$_i$ exchange was assayed as described in the text. Lane 1, enzyme omitted; lane 2, 0.6 μg of wheat germ RNA ligase (ATP-agarose fraction); lane 3, as in 2 except that 5 pmol of termini of 5′-hydroxyl–2′,3′-cyclic phosphate-terminated poly(A) was added. (B) An aliquot of the ATP-agarose fraction was layered onto a 15–35% glycerol gradient that was treated as described in the text. Aliquots (15 μl) of glycerol gradient fractions were assayed for ATP–PP$_i$ exchange activity in the presence of poly(A). The arrow indicates the peak of RNA ligase activity. In lane 16, 0.4 unit of T4 RNA ligase was assayed for ATP–PP$_i$ exchange activity in the absence of RNA. An autoradiogram is shown. The position of authentic ATP, located by ultraviolet illumination, is indicated.

stimulation by various polynucleotide substrates varies, with the preferred substrates for ligation stimulating maximally. The dependence on RNA in the ATP–PP$_i$ exchange reaction makes it possible to use this as an assay for RNA ligase activity during purification, thereby obviating the need to prepare labeled RNA ligase substrates.

Assay of ATP–PP$_i$ Exchange. This assay measures the formation of [^{32}P]ATP from ^{32}PP$_i$. Reaction mixtures (50 μl) contain 20 mM HEPES buffer (pH 7.5), 2 mM DTT, 4 mM MgCl$_2$, 1 mM ATP, 0.2 mM ^{32}PP$_i$ (New England Nuclear, 100–600 cpm/pmol), and enzyme. The mixture is incubated for 60 min at 37° and extracted once with an equal volume of phenol–chloroform (1 : 1) and once with an equal volume of chloroform. An aliquot of the aqueous phase is applied to an PEI-cellulose TLC plate that is developed in 0.75 M KH$_2$PO$_4$ (pH 3.4). Following autoradiography, the species comigrating with authentic ATP and PP$_i$ can be excised and quantitated by liquid scintillation spectroscopy using Econofluor as scintillant.

A sample of an autoradiogram from such an assay is shown in Fig. 4. The RNA dependence of the reaction is demonstrated on Fig. 4A (see legend). In Fig. 4B, a profile of activity separated by glycerol gradient centrifugation is shown. The peak of ATP–PP_i exchange activity corresponds exactly to that of RNA ligase activity.

[41] RNA 3'-Terminal Phosphate Cyclase from HeLa Cells

By WITOLD FILIPOWICZ and OSCAR VICENTE

Introduction

The RNA 3'-terminal phosphate cyclase catalyzes the conversion of a 3'-phosphate to a 2',3'-phosphodiester at the end of RNA. The enzyme was discovered in HeLa cell and *Xenopus* oocyte nuclei extracts that were screened for the presence of RNA ligating enzymes.[1] The HeLa cell cyclase has been subsequently purified and the mechanism of its action studied.[2-4]

Cyclization of the 3'-terminal phosphate in RNA appears to occur in three steps[2-4]:

$$\text{Enzyme} + \text{ATP} \rightarrow \text{Enzyme–AMP} + PP_i \qquad (1)$$
$$\text{RNA-N}^{3'}\text{p} + \text{enzyme–AMP} \rightarrow \text{RNA-N}^{3'}\text{pp}^{5'}\text{A} + \text{Enzyme} \qquad (2)$$
$$\text{RNA-N}^{3'}\text{pp}^{5'}\text{A} \rightarrow \text{RNA-N}_{3'}^{2'} > \text{p} + \text{AMP} \qquad (3)$$
$$\text{Overall reaction: RNA-N}^{3'}\text{p} + \text{ATP} \rightarrow \text{RNA-N}_{3'}^{2'} > \text{p} + \text{AMP} + PP_i$$

It is likely that only steps (1) and (2) require the cyclase. Reaction (3) probably occurs nonenzymatically as the result of a nucleophilic attack by the adjacent 2'-hydroxyl on the phosphorus in the diester linkage, similarly as established for the reaction catalyzed by T4 RNA ligase at the 3'-phosphorylated RNA end.[5] Covalent labeling of the cyclase polypeptide

[1] W. FIlipowicz, M. Konarska, H. J. Gross, and A. J. Shatkin, *Nucleic Acids Res.* **11**, 1405 (1983).

[2] W. Filipowicz, K. Strugala, M. Konarska, and A. J. Shatkin, *Proc. Natl. Acad. Sci. U.S.A.* **82**, 1316 (1985).

[3] D. Reinberg, J. Arenas, and J. Hurwitz, *J. Biol. Chem.* **260**, 6088 (1985).

[4] O. Vicente and W. Filipowicz, *Eur. J. Biochem.* **176**, 431 (1988).

[5] D. M. Hinton, C. A. Brennan, and R. I. Gumport, *Nucleic Acids Res.* **10**, 1877 (1982).

with $[\alpha\text{-}^{32}P]ATP^{2-4}$ and, to a lesser extent, with the other $\alpha\text{-}^{32}P$-labeled ribonucleoside triphosphates[4] has been directly demonstrated. Reaction (2) is deduced from experiments carried out with 3'-phosphorylated RNA substrates containing terminal 2'-deoxy- or 2'-O-methylribose.[2] Incubation of these substrates with cyclase and ATP results in the formation of activated 3'-terminal structures corresponding to $dN^{3'}pp^{5'}A$ or $N^{m3'}pp^{5'}A$.

The physiological function of the RNA 3'-phosphate cyclase is not known. Initially it was suggested[1–3] that the cyclase could be responsible for the generation of the 2',3'-cyclic termini required for ligation of RNA by eukaryotic ligases and, consequently, be involved in tRNA splicing (RNA ligases involved in tRNA splicing in plant, yeast, and animal cells[1,6–10] require a 2',3'-cyclophosphate terminus for ligation; see also [38], [40], and [42] in this volume). It was found, however, that the splicing endonucleases responsible for the excision of introns from tRNA precursors in yeast[11] and $Xenopus$[12] directly produce RNA molecules with cyclic phosphate ends (see [39] and [42] in this volume). Although these findings do not necessarily rule out the involvement of the RNA cyclase in tRNA processing (e.g., the enzyme could be responsible for the regeneration of the cyclic ends in 3'-phosphorylated tRNA half-molecules, formed by the action of decyclizing phosphodiesterases), they make this possibility less attractive. It is possible that activation of the 3' end of RNA by cyclase is required for other, as yet unidentified, types of RNA ligation reactions.

Assay Methods

Materials. Tobacco mosaic virus (TMV) RNA was prepared as described.[13] *Escherichia coli* rRNA and poly(A) were from Boehringer-Mannheim. RNase T1, calf intestine phosphatase (CIP), and T4 RNA ligase were the products of Calbiochem (Sankyo), Boehringer, and Phar-

---

[6] M. Konarska, W. Filipowicz, and H. J. Gross, *Proc. Natl. Acad. Sci. U.S.A.* **79**, 1474 (1982).

[7] R. C. Schwartz, C. L. Greer, P. Gegenheimer, and J. Abelson, *J. Biol. Chem.* **258**, 8374 (1983).

[8] H. Furneaux, L. Pick, and J. Hurwitz, *Proc. Natl. Acad. Sci. U.S.A.* **80**, 3933 (1983).

[9] C. L. Greer, C. L. Peebles, P. Gegenheimer, and J. Abelson, *Cell* **32**, 537 (1983).

[10] W. Filipowicz and A. J. Shatkin, *Cell* **32**, 547 (1983).

[11] C. L. Peebles, P. Gegenheimer, and J. Abelson, *Cell* **32**, 525 (1983).

[12] D. G. Attardi, I. Margarit, and G. P. Tocchini-Valentini, *EMBO J.* **4**, 3289 (1985).

[13] A. Wodnar-Filipowicz, E. Szczesna, M. Zan-Kowalczewska, S. Muthukrishnan, U. Szybiak, A. B. Legocki, and W. Filipowicz, *Eur. J. Biochem.* **92**, 69 (1978).

macia, respectively. [5'-^{32}P]pCp (*pCp, specific activity 3000 Ci/mmol) was obtained from Amersham. Aniline (Fluka) was twice vacuum distilled, brought to pH 5.0 with HCl, diluted to 1 M concentration, and stored at $-20°$ in the dark. Thin-layer chromatography (TLC) solvent A corresponds to saturated $(NH_4)_2SO_4$–1 M sodium acetate–2-propanol (80 : 18 : 2, v/v/v).

Principle. Conversion of the 3'-terminal phosphate in the RNA or oligoribonucleotide substrate to the 2',3'-cyclic form renders this phosphate resistant to the action of alkaline phosphatase. When 3'-terminal phosphate is the only labeled phosphate group in the RNA, generation of the phosphatase-resistant radioactivity can be conveniently measured either by trichloroacetic acid (TCA) precipitation or by retention on activated charcoal. Alternatively, nuclease P1 can be used for digestion of the reaction products,[3] followed by the retention of the nucleotide-linked radioactivity on charcoal. Nuclease P1 has both endonuclease (producing nucleoside 5'-phosphates) and 3'-phosphomonoesterase activities. Hence, it releases the terminal phosphate from the 3'-monoester but not from the cyclic phosphodiester or 2'-monoester forms. Use of nuclease P1 is recommended when fractions assayed for cyclase activity are contaminated with the 2',3'-cyclic nucleotide 3'-phosphodiesterase (CNPase). CNPases convert 2',3'-cyclic phosphate groups, either in mononucleotides or at the end of RNA chains, into 2'-phosphomonoesters (reviewed by Tyc *et al.*[14]).

Assays measuring the generation of phosphatase- or nuclease P1-resistant radioactivity are simple and rapid but require 3'-terminally labeled substrates. When these are not readily available, an alternative procedure can be used. Oligoribonucleotide or RNA is 3'-end-labeled by the T4 RNA ligase-catalyzed ligation[15] of [5'-^{32}P]pNp (*pNp) and used as substrate in the cyclization reaction (4). Digestion of the reaction product with nuclease P1 yields *pN > p, which can be easily identified (and separated from *pN or *pN2'p) by cellulose TLC in solvent A.[1]

$$(Np)_n N^* pN^{3'}p \xrightarrow{\text{cyclase}} (Np)_n N^* pN_{3'}^{2'} > p \qquad (4)$$

Preparation of 3'-Terminally Labeled Substrates

$(Np)_n G^{3'}p.$ The simplest method involves the use of oligoribonucleotides obtained by exhaustive digestion of TMV RNA, ribosomal RNA, or any other easily available RNA, with RNase T1 and alkaline

[14] K. Tyc, C. Kellenberger, and W. Filipowicz, *J. Biol. Chem.* **262**, 12994 (1987).
[15] T. E. England, A. G. Bruce, and O. C. Uhlenbeck, this series, Vol. 65, p. 65.

phosphatase. The resulting oligonucleotides $[(Np)_nG_{OH}]$ are 3'-end-labeled by the T4 RNA ligase-catalyzed ligation[15] of *pCp, and the terminal Cp is removed from the products by treatment with RNase T1, yielding a mixture of $(Np)_nGp*$ oligonucleotides (in the case of TMV RNA, $n = 2$ to 70).

TMV RNA or $E.\ coli$ rRNA (500 μg) is heated at 70° for 2 min in 0.2 ml of 1 mM ethylenediaminetetraacetic acid (EDTA) (pH 7.0). After chilling on ice, N-2-hydroxyethylpiperazine-N'-2-ethanesulfonic acid (HEPES)–KOH (pH 7.6) (final concentration 40 mM) and 15 units of RNase T1 are added, and the sample is incubated at 37° for 90 min. To hydrolyze the 3'-terminal phosphates in the resulting $(Np)_nGp$ oligonucleotides, the sample is supplemented with HEPES–KOH (pH 8.0), MgCl$_2$, sodium dodecyl sulfate (SDS) (final concentrations 60 mM, 3 mM, and 0.1%, respectively), and 5 units of CIP. After incubation for 30 min at 37°, the sample is extracted twice with a phenol–chloroform–isoamyl alcohol mixture (25 : 24 : 1, v/v/v), and oligonucleotides are recovered by ethanol precipitation in the presence of 0.2 M sodium acetate (pH 5.0).

The oligonucleotides are used for the preparation of $(Np)_nG*p$. The reaction mixture (15 μl) contains 70 mM HEPES–KOH (pH 8.3), 10 mM MgCl$_2$, 3 mM dithiothreitol (DTT), 10% glycerol, 10% dimethyl sulfoxide (DMSO), 40 μM ATP, 12 μM *pCp (specific activity 1200 Ci/mmol), 5 μg of oligonucleotide mixture, and 7 units of T4 RNA ligase.[4] After 14 hr at 4°, the sample is diluted with 4 volumes of 50 mM HEPES–KOH (pH 7.6) containing 3 mM EDTA and incubated at 37° for 30 min with 5 units of RNase T1; SDS is then added to a final concentration of 0.1%, and the sample is extracted with a phenol–chloroform–isoamyl alcohol mixture. The aqueous phase is applied to a Sephadex G-25 (fine) column (made in a 10-ml diposable pipet) equilibrated and eluted with 20 mM ammonium acetate (pH 5.5). The peak of radioactivity corresponding to ^{32}P-labeled oligonucleotides is collected, lyophilized, and dissolved in distilled water.

A similar procedure can be used for the terminal labeling of a single defined oligoribonucleotide. For example, the longest RNase T1 fragment of TMV RNA (Ω, 70 nucleotides long) can be purified from the RNase T1 digest by 20% polyacrylamide gel electrophoresis (PAGE, sequencing type) and used for the preparation of the substrate. After *pCp ligation and RNase T1 treatment it is convenient to purify intact Ω-G$^{3'}$p* by PAGE.

$(Ap)_nA^{3'}p$. An alternative procedure[2] for preparation of the 3'-end-labeled substrate involves the use of $(Ap)_nA_{OH}$ as acceptor for *pCp ligation, followed by β-elimination to remove the terminal C residue.[16]

---

[16] H. Fraenkel-Conrat and A. Steinschneider, this series, Vol. 12B, p. 243.

$(Ap)_nA_{OH}$ (average chain length, 30) is obtained by incubation of 0.5 mg of poly(A) in 0.1 ml of 50 mM NaOH for 3 min at 80°. After acid treatment (20 μl of 1.5 M HCl is added to yield ~0.2 M HCl concentration and the sample incubated at 20° for 2 hr) to open the terminal 2',3'-cyclic phosphates, the oligo(A) is recovered by ethanol precipitation, treated with CIP, twice phenol extracted, and purified by Sephadex G-25 gel filtration as described above. The material eluting in the void volume is collected, and a 5-μg portion of it is used as an acceptor for ligation of *pCp (see above). After the ligation reaction, the sample is diluted with 4 volumes of 0.1 M HEPES–KOH (pH 8.0) containing 0.5% SDS and 0.5 units of CIP, and incubation is continued for 30 min at 37°. $(Ap)_nA^*pC_{OH}$ is recovered by phenol–chloroform–isoamyl alcohol extraction and Sephadex G-25 gel filtration as described above and subjected to periodate oxidation and aniline treatment (β-elimination).[16] It is dissolved in 0.1 ml of 50 mM NaIO$_4$ (pH 5.5) (freshly prepared) and incubated for 1 hr at 20° in the dark. Oligonucleotides are then precipitated twice with ethanol and resuspended in 0.1 ml of a solution containing 0.5 M aniline (pH 5.0) and 0.1 M sodium acetate (pH 5.0). After incubation for 3 hr at 20°, $(Ap)_nA^{3'}p^*$ is precipitated 3 times with ethanol, dried, and dissolved in distilled water.

Radioactive substrates are analyzed by digestion with RNase T2, nuclease P1, or CIP, followed by TLC on cellulose plates in solvent A. About 95% of the label is usually found to correspond to the 3'-terminal $N^{3'}p$.

Cyclase Assay

Cyclase activity is assayed by measuring the amount of charcoal-adsorbable, phosphatase-resistant ^{32}P radioactivity, generated after incubation of cyclase fractions with the $(Np)_nGp^*$ or $(Ap)_nAp^*$ substrates. Reaction mixtures (10 μl) contain 30 mM HEPES–KOH (pH 7.6), 187.5 mM NaCl, 4 mM MgCl$_2$, 0.15 mM EDTA, 0.1 mM spermidine, 1.25 mM DTT, 5% glycerol, 0.3 mM ATP, 0.005% Triton X-100, 5 μM phenylmethylsulfonyl fluoride (PMSF), 150–250 fmol of radioactive substrate, and different amounts of cyclase. After 20 min at 25°, 1 μl of 10% SDS and 1 unit of CIP are added, and incubation is continued for 30 min at 37°. Reactions are stopped by dilution with 0.5 ml of 20 mM HCl containing 1% sodium phosphate and addition of 0.2 ml of charcoal suspension (25% charcoal in 80 mM HCl). After 10 min on ice, the charcoal is collected by filtration on glass fiber GF/C (Whatman) filters and radioactivity determined by Cerenkov counting.

One unit of cyclase activity is defined as the amount of protein that renders 1 pmol of labeled termini of the $(Np)_nGp^*$ substrate resistant to phosphatase under the standard assay conditions.

Enzyme Purification

Buffers

Buffer A: 10 mM HEPES–KOH (pH 7.9), 10 mM KCl, 1.5 MgCl$_2$, 0.5 mM DTT

Buffer B: 300 mM HEPES–KOH (pH 7.9), 1.4 M KCl, 30 mM MgCl$_2$

Buffer C: 20 mM HEPES–KOH (pH 7.6), 0.1 mM EDTA, 0.5 mM DTT, 10% (v/v) glycerol, and the concentrations of NaCl as indicated (e.g., buffer C-75 contains 75 mM NaCl)

Buffer D: same as buffer C but containing also protease inhibitors (0.1 mM benzamidine, 0.2 μg/ml aprotinin, 10 μg/ml bacitracin, 0.2 μg/ml leupeptin, 10 $\mu$$M$ PMSF, all obtained from Sigma) and 0.01% (v/v) Triton X-100

Buffer E: same as buffer C but containing also 10 $\mu$$M$ PMSF and 0.01% Triton X-100

Purification Procedure

General Remarks. All operations, except loading and running the Mono-S column (performed at room temperature), are carried out at 0°–4°. Unless indicated otherwise, the samples are frozen in dry ice after each step and stored at −70°. The protein concentration in all fractions, except Blue-Sepharose, is measured by the method of Bradford, using the Bio-Rad reagent and bovine serum albumin (BSA) as standard; the protein concentration of the BS-600 fraction is determined by fluorimetry after reaction with *o*-phthalaldehyde (OPA), using the Mono-S cyclase fraction as protein standard.[4] Dialysis is carried out for 12–24 hr at 0° against 100–500 volumes of buffer, with at least two changes of buffer; if necessary, material precipitated in the dialysis bags is eliminated by centrifugation in a Sorvall SS34 rotor at 15,000 rpm for 10 min.

Poly(A)-Sepharose, Mono-S, and Blue-Sepharose columns (see below) are "precoated" with BSA prior to chromatography in order to minimize losses of cyclase protein. This is done by passing 2 column volumes of a solution of BSA (Biofinex, Praroman, Switzerland; 0.2 mg/ ml in appropriate application buffer) through the column. BSA is eluted with the buffer containing the highest NaCl concentration used in subsequent steps, and the column is finally reequilibrated with the appropriate buffer. Sephacryl or Sephadex columns used for M_r determinations of purified cyclase are subjected to similar treatment. Addition of 0.01% Triton X-100 to the buffers at the final stages of purification increases cyclase recoveries; inclusion of Triton X-100 in the assay reactions greatly improves the reproducibility of activity measurements.

HeLa Cell Culture. HeLa cells are grown in suspension cultures to a density of 5×10^5 cells/ml, in Joklik's modified Eagle's medium (Gibco) containing 5% newborn calf serum (Flow), in 3-liter spinner flasks (Bellco). Cells are collected by centrifugation in an MSE Cool-spin centrifuge (2200 rpm, 10 min at 4°) and washed twice with ice-cold phosphate-buffered saline (PBS) [10 mM sodium phosphate (pH 7.5), 0.15 M NaCl] containing 0.5 mM MgCl$_2$, and the final pellets are frozen in dry ice and stored at $-70°$ until use. Batches of 9 to 12 liters of suspension culture are processed each time.

Preparation of HeLa Cell Postribosomal Supernatant. One hundred milliliters of packed HeLa cells (3.3×10^{10} cells, from 65 liters of suspension culture) is thawed at 4° overnight and washed with buffer A. The cell pellet is resuspended in 3 volumes of buffer A (hypotonic buffer) and the suspension kept on ice. The cells are disrupted in 40-ml aliquots in a glass Dounce homogenizer by 10 strokes of the tight-fitting "B" pestle. Buffer B is then added (one-tenth of the final volume) and the extract centrifuged at 49,000 rpm for 2.5 hr in a Kontron TFT 50.38 or Beckman 60 Ti rotor. The upper four-fifths of postribosomal supernatant is collected, filtered through glass-wool, and dialyzed against buffer C-75.

DEAE-Cellulose Chromatography. One hundred twenty-five milliliters of packed wet DE-52 cellulose (Whatman), equilibrated in buffer C-75, is added to the supernatant, and the suspension is gently stirred for 3 hr. Protein not retained is collected by filtration, the cellulose cake is washed with an additional 100 ml of buffer, and both filtrates are combined. No additional cyclase activity is recovered on washing the DE-52 cake with buffer C-150.

Heparin-Sepharose Chromatography. The filtrate is applied to a heparin-Sepharose CL-6B column (Pharmacia, 2.5×16 cm) equilibrated in buffer C-75. The column is washed with 350 ml of the same buffer, and retained protein is eluted in two steps with buffer C-200 and C-450, respectively. No cyclase activity can be detected in the flow-through and 200 mM NaCl fractions. The fraction eluting with buffer C-450 is collected and dialyzed against buffer C-75. (When the heparin-Sepharose column is developed with a linear gradient of NaCl in buffer C, two peaks of cyclase activity, eluting at 0.25 and 0.33 M NaCl, are consistently obtained. The basis of this separation is not known; both fractions elute at similar positions when analyzed further on Sephacryl S-200 columns and yield identical [α-^{32}P]ATP-labeled ~40-kDa protein bands when analyzed by SDS–PAGE.[2])

Poly(A)-Sepharose Chromatography. The heparin-Sepharose fraction is applied to a column of poly(A)-Sepharose (Pharmacia, 1.6×10 cm) equilibrated in buffer D-75 after being precoated with BSA (see above).

The column is washed with 100 ml of buffer D-150, and retained protein is eluted with a linear gradient of 150–600 mM NaCl in buffer D at 18 ml/hr (2-ml fractions were collected). The fractions with highest cyclase activity, eluting at about 0.5 M NaCl, are pooled, concentrated 5-fold in a dialysis bag by dehydration against dry polyethylene glycol 20,000 (Fluka), and dialyzed against buffer D-75.

Mono-S Chromatography. Dialyzed protein from the previous step is filtered through a 0.22-μm nitrocellulose filter and directly applied to a BSA-precoated Mono-S HR 5/5 column (FPLC, Pharmacia) equilibrated in buffer D-75. The column is washed with 15 ml of the same buffer, and retained protein is eluted with 17 ml of a liner gradient of 75–350 mM NaCl in buffer D at a flow rate of 1 ml/min. Fractions (1 ml each) with cyclase activity, eluting at 250 mM NaCl, are pooled.

Blue-Sepharose Chromatography. The Mono-S fraction, diluted to 8 ml with buffer E-0 (buffer E without NaCl), is applied to a BSA-precoated Blue-Sepharose CL-6B column (Pharmacia, 0.9 × 0.6 cm) equilibrated in buffer E-75. The column is washed with 10 ml of the buffer, and retained protein is eluted stepwise with buffers E-200, E-600, E-1000, and E-2000. Most of the cyclase activity (~70%) is eluted at 600 mM NaCl (fraction BS-600); the fraction eluting at 1 M NaCl (fraction BS-1000) contains ~30% of the recovered cyclase activity. BS-600 and BS-1000 fractions are pooled separately and stored frozen at −70° in aliquots.

Glycerol Gradient Centrifugation. One-half milliliter of the Blue-Sepharose (BS-600) fraction is concentrated about 10-fold in a dialysis bag by dehydration against BioGel concentrator resin (Bio-Rad), dialyzed against buffer E-75, and loaded onto a linear gradient of 15–35% glycerol (3.9 ml) in the same buffer. After centrifugation for 36 hr at 56,000 rpm in a Beckman SW60 rotor, fractions of 100 μl are collected and assayed for cyclase activity. Fractions corresponding to the peak of activity (Fig. 1) are pooled and stored in aliquots at −70°.

Comments and Purity of Cyclase

Cyclase activity, either in crude HeLa cell extracts or in the purified enzyme preparation, is highly resistant to treatment with an excess of proteases (see below). Inclusion of protease inhibitors during purification may therefore not be necessary.

The enzyme can be specifically labeled with [α-^{32}P]ATP, which facilitates the identification (by SDS–PAGE) of the cyclase polypeptide during the final steps of purification. A major contaminating polypeptide of ap-

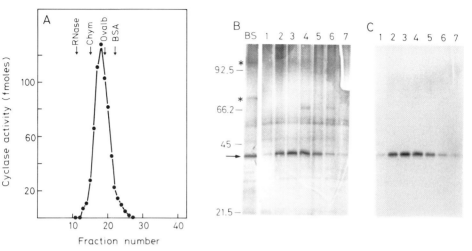

FIG. 1. Analysis of cyclase BS-600 fraction by glycerol gradient centrifugation. (A) Profile of cyclase activity. One-half microgram of the BS-600 fraction was analyzed by sedimentation in a 15–35% glycerol gradient made in buffer E-75. Centrifugation was for 36 hr at 56,000 rpm in a Beckman SW60 rotor. Aliquots of 0.4 μl of each fraction were assayed for cyclase activity with 240 fmol of (Np)$_n$Gp* substrate. The positions of marker proteins are indicated by arrows. (B and C) SDS–10% PAGE analysis of the gradient fractions after labeling with [α-^{32}P]ATP. Aliquots (60 μl) of glycerol gradient fractions 15 (lane 1) and 17–22 (lanes 2 to 7) were incubated with [α-^{32}P]ATP (40 μM, specific activity 2 Ci/mmol) for 3 hr under standard conditions but without the oligonucleotide substrate; lane BS contained 20 ng of BS-600 fraction that was not incubated with [α-^{32}P]ATP. M_r ($\times 10^{-3}$) of marker proteins are shown at left. The band corresponding to cyclase is indicated by an arrow and the contaminating proteins in fraction BS-600 by asterisks. The artifactual bands visible in all gel lines (B) are caused by overstaining with silver. (B) Silver-stained gel. (C) Autoradiogram. [From O. Vicente and W. Filipowicz, *Eur. J. Biochem.* **176**, 431 (1988).]

proximately M_r 40K copurifies with cyclase through the Mono-S step. The Blue-Sepharose column allows the separation of the two proteins: the cyclase elutes from the column at 0.6 and 1 M NaCl, while the contaminating polypeptide remains retained on the column even at 2 M NaCl.

The procedure described above results in about 6000-fold purification of cyclase (Table I).[4] After the Blue-Sepharose step the protein is about 75% pure, as estimated by SDS–PAGE and silver staining. The glycerol gradient centrifugation step removes two contaminating polypeptides of higher M_r (Fig. 1). Purified cyclase does not contain detectable RNase, DNase, ATPase, GTPase, unspecific phosphatase or nucleoside diphosphokinase activities.[4]

TABLE I
PURIFICATION OF RNA 3'-TERMINAL PHOSPHATE CYCLASE[a]

Purification step	Total protein (mg)	Volume (ml)	Specific activity (units/mg)	Total units	Purification (-fold)	Yield (%)
1. Postribosomal super-natant	497	270	170	84,490	1	100
2. DEAE-cellulose	241	344	265	63,865	1.6	76
3. Heparin-Sepharose	38	27	1,120	42,560	6.6	50
4. Poly(A)-Sepharose	0.45	42	60,833	27,375	358	32
5. Mono-S	0.05	3.3	205,882	10,294	1,211	12
6. Blue-Sepharose						
BS-600	0.0035	3.5	1,025,000	3,588	6,029	6
BS-1000[b]	—	8.5	—	1,600	—	

[a] From 100 g (wet weight) of HeLa cells. Adapted from O. Vicente and W. Filipowicz, *Eur. J. Biochem.* **176**, 431 (1988).
[b] Protein concentration of the BS-1000 fraction was not determined.

Properties of Enzyme

Physical Properties. M_r determinations by SDS–PAGE, glycerol density gradient sedimentation, and Sephacryl gel filtration (in the presence of 0.5 M NaCl) all give values of 38,000–40,000, indicating that the cyclase is a single polypeptide enzyme.[4] The Stokes radius of cyclase is 2.66 nm. When Sephacryl gel filtration is performed at 75–100 mM NaCl, cyclase activity elutes from the column in a position corresponding to an apparent M_r of 26,000.[2,4] This anomalous behavior is most likely due to retardation of the protein by ionic interactions with the gel matrix. Adenylation of the enzyme (see below) does not change its mobility in SDS–PAGE (Fig. 1).

Purified cyclase is stable for at least 6 months when stored at −70°, and the activity does not decrease after several cycles of freezing and thawing. The enzyme is inactivated by 50% after 3 days at 4° or after 3–4 hr at room temperature. Incubation at 50° for 2 min inactivates the enzyme completely.

Purified cyclase is strongly resistant to protease digestion. Incubation of the BS-600 fraction with a 100-fold excess of proteinase K or trypsin for 20 min at 37° inhibits the activity by only 34 and 48%, respectively. It has not been tested whether protease treatment changes the size of the polypeptide.

Requirements of Cyclization Reaction. The cyclase requires ATP as cofactor,[1-4] although it can also use other ribonucleoside triphosphates

with lower efficiency.[4] In the presence of GTP, CTP, or UTP the cyclization reaction proceeds 15–20 times slower than in the presence of ATP.
The apparent K_m values for ATP and GTP are 6 and 200 μM, respectively.
It is likely, therefore, that ATP is the physiological cofactor of cyclase.
Covalent modification of cyclase with all four ribonucleotides has been
demonstrated.[4] The [^{32}P]AMP–cyclase complex can be isolated free of
unreacted ATP by Sephacryl gel filtration under nondenaturing conditions, in the presence of 0.5 M NaCl.[17] Incubation of such purified
[^{32}P]AMP–cyclase with unlabeled 3′-phosphorylated substrate results in
the release of [^{32}P]AMP from the complex; similar incubation with a substrate containing a 3′-hydroxyl group does not lead to the hydrolysis of
the complex.[17] These and other data[2–4] indicate that the adenylated cyclase is indeed a physiological intermediate in the cyclization reaction.

No cyclase activity can be detected when ADP or dATP is used in
place of ATP.[2–4] The nonhydrolyzable methylene analogs of ATP,
AMPPCP and AMPCPP, are also inactive, as is the β,γ-imido derivative.[2,3] Adenosine 5′-(γ-thio)triphosphate (ATPγS) effectively replaces
ATP as cofactor. Using ATPγS and partially purified cyclase, Reinberg et
al.[3] have demonstrated that formation of 1 mol of 2′,3′-cyclic phosphate
ends is associated with disappearance of 1 mol of 3′-phosphate termini
and the hydrolysis of 1 mol of ATPγS to AMP and thiopyrophosphate.

The cyclase requires the presence of Mg^{2+} (optimal concentration of
$MgCl_2$ is 4–5 mM) for activity. Only about 5% of activity is observed
when Mg^{2+} is replaced by Ca^{2+} or Mn^{2+}. The optimal concentration of
monovalent cations (Na^+ or K^+) is 150–200 mM. The enzyme has a
slightly alkaline pH optimum of 8.0–9.0; about 70% of maximal activity is
still observed at pH 10.5. The rates of the reaction are similar at 25, 30,
and 37°, decreasing to 60 and 25% at 12 and 0°, respectively.[4] The cyclase
is sensitive to inactivation by sulfhydryl reagents such as N-ethyl-
maleimide and iodoacetamide.[3,4] Vanadyl–ribonucleoside complex is a
very strong inhibitor of cyclase.[3]

Substrate Specificity. No systematic study has been carried out on the
relative efficiency of different RNA substrates in the reaction. AUGp,
tRNA-pNp, 5 S RNA-pCp, pure oligoribonucleotides of different lengths
and sequences, as well as mixtures of homopolymers or natural oligonucleotides are utilized as substrates[1–4] (see also Assay Methods). In the
only comparative study, (Up)$_{10}$Gp was found to be a better substrate than
tRNA^{fMet}-pGp or AUGp; the nature of the 3′-terminal base had no effect
on substrate utilization.[3] Nucleoside bisphosphates (p$^{5'}$N$^{3'}$p) do not act as

---

[17] O. Vicente and W. Filipowicz, unpublished results (1988).

substrates. 2'-Phosphate-terminated oligoribonucleotides are also inactive.[1,3]

Acknowledgments

We would like to thank M. Konarska, K. Strugala, and A. J. Shatkin for contributions to the work discussed in this chapter, and G. Goodall for critical reading of the manuscript. O. Vicente was supported by a fellowship from Fundación Juan March, Madrid, Spain.

[42] Transfer RNA Splicing Endonuclease from *Xenopus laevis*

By D. Gandini-Attardi, I. M. Baldi, E. Mattoccia, and G. P. Tocchini-Valentini

Introduction

Eukaryotic tRNA genes are transcribed as precursors which in turn are processed to become mature tRNA species. The processing reactions include the removal of 5'-leader and 3'-trailer sequences and, in a subset of tRNA gene transcripts, the excision of an intervening sequence (IVS).[1] These reactions are all posttranscriptional and take place in the nucleus. *Xenopus laevis* germinal vesicle (GV) extracts can transcribe cloned tRNA genes and accurately process the primary transcript to form mature tRNA.[2] We identified four enzymatic activities in GV extracts that play a role in the processing of the precursor: the splicing endonuclease (*Xla*I RNase) which precisely exises the IVS; an RNase P-like activity (*Xla*II RNase) which is responsible for the removal of the 5' leader; an endonuclease activity (*Xla*III RNase) responsible for the removal of the 3' trailer; and the ATP-dependent ligase which selectively joins the complementary halves of the tRNA molecules. Results obtained using partially purified preparations of splicing endonuclease[3] lend support to the hypothesis that a single enzyme acts on all the precursors. In *Xenopus* oocytes the activity appears to be soluble. This characteristic was exploited to design the procedure described below.

[1] J. Abelson, *Annu. Rev. Biochem.* **48,** 1035 (1979).
[2] E. Mattoccia, M. I. Baldi, G. Carrara, P. Fruscoloni, P. Benedetti, and G. P. Tocchini-Valentini, *Cell* **18,** 643 (1979).
[3] A. Otsuka, A. De Paolis, and G. P. Tocchini-Valentini, *Mol. Cell. Biol.* **1,** 269 (1981).

Materials and Methods

Plasmids. Plasmid pTO is constructed by subcloning the 2.5-kilobase (kb) *Eco*RI DNA fragment containing the tRNA$_3^{Leu}$ gene from pJB2k[4] into pBR322 at the *Eco*RI site as described by Johnson *et al.*[5] Plasmid pTGEM2 is constructed by placing the yeast tRNA$_3^{Leu}$ gene under the T7 promoter of pGEMI (Promega Biotec). By means of site-directed mutagenesis, a *Bst*NI is inserted at the 3' end of the gene so that the run-off transcript contains the terminal CCA, and the *Xho* site is inserted near the 3' end so that the run-off transcript contains 9 additional nucleotides after the terminal CCA. Plasmid pTGEM2-AC3 is obtained using as a primer on the single-strand DNA of pTGEM2 the *Hpa*I–*Hae*III fragment (47 nucleotides) from the mutant pAC3.[6]

Buffers

Barth's solution: 88 mM NaCl, 1 mM KCl, 0.82 mM MgSO$_4$, 0.4 mM CaCl$_2$, 0.32 mM (CaNO$_3$)$_2$, 2.5 mM NaHCO$_3$, 0.21 mM Na$_2$HPO$_4$, and 0.27 mM KH$_2$PO$_4$

J buffer: 7 mM MgCl$_2$, 0.1 mM ethylenediaminetetraacetic acid (EDTA), 2.5 mM dithiothreitol (DTT), 10% (v/v) glycerol, and 10 mM *N*-2-hydroxyethylpiperazine-*N'*-2-ethanesulfonic acid (HEPES) (pH 7.5)

JB-70: J buffer with 70 mM NH$_4$Cl

JB-150: J buffer with 150 mM NH$_4$Cl

TEMG: 50 mM tris(hydroxymethyl)aminomethane hydrochloride (Tris-HCl) (pH 7.5), 1 mM EDTA, 1.4 mM 2-mercaptoethanol, and 16% (v/v) glycerol

Synthesis of Precursors. To prepare [32P] labeled wild-type pre-tRNA$_3^{Leu}$, plasmid pTGEM2 is cut with *Bst*NI and transcribed utilizing T7 polymerase (Promega Biotec) according to Reyes and Abelson.[7]

Affinity Column. pTGEM2-AC3 DNA is cleaved with the restriction enzyme *Xho*. The DNA is used for multiple assays of transcription with T7 polymerase (Promega Biotec) in the presence of [32P]UTP with a specific activity of 0.1 Ci/mmol. The transcription products are gel purified and stored in ethanol.

[4] J. S. Beckman, P. F. Johnson, J. Abelson, and S. A. Fuhrman, *in* "Molecular Approaches to Eucaryotic Systems" (G. Wilcox, J. Abelson, and C. F. Fox, eds.), p. 213. Academic Press, New York, 1977.

[5] J. D. Johnson, R. Ogden, P. Johnson, J. Abelson, P. Dembeck, and R. Itakura, *Proc. Natl. Acad. Sci. U.S.A.* **77,** 2564 (1980).

[6] M. I. Baldi, E. Mattoccia, S. Ciafrè, D. Gandini-Attardi, and G. P. Tocchini-Valentini, *Cell* **47,** 965 (1986).

[7] V. M. Reyes and J. Abelson, *Anal. Biochem.* **166,** 90 (1987).

Six nanomoles of pAC3 pre-tRNA (250 μg) is resuspended in 0.5 ml of 0.1 M sodium acetate (pH 5.0) and oxidized in the dark with 10 mM sodium periodate at 23° for 1 hr.[8] Excess periodate is removed by washing the pre-tRNA with 0.1 M sodium acetate (pH 5.0) several times and by repeated precipitation with ethanol. To evaluate the efficiency of the oxidation the treated pre-tRNA is used as a substrate in a RNA ligase reaction with cytidine 3',5'-[5'-^{32}P]biphosphate.[9] More than 95% of the pre-tRNA should be oxidized.

Agarose-adipic acid dihydrazide (Sigma, St. Louis, MO) is suspended in 0.2 M NaCl and washed thoroughly with at least 100 volumes of 0.2 M NaCl. The resin is then washed several times with 0.1 M sodium acetate (pH 5.0) and stored at 4°. The oxidized pre-tRNA is resuspended in 0.4 ml of 0.1 M sodium acetate (pH 5.0) and coupled to 0.4 ml of agarose-adipic acid dihydrazide by mixing the two preparations and agitating gently by end-over-end rotation in a plastic tube at 4° for 12 hr.[10] The pre-tRNA-agarose is packed in a small column (0.6 × 0.5 cm) and washed with 2 M KCl, 20 mM Tris-HCl (pH 7.5), 5 mM EDTA at 4° until no radioactive pre-tRNA is released. Coupling efficiency is calculated on the basis of the amount of radioactivity recovered in the washes. More than 90% of the pre-tRNA should be coupled to the resin. The column is finally equilibrated with binding buffer JB-70. After it is used the column is washed with 2.0 M KCl, 20 mM Tris-HCl (pH 7.5), 5 mM EDTA, reequilibrated in binding buffer, and stored at 4°.

Assays

Principle. The splicing endonuclease acts in two steps: (1) specific recognition and binding of the precursors and (2) cleavage at the two sites with the consequent formation of the two half-molecules and of the intron.[6] Two different assays are used to test for binding and cleavage.

Electrophoretic-Pre-tRNA Binding Assay. Binding reactions contain 3 fmol of labeled pre-tRNA and 10 μl of purified endonuclease (from Step 6, see below) in a total volume of 20 μl. Assay conditions are identical to those used in the endonuclease cleavage assay.[11] Reactions are initiated by addition of the enzyme and incubated for 10 min at 0°, after which 2 μl of electrophoresis buffer containing bromphenol blue is added. The entire reaction is loaded onto a gel in the cold room (4°) with running voltage

[8] S. R. Fahnestock and M. Nomura, *Proc. Natl. Acad. Sci. U.S.A.* **69,** 333 (1972).
[9] T. L. Marsh and N. R. Pace, *Science* **229,** 79 (1985).
[10] H. R. Burell and J. Horowitz, *Eur. J. Biochem.* **75,** 533 (1977).
[11] D. Gandini-Attardi, I. Margarit, and G. P. Tocchini-Valentini, *EMBO J.* **4,** 3289 (1985).

applied. The addition of bromphenol blue to the binding reaction does not influence the stability of the complex. The gel ($15 \times 30 \times 0.1$ cm) contains 6% polyacrylamide (bisacrylamide–acrylamide, 1:29) in a buffer of 50 mM Tris–acetate (pH 7.6), 1 mM Mg^{2+}. Buffer is recirculated between the two reservoirs. The gel is electrophoresed until the dye migrates 8–10 cm and then autoradiographed.

Cleavage Assay. The standard reaction mixture contains 2–20 fmol of ^{32}P precursor and 10 μl of enzyme in 50 μl of JB-70. To identify the chromatographic fractions with endonuclease activity, the assay contains 10 μl of each column fraction in a volume of 60 μl of JB-70 so that the salt concentration of the fraction will not inhibit enzyme activity. Reactions are incubated for 2 hr at 22° and stopped by addition of 5 μl of 5% sodium dodecyl sulfate (SDS) and 5 μl of 1 mg/ml proteinase K. After incubation for 30 min at room temperature and addition of 1 μg of yeast tRNA and NaCl to 0.2 M, the samples are precipitated with 3 volumes of ethanol at −70° for 30 min.

Quantitation of endonuclease activity is performed in 50 μl of JB-70 containing 20 fmol of ^{32}P-labeled precursor and 10 μl of purification fractions. All fractions are assayed in the same experiment: the products are analyzed by gel electrophoresis as described, and the reactants and products are Cerenkov counted. One unit of endonuclease is defined as the amount of enzyme that cleaves 0.1 fmol of precursor in 2 hr at 22°.

Isolation of Germinal Vesicles

Our procedure is derived from the method described by Schalenghe *et al.*[12]

Oocyte Isolation. Oocytes free of follicle cells are obtained from ovaries of *X. laevis* females by collagenase digestion. For a typical preparation, 150 ml of ovaries are collected in Barth's solution, rinsed, and transferred to a beaker containing 1.5 g of collagenase (type I, Sigma, 300 units/mg) with excess solution wiped off on tissue paper. Digestion is carried out for 2–3 hr at 30° with occasional gentle shaking. Oocytes are thoroughly rinsed with Barth's solution; stage 6 oocytes are selected by sieving and kept in ice until use.

Pronase Digestion. Oocyte batches (35 ml) are incubated in ice with 20 ml of Barth's solution containing 9 mg/ml of Pronase (Calbiochem 72600 proteolytic units (P.U.K.)/g powder) for 15–20 minutes. At this point

[12] F. Schalenghe, M. Buscaglia, C. Steinbeil, and M. Crippa, *Chromosoma* **66,** 299 (1978).

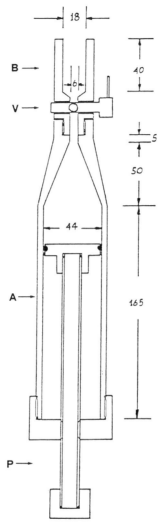

FIG. 1. Syringelike device used for germinal vesicle preparation. The main chamber (A), the upper chamber (B), the valve (V), and the piston (P) are indicated. Measurements are given in millimeters.

oocytes are fragile and have to be handled gently. Pronase is removed by successive rinsing with Barth's solution (200-ml aliquots, 20 times).

Treatment with Nonidet P-40 in JB Buffer. The oocyte suspension is poured into the main chamber (A) of a syringe-like device (see Fig. 1) with valve (V) closed. After the oocytes have settled, the excess buffer is withdrawn. Then 120 ml of glycerol-free JB-70 containing 0.2% Nonidet P-40 (NP-40) is added to the oocytes. The piston (P) is gently screwed onto the syringe so that it can be placed in a vertical position with the piston at the bottom. The valve (V) is opened, and pressure is exerted on the piston to push the oocyte–JB-70 mixture upward until only the upper-most, cone-shaped part of the syringe remains filled with air. The valve (V) is then shut, and the syringe is placed horizontally and rolled gently over ice for 10–15 min. This can be easily accomplished with the aid of an electric motor fastened to the syringe via a rubber cork inserted in the upper, valve-containing chamber (B) of the device. During this stage the oocytes lose their pigment and appear compact. At this point the syringe is carefully returned to a vertical position and the oocytes are allowed to sediment on the piston. The upper chamber (B) is unscrewed from the top and the pigmented buffer is withdrawn and replaced with 120 ml Barth's solution. The upper chamber (B) is screwed back on and the syringe is once again rolled on ice as described above. After a few minutes the germinal vesicles start to separate from the cytoplasm.

Complete lysis of the cytoplasm is usually obtained after 15–20 min. The syringe is once again put in a vertical position, and the GVs are allowed to rise to the surface. After a few minutes, when all the vesicles have gathered at the top of the solution, the valve (V) is opened and the piston is used to push the vesicles upward all the way into the upper chamber (B), where they are collected using a Pasteur pipet with a wide opening. About 2 ml of vesicles is then carefully layered on 20 ml of JB-70 in 30 ml Corex tubes kept in ice. The GVs are mixed gently with the uppermost one-third of the buffer to facilitate sedimentation. After a few minutes the GVs slowly start to sediment, while most of the cytoplasmic material remains in the upper part of the buffer. When all the GVs have reached the bottom of the tube, the buffer is carefully removed without disturbing the GV pellet.

Splicing Endonuclease Purification

The procedure described below allows for a 1000-fold purification of splicing endonuclease from GVs (see Table I).

Step 1: Nuclear Extract. Germinal vesicles are collected from 350 ml of collagenase-treated stage 6 oocytes in batches of 2–3 ml of JB-70 and kept in Corex tubes at $-70°$. All procedures are carried out at $4°$. The nuclear extract is prepared by disrupting the germinal vesicles by means of several passages (5–7) through a Gilson plastic tip fastened to a 5-ml pipet. The material is collected into two 30-ml Corex tubes, and the non-soluble material is removed by centrifugation at 755 g for 10 min in the

TABLE I
PURIFICATION OF RNase XlaI

Step	U/ml	Volume (ml)	Total units	Yield (%)	Protein (mg)	Specific activity (U/mg)
1. Nuclear extract	7,800	50	370,000	100	260	1,400
2. Phosphocellulose	2,100	160	331,000	90	32	10,300
3. Heparin-Sepharose	8,900	15	133,500	36	2	60,700
4. DEAE-Sephadex	5,460	15	82,000	22	1	91,000
5. CM-Sephadex	11,200	2.5	27,000	7	0.120	220,000
6. Sucrose gradient	1,830	3.6	6,600	2	0.004	1,600,000

fixed-angle SS34 Sorvall rotor. The supernatant is dialyzed against 60 volumes of TEMG buffer for 3 hr with two buffer changes during the dialysis. The dialyzed extract is divided into two equal batches and kept at −70°.

Step 2: Phosphocellulose Chromatography. Each batch of extract is applied to a 2 × 14 cm phosphocellulose column (P-II, Whatman) equilibrated overnight with TEMG. The column is washed with the equilibration buffer and eluted with a 600-ml nonlinear gradient from 0 to 1.5 M KCl in the same buffer (flow rate 50 ml/hr). The endonuclease activity elutes between 0.25 and 0.40 M KCl.

Step 3: Heparin-Sepharose. The pool of the active fractions derived from the two phosphocellulose columns is dialyzed against TEMG buffer and divided into two batches. Each batch is applied separately to a 1.5 × 2.5 cm heparin-Sepharose CL-6B column (Pharmacia) equilibrated with TEMG buffer. The column is washed and then eluted with a 60-ml linear gradient from 0 to 0.8 M NaCl in the same buffer (flow rate 6 ml/hr). The endonuclease activity elutes between 0.4 and 0.5 M NaCl.

Step 4: DEAE-Sephadex. The pool of the active fractions derived from the two heparin-Sepharose columns is dialyzed against TEMG buffer (pH 8.0) and applied to a 1.5 × 2.5 cm DEAE-Sephadex A-25 column (Pharmacia) equilibrated with the same buffer. The column is washed and then eluted with a 60-ml linear gradient from 0 to 0.5 M KCl in the same buffer (flow rate 6 ml/hr). The endonuclease activity elutes between 0.1 and 0.2 M KCl.

Step 5: CM-Sephadex. The active fractions are dialyzed against TEMG buffer and applied to a 1.5 × 6 cm CM-Sephadex G-25 column (Pharmacia) equilibrated with the same buffer. The column is washed and then eluted with a 160-ml linear gradient from 0 to 1 M KCl in the equilibration buffer (flow rate 6 ml/hr). The active fractions around 0.25 M KCl are dialyzed against JB-70 and stored in aliquots at $-70°$.

Step 6: Sucrose Gradient Sedimentation. Aliquots (0.3 ml) of active fractions are applied to 5-ml sucrose gradients (5–20% sucrose in JB-150) and centrifuged for 22 hr at 48 rpm in a SW55.1 rotor at 3°. A total of 32 fractions are collected from the bottom of the tube. The standard proteins used as markers are bovine immunoglobulin G (160,000), bovine serum albumin (66,000), ovalbumin (45,000), and myoglobulin (17,000). The endonuclease activity cosediments with the bovine immunoglobulin G marker.

Gel Filtration with Superose 12. As an alternative procedure to sucrose gradient sedimentation, gel filtration with Superose 12 HR 10/30 (Pharmacia FPLC) is used. A 2-ml aliquot of the fraction from Step 5 prior to dialysis is concentrated by centrifugation with a Centricon 30 (Amicon) to 0.2 ml and applied to a Sepharose 12 column equilibrated with 0.15 M NaCl, 50 mM phosphate buffer (pH 7.5) and eluted with the same buffer at a flow rate of 0.5 ml/min. Fractions of 0.5 ml are collected into glass tubes containing 0.250 ml of 30 mM HEPES (pH 7.5), 21 mM MgCl$_2$, 0.3 mM EDTA, 7.5 mM DTT, and 30% (v/v) glycerol. Endonuclease activity elutes in fractions of molecular weight 160,000, as determined by comparison with the standard proteins bovine immunoglobulin G (160,000) and bovine serum albumin (66,000) separated on Superose 12 under identical conditions.

Affinity Chromatography. Affinity chromatography can be applied. The ligand used is pAC3, a mutant precursor of yeast pre-tRNA$_3^{Leu}$ characterized by two blocks of base substitutions at positions 35, 36, and 37 (CAA $\rightarrow$ GUC) and 46, 47, and 48 (UUG $\rightarrow$ GAC). This mutant pre-tRNA is recognized by the splicing endonuclease in a binding assay, but is not cleaved.[6]

The fractions from Step 6 are dialyzed against JB-70 and loaded on the pre-tRNA-agarose column. The sample is recirculated over the column for 2 hr using a peristaltic pump at 6 ml/hr. The recirculated sample contains residual endonuclease activity. Although longer recirculation times increase the binding of the enzyme to the column, these conditions minimize inactivation of the enzyme. The column is then washed with binding buffer and eluted with a linear gradient (5.0 ml) of 0.07–0.70 M NH$_4$Cl in J buffer at 6 ml/hr. Fractions of 150 μl are collected. The salt concentration of the collected fractions is calculated on the basis of the conductivity. Twenty microliters from each fraction is assayed for endonuclease activity. The salt concentration in the assays is brought within the range of 70–100 mM NH$_4$Cl by dilution with J buffer. The peak of endonuclease activity elutes around 200 mM NH$_4$Cl.

Section IV

Ribozymes

[43] Genetic and Molecular Analysis of RNA Splicing in *Escherichia coli*

By MARLENE BELFORT, KAREN EHRENMAN, and P. SCOTT CHANDRY

Introduction

The study of splicing is greatly facilitated in organisms with well-characterized genetic systems. Such systems make it possible to isolate and characterize mutations that affect splicing. For example, the availability of mutants in *Saccharomyces cerevisiae* is helping to unravel the complexities of the spliceosome involved in nuclear pre-mRNA splicing,[1,2] while in *Neurospora crassa* mutational approaches are defining accessory factors in the group I splicing process.[3] The demonstration that a eukaryotic group I intron can be accurately processed in *Escherichia coli*[4,5] and the discovery of introns in the T-even phages[6–8] have opened up bacterial genetic systems for the analysis of RNA splicing.

With its well-developed genetic systems and powerful selection techniques, *E. coli* allows analysis of rare molecular events and is a particularly suitable host for the study of group I introns. First, *E. coli* is readily transformable by plasmids, allowing *in vitro* DNA manipulations to be coupled to the natural genetics of the host cell. Second, *E. coli* is host to a variety of bacteriophages, which are useful both as recombinant vectors and as carriers of naturally occurring group I introns. Third, the phenotypic selection and screening procedures available in *E. coli* are ideal for isolation of mutants and pseudorevertants as well as for genetic mapping. Fourth, the coupling of transcription and translation allows protein products of intron-containing genes to be readily monitored as a reflection of splicing events. It is of particular advantage to assay stable polypeptide

[1] A. J. Lustig, R.-J. Lin, and J. Abelson, *Cell* **47**, 953 (1986).
[2] R. Parker, P. G. Siliciano, and C. Guthrie, *Cell* **49**, 229 (1987).
[3] R. A. Akins and A. M. Lambowitz, *Cell* **50**, 331 (1987).
[4] J. V. Price and T. R. Cech, *Science* **228**, 719 (1985).
[5] R. B. Waring, J. A. Ray, S. W. Edwards, C. Scazzocchio, and R. W. Davies, *Cell* **40**, 371 (1985).
[6] F. K. Chu, G. F. Maley, F. Maley, and M. Belfort, *Proc. Natl. Acad. Sci. U.S.A.* **81**, 3019 (1984).
[7] J. M. Gott, D. A. Shub, and M. Belfort, *Cell* **10**, 81 (1986).
[8] B.-M. Sjoberg, S. Hahne, C. Z. Mathews, C. K. Mathews, K. N. Rand, and M. J. Gait, *EMBO J.* **5**, 2031 (1986).

products of a split gene since RNA turnover is very rapid in prokaryotes (the average mRNA half-life is 2 min).

The model system used throughout this chapter is the phage T4 intron-containing td gene that encodes thymidylate synthase (TS).[6,9] First, principles underlying the genetic methodologies are presented. These genetic approaches may be applied to any prokaryotic or eukaryotic intron that can be processed in E. coli, if the intron is fused to a gene with distinctive and, preferably, selectable phenotypes. Thereafter, widely applicable molecular techniques that are of particular value in studying the RNA splicing process in E. coli are presented.

Strategies for Isolating Nondirected Mutations Causing Splicing Defects

Bacterial genetic methods may be applied to the study of introns that can be processed in E. coli and that are associated, either naturally or by artificial DNA fusion, with genes which affect growth properties on bacteriological media. For example, the natural phenotypes of the td gene made possible the isolation of two libraries of nondirected splicing-defective td⁻ mutants, one in the phage,[10] the other using a plasmid-borne td gene.[11,12] As another example, the in-frame lacZ fusion with the Tetrahymena DNA encoding the large rRNA intron has facilitated mutational analysis of this intron utilizing β-galactosidase phenotypes.[4,5]

The usefulness of employing a variety of mutant isolation strategies is underscored by the different spectra of td⁻ mutations isolated in the phage and plasmid systems.[10–12] The scheme for both td⁻ mutant identification and screening for splicing-defective variants differs somewhat between the phage and plasmid-borne mutations, as outlined in Fig. 1. First, after mutagenesis, T4td⁻ phage are screened by plaque morphology, while td⁻ plasmids are isolated directly on selective media. Second, whereas splicing-defective plasmid mutants are readily screened by a functional assay for imbalances of td protein products [the NH_2-TS (amino-terminal product of td gene encoded by exon I) assay described below], these products are obscured after T4 phage infection. T4td⁻ mutations are therefore screened by genetic mapping to identify those mutations located within or near the intron. The dot-blot RNA splicing assay

⁹ F. K. Chu, G. F. Maley, D. K. West, M. Belfort, and F. Maley, Cell 45, 157 (1986).
¹⁰ D. H. Hall, C. M. Povinelli, K. Ehrenman, J. Pedersen-Lane, F. Chu, and M. Belfort, Cell 48, 63 (1987).
¹¹ P. S. Chandry and M. Belfort, Genes Dev. 1, 1028 (1987).
¹² M. Belfort, P. S. Chandry, and J. Pedersen-Lane, Cold Spring Harbor Symp. Quant. Biol. 52, 181 (1987).

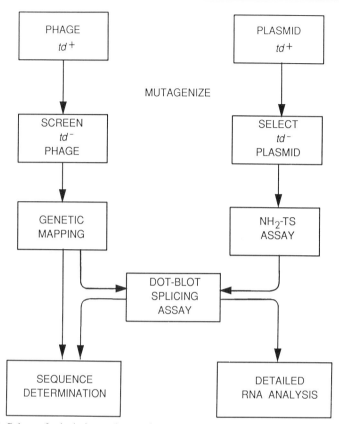

Fig. 1. Scheme for isolation and screening of splicing-defective *td* mutants. Details of the procedures for isolation of T4*td⁻* phage and *td⁻* plasmid mutants are described in the text.

with oligonucleotide probes (described below) is subsequently used as the definitive screen for splicing defects in both plasmids and phage. Sequence determination is then coupled with detailed molecular analysis of the RNA products to establish structure–function correlates for the catalytic RNA.

Genetic Methods

Nondirected Mutagenesis

Nontargeted mutagenesis approaches based on phenotypic selections are useful alternatives to site-directed mutagenesis. Since these nondirected approaches need not be based on *a priori* assumptions, they are of particular value when a good model for higher order structure of an RNA molecule is not available. Positive selections, or at least screening procedures, for gene function and dysfunction open up extensive mutagenesis options, as described below. It is also important to consider the type of mutational event desired, as in splicing the effective change is at the RNA level. As will be considered in more detail, the molecular consequences of transition versus transversion events for an RNA molecule, as opposed to a protein, can be great.

Consider, for example, a critical G-C Watson–Crick base pairing interaction in an RNA structure. Transversion of the G to a C would be useful in verifying this pairing, since after isolation of intragenic suppressors, second-site revertants would convert a G-C pairing in the RNA structure to a C-G pairing. In contrast, transition of the G in this same G-C pairing to an A, followed by the compensatory transition of the C to a T, would result in substitution of a G-C pairing with an energetically much less favorable A-U pairing. Restoration of phenotype in such cases would depend on the ability of the less stable RNA structure to function efficiently. In some cases unidirectional transition mutagens may also be useful (even advantageous), as when considering generating second-site revertants. Hydroxylamine (HA), for example, causes only one-way G-to-A or C-to-T changes. If one treats a primary transition mutant (say G-to-A, disrupting a G-C pair) with HA, only pseudorevertants (i.e., C-to-T compensatory changes) and not true revertants (A-to-G) will be induced. Selection for restoration of gene function may then provide a population enriched for pseudorevertants above the background of spontaneous true revertants. The usefulness of this approach, therefore, lies in the directional specificity of the mutagen.

Transitions are readily induced by chemical base modifiers such as HA (G-C → A-T, unidirectional) and nitrous acid (G-C ↔ A-T, bidirectional). There are, however, no chemical agents that efficiently induce transversions, although such changes may occur spontaneously at low frequency, owing to errors of DNA replication or misrepair (reviewed by Glass, 1982).[13] UV irradiation enhances misrepair and therefore transversion frequency, and, furthermore, there are mutator strains of *E. coli* that are transversion specific.[14] However, flexibility in terms of the choice of base substitution is greatest when inducing transversion *in vitro,* with *E. coli* DNA polymerase I (pol I).[15] In this case a biased dNTP pool used in

[13] R. E. Glass, "Gene Function," p. 110. University of California Press, Berkeley and Los Angeles, 1982.
[14] Y. Nghiem, M. Cabrera, C. G. Cupples, and J. H. Miller, *Proc. Natl. Acad. Sci. U.S.A.* **85,** 2709 (1988).
[15] D. Botstein and D. Shortle, *Science* **229,** 1193 (1985).

gap repair will favor misincorporation. For example, limiting dCTP and adding α-thio-dGTP (α-thiophosphate derivatives defy excision by the proofreading exonuclease of pol I) will promote C-to-G transversions. Region-specific saturation mutagenesis with oligonucleotides containing low-frequency random errors[16] can also be a viable approach for genetic probing of regions within a gene for which selectable phenotypes are available.

Phenotypic Selection and Screening

The thymidylate synthase (TS) phenotype has been instrumental in isolating td^- mutations in plasmid and phage. A *thyA* host that lacks the cellular TS is unable to grow in the absence of exogenous thymine or thymidine ($-$THY media). The cloned *td* gene can complement *thyA* cells for growth on $-$THY media. On the other hand, cells in which TS is limiting or absent are at a selective advantage on media containing folate analogs (such as trimethoprin) and thymine (TTM medium). TTM medium therefore selects for td^- (TS$^-$) mutants in a *thyA* host,[11,12] whereas TS$^+$ revertants (both true revertants and second-site suppressors) can be selected on minimal media lacking thymine.

Plaque phenotypes associated with *td* function have been adapted to the study of splicing-defective mutants in the laboratory of Hall.[10] Phage T4*td*$^-$ mutants can be recognized on the basis of their small-plaque phenotype on a *thyA* host when thymine is limiting.[17] Alternatively, td^- mutant plaques can be identified by their white-halo phenotype on cells requiring uracil.[18] Revertants form large plaques under the former plating conditions and lack halos under the latter conditions.

Minimal $-THY$ and TTM Media. Several minimal media (e.g., M63 or M9, with 0.4% dextrose as carbon source)[19] serve as satisfactory basal media for the selections. The $-$THY media are supplemented with 0.1% Norit A-treated casamino acids, the nutritional requirements of the host cell, and 1.5% Bacto-agar. [Norit A decolorizing carbon (Fisher) is used to remove heterocyclic compounds, including thymine, from a stock solution of 10% casamino acids (Difco). The charcoal (15 g/liter) decolorizes the solution within 30 min, with stirring, after which it is removed by filtration and the straw-colored solution autoclaved.] TTM plates contain in addition 20 μg/ml trimethoprim and 50 μg/ml thymine. $-$THY medium

[16] K. M. Derbyshire, J. J. Salvo, and N. D. F. Grindley, *Gene* **46**, 145 (1986).
[17] E. H. Simon and I. Tessman, *Proc. Natl. Acad. Sci. U.S.A.* **50**, 526 (1963).
[18] D. H. Hall, I. Tessman, and O. Karlstrom, *Virology* **31**, 442 (1967).
[19] J. H. Miller (ed.), "Experiments in Molecular Genetics," p. 431. Cold Spring Harbor Laboratory, Cold Spring Harbor, New York, 1972.

with 1.1% Bacto-agar is suitable for detecting the small versus large plaque phenotypes. The top agar (3 ml with 0.7% Bacto-agar) contains 0–8 μg thymidine (depending on the genetic background of phage and host) as the limiting condition for T4td^- plaque formation and 100 μg for thymidine excess.

Genetic Mapping

The ability to recognize TS^+ recombinants in the phage or plasmid enables the use of marker rescue to map point mutations.[10,12,20] Nested sets of *td* subfragments, cloned into a suitable replicon (M13 or compatible plasmid), are used as donors of wild-type sequence during recombination events with the *td*$^-$ T4 phage or *td*$^-$ mutant plasmid (Fig. 2). Cells containing the plasmid-borne *td* tester subfragment are infected with T4td^- phage carrying the point mutation to be mapped. Alternatively, F^+ cells can be coinfected with T4td^- point mutants and M13–*td* hybrids. In both cases positive marker rescue is then scored by the appearance of large T4td^+ plaques on F^- *thyA* cells on media containing limiting thymidine.[10] When the point mutation is carried on a plasmid (e.g., pBR322, a ColEI derivative), marker rescue can be achieved by introducing a member of the fragment library on a compatible plasmid (e.g., pACYC184, a p15A derivative). Alternatively, the F^+ cell carrying the point mutation on a plasmid can be infected with an M13–*td* hybrid. In either case positive marker rescue is scored by the appearance of TS^+ colonies on media lacking thymine, with the mutation mapping to the smallest subfragment capable of rescue. Recombination is detectable with the point mutation as close as 10 nucleotides from the deletion endpoint.

Molecular Analyses

NH$_2$-Terminal Polypeptide (NH$_2$-TS) Screening Assay for Splicing

The NH$_2$-terminal polypeptide assay for *td* RNA splicing (the NH$_2$-TS assay) is based on the ratio of NH$_2$-TS, the translation product of the pre-mRNA, to TS, the product of the mature mRNA. This ratio is elevated when splicing is defective.[11,12] The suitability of *E. coli* as the host for this rapid assay rests on its meeting three criteria. First, *E. coli* is readily transformed by mutant constructs generated *in vitro*. Second, transcriptional amplification of the cloned split gene is possible. Third, transcription, RNA processing, and translation all occur in the same cellular com-

[20] M. Belfort and J. Pedersen-Lane, *J. Bacteriol.* **140**, 371 (1984).

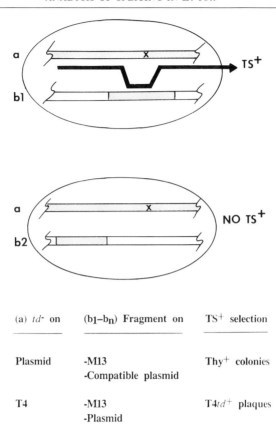

(a) *td⁻* on	(b₁–bₙ) Fragment on	TS⁺ selection
Plasmid	-M13 -Compatible plasmid	Thy⁺ colonies
T4	-M13 -Plasmid	T4*td⁺* plaques

Fig. 2. Marker rescue mapping of *td⁻* point mutations. The recombination events that occur between the *td⁻* allele on either plasmid or T4 phage (a) and the rescuing fragment on either M13 or a compatible plasmid (b₁, b₂, . . . , bₙ) are depicted by a heavy line. The site of the point mutation is marked in the *td* sequence (shaded) by an X.

partment. This latter requirement confines the NH_2-terminal polypeptide assay to bacterial systems. The assay is suitable for screening splicing-defective mutants of all three split T4 genes (*td, nrdB,* and *sunY*),[21] because all three express stable amino-terminal polypeptides from the pre-mRNA, in addition to the extended translation products from the ligated exons. The assay is not, however, applicable to intron–*lacZ* fusions. This is because the site of fusion is usually within the first 20 amino acids of β-

[21] D. A. Shub, J. M. Gott, M.-Q. Xu, B. F. Lang, F. Michel, J. Tomaschewski, J. Pedersen-Lane, and M. Belfort, *Proc. Natl. Acad. Sci. U.S.A.* **85,** 1151 (1988).

galactosidase, resulting in NH_2-terminal polypeptides that are too small for detection by rapid methods.

The NH_2-TS assay is shown in Fig. 3. Translation of the pre-mRNA yields the 25-kDa NH_2-TS product of exon I, as well as a downstream product of molecular weight 11.4K (X), whose coding sequence is co-transcribed with the *td* gene. The mature mRNA encodes TS (M_r 32,000) and X. Whereas wild-type clones have very high levels of TS relative to NH_2-TS, for splicing-defective mutants the reverse is true and the NH_2-TS/TS ratio is high. Product X is a useful indicator of an intact primary transcript, as it is absent in mutants affecting transcription and RNA stability.

Procedure. The method was worked out for pKK-TdΔ1-3 and its derivatives. The parent plasmid contains the splicing-proficient *td*Δ1-3 construct under p*tac* control in the vector pKK223-3.[12] The host cell is D1210*thyA*, a *lacI*^{q} *thyA* derivative of *E. coli* strain HB101.

1. Transformants are grown in TBYETA (1.0% Bacto-tryptone, 0.5% NaCl, 0.5% yeast extract, 50 μg/ml thymine, and 100 μg/ml ampicillin) at 37° to an OD_{650} of 0.2.

2. Isopropylthio-β-galactoside is added to a final concentration of 2.5 mM to induce transcription from p*tac*. Vigorous aeration at 37° is continued for 3.5 hr.

3. Aliquots of 1.5 ml are chilled and centrifuged at 16,000 g in a microcentrifuge for 3 min at 4°.

4. The cell pellet is washed in ice-cold TM [10 mM tris(hydroxymethyl)aminomethane hydrochloride (Tris-HCl) (pH 7.4), 10 mM $MgCl_2$], and the cells are resuspended in 75 μl of cracking buffer [150 mM Tris-HCl (pH 6.8), 6% sodium dodecyl sulfate (SDS), 8% glycerol, 2 M 2-mercaptoethanol and 0.005% bromphenol blue] by vigorous agitation in a Vortex mixer.

5. The samples are heated for 3 min in a boiling water bath, then centrifuged at 16,000 g at room temperature for 10 min.

6. Five to ten microliters of the upper phase is loaded directly onto a 15% bisacrylamide slab gel,[22] taking care not to disturb the DNA that has sedimented to the bottom of the tube.

7. The bands are visualized by staining for 30 min at 60° in a 0.25% Coomassie Brilliant Blue R solution in 45% (v/v) methanol–9.2% (v/v) glacial acetic acid. Destaining is in 10% ethanol–10% acetic acid at room temperature with multiple changes, until the stained bands are clearly visible.

---

[22] U. K. Laemmli, *Nature (London)* **227**, 680 (1970).

A

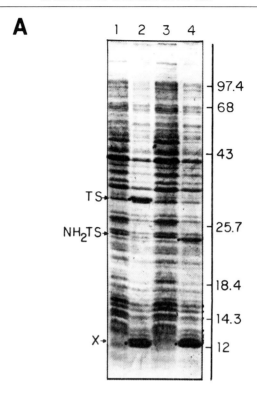

B

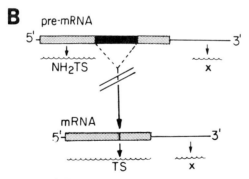

FIG. 3. NH₂-TS assay. (A) Cells (D1210*thyA*) containing pKK223-3 vector (lane 1), the wild-type TS⁺ parent pKK-TdΔ1-3 (lane 2), and TS⁻ derivatives defective in transcription (lane 3) and splicing (lane 4). Transcription was induced in the transformants and proteins separated from a crude cell extract as described in the text. The sequences coding for the products TS, NH₂-TS, and X are depicted in (B). Exons are shaded, the intron is a solid bar, and downstream sequences are represented by a thin line.

Dot-Blot RNA-Oligodeoxynucleotide Splicing Assay

Junction-specific oligodeoxynucleotide hybridization probes have been valuable tools for screening and confirming splicing defects. These probes have been used successfully with RNA isolated from phage-infected cells and after transcriptional induction of transformed *E. coli*. The RNA extraction procedure is similar to that used for all subsequent RNA analysis and is described below. The dot-blot assay is equally appropriate for studying RNA products in eukaryotic systems.

RNA Extraction. The lysozyme freeze–thaw method followed by phenol extraction was found to be superior to other RNA extraction protocols (e.g., the hot phenol method) for producing intact RNA. The procedure has been adapted from McKenney *et al.*,[23] with modifications dependent on subsequent use of the RNA.

Reagents

Solution A: 10 mM Tris-HCl (pH 7.5), 1 mM ethylenediaminetetraacetic acid (EDTA), 10 mM dithiothreitol (DTT), 200 U/ml RNasin (placental ribonuclease inhibitor from Promega Biotec, Madison, WI), 0.25 mg/ml lysozyme (freshly prepared)

Solution B: 200 mM magnesium acetate, 1 mg/ml DNase (RNase-free, from Bethesda Research Laboratories, Gaithersburg, MD), 200 U/ml RNasin

Solution C: 0.1 M acetic acid, 5% SDS

Procedure

1. Cells, grown in TBYE (1.0% Bacto-tryptone, 0.5% NaCl, 0.5% yeast extract) with indicated supplements to an OD_{650} of about 0.2, are infected with phage or transcriptionally induced as appropriate. One minute prior to harvest, chloramphenicol is added to a concentration of 100 μg/ml to stabilize transcripts. Cells are rapidly chilled to 4° by swirling in an ice–ethanol bath (30 sec for a 30-ml culture) and sedimented by centrifugation for 5 min at 12,000 g (10,000 rpm in a Sorvall SS34 rotor) at 4°. Pellets are resuspended in 1.0 ml Tris-HCl (pH 7.5) and the cells transferred to 1.5-ml Eppendorf tubes. After centrifugation for 3 min at 16,000 g at 4° (14,000 rpm in a Brinkman microcentrifuge) the supernatant is removed and the pellets frozen at −80°. Although pellets may be stored at this temperature for several weeks before RNA extraction, the shorter the period of freezing, the greater the RNA yields.

---

[23] K. McKenney, H. Shimatake, D. Court, U. Schmeissner, C. Brady, and M. Rosenberg, *in* "Gene Amplification and Analysis" (J. S. Chirikjian and T. S. Papas, eds.), p. 383. Elsevier, Amsterdam, 1981.

2. A cell pellet from a 30-ml culture is thawed on ice and resuspended in 160 μl of ice-cold solution A. Three freeze–thaw cycles are performed by placing tubes alternatively into a dry ice–ethanol bath and 25° water bath (until almost thawed) with thorough mixing before refreezing.

3. After the addition of 20 μl of solution B and mixing, the tubes are incubated on ice for 45 min.

4. After the addition of 20 μl of solution C and mixing, the tubes are incubated at room temperature for 5 min.

5. Phenol extraction is performed with 300 μl of saturated phenol,[24] after which the upper aqueous phase is reextracted with an equal volume of phenol–chloroform–isoamyl alcohol (25 : 24 : 1, v/v), and then with chloroform–isoamyl alcohol (24 : 1). The aqueous phase may be used at this stage for qualitative dot-blot assays without further treatment of the RNA. The subsequent steps are recommended for all quantitative RNA analysis procedures.

6. The aqueous phase is ethanol precipitated after the addition of 1/100 volume of 1 M magnesium acetate, 1/10 volume of 3 M NaOAc (pH 5.2), and 2.5 volumes 100% ethanol (−20°).

7. After incubation at −20° overnight or −80° for 30 min, the RNA pellet is collected by centrifugation at 16,000 g for 10 min at 4°. Pellets are washed in 70% ethanol, resedimented, and dried in a vacuum desiccator for 10 min.

8. The RNA pellet is resuspended in 50 μl ice-cold water treated with diethyl pyrocarbonate (DEPC) and stored at −80°. Preparations are stable for several months at this temperature. RNA concentration, as determined by OD_{260} readings of a 1/500 dilution of the preparation, range from 8 to 20 mg/ml.

Probes. Probes are designed to detect precursors or products by hybridizing to junctions that are specific to one or the other. Probes to detect precursors are directed at the 5′ splice site (exon I–intron) and 3′ splice site (intron–exon II), while probes to detect splice products target the exon ligation junction or intron cyclization junction (Fig. 4, probes 1, 2, 3, and 4, respectively).[10–12] Intron and exon probes will detect both precursors and products and are used to normalize the assay. Junction-specific probes are typically 23–27 nucleotides in length, with approximately one-half of the probe hybridizing to each of the two RNA segments. A C + G content of the probes ranging from 30 to 50% has been found to give satisfactory results.

[24] T. Maniatis, E. F. Fritsch, and J. Sambrook, "Molecular Cloning: A Laboratory Manual," p. 458. Cold Spring Harbor Laboratory, Cold Spring Harbor, New York, 1982.

pre-mRNA

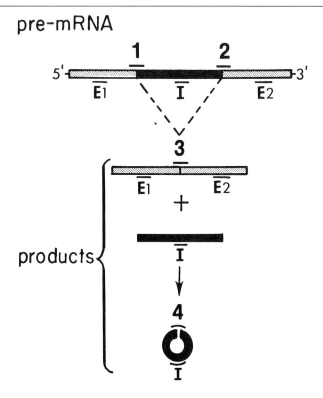

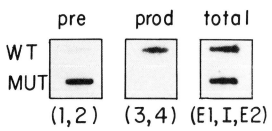

FIG. 4. Dot-blot RNA–oligodeoxynucleotide splicing assay. The probes that specifically target precursors (1 and 2), ligated exons (3), and intron cyclization products (4) are represented by bold numbers above the RNAs. Exon (E1 and E2) and intron (I) probes, shown below the RNAs, help quantitate the total amount of RNA present. Representative slot-blot data below show RNA from the wild-type construct pKK-TdΔ1-3 (WT) and a splicing-defective derivative (MUT). Oligodeoxynucleotides 1, 3, and I were used to probe precursor (pre), product (prod), and total RNA in the three blots shown. Probes giving similar results are grouped in parentheses below the blots.

Labeling. Oligonucleotide (20 pmol) is incubated at 37° in a total volume of 10 μl for 45 min with 40 μCi of [γ-^{32}P]ATP (3,000 Ci/mmol) and 10 U of polynucleotide kinase in 0.1 M Tris-HCl (pH 8.0) containing 5 mM DTT, 10 mM MgCl$_2$, and 0.2 mM spermidine. After heat inactivation at 90° for 3 min, 25 μl of water and 200 μl of low-salt buffer [0.2 M NaCl, 20 mM Tris-HCl (pH 7.4), 1.0 mM EDTA] are added. Unincorporated [^{32}P]ATP is removed by passage through an RPC-5 column (Elutip-D, Schleicher and Schuell, Keene, NH) according to the instructions of the supplier. Labeled oligonucleotide is eluted with high-salt buffer [1.0 M NaCl, 20 mM Tris-HCl (pH 7.4), 1 mM EDTA] and added directly to the prehybridization solution.

Blotting and Hybridization Protocol

1. Phenol-extracted cellular RNA (100 ng–1 μg per spot) may be applied manually to a nitrocellulose or nylon (Hybond-N, Amersham, Arlington Heights, IL) membrane. In this case, denaturation is achieved by heating the RNA preparation to 95° for 3 min and rapid chilling. The RNA is then applied to the membrane with a micropipet in 1–2 μl of DEPC-treated water. Alternatively, the RNA may be applied using a filtration manifold. In this case a slot-blot apparatus is most suitable for subsequent quantitation. The RNA is brought up to 50 μl with DEPC-treated water, diluted to 200 μl with 6.15 M formaldehyde–10× SSC [1.5 M NaCl, 0.15 M trisodium citrate (pH 7.5)], and denatured at 65° for 15 min. The membrane is prepared by soaking in DEPC-treated water and then in 10× SSC, before being placed on the filtration apparatus. The wells are rinsed with 10× SSC under vacuum, the samples applied, and the wells washed with 400 μl of 10× SSC. RNA fixation to the nitrocellulose entails baking the air-dried filter for 2 hr at 80° under vacuum. In contrast, the air-dried Hybond membrane is wrapped in plastic wrap and exposed to a long-wave (305 nm) UV transilluminator (RNA side down) for 5 min.

2. After prehybridization in "Wallace buffer" [0.9 M NaCl, 90 mM Tris-HCl (pH 7.5), 6 mM EDTA, 5× Denhardt's solution (0.1% Ficoll, 0.1% polyvinylpyrrolidone, 0.1% bovine serum albumin, BSA) 0.5% SDS, and 50 μg/ml yeast tRNA][25] at 37° for 15–30 min, [γ-^{32}P]oligonucleotide is added at 10^6 cpm/ml. After thorough mixing, incubation is continued at 37° for 16 hr.

3. After washing the filters 3 times at 23° for 30 min in 6× SSC, a good differential signal is usually obtained. If nonspecific probe binding is a

[25] R. B. Wallace, M. Schold, M. J. Johnson, P. Dembek, and K. Itakura, *Nucleic Acids Res.* **9**, 3647 (1981).

problem, filters may be washed at higher temperatures [~10° below the calculated melting temperature (T_m) for an oligonucleotide; as a rule we have found that 45° is suitable for probes of the indicated length and composition] for 30 min in 6× SSC.

4. Excess moisture is blotted off, without allowing the filter to dry. Filters are wrapped in plastic wrap before being subjected to autoradiography.

Filters may be reused after stripping with probe removal buffer at 65° for 1–2 hr as per Thomas.[26]

Quantitation of Splice Products in Vivo

The short half-life of mRNA in *E. coli* makes quantitation of the RNA difficult. Nevertheless, the methods described below have been successful in identifying precursors and products, and in measuring their relative levels in the wild type and in mutants. It must be kept in mind, however, that RNA abundance reflects both its synthesis and degradation and that the relative levels of two species in the same preparation is not necessarily a reflection of relative synthesis. For example, the *td* intron is 7- to 10-fold more stable than the *td* mRNA, and is present at 20–50 times the level of the messenger in cellular RNA preparations, despite their being generated in equimolar amounts in the normal splicing reaction.[27] These considerations notwithstanding, the various quantitative hybridization methods and primer-extension analyses are of great utility in comparing RNA products of the wild type and splicing-defective mutants.

Quantitative Hybridization. Northern analysis or dot-blot hybridization, with the probe in molar excess, may be used to quantitate RNA species. In these cases signal intensity is measured by densitometric analysis of autoradiograms. Northern analysis is an effective means to identify specific RNA species, although diffuse bands arising from RNA instability may present problems in resolution and quantitation. To some extent these problems may be obviated by quantitative dot-blot assay using multiple probes or by S1 protection analysis with oligonucleotide probes (below). With both methods a comparison of signal intensities is valid only for the same probe, because of variable binding affinities of the different probes. The S1 protection assay has been useful for time-course

---

[26] P. S. Thomas, this series, Vol. 100, p. 255.
[27] W. K. Y. Chan, G. Belfort, and M. Belfort, *Gene* **73,** 295 (1988).

analysis of precursor and mRNA levels after transcriptional induction and phage infection.[28,29]

S1 Protection Analysis

1. Deproteinized cellular RNA (10–100 μg) is mixed with 0.4 pmol of 5'-^{32}P-labeled oligonucleotide (~2 × 10^6 cpm) that has been purified through an Elutip-D column as indicated above. The total amount of RNA is made up to 100 μg with yeast tRNA. One-half volume of 1 M sodium acetate (pH 5.4) and 2 volumes of ice-cold ethanol are added, and precipitation is allowed to proceed at $-20°$ overnight or at $-80°$ for 15 min.

2. After washing the precipitate in 70% ethanol, the pellet is dried for 5 min in a vacuum desiccator and resuspended in 30 μl of oligonucleotide hybridization buffer [40 mM piperazine-N, N'-bis(2-ethanesulfonic acid) PIPES (pH 6.7), 1 mM EDTA, 400 mM NaCl]. Annealing is performed for 30 min at 5–10° below the calculated T_m for the probe.

3. S1 nuclease digestion is performed at 23° for 30 min after the addition of 300 μl of S1 buffer [0.25 M NaCl, 50 mM sodium acetate (pH 4.6), 4.0 mM ZnSO$_4$, 20 μg/ml single-stranded DNA, 5% glycerol] and 200 U/ml enzyme.

4. The reaction is stopped with the addition of EDTA to 12.5 mM. After the addition of 50 μg yeast tRNA and ammonium acetate (to 0.5 M), the reaction products are precipitated with 2 volumes of ethanol.

5. The protected oligomer is separated on a 20% acrylamide–8 M urea sequencing gel. Quantitation is either by densitometric scanning of an autoradiogram or by cutting out a gel slice containing the radioactive band and counting in a liquid scintillation counter.

Primer-Extension Analysis. cDNA synthesis is useful for comparing the levels of processing products within an RNA preparation. Since both precursor and product RNAs are quantitated using the same primer, relative amounts of different species can be compared, bearing in mind, however, that steady-state levels are a function of both synthesis and stability of the particular species. Primer-extension analysis can take one of two forms: (1) Long-distance cDNA synthesis generates DNA copies which reflect both the length and the amount of the RNA template (Fig. 5). Samples are separated on gels that allow sizing of the cDNAs. Though

[28] M. Belfort, J. Pedersen-Lane, K. Ehrenman, F. K. Chu, G. F. Maley, F. Maley, D. S. McPheeters, and L. Gold, *Gene* **41**, 93 (1986).
[29] D. K. West, M. Belfort, G. F. Maley, and F. Maley, *J. Biol. Chem.* **261**, 13446 (1986).

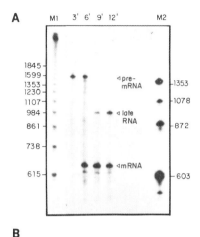

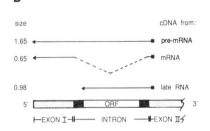

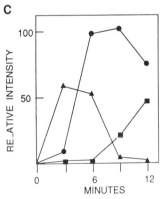

FIG. 5. Primer-extension analysis. cDNAs corresponding to pre-mRNA, mRNA, and a late transcript of the *td* gene were monitored at the indicated times (3–12 min) after T4 infection. M1 and M2 in (A) are DNA size markers with numbers of residues indicated. The cDNAs of the RNA species that are labeled in (A) are diagrammed in (B) and quantitated in (C) (▲, pre-mRNA; ●, mRNA; ■, late RNA). The primer (solid box in B) anneals to the downstream exon, 50 nucleotides from the 3′ splice site. Quantitation in (C) was by densitometry of lighter exposures of the autoradiogram shown in (A). [From J. M. Gott, A. Zeeh, D. Bell-Pedersen, K. Ehrenman, M. Belfort, and D. A. Shub, *Genes Dev.* **2**, 1791 (1988).]

RNA instability considerations cannot be ignored, the resolution of bands is much sharper than in Northern hybridization analysis, because the 3' end is fixed by a primer rather than being heterogeneous. (2) Short-distance cDNA synthesis conducted near intron–exon junctions in the presence of ddNTPs can reveal the relative amounts of precursor and splice products.[10–12] Reactions are separated on sequencing gels to reveal junction sequences, allowing inference of the relative amounts of precursor and splice products. Procedures have been adapted from Zaug *et al.*[30] and McPheeters *et al.*[31]

1. Phenol-extracted labeled oligonucleotide primer (0.4 pmol, 5×10^6 cpm/pmol) is annealed to deproteinized RNA (8–16 μg) at 60° in 10 μl of annealing buffer [50 mM Tris-HCl (pH 8.3), 60 mM NaCl, 10 mM DTT] for 3 min and quick-frozen in a dry ice–ethanol slurry.

2. Magnesium acetate is added to a final concentration of 6 mM, and cDNA synthesis is performed in the presence of 400 μM dNTPs and 0.1 U of reverse transcriptase at 48° for 30 min. The reaction is terminated by the addition of 1.5 volumes of formamide–dye mix (0.1% xylene cyanol, 0.1% bromphenol blue, 10 mM EDTA, and 95% v/v deionized formamide). After heating in a boiling water bath for 3 min, samples are separated on denaturing gels containing 4–5% acrylamide and 8 M urea.

3. For primer-extension sequence analysis the synthesis reaction is split into five aliquots, and one of each of the ddNTPs (10 μM) is added to each of four tubes, with one tube not receiving any ddNTP but instead an equal volume of water. Reverse transcriptase (0.02 U) is added to the individual reactions and synthesis allowed to proceed as above. Reaction products are separated on an 8% acrylamide–8 M urea sequencing gel.

GTP Labeling to Screen RNA Populations for Group I Introns

The occurrence of group I self-splicing introns in phage-infected cells has been successfully probed by labeling mixed RNA populations with [^{32}P]GTP under self-splicing conditions (Fig. 6A).[7] This approach, which is based on the covalent attachment of the labeled GTP to the 5' end of the intron in the first step of splicing, has led to the identification of several phage introns and to studies demonstrating the variable occurrence of these introns in the T-even phage family.[32] The labeling protocol (below) was adapted from Garriga and Lambowitz.[33] RNA extracted 6–12 min

[30] A. J. Zaug, J. R. Kent, and T. R. Cech, *Science* **224**, 574 (1984).
[31] D. S. McPheeters, A. Christensen, E. T. Young, G. Stormo, and L. Gold, *Nucleic Acids Res.* **14**, 5813 (1986).
[32] J. Pedersen-Lane and M. Belfort, *Science* **237**, 182 (1987).
[33] G. Garriga and A. M. Lambowitz, *Cell* **38**, 631 (1984).

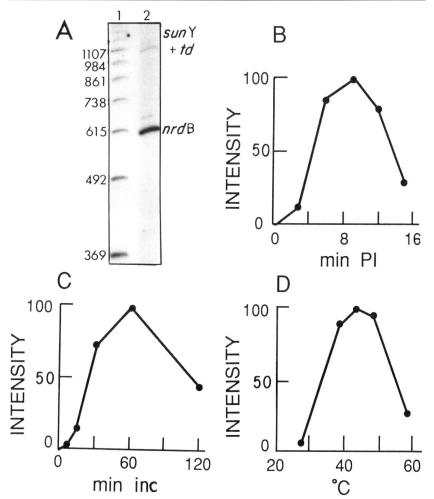

Fig. 6. GTP labeling of group I introns in T4 RNA. The *td* and *sun*Y introns (comigrating at ~1 kb) and the *nrdB* intron (at ~0.6 kb) are shown in (A) (lane 2) next to DNA size markers, in nucleotides (lane 1). The labeling of the *nrdB* intron (representative of the family of three introns) was measured under reaction conditions described in the text. The labeling intensity was monitored at various times postinfection (PI) in a 1-hr reaction at 38° (B). The length of incubation (inc) was varied at 45°, with RNA being extracted 9 min after infection (C). The effect of temperature on the reaction was measured in a 1-hr incubation, again with RNA being extracted 9 min after infection (D).

postinfection was optimally active in the GTP labeling reaction (Fig. 6B), with maximal incorporation after 1 hr of incubation (Fig. 6C) and a broad temperature optimum of 38–48° (Fig. 6D).

Procedure. Ten micrograms of RNA is incubated with 10 μCi [α-^{32}P]-GTP (3000 Ci/mmol) in buffer containing 50 mM Tris-HCl (pH 7.5), 30 mM MgCl$_2$, 50 mM NH$_4$Cl, 1 mM spermidine, 5 mM DTT, and 10 U of RNasin in a total volume of 10 μl at 38–42°. After 1 hr, 150 μl of chilled stop solution is added (2.5 M ammonium acetate, 25 mM EDTA, and 300 μg/ml yeast tRNA), and the labeled RNA is ethanol precipitated with 2.5 volumes of ice-cold 100% ethanol. The precipitate is washed with 70% ethanol, dried under vacuum, and dissolved in 10 μl of DEPC-treated water. After the addition of 1.5 volumes of formamide–dye mix (see above) and heating at 65° for 15 min, the RNA is resolved on an agarose–formaldehyde or acrylamide–urea gel. The labeled RNA may also be used as a hybridization probe for mapping the introns in the phage genome.[7]

Acknowledgments

We thank all members of our laboratory for help in refining the various methodologies, and particularly William Chan for working out the slot-blot protocol. The role of Dwight Hall in adapting phage techniques for the splicing studies is thankfully acknowledged. Special thanks are due to Debbie Bell Pedersen, Jill Galloway Salvo, and Sue Quirk for critically reading the manuscript. The methods described were developed while performing work supported by National Science Foundation Grants DMB 8502961 and DMB 850552 and National Institutes of Health Grants GM 33314 and GM 39422 to M.B.

[44] Genetic Analysis of RNA Splicing in Yeast Mitochondria

By PHILIP S. PERLMAN

Introduction

The mitochondrion of bakers' yeast has been studied intensively since the 1950s as a model system for various processes, including oxidative phosphorylation, protein targeting, mitochondrial DNA (mtDNA) structure, transcription, and RNA processing. Both nuclear and mitochondrial mutations are available as variables in studying those processes, and both have been used, especially in the 1980s, very effectively. This chapter summarizes genetic methods that have contributed to our knowledge of RNA splicing mechanisms in yeast mitochondria. Many of the basic approaches were reviewed earlier in this series,[1] so this chapter briefly summarizes key points needed for completeness and provides more detailed descriptions of new strategies that have appeared since 1983.

Introns in Yeast Mitochondrial DNA

Mitochondrial DNA of bakers' yeast, *Saccharomyces cerevisiae*, is a circular molecule containing about 75,000 base pairs (bp). It contains genes for 2 rRNAs, 24 tRNAs, 7 subunits of the oxidative phosphorylation system, 1 ribosomal protein, and a small RNA involved in 5'-end-processing of tRNAs.[2] All of the coding sequences are on the same strand of the genome and are transcribed from a limited number of promoters using an RNA polymerase whose subunits are encoded by nuclear DNA. Virtually all primary transcripts are polygenic, including groups of adjacent coding sequences, requiring processing at several types of internal sites to yield individual, functional RNAs.

Three genes contain introns in some or all isolates of *Saccharomyces* analyzed to date (Fig. 1). The *coxI* gene, encoding subunit I of cytochrome-*c* oxidase, contains up to ten different introns[3–7]; the *cob* gene, encoding the apoprotein of cytochrome *b*, contains up to five introns;[8–10] and the 21 S rRNA gene contains one intron.[11] Most of those introns are optional so far as the function of their host gene is concerned since some natural isolates lack various introns. Our recent survey of the intron configurations of these genes in yeast isolates reveals that only aI3 and bI5 are ubiquitous.[7] However, intronless *cob*[12] and *coxI*[13] genes have been made, and each appears to be fully functional *in vivo*.

[1] P. S. Perlman and H. R. Mahler, this series, Vol. 97, p. 374.
[2] A. Tzagoloff and A. Myers, *Annu. Rev. Biochem.* **55**, 249 (1986).
[3] S. G. Bonitz, G. Coruzzi, B. E. Thalenfeld, A. Tzagoloff, and G. Macino, *J. Biol. Chem.* **255**, 11927 (1980).
[4] L. A. M. Hensgens, L. Bonen, M. de Haan, G. van der Horst, and L. A. Grivell, *Cell* **32**, 379 (1983).
[5] Z. Kotylak, J. Lazowska, D. C. Hawthorne, and P. P. Slonimski, *in* "Achievements and Perspectives of Mitochondrial Research, Volume II: Biogenesis" (E. Quagliariello, E. C. Slater, F. Palmieri, C. Saccone, and A. M. Kroon, eds.), p. 1. Elsevier, Amsterdam, 1985.
[6] D. Ralph, Ph.D. dissertation, The Ohio State University, Columbus, Ohio, 1986.
[7] S. Dib-Hajj, D. Ralph, S. Boulanger, and P. S. Perlman, in preparation.
[8] F. G. Nobrega and A. Tzagoloff, *J. Biol. Chem.* **255**, 9828 (1980).
[9] J. Lazowska, C. Jacq, and P. P. Slonimski, *Cell* **22**, 333 (1980).
[10] J. Lazowska, A. Gargouri, and P. P. Slonimski, *in* "Mitochondria 1983" (R. Schweyen, K. Wolk, and F. Kaudewitz, eds.), p. 405. de Gruyter, Berlin, 1983.
[11] B. Dujon, *Cell* **20**, 185 (1980).
[12] M. Labouesse and P. P. Slonimski, *EMBO J.* **2**, 269 (1983).
[13] B. Seraphin, A. Boulet, M. Simon, and G. Faye, *Proc. Natl. Acad. Sci. U.S.A.* **84**, 6810 (1987).

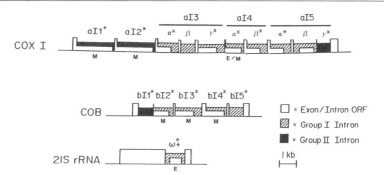

FIG. 1. Introns in yeast mitochondrial genes. The three intron-containing genes of yeast mtDNA are shown diagrammatically. Each intron is designated as group I or group II; introns known to self-splice are marked with an asterisk; introns that do not appear to self-splice under conditions suitable for detecting reactions with the other yeast introns are identified by an x; and intron open reading frames encoding known functions (M, maturase; E, endonuclease) are also labeled. The potentially confusing numbering of the *coxI* gene introns results from the discovery of additional introns after the sequence of the five-intron version of the gene was first published.[3] That form contains introns 1, 2, 3 (now 3α), 4 (now 4α), and 5 (now 5γ). Next, two more introns were found (together with the original five) in strain KL14-4A[4] (and others); since they are between aI4 and aI5, they were called aI5α and aI5β and the original aI5 was renamed aI5γ. Using that same strategy, two new introns between aI3α and aI4α and one between aI4α and aI5α were named appropriately[5–7] [see also J. M. Wenzlau, R. J. Saldanha, R. A. Butow, and P. S. Perlman, *Cell* **56**, 421 (1989)]. To date the largest *coxI* gene found in a natural isolate of *Saccharomyces* contains either of two different combinations of seven of the ten introns.[6,7] Either of two forms of the *cob* gene is commonly found in standard laboratory strains of yeast: the first has two introns[8] and the second has those plus three others in exon 1 of the two-intron form[9]; introns 1 and 2 of the former are introns 4 and 5 of the latter. The single intron of the 21 S rRNA gene is denoted by a greek letter because its presence was first detected as a "genetic determinant" that influences the transmission of nearby point mutations in crosses; that determinant exists in two forms, + and −. Later it was determined that the ω⁺ allele denotes the presence of the intron while the ω⁻ allele defines the form of the gene lacking the intron.

Two very different kinds of introns are present in yeast mtDNA, each being assigned to either group I or group II. Recent reviews define the basic features distinguishing those two intron types and their splicing mechanisms.[14–16] Each is basically self-splicing, though by somewhat different mechanisms; each type of intron contains numerous sequences that contribute to a relatively conserved secondary and tertiary structure that is essential for splicing. Mitochondrial genomes of most other fungi ana-

[14] T. R. Cech, *Gene* **73**, 259 (1988).
[15] F. Michel, K. Umesono, and H. Ozeki, *Gene* **82**, 5 (1989).
[16] P. S. Perlman, C. L. Peebles, and C. Daniels, in "Intervening Sequences in Evolution and Development" (E. M. Stone and R. J. Schwartz, eds.), in press. Oxford Univ. Press, New York, 1990.

lyzed to date also contain introns and usually have representatives of both types. Interestingly, group I introns are also found in the nuclear genomes of some lower eukaryotes, in bacteriophage genomes, and in chloroplast DNAs of some plants.[16] Group II introns are quite prevalent in chloroplast DNAs but are also found in mitochondrial DNA of some plants.[16]

Mitochondrially Inherited Splicing Defects

Types of Splicing-Defective Mitochondrial Mutants

Point and small deletion mutants of mtDNA are obtained readily using methods described previously.[1] For structural genes, and their introns, all such mutants yield cells with the same growth deficiency, namely, no growth on glycerol medium but normal growth on glucose medium. Large sets of mutants can be obtained fairly rapidly and screened to the subset mapping to any given gene or intron using petite deletion strains as mapping reagents (see Ref. 1). Introns comprise most of the sequence of the *cob* and *coxI* genes, and quite early in the analysis of mitochondrial point mutants it became clear that mutations of those introns are not rare. No splicing-defective mutants located in the single intron of the 21 S rRNA gene have been reported, presumably because such mutants would lack mitochondrial translation and thereby become petite mutants.

A transient complementation test is available,[1] that permits nearly all mutants in the *cob* and *coxI* genes to be divided into two classes. Those that fail to complement other mutants mapping elsewhere in the same gene, affecting either an exon sequence or a splicing signal within the intron, are cis-dominant mutants. Those that are complemented by most downstream mutants and by cis-dominant mutants in the same intron are trans-recessive. Early genetic studies of mitochondrial point mutations in two group I introns (bI2[9] and bI4[17–19]) provided clear evidence that their reading frames are translated efficiently in appropriate mutants and that, in wild-type strains, their translation products are required for splicing. Such proteins were called "maturases," and the complementation data showed that each maturase-encoding intron is a "gene within a gene." Cis-dominant mutants accumulate the partially spliced RNA that encodes the maturase and, so, usually overproduce maturase protein to such an extent that it is one of the most abundant products of the mitochondrial translation system.

[17] H. de la Salle, C. Jacq, and P. P. Slonimski, *Cell* **28,** 721 (1982).
[18] P. Q. Anziano, D. K. Hanson, H. R. Mahler, and P. S. Perlman, *Cell* **30,** 925 (1982).
[19] B. Weiss-Brummer, G. Rodel, R. J. Schweyen, and F. Kaudewitz, *Cell* **29,** 527 (1982).

Some so-called cis–trans mutants have been reported.[19,20] Such mutants fail to complement cis-dominant mutants of the same gene (whether in exons or in the same or different intron) and also fail to complement trans-recessive mutants of the same intron. A specific cis–trans mutant of bI4 alters a crucial splicing signal but also truncates the intron reading frame. Interestingly, there are a few examples of cis-dominant mutants of bI4 that complement trans-recessive mutants even though they have a premature termination signal in the intron open reading frame (ORF).[18,21] Since no maturase protein is detected in those mutants, it is likely that the termination signals are "leaky," permitting sufficient read-through to accumulate sufficient maturase to be detected in complementations. Leakiness of frameshift mutations of another gene of mtDNA has been reported previously,[22] but so far no hypothesis permits prediction of which mutants of that sort will be leaky or tight.

Genetic Studies of Intron-Encoded Proteins

As indicated schematically in Fig. 1, ten of the twelve group I introns and two of the four group II introns can encode a protein. Thorough genetic analysis defines a maturase function encoded by three group I introns (bI2,[9] bI3,[23] and bI4[17–19]) and both group II introns with reading frames (aI1[20] and aI2[24]). In addition, aI4 does not encode a maturase in wild-type strains; instead it "borrows" the bI4-encoded protein as its maturase. With that exception, the other maturases appear to be intron-specific.

Further evidence supporting the maturase model, at least for individual introns, is worth summarizing. First, where an intron-encoded protein is required for splicing the intron that encodes it, it follows that mutations which alter translation should block splicing of maturase-dependent introns. For example, maturase-encoding bI4 does not splice in petite mutants, totally lacking in mitochondrial translation.[25] Also, mutants that block splicing of bI1 abolish splicing of the three maturase-requiring in-

[20] G. Carignani, O. Groudinski, D. Frezza, E. Schiavon, E. Bergantino, and P. P. Slonimski, Cell 35, 733 (1983).

[21] M. R. Lamb, P. Q. Anziano, K. R. Glaus, D. K. Hanson, H. J. Klapper, P. S. Perlman, and H. R. Mahler, J. Biol. Chem. 258, 1991 (1983).

[22] T. D. Fox and B. Weiss-Brummer, Nature (London) 288, 60 (1980).

[23] J. Lazowska, M. Claisse, A. Gargouri, Z. Kotylak, A. Spyridakis, and P. P. Slonimski, J. Mol. Biol. 205, 275 (1989).

[24] K. L. Mecklenburg, Ph.D. dissertation, The Ohio State University, Columbus, Ohio, 1986.

[25] S. Bonitz, G. Homison, B. E. Thalenfeld, and A. Tzagoloff, J. Biol. Chem. 257, 6268 (1981).

trons located downstream of it (bI2–4) but not the intron that lacks an ORF (bI5).[26] These "polar" effects on splicing support the interpretation that each maturase is initially translated as a chimeric protein encoded by the spliced upstream exons and the entire intron reading frame.

Splicing-defective mitochondrial mutants are relatively easy to characterize biochemically for several reasons. First, unlike nuclear pre-mRNA introns, the major precursor RNA that accumulates owing to a splicing block is quite abundant.[21,26–28] Second, because that precursor contains the 5' end of the mature mRNA, it is translated to such an extent that intron-encoded proteins are prominent products of mitochondrial translation in such mutants.[9,20,21]

It appears that bI2 splicing requires a 54-kDa maturase that is encoded by *cob* exons 1 and 2 and the intron ORF; such a protein is detected among mitochondrial translation products of cis-dominant mutants and missense trans-recessive mutants; appropriately shortened forms are detected in trans-recessive mutants that have stop codons in the intron ORF.[9] *cob* intron 4 appears to encode a 65-kDa protein that is processed proteolytically to yield 28- and 35-kDa proteins. The former (p28) has been shown to be wholly intron encoded[18,29] and to be responsible for maturase activity[30,31] (see also below). The latter is clearly the amino-terminal fragment.[18] More circumstantial evidence exists identifying the sizes of maturases encoded by bI3[23] and two group II introns aI1[20,32] and aI2.[24]

Recent studies of mutants in other introns yielded new insights to intron-encoded functions. The protein encoded by the intron of the 21 S rRNA gene plays a key role in a special recombination mechanism involving that intron in crosses with strains lacking the intron.[33,34] That intron splices efficiently in petite mutants, indicating that the protein encoded by

[26] A. Haid, G. Grosch, C. Schmelzer, R. J. Schweyen, and F. Kaudewitz, *Curr. Genet.* **1**, 155 (1980).

[27] C. Schmelzer, A. Haid, G. Grosch, R. J. Schweyen, and F. Kaudewitz, *J. Biol. Chem.* **256**, 7610 (1981).

[28] L. A. M. Hensgens, G. van der Horst, H. L. Vos, and L. A. Grivell, *Curr. Genet.* **8**, 457 (1984).

[29] C. Jacq, J. Banroques, A.-M. Becam, P. P. Slonimski, N. Guiso, and A. Danshin, *EMBO J.* **3**, 1567 (1984).

[30] J. Banroques, A. Delahodde, and C. Jacq, *Cell* **46**, 837 (1986).

[31] A. Delahodde, V. Goguel, A. M. Becam, F. Creusot, J. Perea, J. Banroques, and C. Jacq, *Cell* **56**, 431 (1989).

[32] G. Carignani, P. Netter, E. Bergantino, and S. Robineau, *Curr. Genet.* **11**, 55 (1986).

[33] I. Macreadie, R. Scott, A. Zinn, and R. Butow, *Cell* **41**, 395 (1985).

[34] A. Jacquier and B. Dujon, *Cell* **41**, 383 (1985).

it does not also function in splicing.[35] That conclusion is further supported by the finding that mutants with stop condons in the ORF lack the endonuclease function but splice normally.[33,34] Similarly, aI4a encodes a recombinogenic endonuclease[31,36]; in wild-type strains aI4a splicing employs the bI4 maturase, and chain termination mutants of the aI4a ORF affect splicing only under special circumstances.

Thus, some mitochondrial ORFs encode maturase functions whereas others encode functions unrelated to splicing. Only three of the eight ORF-containing introns of the *coxI* gene (aI1, aI2, and aI4) have been analyzed in sufficient detail to determine whether they encode a maturase. Available data indicate that some of the others do not. Nearly all ORF-containing group I and and group II introns of other organisms have not been studied genetically so that it is uncertain whether any of those ORFs encode splicing functions. In this regard, recent studies indicate that a number of other group I intron ORFs encode recombinogenic endonucleases[37–39] and that this is presently the easiest-to-detect intron-encoded function. There is no evidence whatsoever that any intron ORF of an organism other than yeast participates in splicing; the best case that can be made so far is based on sequence homologies among introns of different organisms. For example, bI3 of yeast mtDNA encodes a maturase function, and the single intron of the *cob* gene of *Aspergillus nidulans* occupies the same exonic site and can encode a very similar protein.[24]

At the moment, the best system for further studying a group I intron-encoded maturase has been developed by Claude Jacq and co-workers. They used repeated cycles of site-directed mutagenesis to prepare a version of the sequences encoding the bI4 maturase function that has been purged of all codons which are peculiar to the mitochondrial genetic code, thereby producing a so-called universal code equivalent gene.[29] That gene was first expressed in *Escherichia coli* to permit various studies of the protein. More recently, they made a further derivative of that gene capable of expression in the yeast nucleus.[30] They have defined conditions in which the nuclear-situated maturase gene is expressed and the maturase function is restored to the mitochondria. In this way, it is possible to think about more detailed *in vivo* studies of the protein. It should be noted that missense mutations of the bI4 ORF (and other intron ORFs) are quite rare

---

[35] J. Locker and M. Rabinowitz, *Plasmid* **6**, 302 (1981).
[36] J. M. Wenzlau, R. J. Saldanha, R. A. Butow, and P. S. Perlman, *Cell* **56**, 421 (1989).
[37] S. M. Quirk, D. Bell-Pedersen, and M. Belfort, *Cell* **56**, 455 (1989).
[38] D. Muscarella and V. Vogt, *Cell* **56**, 443 (1989).
[39] B. Dujon and L. Belcour, *in* "Mobile DNA" (D. E. Berg and M. M. Howe, eds.), p. 861. ASM Press, Washington, D.C. 1989.

using methods of forward genetics;[17–19] studies of truncated versions of maturase proteins (e.g., most examples in Ref. 21) do not provide much insight to the structure or mechanism of action of the protein.

Genetic Analysis of Splicing Signals

Cis-dominant point mutants in group I introns have proved to be very valuable for understanding splicing mechanisms. Such mutants have defined specific residues as essential for splicing *in vivo*. Simultaneously, studies of site-directed mutants (generally of group I introns from other organisms) showed that most of those same sites are also needed for *in vitro* self-splicing. Yeast mutants have been reported affecting the internal guide sequence and parts of the 5' exon with which it base-pairs (both arms of P1), and both arms of P3 and P7. The effects of cis-dominant mutations of maturase-dependent introns are known only *in vivo* as those introns do not self-splice *in vitro*. There are a few cases where the consequences of specific changes are known both *in vivo* and *in vitro;* these include two mutations of bI5[40] and two of aI3.[41] The only group I introns for which a given mutation can be analyzed easily under both situations are the T-even coliphage introns.[42]

A few cis-dominant mutants are available in each of the four group II introns[20,24,32,43–45]; these are quite useful since all four introns self-splice so that effects of sequence changes can be assayed under protein-dependent (*in vivo*) and protein-independent (*in vitro*) conditions. Interestingly, while virtually all *in vivo* mutants of group I introns demonstrate participation of conserved sequences/structures, many of the group II mutations identify intron sequences that are not highly conserved. Several group II mutants, though having tight splicing defects *in vivo*,[45] are quite reactive *in vitro*, at least under some reaction conditions. Others represent the more expected situation where splicing is blocked under both assay conditions.[44]

Revertants and Pseudorevertants of Mitochondrial Mutants

While descriptive studies of mitochondrial mutants have yielded much useful information about splicing, detailed genetic analyses of revertants

---

[40] A. Gampel and A. Tzagoloff, *Mol. Cell. Biol.* **7**, 2545 (1987).
[41] P. Hoffman, Ph.D. dissertation, The Ohio State University, Columbus, Ohio, 1989.
[42] M. Belfort, K. Ehrenman, and P. S. Chandry, this volume.
[43] C. Schmelzer, C. Schmidt, K. May, and R. J. Schweyen, *EMBO J.* **2**, 2047 (1983).
[44] C. Schmelzer and R. J. Schweyen, *Cell* **46**, 557 (1986).
[45] S. Belcher, R. Dietrich, and P. S. Perlman, in preparation.

have been equally valuable, and sometimes essential for defining the sequence change responsible for loss of splicing in individual mutant strains. As outlined in Fig. 2, a number of pathways are possible for reversion, most of which involve suppression of the retained primary defect. All of those modes of reversion have been demonstrated, and this section discusses each.

Backmutations

A peculiar feature of the use of $MnCl_2$ to mutagenize mtDNA *in vivo*[1] is the finding that many mutants so induced contain multiple mutations, all or most of which are clustered in a short region within or near the map location of the mutation. In general, organisms having multiple mutations have complex phenotypes that can be difficult, if not impossible, to dissect. Yet, in our experience, most such mutants in yeast mtDNA readily yield spontaneous revertants, suggesting that just one of the changes is responsible for the *in vivo* phenotype. The ability to map the site of the base change(s) responsible for the gly$^-$ phenotype quite accurately prior to DNA sequence analysis often provides a partial solution to this problem: sequence changes outside of the mapped area of the gene or intron can be disregarded.

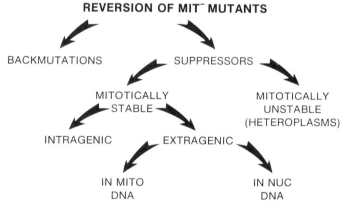

FIG. 2. Different types of revertants of mitochondrial splicing-defective mutants. All splicing-defective mutations of yeast mtDNA produce the same growth defect, loss of ability to grow on glycerol media without affecting growth on glucose medium. Derivatives of mutant strains that regain full or partial respiratory ability result from a number of different processes, here indicated diagrammatically. With the exception of backmutations, all of the other classes are "pseudorevertants," most of which still retain the original mutant site. The exception is a special class of mitotically stable, intragenic suppressors in which the mutant intron is excised from the genome, thus avoiding the splicing block. Examples of that case and all of the other classes are described in the text.

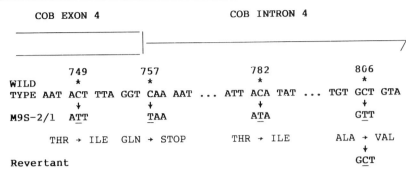

FIG. 3. Full revertants resolve a complex mutation. A portion of the *cob* gene is shown, including the 5' splice junction of bI4 and portions of the upstream exon and the intron. Part of the wild-type base sequence is shown with nucleotide positions noted to correspond with the published DNA sequence.[8] Mutant M9S-2/1 blocks splicing of bI4 and contains four C → T point mutations in the region shown.[21] Despite having a complex change of the wild-type sequence, the mutant strain yields spontaneous revertants that grow as well on glycerol as does the parent strain. Several such "full" revertants were sequenced and found to have regained the wild-type sequence at position 806 while retaining the mutant sequence at the other three sites.[46] Note that the mutation at position 757 alters the first intron base, creating a stop codon; since this mutant expresses the *coxI* gene,[21] it appears that chain termination at the site is leaky so that enough maturase to splice aI4 is made but bI4 fails to splice because of the cis-dominant mutation at position 806.

In several cases, analysis of revertants of multiple point mutants has identified an unexpected sequence as involved in splicing. For example, mutant M9S-2/1 contains four point mutations near the 5' splice junction of bI4 (see Fig. 3).[21,46] When such data were first obtained, we hypothesized that the alteration of the first intron base would prove to be the culprit; however, a number of revertants having complete restoration of splicing function all proved to have regained the wild-type sequence at the most downstream mutant site while remaining mutant at the other three sites.[46] That analysis defined the first example of a mutation in what is now known as the 5' arm of P3.

Mitochondrial Second-Site Suppressors

Revertant analysis is not, of course, limited to backmutations. Second-site suppressors (present in "pseudorevertant strains") are possible and, given the very powerful selection used (regaining the ability to grow on glycerol medium), are not difficult to obtain. Typically, suppressors do not restore full function; revertants with temperature-sensitive or slowed

---

[46] M. L. Haldi, Ph.D. dissertation, The Ohio State University, Columbus, Ohio, 1985.

growth on glycerol media are readily identified and are often used to avoid characterizing repeated examples of backmutations.

Revertants with conditional or partial restoration of glycerol growth are quite valuable since the location of the second mutation can be characterized genetically. For example, mating a second-site revertant strain to a wild-type one will yield mutant segregants at a frequency related to the distance between the original mutant site and the suppressor mutation. Since in this system mutations only 100 bp apart yield about 1% recombinants, this test is very sensitive. Second-site suppressors located elsewhere in the genome are detected quite readily using this test as it yields at least 10% mutant progeny.

In our experience, some suppressor mutations permit "normal" growth on glycerol yet inefficiently restore splicing.[45] These can be distinguished from backmutants by scoring blots of RNA samples from revertant strains for increased levels of precursor RNAs relative to control samples. If the initial mutation provides a restriction site polymorphism, second-site revertants having the same growth characteristics as true revertants can be detected using DNA blots.[46] Related to this, many intron mutants yield revertants with wild-type growth characteristics by cleanly deleting the mutant intron[47] (see below).

The first two examples of intragenic second-site suppression of intron mutations illustrate the value of this approach. Schweyen's group obtained second-site revertants of bI4 mutants affecting sequences known as E (3' arm of P3)[48] and S (3' arm of P7).[49] The splicing defect of the former was restored by a mutation in the 5' arm of P3 (E') providing strong genetic evidence that the E and E' sequences, clearly capable of base pairing, probably do so and that their pairing helps to define whether the precursor RNA is spliceable *in vivo*. Interestingly, the E' mutation defined by our studies of mutant M9S-2/1 is the same change there that suppresses the E mutation[21]; thus, it is quite clear that both single changes block splicing *in vivo* and that combining the compensating single mutations restores splicing. Similarly, a change in the 5' arm of P7 (R) compensated for the mutation in the S element. Subsequent *in vitro* studies using site-directed mutations of those elements in self-splicing introns indicate

[47] C. Jacq, P. Pajot, J. Lazowska, G. Dujardin, M. Claisse, O. Groudinsky, H. de la Salle, C. Grandchamp, M. Labouesse, A. Gargouri, B. Guiard, A. Spyridakis, M. Dreyfus, and P. P. Slonimski, *in* "Mitochondrial Genes" (P. Slonimski, P. Borst, and G. Attardi, eds.), p. 155. Cold Spring Harbor Laboratory, Cold Spring Harbor, New York, 1982.

[48] B. Weiss-Brummer, J. Holl, R. J. Schweyen, J. Rodel, and F. Kaudewitz, *Cell* **33**, 195 (1983).

[49] J. Holl, G. Rodel, and R. J. Schweyen, *EMBO J.* **4**, 2081 (1985).

that those pairings are important for at least some other members of the group.[14,50]

Suppressors of mitochondrial splicing defects located in other genes of mtDNA are also possible, and a few cases have been reported. The best characterized is the MIM2 mutation in which a missense mutation in the reading frame of aI4 suppresses the absence of the bI4 maturase owing to a chain termination mutation in the bI4 ORF.[51] In that case it appears that the single amino acid change in the aI4-encoded protein activates its latent maturase function. Note that the wild-type form of the protein is a potent recombinogenic endonuclease and that the mim2 form of it retains endonuclease function.[31] Some other extragenic suppressors are mutations in rRNA genes that evidently affect the fidelity of translation sufficiently to suppress nonsense or frameshift mutations.[52,53]

Unstable Pseudorevertants

Suppression of maturase deficiencies is quite readily detected. It is clear that very little of each maturase protein is made in wild-type cells[29]; it is also becoming evident that the amount of cyt *b* and coxI protein is not limiting for growth in the wild-type.[54] So, even a very inefficient suppressor may provide enough maturase function to permit sufficient splicing for robust growth on glycerol medium.

Some quite unusual types of maturase suppressors are now known. Our laboratory has characterized two examples of unstable pseudorevertants of maturaseless mutants.[55] In one, a frameshift mutant of bI4 is suppressed by petite genomes containing in-frame fusions of the upstream part of bI4 to the downstream part of aI4. Genomes containing such chimeric introns cannot recombine with the original mutant genome to restore splicing function, but they suppress the defect evidently by encoding a hybrid protein that has at least some maturase function. Only cells that contain both defective genomes can grow on glycerol, and standard features of mtDNA segregation render such heteroplasmic cells vegetatively unstable. Provided that revertant cells are maintained on glycerol media however, such strains can be maintained.

The petite genome that encodes the novel maturase can be obtained in homoplasmic segregants and can be used to reconstruct the original hetero-

[50] J. M. Burke, *Gene* **73**, 273 (1988).
[51] G. Dujardin, C. Jacq, and P. Slonimski, *Nature (London)* **298**, 628 (1982).
[52] G. Dujardin, P. Pajot, O. Groudinsky, and P. Slonimski, *Mol. Gen. Genet.* **179**, 469 (1980).
[53] O. Groudinsky, G. Dujardin, and P. Slonimski, *Mol. Gen. Genet.* **184**, 493 (1981).
[54] H. Conrad-Webb, P. Perlman, H. Zhu, and R. A. Butow, submitted for publication (1990).
[55] P. Q. Anziano, J. Moran, D. Gerber, and P. Perlman, submitted for publication (1990).

plasm by mating the petite strain to the original mutant strain. Given considerable current interest in the proteins encoded by wild-type forms of aI4 and bI4, these proteins encoded by parts of both intron ORFs are potentially quite useful.

Nuclear Suppressors of Mitochondrial Mutations

In principle, selection for restoration of growth on glycerol can yield second-site revertants having suppressor mutations in nuclear DNA. Dujardin *et al.* first reported the existence of a class of mutants, so called *NAM* mutants for nuclear accommodation of mitochondria,[52] and since then they and others have described a number of examples in detail, some of which involve splicing.

The best characterized is the *NAM2* gene[52,53,56,57]; mutations there are dominant suppressors of bI4 maturaseless mutants. While most of the mutant alleles which are suppressed are chain terminations, it is quite clear that the mutant *NAM2* protein does not alter the expression of the bI4 reading frame. Instead, it requires the integrity of the aI4 ORF for suppression and appears to "activate" the latent maturase function of the protein encoded by that protein. As noted above, the *NAM2* gene encodes an essential[56] tRNA sythetase activity[57,58] that may be the yeast cognate of the *CYT-18* gene of *Neurospora crassa.*

Mitochondrial Reverse Genetics

Until recently, studies of splicing in yeast mitochondria have been limited by the inability to test specific hypothesis using reverse genetics. Whatever mutants could be obtained could be analyzed quite thoroughly using approaches described in this chapter and previously[1]; however, intron sequences not yet represented in the mutant set could not be studied *in vivo* at all. Related to this, the demonstration that four group I introns [40,59–61] and four group II introns[44,62–64] of yeast mtDNA self-splice *in vitro* set the stage for rapid progress studying cis-acting intron se-

[56] M. Labouesse, G. Dujardin, and P. P. Slonimski, *Cell* **41**, 133 (1985).
[57] C. J. Herbert, M. Labouesse, G. Dujardin, and P. P. Slonimski, *EMBO J.* **7**, 473 (1988).
[58] A. Tzagoloff, A. Akai, M. Kurkulos, and B. Repetto, *J. Biol. Chem.* **263**, 850 (1988).
[59] G. van der Horst and H. Tabak, *Cell* **40**, 759 (1985).
[60] H. Tabak, G. van der Horst, A. Kamps, and A. Arnberg, *Cell* **48**, 101 (1987).
[61] S. Partono and A. Lewin, *Mol. Cell. Biol.* **8**, 2562 (1988).
[62] C. Peebles, P. Perlman, K. Mecklenburg, M. Petrillo, J. Tabor, K. Jarrell, and H. L. Cheng, *Cell* **44**, 213 (1986).
[63] R. van der Veen, A. C. Arnberg, G. van der Horst, L. Bonen, H. F. Tabak, and L. A. Grivell, *Cell* **44**, 225 (1986).
[64] S. K. Hebbar, S. Belcher, and P. S. Perlman, submitted for publication (1990).

quences *in vitro*. But, for the self-splicing introns few *in vivo* cis-acting mutations have been reported.

So, it has become quite a pressing need to be able to analyze site-directed mutants with known *in vitro* consequences under *in vivo* conditions. Fortunately, just as that need was becoming acute, the laboratories of Ron Butow and Tom Fox independently demonstrated that cloned copies of yeast mitochondrial DNA sequences can be transformed into the organelles. Both used an unusual DNA delivery system developed initially to transform various types of plant nuclear genomes.[65] The system involves coating tungsten microspheres with appropriate DNA and literally shooting those projectiles at living yeast cells.

Butow's group reported a novel selection system in which the recipient yeast cells contain a nonreverting partial deletion of two introns of the *coxI* gene and the transforming DNA contains a cDNA version of that region of the gene.[66] Cells are transformed with a mixture of plasmids, one containing the yeast cDNA and the other a selectable nuclear marker. Nuclear transformants are selected on appropriate media and mitochondrial cotransformants counterselected for regaining the ability to grow on glycerol media. While the reported frequency of such transformation events was low, by yeast nuclear transformation standards, it was clear that they had achieved this important technical breakthrough. Simultaneously, it was reported that the same basic approach permits transformation of chloroplast genes in *Chlamydomonas*;[67] that system also contains group I and group II introns, and the genetic analysis of splicing processes there should also benefit from this development. The transformants obtained by Johnston *et al.* were, in effect, transplacements, in which the incoming DNA replaced the host mutation, presumably by a double cross-over or gene conversion event; vector sequences were not incorporated into mtDNA, nor were they retained as separate DNA molecules in those experiments.

Fox's group demonstrated a different assay system though using the same technology.[68] They used a plasmid containing the entire *oxiI* gene (encoding subunit II of cytochrome-*c* oxidase) as the donor sequence; a plasmid containing a selectable nuclear marker gene was also present. In

[65] T. M. Klein, E. D. Wolf, R. Wu, and J. C. Sanford, *Nature (London)* **327**, 70 (1987).
[66] S. A. Johnston, P. Q. Anziano, K. Shark, J. C. Sanford, and R. A. Butow, *Science* **240**, 1538 (1988).
[67] J. E. Boynton, N. W. Gillham, E. H. Harris, J. P. Hosler, A. M. Johnson, A. R. Jones, B. L. Randolph-Anderson, D. Robertson, T. M. Klein, K. B. Shark, and J. C. Sanford, *Science* **240**, 1534 (1988).
[68] T. D. Fox, J. C. Sanford, and T. W. McMullin, *Proc. Natl. Acad. Sci. U.S.A.* **85**, 7288 (1988).

one experiment they used an *oxi1* mutant strain as recipient and carried out the same type of two-step selection of transformants. In another type of experiment, they used a petite strain totally lacking mitochondrial DNA sequences as recipient. They isolated cells transformed for the nuclear marker and then obtained the desired mitochondrial transformants by screening for individuals that had acquired the ability to recombine with mitochondrial mutants in the *oxi1* gene to yield glycerol-growing progeny. Transformed cells are rho⁻ mutants with novel mtDNAs, containing bona fide mtDNA sequences joined to vector sequences. Yet, those genomes can engage in recombination in crosses so that transformed sequences can be replaced into an otherwise intact mtDNA.

The complementation approach permits introduction of manipulated DNA sequences into mitochondria so long as they are compatible with respiratory growth. The second approach, as reported, is more general since mutant versions of mitochondrial genes can be transformed into cells without regard for phenotypic consequences. The mutations can be placed in an otherwise rho⁺ configuration by simple recombination. As published, both approaches are quite inefficient; it is likely that improved transformation and selection/screening methods will be reported soon.

Nuclear Mutants Affecting Mitochondrial Splicing

Direct Screening for Nuclear Genes Involved in Mitochondrial Splicing

pet⁻ mutants are valuable genetic resources for studying mitochondrial processes. Basic strategies for obtaining and analyzing such mutants were described previously.[1] *pet⁻* mutants are usually recessive, respiration-deficient mutants, viable on fermentable media but unable to grow on nonfermentable ones. It has been estimated that at least 20% of yeast nuclear genes are involved in mitochondrial biogenesis and/or function, and the vast majority of those genes can be disrupted (at least individually) without affecting viability on fermentable carbon sources. Since few yeast genes are present in multiple copies, the use of *pet* mutants makes *most* mitochondrial processess amenable to genetic dissection.

There are some exceptions. A few genes have been identified that encode protein(s) with functions in the mitochondrion and in another cell compartment. The gene encoding histidine tRNA synthetase encodes both the mitochondrial and cytoplasmic enzyme so that a null mutation blocks even growth on glucose.[69] Several genes encoding enzymes that

---

[69] G. Natsoulis, F. Hilger, and G. Fink, *Cell* **46**, 235 (1986).

modify nucleotides in mitochondrial and cytoplasmic tRNAs are nonessential so that they are not present in *pet* mutant collections.[70]

In the past decade, a number of groups have screened sets of *pet* mutants to define genes involved in mitochondrial RNA processing.[71–73] For example, Tzagoloff's group isolated a large set of *pet⁻* mutants and screened representatives of each complementation group, using RNA blots, for alterations in splicing of *cob* gene transcripts.[71] From that undertaking one clear-cut example of a nuclear gene needed for mitochondrial splicing was obtained, the *CBP2* gene.[74] The original mutation defining that gene and subsequent disruption mutations obtained by reverse genetics block splicing of the terminal intron of the *cob* gene [bI5 of Fig. 1, but called bI2 when dealing with strains containing only two introns (bI4 and bI5) in that gene]. Strains having the *cbp2* mutation but lacking bI5 do not have a respiration defect so that it is clear that the *cbp2* gene product is intron specific (see below).

That mutant screen also yielded mutants lowering *cob* mRNA content for other reasons. The *cbp1* mutation destabilizes the *cob* mRNA but has no effect on splicing;[71] it appears to recognize some feature of the untranslated 5′ end of the mRNA.[75] A number of other complementation groups were identified that affect bI4 (or bI2, in strains having only two *cob* introns) splicing indirectly.[76] Those genes encode mitochondria-specific tRNA synthetases[77,78]; leaky mutations in them deplete mitochondria for some proteins more than others, presumably owing to differences in codon utilization between genes and, so, appear to block maturase synthesis. Tight mutations in those genes (e.g., disruptions) abolish mitochondrial translation and result in the loss of wild-type mtDNA.[56,79]

Studies of temperature-sensitive alleles of the nuclear *CYT-18* gene of *Neurospora crassa* showed that the wild-type gene product is required

[70] A. K. Hopper, A. H. Furuhawa, H. D. Pham, and N. C. Martin, *Cell* **28**, 543 (1982).
[71] C. Dieckmann, L. Pape, and A. Tzagoloff, *Proc. Natl. Acad. Sci. U.S.A.* **79**, 1805 (1982).
[72] T. Pillar, B. F. Lang, I. Steinberger, B. Vogt, and F. Kaudewitz, *J. Biol. Chem.* **258**, 7954 (1983).
[73] J. E. McEwen, C. Ko, B. Kloeckener-Gruissem, and R. O. Poyton, *J. Biol. Chem.* **261**, 11872 (1986).
[74] P. McGraw and A. Tzagoloff, *J. Biol. Chem.* **258**, 9459 (1983).
[75] C. L. Dieckmann, T. J. Koerner, and A. Tzagoloff, *J. Biol. Chem.* **259**, 4722 (1984).
[76] C. L. Dieckmann, S. G. Bonitz, J. Hill, G. Homison, P. McGraw, L. Pape, B. E. Thalenfeld, and A. Tzagoloff, *in* "Mitochondrial Genes" (P. Slonimski, P. Borst, and G. Attardi, eds.), p. 213. Cold Spring Harbor Laboratory, Cold Spring Harbor, New York, 1982.
[77] L. K. Pape, T. J. Koerner, and A. Tzagoloff, *J. Biol. Chem.* **260**, 15362 (1985).
[78] A. M. Myers and A. Tzagoloff, *J. Biol Chem.* **260**, 15371 (1985).
[79] A. M. Myers, L. K. Pape, and A. Tzagoloff, *EMBO J.* **4**, 2087 (1985).

for splicing the intron of the large rRNA gene[80] (and some others) and revealed the unexpected involvement of a tRNA synthetase in group I intron splicing. The *CYT-18* gene encodes the mitochondrial tyrosine tRNA synthetase that also participates directly in splicing[81,82]; the latter activity is evident from *in vivo* studies and has also been assayed directly *in vitro*. It is possible that one or more yeast synthetases might play a role in splicing beyond being essential for translation of maturase mRNAs (see above).

A New Screen for Nuclear Mutants Affecting Mitochondrial Splicing

While many hundreds of yeast nuclear mutants have been screened directly for splicing defects, at most a few genes having a direct role in splicing have been identified. Recently a new strategy for screening mutants has been developed that should have a real impact on this issue.[13] It takes advantage of the optional nature of most (now all) mitochondrial introns.

This approach begins with the construction of a library of *pet*⁻ mutants using a strain whose mtDNA contains some or all of the known introns. Based on previous studies of such mutant collections, only a subset of the mutants is expected to affect splicing, and some of those probably do so indirectly. The basic idea is to replace the intron-containing mtDNA in each candidate strain with one lacking introns using established cytoduction methods.[1] These involve taking advantage of *kar1* mutants that block nuclear fusion in zygotes so that mating creates a transient heterokaryon; if intron-containing mtDNA is first eliminated from a *pet* mutant and that petite strain then mated to a *kar1* strain with mtDNA lacking introns, segregants (cytoductants) from those zygotes having the *pet* mutation and the intronless mtDNA are formed. Any cytoductant that regains the ability to grow on a nonfermentable medium should contain a nuclear mutation that is "suppressed" by the absence of introns. In principle, this strategy could be carried out to screen for nuclear genes needed for splicing (or some other form of accommodation of the presence of the intron) any given intron or all of them, depending on the intron configuration of the original and *kar1* tester strains used.

[80] H. Bertrand, P. Bridge, R. A. Collins, G. Garriga, and A. M. Lambowitz, *Cell* **29,** 517 (1982).
[81] R. A. Akins and A. M. Lambowitz, *Cell* **50,** 331 (1987).
[82] A. L. Majumder, R. A. Akins, J. G. Wilkinson, R. L. Kelley, A. J. Snook, and A. M. Lambowitz, *Mol. Cell. Biol.* **9,** 2089 (1989).

Recently, the most general form of this experiment was reported by Seraphin *et al.*[13] It was necessary to first construct a respiration-sufficient strain with mtDNA lacking all of the introns. Those workers took advantage of earlier findings in Slonimski's laboratory that some intron mutants revert by losing the mutant intron from mtDNA[47]; remarkably, many such events result in the simultaneous loss of the mutant intron and one or more nearby introns.[13,47,83] In this way it was possible to construct intronless forms of the *cob*[12] and *coxI*[13] genes in several steps; intronless forms of the 21 S rRNA gene occur naturally and are readily available. Importantly, each intronless gene is functionally equivalent to the intron-containing form, at least under certain conditions that were met.

Combining the three intronless genes into a single mitochondrial genome was a straightforward application of mitochondrial genetic principles[13]; Ref. 1 discusses methods for strain constructions involving multiple mutations.[1] It was found that the intronless genome is compatible with respiratory growth, thus completing the setup of the basic screen.

A surprisingly large fraction of the *pet*⁻ mutants screened by Seraphin *et al.* were suppressed by the intronless mtDNA.[13] It was estimated that there are about 18 genes devoted to mitochondrial splicing. Two proved to be allelic to known mutations affecting RNA splicing, thus validating the basic approach. One of those, defining a new allele of the *MSS116* gene, provides a first insight to a potentially important class of splicing functions.[84] Mutants in the gene block, or greatly reduce, the splicing of four introns, including representatives of both types. The gene encodes a protein with homology to several types of known proteins, perhaps most importantly to DNA helicases. It was suggested that *MSS116* encodes an RNA helicase involved in splicing and possibly other processes in mitochondria.

The basic approach, in various versions noted above, should suffice to identify a significant set of nuclear genes whose products are involved in splicing in mitochondria and have no essential nuclear function. It was noted above that some genes will not be present in sets of *pet* mutants so that such genes will not be detected efficiently using the above screening method. Collections of conditional *pet*⁻ mutants exist and could be used to screen for such genes. Alternatively, second-site suppressors of missense alleles of nuclear splicing genes may identify genes encoding multifunctional proteins needed for splicing in mitochondria.

Since a number of nuclear genes are involved in splicing mitochondrial introns, mention of a special feature of the yeast mitodchondrial system

[83] A. Gargouri, J. Lazowska, and P. Slonimski, *in* "Mitochondria, 1983" (R. Schweyen, K. Wolf, and F. Kaudewitz, eds.), p. 259. de Gruyter, Berlin, 1983.

[84] B. Seraphin, M. Simon, A. Boulet, and G. Faye, *Nature (London)* **337,** 84 (1989).

germane to defining the splicing defect of each mutant is in order. Because, as noted above, blocking splicing of the first intron of the *cob* and *coxI* genes has a polar effect on downstream intron splicing, any *pet* mutant blocking splicing of either of those upstream introns will necessarily block splicing of other introns. Distinguishing between direct involvement of the *pet* gene in splicing each affected intron and polar effects on splicing can be accomplished by taking advantage of the optional nature of mitochondrial introns. Some alternate forms of the two genes have been found in other *Saccharomyces* species, and others have been constructed using reversion by intron excision. Cytoduction of mtDNAs having several combinations of introns into *mss116* mutant cells helped define which introns need the *MSS116* gene product for splicing and which are blocked in the original isolate for secondary reasons.[84] Such intron configuration variants were used earlier to characterize the phenotypes of other *pet* genes.[75]

Revertants of Nuclear Mutants

Many *pet⁻* mutants regain the ability to grow on glycerol medium spontaneously (or upon mutagen treatment). Such revertants can result from backmutations, intragenic suppressors, and extragenic suppressors. These are readily distinguished by mating each revertant to a *PET⁺* tester strain with wild-type mtDNA and scoring meiotic products for glycerol growth. A reversion event at the site of the original mutation will result in all tetrads being 4 gly⁺/0 gly⁻. In the case of an extragenic suppressor, the original mutant phenotype will reappear among the spores at a frequency indicating its location (on the same or different chromosome). Intragenic second-site suppressors will yield gly⁻ spores but at quite low frequencies since intragenic crossingover or gene conversion events are needed to separate the two linked mutations. So far, nuclear extragenic suppressors of *pet* mutants affecting mitochondrial splicing have not been reported. However, other complex nuclear-encoded processes have benefitted greatly from some genetic studies.[85–87] Available alleles of *cbp2* and other genes involved in splicing may not be suitable for suppressor studies, but other alleles (e.g., missense mutants yielding temperature-sensitive, cold-sensitive, slow-growing phenotypes) can be found or made by reverse genetics methods quite readily in this system.

[85] J. Jarvik and D. Botstein, *Proc. Natl. Acad. Sci. U.S.A.* **72**, 2738 (1975).
[86] J. H. Thomas, P. Novick, and D. Botstein, *in* "Molecular Biology of the Cytoskeleton" (G. G. Borisy, D. W. Cleveland, and D. B. Murphy, eds.), p. 153. Cold Spring Harbor Laboratory, Cold Spring Harbor, New York, 1984.
[87] C. Weil, C. E. Oakley, and B. R. Oakley, *Mol. Cell. Biol.* **6**, 2963 (1986).

Mutations of mtDNA can suppress *pet* mutations; these are a special class of extragenic suppressors. They are detected genetically by meiotic analysis of a number of diploids of the test cross noted above. Since mitochondrial markers segregate mitotically, those diploids that received the mitochondrial suppressor mutation will yield meiotic data compatible with a backmutation in the nuclear gene; however, those diploids that received the mtDNA from the tester strain lack the suppressor mutation and will yield 2 gly$^+$/2 gly$^-$ spores. Also cytoduction tests can detect revertants strains that depend on the resident mtDNA for ability to grow on glycerol.

Several interesting situations in which a mitochondrial mutation suppresses a nuclear-encoded splicing defect have been analyzed. First, the clean deletion of bI5 suppresses the *cbp2* mutation[88]; this finding provided an independent demonstration that the *CBP2* gene product is specific for splicing the intron. Second, Seraphin *et al.* studied revertants of the a mutant of the *MSS18* gene, apparently involved in splicing aI5β.[89] They found mitochondrial suppressors that restored splicing to aI5β owing to various changes in the exon preceding that intron. While it is generally accepted that the exon bases immediately adjacent to group I intron 5′ splice junctions are essential for splicing (by forming structure P1 with the internal guide sequence), those suppressor mutations were some bases upstream of the sequence implicated in P1 formation for that intron.

Acknowledgments

I thank my students for thoughtful comments on drafts of the manuscript. Research was supported by grants from the National Institutes of Health.

[88] J. Hill, P. McGraw, and A. Tzagoloff, *J. Biol. Chem.* **260**, 3235 (1985).
[89] B. Seraphin, M. Simon, and G. Faye, *EMBO J.* **7**, 1455 (1988).

[45] Self-Splicing and Enzymatic Cleavage of RNA by a Group I Intervening Sequence

By John A. Latham, Arthur J. Zaug, and Thomas R. Cech

Introduction

The nuclear rRNA precursor from *Tetrahymena thermophila,* which contains a 413-nucleotide intervening sequence (IVS), is capable of undergoing self-splicing. In the absence of any protein or external energy source, the IVS catalyzes a series of transesterification reactions that result in its excision from the rRNA precursor and concomitant ligation of the flanking sequences or exons.[1] A number of related group I IVSs have been shown to undergo self-splicing by the same basic pathway.[1] We have exploited the catalytic activity by using the *Tetrahymena* IVS as an enzyme to mediate cleavage–ligation reactions on both small and large RNA substrates; these reactions occur with multiple turnover.[2,3]

One use of the IVS RNA enzyme (ribozyme) is to cleave large RNA molecules in a sequence-specific manner.[2] The IVS-mediated cleavage occurs after the sequence CUCU and related sequences such as CCCU. In the presence of denaturants such as urea or formamide, the specificity of cutting increases.[2] There is no loss in activity for the cognate sequence CUCU using up to 2.5 *M* urea; substantial decrease in cleavage is observed with other sites. When the substrate binding site on the IVS is altered using site-specific mutagenesis, there is a corresponding change in the sequence specificity of cleavage. Thus, a set of ribozymes can be synthesized to cleave a variety of substrate sequences, making available the capacity to cleave selectively large RNA molecules with specificity approaching that of DNA restriction endonucleases.

Sequence-specific cleavage of large RNA molecules is a highly desirable tool. The highest level of specificity until recently has been at the nucleotide level using base-specific endoribonucleases (e.g., RNase T1 cutting after G, RNase U2 cutting after A). Schultz and co-workers have broadened this option by making use of an oligonucleotide–enzyme hybrid of staphylococcal nuclease to exhibit sequence recognition.[4] The cleavage location and specificity are controlled by the oligonucleotide sequence coupled to the nuclease. RNase H is a second option in this line, making use of RNA–DNA hybrid recognition. In both cases, extensive knowledge of flanking sequences is required for specific recognition and construction.

In this chapter we describe a number of procedures for studying the self-splicing *Tetrahymena* pre-rRNA that are useful for studying the entire set of self-splicing group I IVSs. In addition, we describe the method for using the *Tetrahymena* ribozyme as a sequence-specific endoribonuclease. Most of these techniques were adopted from those published by others and have undergone modification to enhance their application to problems under study in our laboratory.

[1] T. R. Cech, *Cell* **44**, 207 (1986).
[2] A. J. Zaug, M. Been, and T. R. Cech, *Nature (London)* **324**, 429 (1986).
[3] A. J. Zaug and T. R. Cech, *Science* **231**, 470 (1986).
[4] R. N. Zuckermann, D. R. Corey, and P. G. Schultz, *J. Am. Chem. Soc.* **110**, 1614 (1988).

DNA Templates

Two types of DNA templates are used to produce large quantities of RNA using phage T7 RNA polymerase. In the case of smaller RNA species up to tRNA in size, a system using synthetic DNA containing a double-stranded T7 promoter and adjacent single-stranded coding sequence is used (see this series, Vol. 180, p. 51). The second type of template is linearized plasmid DNA where the T7 promoter is adjacent to the sequence of interest. Commonly, the latter system is used to synthesize RNA greater than 80 nucleotides in length. Plasmids pBGST7 and pTT1A3 have been particularly useful to study self-splicing of the *Tetrahymena* pre-rRNA.[5,6] Each construct contains the coding region for the IVS and portions of the flanking exon sequences. pT7L-21 is a construct that is used to synthesize large quantities of the *Tetrahymena* ribozyme for use as a sequence-specific endoribonuclease.[7]

RNA Synthesis

Large quantities of homogeneous RNA can be prepared using the phage T7 RNA polymerase transcription system. T7 RNA polymerase is isolated using the procedure of Davanloo *et al.*[8] Reaction volumes can range from 20 μl (used to prepare small quantities of RNA with high specific radioactivity) to 4 ml. Under optimal conditions as many as 700 full-length RNA molecules per template DNA molecule have been obtained.

Nucleotides

Individual stock solutions of each nucleoside triphosphate (Pharmacia) are prepared in water. One hundred milligrams of each is dissolved in 3 ml of water, and the pH is adjusted to 7 using NaOH. (The pH is monitored using pH paper.) The final concentration of each nucleotide is then determined spectrophotometrically. Final solution concentrations range from 60 to 70 mM. A 10× nucleotide mix is prepared by dilution of the initial stock solutions to 10 mM in each nucleotide using water. This is stored at −20°. If high specific activity ^{32}P-labeled RNA is required, the corresponding unlabeled nucleotide is lowered to 1.25 mM in this mix.

[5] M. Been and T. R. Cech, *Cell* **47**, 207 (1986).
[6] J. V. Price and T. R. Cech, *Science* **228**, 719 (1985).
[7] A. J. Zaug, C. Grosshans, and T. R. Cech, *Biochemistry* **27**, 8924 (1988).
[8] P. Davanloo, A. H. Rosenberg, J. J. Dunn, and F. W. Studier, *Proc. Natl. Acad. Sci. U.S.A.* **81**, 2035 (1984).

Transcription

For a 1-ml transcription reaction, 100 μl of 10× transcription buffer [400 mM tris(hydroxymethyl)aminomethane hydrochloride (Tris-HCl) (pH 7.5), 120 mM MgCl$_2$, 100 mM dithiothreitol (DTT), 20 mM spermidine], 100 μl of 10× nucleotide mix, and linearized template DNA (10–50 μg/ml depending on the template) are pooled and brought up to the corrected volume, all performed on ice. RNA synthesis is initiated by the addition of T7 polymerase (100–500 U/μg DNA), and the reaction proceeds at 37° for 2 hr.

RNA Purification

Small-scale transcription solutions are directly loaded onto a polyacrylamide gel to isolate the RNA of interest. For RNA transcripts greater than 200 bases in length, a 4% polyacrylamide (29 : 1 acrylamide–bisacrylamide)–8 M urea gel and TBE buffer [0.1 M Tris base, 83 mM boric acid, 1 mM ethylenediaminetetraacetic acid (EDTA) (pH 8.0–8.3)] are used. For small RNA molecule purification, higher percentage (8–20%) polyacrylamide gels are required in order to resolve the transcription products.

Large-scale transcriptions must first be precipitated. One-third the reaction volume of 6 M ammonium acetate is added to the transcription solution. The use of this salt minimizes the precipitation of mononucleotides and small oligonucleotides. Three volumes of absolute ethanol are added to the solution, which is mixed and placed at −70° for 30 min. The nucleic acid is pelleted by centrifugation at 10,000 rpm at 4° in a swinging-bucket rotor (e.g., Beckman JS-13 or Sorvall HB4), and the resulting pellet is dried *in vacuo*.

To prepare the sample for gel purification, the pellet is resuspended in 100–200 μl of water, and an equal volume of loading buffer (0.22 g urea in 200 μl 0.1× TBE: 0.01% bromphenol blue, 0.025% xylene cyanol, 20% sucrose) is added. The sample is loaded onto the appropriate gel and run. The gel is disassembled and transferred to plastic wrap. Visualization is performed by laying the wrapped gel on top of an intensifying screen (e.g., Du Pont Cronex lightning plus) and UV shadowing (254 nm). (The fluorescence of the screen is quenched by the nucleic acid.) Exposure to UV radiation should be minimized to avoid damage to the RNA. The band of interest is a dark region in a bright background. It is cut out using a flamed razor blade.

The gel slice is frozen at −70°. While thawing the acrylamide is pulverized using a sterile, siliconized glass rod. One to two milliliters of TE [10 mM Tris-HCl (pH 7.5), 1 mM EDTA] with 0.25 M NaCl is then added and the solution refrozen. On thawing the RNA is extracted at 4° overnight in the case of large RNA products. In the case of RNA small species, extraction may be performed at room temperature for the same period of time. At the end of the extraction period, the acrylamide is pelleted via centrifugation. The resulting supernatant is removed, and the RNA is precipitated overnight at −20° with 3 volumes of absolute ethanol.

If further purification is required, the RNA can be applied to a gel filtration column using the appropriate Sephadex support needed to place the product RNA at the void volume. For large RNA molecules (>100 nucleotides), Sephadex G-50 or G-100 is routinely used. The column is prepared by plugging the end of a 10-ml serological pipet (Falcon) with siliconized glass wool. Resin equilibrated in TE with 0.25 M NaCl is packed to 10 ml total volume. The RNA is loaded onto this column, eluted, and the fractions analyzed. Fractions containing RNA are pooled and the RNA precipitated according to the format described above.

End Labeling of Tetrahymena IVS

A number of different forms of the IVS are readily end-labeled. End-labeled RNA has been prepared for use in studies of RNA structure and reaction kinetics.[2,9,10] Radiolabel incorporation is accomplished either by self-processing or by use of commercially available enzymes.

5'-End-Labeling of L IVS

The full-length linear IVS RNA is readily 5'-end-labeled by splicing of cold RNA precursor (prepared using the T7 polymerase system described above) with [α-^{32}P]GTP. The reaction is performed with only a fraction of the precursor synthesized; the remaining RNA is routinely stored at −20° until it is required.

A typical experiment is described below (all additions are done while on ice): 156 pmol [α-^{32}P]GTP [800 Ci/mmol, 12.5 μl; New England Nuclear (NEN)] is dried down in a 1.5-ml Eppendorf tube using a Savant Speedvac. Then 12 μl precursor RNA (241 pmol) is added to the tube. Five microliters of 5× splicing buffer [250 mM Tris (pH 7.5), 25 mM MgCl$_2$, 2000 mM NaCl] is added and the reaction brought to 25 μl total volume with sterile distilled water. After the splicing reaction is run at 30° for 30 min, 3 μl of 0.25 M EDTA is added to stop the reaction. The

---

[9] N. K. Tanner and T. R. Cech, *Nucleic Acids Res.* **13**, 7759 (1985).

[10] T. R. Cech, N. K. Tanner, I. Tinoco, Jr., B. R. Weir, M. Zuker, and P. S. Perlman, *Proc. Natl. Acad. Sci. U.S.A.* **80**, 3903 (1983).

mixture is diluted with an equal volume of acrylamide gel loading buffer and heated to 65° for 2 min. The RNA is then applied to a 4% polyacrylamide–8 M urea gel, and the xylene cyanol (XC) dye front is run to the end of the gel. Labeled L IVS is visualized by autoradiography. The band is cut out with a flamed razor and frozen on dry ice. The gel is pulverized while thawing using a glass rod (sterile and siliconized). Then 1 ml of TE with 0.25 M NaCl is added, and the mixture is vortexed and refrozen on dry ice. The material is thawed at room temperature, and extraction proceeds at 4° overnight with or without agitation. Acrylamide is pelleted via centrifugation, and the supernatant is removed. The pelleted acrylamide is washed with 0.25 ml of additional TE with 0.25 M NaCl and respun. Pooled supernatants are then precipitated with 3 volumes of ethanol on dry ice for 30 min. The precipitate is collected via centrifugation for 30 min at 10,000 rpm in a cold (4°) swinging-bucket rotor (Beckman JS-13 or Sorvall HB4). The pellet is washed with 70% ethanol (stored at −20°) and recentrifuged under the same conditions. The ethanol is discarded, and the pellet is dried in a vacuum oven for 15 min. The pellet is resuspended in 100 μl of TE, and portions are removed for specific activity determination via spectrophotometry and liquid scintillation counting. Typical specific activity levels are in the range of 700–800 Ci/mmol.

L-21 ScaI RNA Labeling

The *Sca*I form of the IVS is readily 3′- and 5′-end-labeled using commercially available enzymes. 5′-End-labeling is accomplished by removal of the 5′-triphosphate of the transcript with calf intestinal phosphatase (CIP) followed by phosphorylation using T4 polynucleotide kinase and [γ-^{32}P]ATP. 3′-End-labeling is achieved with [α-^{32}P]cordycepin triphosphate (3′-dATP) and poly(A) polymerase.

CIP Treatment of L-21 ScaI RNA

L-21 *Sca*I IVS (270 pmol) is treated with 0.01 U/pmol CIP (NEN) in 50 mM Tris-HCl (pH 8.0), 10 mM MgCl$_2$. Five μl of 5× buffer is mixed with the L-21 *Sca*I RNA and the volume adjusted to 22 μl (stored on ice). Three microliters of 1 U/μl CIP is then added and the solution placed at 37°. At 30 min, a second 3 μl of 1 U/μl CIP is added. After a total of 1 hr the reaction is diluted with 150 μl TE and extracted 3 times with 100 μl of TE-saturated phenol and once with an equal volume of water-saturated ether. The solution is brought to 0.25 M NaCl and precipitated with 3 volumes of ethanol on dry ice for 15 min. The RNA is centrifuged in an

Eppendorf centrifuge (4°) for 15 min, and the resulting pellet is washed with cold ($-20°$) 70% ethanol and recentrifuged for 15 min. The pellet is dried, resuspended in TE, and requantified using UV spectrophotometry. The concentration is determined using the extinction coefficient at 260 nm (ε_{260}) = 3.2 × 10^6 M^{-1} cm^{-1}.

5'-End-Labeling of L-21 ScaI RNA

Standard T4 polynucleotide kinase conditions are used to end-label the CIP-treated RNA. CIP-treated ScaI IVS (38 pmol) is mixed with 1 µl of 10× T4 polynucleotide kinase buffer [500 mM Tris (pH 9), 100 mM MgCl$_2$, 50 mM DTT, 10 mM spermidine], and the volume is adjusted to 8 µl with sterile distilled water. One microliter (21 pmol, of 7000 Ci/mmol, ICN) of [γ-^{32}P]ATP is added. The solution is thoroughly mixed, and the reaction is initiated by the addition of 10 U (1 µl) of T4 polynucleotide kinase (U.S. Biochemicals, Cleveland, OH). The labeling is performed at 37° for 30 min. The reaction is diluted with 15 µl of TE and 25 µl of acrylamide loading buffer. The sample is applied to a 4% polyacrylamide–8 M urea gel which is run until the XC is at the lower edge of the gel. The 5'-end-labeled RNA is located by autoradiography. All subsequent operations are identical to those described above for the isolation of [5'-^{32}P]L IVS.

3'-End-Labeling of L-21 ScaI IVS

[α-^{32}P]3'-dATP (66 pmol) is dried down in a 1.5-ml Eppendorf tube. Two microliters of 5× buffer [250 mM Tris (pH 8.0), 50 mM MgCl$_2$, 1.25 M NaCl] is added. Then 1 µl of MnCl$_2$ (25 mM) and 17.9 pmol of RNA are added, and the volume is adjusted to 9 µl. The components are thoroughly mixed, and the reaction is initiated by the addition of 1 µl (10 U) of poly(A) polymerase (Pharmacia, Piscataway, NJ). The reaction is run at 37° for 30 min. It is then diluted with 10 µl of TE and 20 µl of polyacrylamide gel loading buffer and applied to a 4% polyacrylamide–8 M urea gel. The XC is run to the lower edge of the gel, and the labeled RNA is located via autoradiography. Subsequent operations are identical to the isolation procedures described above.

RNA Processing Reactions

RNA synthesized by T7 transcription of plasmids pBGST7 and pTT1A3 can be used to study self-splicing reactions.[5,6] The excision of the IVS and its subsequent circularization, exon ligation, and site-specific

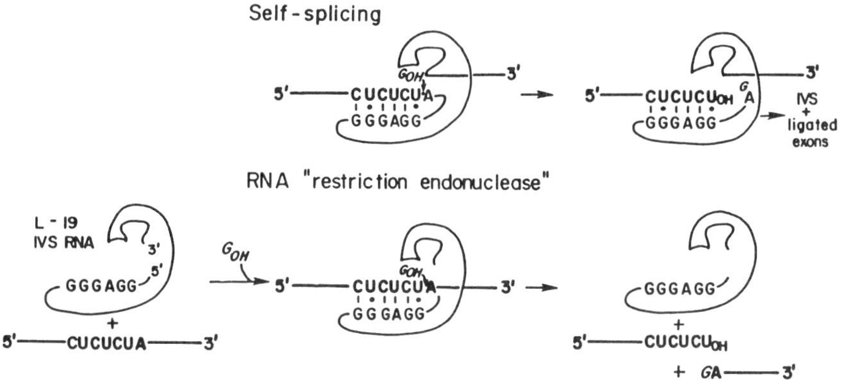

Self - splicing

FIG. 1. Model for the endoribonuclease activity of the L-21 ScaI ribozyme. The mechanism is an intermolecular version of the first step of pre-rRNA self-splicing. Thin letters and lines, IVS; bold letters and thick lines, exon sequences (top) or substrate (bottom); italicized G, free guanosine nucleotide or nucleoside. (Reprinted from Ref. 2 with permission.)

hydrolysis at the splice sites can be analyzed using the appropriate reactions conditions. Standard splicing reactions contain 50 mM N-2-hydroxyethylpiperazine-N'-2-ethanesulfonicacid (HEPES) (pH 7.5), 100 mM $(NH_4)_2SO_4$, 5 mM $MgCl_2$, and 0.1 mM GTP at 30° with 1–50 nM pre-rRNA. Tris buffer can be substituted for the HEPES. The reaction is stopped by the addition of 10 μl of loading buffer, and the solution is applied to a 4% polyacrylamide–8 M urea gel (10 cm wide × 20 cm long). XC is run 10 cm into the gel. The reaction is analyzed by autoradiography in the case of radiolabeled RNA or by UV shadowing when unlabeled RNA is used. To promote circularization, the $MgCl_2$ concentration is increased to 10 mM and the reaction temperature to 42°. Site-specific hydrolysis is studied using circularization conditions but in the absence of GTP; the reaction rate increases with increasing pH.[11]

Sequence-Specific Cleavage of Large RNA

Sequence-specific cleavage is mechanistically similar to the first step in self-splicing (Fig. 1). GTP is a required cofactor, and it becomes covalently attached to the RNA downstream of the cut site during the cleavage reaction. Thus, one cleavage product is automatically 5'-end-labeled when [^{32}P]GTP is used as the cofactor, analogous to the labeling of L IVS during self-splicing described earlier. Mapping the position of cutting in

[11] A. J. Zaug, J. R. Kent, and T. R. Cech, *Biochemistry* **24,** 6211 (1985).

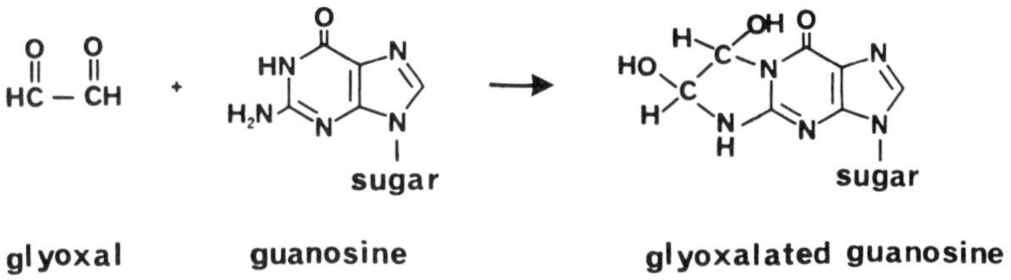

FIG. 2. Glyoxalation of guanosine as described by Carmichael and McMaster.[14] Glyoxal adds to N-1 and exocyclic amine of guanine. The reaction is reversible only at high pH and temperature.

the RNA is then possible using this end-labeled RNA and the enzymatic sequencing techniques developed by Donis-Keller et al.[12,13]

CUCU sites that are located in duplex regions are not available for ribozyme cleavage. Therefore, to obtain complete cutting of large RNA molecules using the Tetrahymena ribozyme, all secondary and tertiary structure must be removed. Glyoxal denatures the substrate RNA without interfering with its processing by the ribozyme.* Once cutting is complete, this chemical modification must be removed in order to analyze the processing products.

Glyoxalation involves chemical modification of guanosine residues in the RNA via covalent attachment to N-1 and the exocyclic amine (Fig. 2).[14] The modified guanosine is no longer capable of hydrogen bond interactions used to form the Watson–Crick GC base pair. Glyoxal is commercially available as a 30 or 40% aqueous solution (Sigma, St. Louis, MO). Before use this reagent stock solution must be deionized. This is accomplished by stirring the solution in the presence of Bio-Rad AG 501-X8(D)

---

* With an altered ribozyme that recognized a sequence containing a G, glyoxalation would presumably prevent cleavage necessitating the use of an alternative denaturant.

[12] H. Donis-Keller, A. M. Maxam, and W. Gilbert, Nucleic Acids Res. 4, 2577 (1977).
[13] H. Donis-Keller, Nucleic Acids Res. 8, 3133 (1980).
[14] G. G. Carmichael and G. K. McMaster, this series, Vol. 65, p. 380.

FIG. 3. Site-specific cleavage of a highly structured RNA substrate by the Tetrahymena ribozyme. (A) Uniformly labeled, glyoxalated M1 RNA (the catalytic subunit of RNase P[15]) was incubated with the Tetrahymena ribozyme as described in text, with varying urea concentrations. As the urea concentration is decreased, specificity of the ribozyme for its cognate sequence CUCU decreases and the substrate RNA is multiply cleaved. (B) Secondary structure of M1 RNA near the site of cleavage (arrowhead).[16] The CUCU sequence that must be recognized by the ribozyme is shaded. Hairpins with CUUCGG such as this are extraordinarily stable,[17] so it is not surprising that the ribozyme does not cleave this substrate without prior glyoxalation.

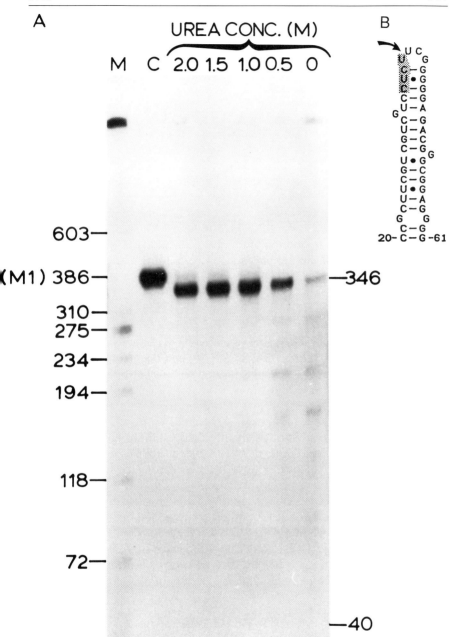

mixed-bed resin for 1 hr at 4°. If the resin is depleted (as indicated by the color change from blue to yellow), it is removed via filtration and additional resin added. The final conductivity of the glyoxal solution should be at most 5 μS/cm. Deionized stock glyoxal solutions are stored in small aliqouts (0.5 ml) in Eppendorf tubes at $-20°$. A standard glyoxalation reaction can be performed on as much as 50 μg of RNA in 200 μl final volume.

Glyoxalation is achieved by incubation of the RNA in 10 mM sodium phosphate (pH 7.0), 50% dimethyl sulfoxide (DMSO), 6% deionized glyoxal at 50°C for 10 min. The reaction is immediately applied to a 10-ml Sephadex G-100 column equilibrated in TE with 250 mM NaCl. Fractions are collected and analyzed via scintillation counting in the case of radiolabeled RNA or spectrophotometry in the case of unlabeled material. RNA-containing fractions are pooled and precipitated with 3 volumes of absolute ethanol. After precipitation and drying, the RNA is resuspended in sterile water. This can be difficult and, in certain instances, may take a long period of time.

Ribozyme Cutting of Large RNA

Cutting has been possible using small amounts of high specific activity RNA or up to 1 μg quantities of cold RNA when the guanosine cofactor has been [α-^{32}P]GTP. RNA cleavage reactions are performed in 50 mM Tris-HCl (pH 7.5), 10 mM MgCl$_2$, 2.5 M urea, 0.25 mM GTP (or 0.1 mM when using [^{32}P]GTP) and 0.18 μM ribozyme. To obtain end-labeling of the downstream cleavage fragment, 62.4 pmol of [α-^{32}P]GTP (800 Ci/mmol, 5 μl; NEN) is substituted for the cold GTP normally used. The reaction is incubated for 20 min at 50°. If labeled GTP is used, unincorporated material may be removed via gel filtration chromatography using the conditions described to remove unreacted glyoxal. The cut RNA is then precipitated with 3 volumes of absolute ethanol, washed with 70% ethanol, respun, and dried. The RNA is then resuspended in water. If cleavage is to be analyzed by gel electrophoresis, the RNA does not need to be deglyoxalated (Fig. 3).[15–17] Markers that are run for size comparison during the gel analysis need to be modified with glyoxal in order to obtain valid molecular weight estimates. *Hae*III digests of ϕX-174 DNA treated

[15] C. Guerrier-Takada, K. Gardiner, T. Marsh, N. Pace, and S. Altman, *Cell* **35**, 849 (1983).
[16] B. D. James, G. J. Olsen, J. Liu, and N. Pace, *Cell* **52**, 19 (1988).
[17] C. Tuerk, P. Gauss, C. Thermes, D. R. Groege, N. Guild, G. Stormo, M. Gayle, Y. d'Aubenton-Carata, O. C. Uhlenbeck, I. Tinoco, Jr., E. N. Brody, and L. Gold, *Proc. Natl. Acad. Sci. U.S.A.* **85**, 1364 (1988).

with glyoxal in the same fashion as the RNA have proved to be an adequate standard system for this comparative purpose. On the other hand, if enzymatic sequencing is to be performed to determine the cut site location, the RNA must be deglyoxalated.

Glyoxal Removal

Removal of glyoxal modifications requires high temperature and pH in order for its removal, but care must be taken to avoid alkaline hydrolysis. The RNA is resuspended in 0.3 M Tris base (pH not adjusted; pH 9.5) and incubated for 2 hr at 50°. The RNA is then precipitated with 3 volumes of absolute ethanol. Removal of the glyoxal is monitored by polyacrylamide gel electrophoresis and staining the RNA with ethidium bromide. Glyoxalated RNA exhibits severely decreased fluorescence and slightly retarded electrophoretic mobility, both of which return to normal on complete deglyoxalation.

[46] Preparation and Characterization of RNase P from *Escherichia coli*

By MADELINE F. BAER, JOHN G. ARNEZ, CECILIA GUERRIER-TAKADA, AGUSTIN VIOQUE, and SIDNEY ALTMAN

Introduction

RNase P from *Escherichia coli* is an endoribonuclease which cleaves precursor tRNAs to generate the 5' termini of mature tRNAs. The enzyme also cleaves the precursor to 4.5 S RNA, an RNA that is thought to be involved in protein synthesis,[1] to generate the 5' terminus of the mature molecule. RNase P is a ribonucleoprotein. It is composed of an RNA called M1 which is 377 nucleotides long and a very basic protein of M_r 13,800 called C5.[2] Both the gene for the RNA, *rnpB*,[3] and the protein, *rnpA*,[4] have been cloned, and their nucleotide sequences have been determined.

[1] D. B. Bourgaize and M. J. Fournier, *Nature (London)* **325,** 281 (1987).

[2] S. Altman, M. Baer, C. Guerrier-Takada, and A. Vioque, *Trends Biochem. Sci.* **11,** 515 (1986).

[3] R. E. Reed, M. F. Baer, C. Guerrier-Takada, H. Donis-Keller, and S. Altman, *Cell* **30,** 627 (1982).

[4] F. G. Hansen, E. B. Hansen, and T. Atlung, *Gene* **38,** 85 (1986).

The subunits are present in a molar ratio of 1 : 1 in the holoenzyme.[5] Although both subunits are required for viability *in vivo*,[6] in buffers that contain a high concentration of magnesium ions, M1 RNA functions *in vitro* as a true enzyme, accurately and efficiently cleaving the 5' end of precursor tRNA (i.e., pre-tRNA) molecules. The addition of C5 protein enhances the rate of cleavage of pre-tRNA *in vitro*.[7] In this chapter we describe the methodology for isolation of the RNase P holoenzyme from *E. coli*, preparation of the individual RNA and protein subunits, reconstitution of the holoenzyme from purified subunits, and detection of the enzymatic activity of RNase P.[8,9]

Materials

Buffers

Buffer A: 50 m*M* tris(hydroxymethyl)aminomethane hydrochloride (Tris-HCl) (pH 7.5), 10 m*M* Mg(CH₃COO)₂, 60 m*M* NH₄Cl

Buffer B: 50 m*M* Tris-HCl (pH 7.5), 10 m*M* Mg(CH₃COO)₂, 500 m*M* NH₄Cl

Buffer C: 50 m*M* CH₃COONa (pH 7.2), 10 m*M* MgCl₂, 7 *M* urea

Buffer D: 50 m*M* Tris-HCl (pH 7.5), 10 m*M* MgCl₂, 100 m*M* NH₄Cl

Buffer E: 15 m*M* NaCl, 1.5 m*M* sodium citrate (pH 7.0), 1 m*M* ethylenediaminetetraacetic acid (EDTA)

Buffer F: 50 m*M* Tris-HCl (pH 7.5), 100 m*M* NH₄Cl, 100 m*M* MgCl₂, 5% (w/v) polyethyleneglycol (PEG)

Buffer G: 10 m*M* Tris-HCl (pH 7.5), 1 m*M* EDTA

Enzymes

T7 RNA polymerase, restriction endonucleases (New England Biolabs, Beverly, MA), RNase-free DNase I (Cooper Biomedical, Malvern, PA, formerly Worthington)

Reagents

Isopropyl-β-D-thiogalactopyranoside (IPTG), phenylmethylsulfonyl fluoride (PMSF), ampicillin, chloramphenicol, thiamin, alumina, Tris (Trizma base) (Sigma, St. Louis, MO)

---

[5] A. Vioque, J. Arnez, and S. Altman, *J. Mol. Biol.* **202,** 835 (1988).

[6] R. Kole, M. F. Baer, B. C. Stark, and S. Altman, *Cell* **19,** 881 (1980).

[7] C. Guerrier-Takada, K. Gardiner, T. Marsh, N. Pace, and S. Altman, *Cell* **35,** 849 (1983).

[8] R. Kole and S. Altman, *Proc. Natl. Acad. Sci. U.S.A.* **76,** 3795 (1979).

[9] S. Altman, E. J. Bowman, R. L. Garber, R. Kole, R. A. Koski, and B. C. Stark, *in* "Transfer RNA: Biological Aspects" (D. Söll, J. N. Abelson, and P. R. Schimmel, eds.), p. 71. Cold Spring Harbor Laboratory, Cold Spring Harbor, New York, 1980.

Sepharose and Sephadex chromatography resins (Pharmacia, Piscataway, NJ)
ATP, GTP, CTP, UTP (Pharmacia)
Ultrapure urea (Schwarz-Mann, Cleveland, OH)
Acrylamide, bisacrylamide, $N,N,N'N'$-tetramethylethylenediamine (TEMED), agarose (Bio-Rad, Richmond, CA)
All chemicals are reagent grade.

Purification of the RNase P Holoenzyme[10–12]

Two hundred grams of frozen *E. coli* MRE600 cells is mixed with 400 grams of alumina that have previously been chilled at $-20°$ in a mortar. This procedure can be scaled down with good results, but the use of larger amounts of cells (i.e., over 1 kg) results in relatively greater losses of enzymatic activity. The cells are ground by hand with a precooled $(-20°)$ pestle in the mortar, placed on ice, until a paste is formed. Then 200 ml of buffer A is added and mixed with the paste. It is worth noting here that other methods of cell lysis do not give as high a yield of active holoenzyme as does lysis by grinding. Two hundred microliters of 10 mM PMSF and 200 μl of DNase I (1 mg/ml) are then added to the slurry. All subsequent steps are carried out on ice or at $4°$.

After 30 min on ice the mixture is centrifuged for 10 min at 10,000 rpm (12,000 g) in 250-ml polypropylene centrifuge bottles (in a GSA rotor and a Sorvall RC-5 centrifuge) to remove the alumina and cell debris. The supernatant (called crude extract) is then centrifuged for 30 min at 13,500 rpm (30,000 g) as above. The RNase P activity is in the supernatant (S30). The S30 is diluted 4-fold with buffer A and loaded onto a DEAE-Sephadex A-50 column (2 liter bed volume) equilibrated with the same buffer. The column is eluted stepwise with 2 liters each of 0.2, 0.3, 0.39, and 0.5 M NH$_4$Cl in buffer A. RNase P activity is detected in the fractions eluted with 0.5 M NH$_4$Cl and is precipitated with ammonium sulfate (55 g/100 ml solution). The precipitate is resuspended in buffer A that contains 0.5 M NH$_4$Cl (buffer B) and dialyzed against the same buffer.

The next step in the purification involves gel filtration on a Sepharose 4B column (2.5 × 90 cm) with buffer B as the elution buffer. Fractions of 4 ml are collected at a rate of 16 ml/hr. RNase P elutes after the void

[10] H. D. Robertson, S. Altman, and J. D. Smith, *J. Biol. Chem.* **247**, 5243 (1972).
[11] B. C. Stark, R. Kole, E. J. Bowman, and S. Altman, *Proc. Natl. Acad. Sci. U.S.A.* **75**, 3717 (1978).
[12] R. Kole and S. Altman, *Biochemistry* **20**, 1902 (1981).

volume, between the peaks of rRNA and 4 S RNA. The fractions containing RNase P activity are pooled and precipitated with ammonium sulfate as described above. At this point the preparation, although highly enriched for RNase P, is still not homogeneous.

Several other chromatographic procedures (e.g., Sephadex G-200 and *n*-octyl-Sepharose) have been tested in attempts to purify RNase P from *E. coli* to homogeneity. However, the yields and the specific activities of the enzyme obtained by such methods are low. The best method for obtaining a homogeneous preparation of the RNase P holoenzyme is to purify the individual RNA and protein subunits separately and then to reconstitute the holoenzyme as described below.

Purification of Protein Subunits of RNase P from *Escherichia coli*

Preparation of C5 Protein from RNase P Holoenzyme from E. coli

The preparation of RNase P obtained by chromatography on Sepharose 4B is resuspended in buffer C and dialyzed against the same buffer. The dialysate is then loaded onto a CM-Sephadex C-50 column (1 × 3 cm) equilibrated with buffer C (any RNAs present in the loaded sample flow through the column). C5 protein is eluted with a linear gradient of 0 to 0.5 M NaCl in buffer C.[12] Fractions containing C5 protein are detected by reconstitution assays with the RNA subunit of RNase P (see below). This procedure yields highly purified C5 protein. Minor contaminants can be removed as described below.

To improve the purification of C5 protein preparative polyacrylamide gel electrophoresis has been used with some success.[7] This procedure involves electroelution of C5 protein from the gels followed by precipitation with acetone in order to remove traces of sodium dodecyl sulfate (SDS) from the preparation. Unfortunately, this procedure, even though it yields a homogeneous band of protein, is not reproducible and is, therefore, not recommended. Another method for preparing C5 protein from the RNase P holoenzyme was developed by Vioque and Altman[13]: 1 volume of the preparation of RNase P purified through the Sepharose 4B step is mixed with 2 volumes of 10.5 M LiCl. EDTA (adjusted to pH 8.0 with NaOH) is added to a final concentration of 10 mM. The mixture is gently agitated overnight at 4° and then centrifuged at 100,000 g for 10 hr. The supernatant, which contains the proteins, is dialyzed against buffer D and can then be used to generate the RNase P holoenzyme.

[13] A. Vioque and S. Altman, *Proc. Natl. Acad. Sci. U.S.A.* **83**, 5904 (1986).

Active C5 protein has also been purified by affinity chromatography. A crude preparation of proteins (S30) is subjected to the LiCl extraction procedure described above, and C5 is purified by affinity chromatography on a column of agarose-adipic acid dihydrazide to which M1 RNA has been covalently linked. C5 protein binds specifically to the immobilized M1 RNA and can be eluted from it.[13] However, the amounts of pure protein obtained are small, and contamination of the protein with M1 RNA, leached from the solid support, can sometimes occur.

Purification of C5 Protein from Cells Harboring Plasmids Carrying the rnpA Gene

Large-scale purification of the C5 protein has been difficult to achieve for the following reasons: (1) RNase P is present at very low levels in the cell. No more than a few hundred molecules of the enzyme are present in each *E. coli* cell.[14] (2) C5 protein is very unstable in the absence of RNA and in dilute solution. (3) In the absence of urea or in a low-salt solution the protein precipitates out.[14] (Dilutions of up to 50 μg/ml remain in solution.)

Recently a genetic construct for the overexpression of C5 protein has been developed, and large-scale purification of C5 protein is now feasible. Overexpression has been achieved using a vector in which the gene for the C5 protein is placed under the control of a T7-specific promoter.[5] A T7 terminator is located downstream from the C5 protein. The ribosome-binding site is that of gene 10 of the T7 phage, i.e., the gene that encodes the coat protein of the bacteriophage.[15] This plasmid, pARE7,[5] is carried by *E. coli* strain BL21(DE3)[16] which also carries a chromosomal copy of the gene for T7 RNA polymerase under the control of the *lac* promoter.[15,16]

Induction and Overproduction of C5 Protein in E. coli. Cells are grown overnight in M9 medium supplemented with glucose (2 mg/ml), thiamin (1 μg/ml), and ampicillin (0.1 mg/ml; M9A). One liter of fresh M9A medium in a 2-liter flask is then inoculated with the overnight culture such that the initial density of cells is low, i.e., the absorbance at 550 nm is about 0.05. The culture is incubated with vigorous shaking at 37° until the absorbance at 550 nm reaches 0.4. At this point IPTG is added to a

[14] A. Vioque and S. Altman, unpublished results.
[15] A. K. Rosenberg, B. N. Lade, D. Chui, S. W. Lin, J. J. Dunn, and F. W. Studier, *Gene* **56**, 125 (1987).
[16] F. W. Studier and B. A. Moffat, *J. Mol. Biol.* **189**, 113 (1986).

final concentration of 2 mM and the incubation is continued. After 3 hr the cells are harvested. Longer incubations do not increase the yield of the protein, and the number of cells that carry the plasmid can be exceeded by the number of other cells that do not carry the plasmid or that carry nonfunctional plasmids.

Before proceeding with the purification it is essential that the degree of overexpression be checked by analysis of total cellular protein on a 12% polyacrylamide, SDS–urea gel.[7] If the overexpression of C5 protein has been successful, a band corresponding to a molecular weight of 15,500, as determined by a comparison with the mobility of molecular weight markers, is very intense, usually representing between 10 and 15% of the total cellular protein. This band corresponds to C5 protein, which has aberrant mobility in SDS gels. The level of C5 in the preparation must always be checked initially because cells carrying the target plasmid grow slowly, and they can be easily overgrown by cells without the plasmid. Cells without the plasmid can grow if enough β-lactamase is secreted into the medium by the plasmid-carrying cells. In such cases, C5 protein is not amplified.

After the induction period, the cells are harvested by centrifugation at low speed in a Sorvall RC-5 centrifuge. The cell pellet is recovered, weighed, and frozen on dry ice. About 2 g of wet cells is usually recovered from 1 liter of induced culture. The frozen cells can be stored at $-70°$ for long periods.

Purification of C5 Protein. All operations are performed on ice or in the cold room (4°). The crude extract is prepared as described above for purification of the holoenzyme, although some changes have been made in the procedure to account for the behavior in crude extracts of excess C5 protein.

The cell pellet is placed in a prechilled mortar, on ice, and mixed with twice its weight of cold alumina. The cells are ground by hand with a chilled pestle. When a homogeneous paste is obtained, 10 ml of buffer A is added and mixed with the paste. Then 10 μl of 50 mM PMSF and 10 μl of DNase I (1 mg/ml) are added. The extract is incubated for 30 min on ice with occasional stirring. The extract is centrifuged for 10 min at 10,000 rpm in the SS34 rotor of a Sorvall RC-5 centrifuge to remove alumina and cell debris. The supernatant is centrifuged again for 30 min at 15,500 rpm (30,000 g). About 90% of the overproduced C5 protein is in the supernatant (S30).

The S30 is centrifuged for 2 hr at 41,000 rpm (100,000 g) in a Beckman Ti50 rotor. The supernatant is removed and the ribosomal pellet, which contains the C5 protein, is resuspended in 10 ml of buffer A that contains 1 M NH$_4$Cl. The suspension is gently agitated for 2 hr at 4° and centri-

fuged again for 2 hr at 41,000 rpm in a Beckman Ti50 rotor. The supernatant (ribosomal wash) contains most of the C5 protein.

The ribosomal wash (about 10 ml) is dialyzed for 16 hr against 2 liters
of buffer D, which is changed twice during that time. A white precipitate
is formed in the dialysis bag. The contents of the bag are transferred to a
30-ml Corex tube, and the bag is carefully rinsed with buffer D to maximize recovery of the precipitate. The suspension is centrifuged for 30 min
at 15,500 rpm in a Sorvall SS34 rotor. The supernatant is discarded, and
the precipitate is resuspended in 5 ml of buffer C. To resuspend the
precipitate completely, it is sometimes necessary to stir the suspension
gently in the cold room for 1–2 hr. At this point C5 protein is 80–90% pure
as judged by analysis on SDS gels stained with Coomassie blue.

The solution of C5 protein is applied to a 3-ml column of CM-Sephadex C-50 equilibrated with buffer C. The column is washed with 10 ml of
buffer C at a rate of 0.5 ml/min. It is then eluted with 40 ml of a linear
gradient of NaCl (0 to 0.5 M) in buffer C. Fractions of 1 ml are collected.
C5 protein is usually eluted by approximately 0.3 M NaCl. The 6 to 8
fractions encompassing the peak of C5 protein are pooled and stored in
small aliquots at −70°. Typically, 1 to 2 mg of pure C5 protein is obtained
in this way. The protein should be stored at a concentration of 0.2 mg/ml
or higher. If necessary, the protein can be concentrated severalfold with
an Amicon Centricon 10 concentrator, although a substantial amount of
protein can be lost during such a procedure.

C5 protein is very unstable once it has been diluted. Frozen aliquots
should be thawed and diluted immediately prior to use, and they should
never be refrozen. Urea and salt present in the concentrated solution of
protein cause no interference in most experiments because large dilutions
(1 : 1000) of the preparation are usually made for most experiments.

The complete purification procedure, after harvesting of the cells, can
be accomplished in less than 48 hr. If care is taken to avoid exposure of
the protein fractions to temperatures higher than 4–6° and delays are
avoided, highly active protein can be obtained as determined by a reconstitution assay with M1 RNA (see below). The protein prepared by this
procedure is about 95% pure.

Preparation of C5 Protein Containing the A49 Mutation

The gene for the A49 mutant[17] contains a single nucleotide change that
results in a substitution of His for Arg at position 46 in the C5 protein.[18]

[17] P. Schedl and P. Primakoff, *Proc. Natl. Acad. Sci. U.S.A.* **70,** 2091 (1973).
[18] L. Kirsebom, M. F. Baer, and S. Altman, *J. Mol. Biol.* **204,** 879 (1988).

The gene with this mutation has been cloned behind the T7 promoter in a construction identical to that described above for the wild-type gene. This plasmid, pMBA49, was transformed into strain BL21-A49 which carries the C5 gene with the A49 mutation on its chromosome as well as the gene for T7 RNA polymerase under the control of the *lac* promoter. The host strain is not viable at 43° unless it is carrying the pMBA49 plasmid. Cells are grown at 37°, induced with IPTG, and harvested as described above.

Preparation of the crude extract is performed in the same way as for the wild-type protein through the 30,000 *g* centrifugation. After this step the A49 protein is found in the pellet (P30) rather than in the supernatant. The protein may be recovered from the bulk of the proteins in the P30 by gentle resuspension in buffer A that contains 4 *M* urea. The protein solution is gently agitated at 4° for 45 min and centrifuged again at 30,000 *g* for 30 min. The supernatant (P30 wash) now contains at least one-half of the total mutant C5 protein. Vigorous agitation or use of a higher concentration of urea releases too many contaminating proteins into solution and leads to a low yield of less active mutant C5 protein.

The P30 wash is then dialyzed at 4° for 16 hr against 2 liters of buffer D with 2 changes as described above. The white precipitate which appears is then treated in the same manner as in the procedure for the wild-type protein.

Purification of RNA Subunit of RNase P from *Escherichia coli*

Purification of M1 RNA from Total RNA

Total RNA is prepared from 1 liter of a culture of *E. coli* MRE600 by the procedure described by Reed.[19] Five grams of frozen cells is resuspended in 10 ml of buffer E and stirred with 10 ml of phenol, which has been saturated with the same buffer, for 1 hr at 4°. The solution is then centrifuged at 10,000 *g* for 10 min. The aqueous phase is poured into a beaker. The phenol phase is extracted with an additional 5 ml of buffer E for 15 min at 4° and centrifuged as before. Both aqueous phases are pooled and extracted with an equal volume of phenol, 0.1 volume of 20% SDS, and 0.1 volume of 3 *M* sodium acetate. After centrifugation the aqueous phase is removed, and 3 volumes of 95% ethanol is added slowly so that 2 phases are created. A glass rod is then used to spool out any DNA present in the precipitate; this step can be omitted if freshly harvested rather than frozen cells are used initially. The remaining solution is

[19] R. E. Reed, Ph.D. thesis, Yale University, New Haven, Connecticut, 1984.

mixed thoroughly, placed at −20° for 2 hr or at −70° for 5 min and then centrifuged to pellet the RNA.

The pellet of total cellular RNA is dissolved in a loading buffer and subjected to electrophoresis on a 5% nondenaturing polyacrylamide gel, and the gel is then stained with ethidium bromide. The band of M1 RNA is excised and eluted by crushing and soaking the gel slice with buffer G and centrifuging the suspension of gel at 4° in an Eppendorf centrifuge.[20] After addition of sodium acetate to a concentration of 0.3 *M* and 2.5 volumes of 95% ethanol, the mixture is incubated at −70° for 5 min to precipitate the eluted RNA. After centrifugation, the RNA pellet is washed twice with 75% ethanol and lyophilized; it is then resuspended in a 10 *M* urea solution that contains 0.05% xylene cyanol and 0.05% bromphenol blue and is further purified on a 5% polyacrylamide–7 *M* urea gel. The gel is stained with ethidium bromide, and the M1 RNA is excised and eluted as before. The RNA purified by this procedure can be used successfully in reconstitution experiments (see below).

Purification of M1 RNA from Partially Purified RNase P

An aliquot of the holoenzyme, purified through the Sepharose 4B step, is mixed with Pronase to give a final concentration of 1 mg Pronase/ml and SDS to a final concentration of 1%. The mixture is incubated at 37° for 30 min and an equal volume of phenol, presaturated with buffer G, is added. After mixing and centrifugation, the aqueous phase is removed. The phenol phase is extracted with 0.5 volumes of buffer G that contains 1% SDS. After mixing and centrifugation, the aqueous layer is again removed. The combined aqueous layers are reextracted with 0.5 volumes of phenol. The aqueous layer is removed as before and transferred to a glass test tube. After addition of an equal volume of ether, presaturated with water, the two phases are mixed and then allowed to settle. The ether layer is removed and discarded. The extraction is repeated.

After the second ether extraction, 0.1 volume of 20% potassium acetate is added to the aqueous layer. The solution is mixed and put on ice for 15 min, then centrifuged to pellet the SDS that has precipitated. The supernatant is removed and combined with 2.5 volumes of 95% ethanol; the resulting mixture is frozen at −70° for 5 min to precipitate the RNA. After centrifugation of the mixture, the pellet is collected, washed twice with 75% ethanol, and dried. This RNA fraction may be used to reconstitute the RNase P holoenzyme (see below). If further purification is required, the RNA pellet may be resuspended with the appropriate buffer

[20] C. Guerrier-Takada and S. Altman, *Biochemistry* **23,** 6327 (1984).

for loading on a polyacrylamide gel, either native or denaturing, and then subjected to electrophoresis.

Preparation of M1 RNA from Cells Harboring Plasmids Carrying the rnpB Gene

This procedure was adapted from that of Neidhardt *et al.*[21] Strains of HB101 harboring a plasmid that carries the *rnpB* gene[19,22] are grown in LB medium supplemented with ampicillin (100 μg/ml; LBA) until late log phase. Chloramphenicol (250 μg/ml) is added to the culture, and after an overnight incubation the cells are harvested, washed twice with fresh LBA medium, and resuspended in LBA medium. The cells are incubated for 2–3 hr and the total cellular RNA is extracted as described above (Purification of M1 RNA from Total RNA). The RNA is purified only once on a 5% nondenaturing polyacrylamide gel. If further purification is necessary, the preparation may be fractionated by HPLC as described below.

Preparation of M1 RNA Using T7 RNA Polymerase

M1 RNA may be synthesized *in vitro* by transcription by T7 RNA polymerase of the plasmids pJA2 or pJA2' cut with *Fok*I restriction endonuclease. These plasmids are derivatives of pUC19 and contain the *rnpB* gene spliced to a synthetic promoter specific for T7 RNA polymerase such that transcription begins precisely with the first nucleotide (a guanosine residue) of authentic M1 RNA. In addition, the construct harbors a new *Fok*I site inserted 7 nucleotides from the 3' end of the gene in the orientation that produces the authentic 3' terminus after digestion with *Fok*I and subsequent transcription by T7 RNA polymerase.[5] Plasmid DNA from pJA2 (or pJA2') is prepared in large quantities[23] to produce the template for transcription *in vitro*. The restriction mixtures are checked by electrophoresis on agarose gels to monitor the extent of the digestion. When digestion is complete, the digests are extracted once with 1 volume of phenol–chloroform (1:1), twice with 1 volume of chloroform, and then precipitated with 2 volumes of 95% ethanol after the addition of sodium acetate to a final concentration of 0.3 *M*. After centrifugation the pellets are washed twice with 75% ethanol, dried, and resuspended in 0.1× buffer G.

[21] F. C. Neidhardt, R. Wirth, M. W. Smith, and R. Van Bogelen, *J. Bacteriol.* **143**, 535 (1980).
[22] N. P. Lawrence and S. Altman, *J. Mol. Biol.* **190**, 163 (1986).
[23] C. Guerrier-Takada and S. Altman, *Cell* **45**, 177 (1986).

Rather than one large reaction, it is preferable that a series of small reactions (1 ml each) are carried out to produce sufficient quantities of M1 RNA. The optimal ratio of T7 RNA polymerase to DNA has been found to be 20 U/μg. The reaction mixture, based on conditions described by Milligan *et al.*,[24] consists of 40 mM Tris-HCl (pH 7.5), 24 mM MgCl$_2$, 2 mM spermidine, 10 mM DTT, 0.1% Triton X-100, 3.8 mM ATP, 5.3 mM GTP, 4.1 mM CTP, 2.5 mM UTP (these concentrations of nucleoside triphosphates correspond to the ratio of nucleotides found in M1 RNA), 40 μg of the DNA template, and 800 units of T7 RNA polymerase. The reactions are incubated overnight at 37°. A white precipitate appears as the reaction progresses.

The reaction mixtures are treated with 5 μl of 1 mg/ml RNase-free DNase I and incubated at 37° for 30 min. To dissolve the precipitate, EDTA is added to each tube to a final concentration of 40 mM (about twice the concentration of magnesium ions). Aliquots of the transcription reactions should be analyzed by electrophoresis on a 3% agarose gel or on a 5% polyacrylamide–7 M urea gel, at the same time as the reaction mixtures are pooled and extracted with phenol–chloroform and chloroform as described above. The resulting aqueous mixture is free of proteins, but it still contains shorter polynucleotide fragments and oligonucleotides as well as residual products of abortive transcription. Smaller fragments can be easily eliminated by loading the aqueous preparation onto a spun column of Sephadex G-50.[25] These columns are packed in syringes (5 ml) and spun at 3500 rpm, then washed 3 times with buffer G. About 300 μl of the mix can be purified through each column, by centrifugation for 20 min at 3000 rpm.

Finally, the transcripts are precipitated with 2 volumes of 95% ethanol after the addition of sodium acetate to a final concentration of 0.3 M. The pellets are washed twice with 75% ethanol and dried. The dried pellets can be resuspended in water or an appropriate buffer, and the absorbance at 260 nm is then measured. The yield is approximately 2 mg of RNA per 1-ml reaction mixture.

In order to obtain very pure preparations of M1 RNA, the transcript (or material extracted from cells) may be fractionated by HPLC, using a Macherey-Nagel (Düren, FRG) Nucleogen-DEAE 500-7 column (6 × 125 mm). The pellet containing the transcript is resuspended in 20 mM potassium phosphate (pH 6.75), 5 M urea, and loaded on the column which has

[24] J. Milligan, D. Groebe, G. W. Witherel, and O. C. Uhlenbeck, *Nucleic Acids Res.* **15**, 8783 (1987).
[25] T. Maniatis, E. F. Fritsch, and J. Sambrook, "Molecular Cloning: A Laboratory Manual." Cold Spring Harbor Laboratory, Cold Spring Harbor, New York, 1982.

been equilibrated with the same buffer. The transcript is eluted with a gradient of 0 to 1.5 M KCl in the same buffer. Under these conditions M1 RNA elutes in approximately 0.6 M KCl. The fractions containing M1 RNA are collected and precipitated with 0.3 M sodium acetate and ethanol, and the pellet is washed with 75% ethanol and dried as previously described. Use of HPLC does not adversely affect the yield of M1 RNA.

Reconstitution of RNase P Holoenzyme and Assays of Cleavage Activity of RNase P

Reconstitution may be performed either by dialysis or by direct mixing of subunits.[7]

Dialysis

Purified RNA and protein subunits are mixed or resuspended separately in buffer C that contains 0.5 M NaCl and dialyzed for 2 hr against buffer B. Aliquots are then assayed for RNase P activity by incubation in buffer D, in the presence of radioactively labeled pre-tRNA substrate at 37°. A similar dialysis of either the purified RNA or protein subunit alone followed by incubation of aliquots in buffer D as described above should show no cleavage activity. The activity of M1 RNA alone may be similarly assayed by incubation in buffer F.

Direct Mixing

RNA and protein subunits can be incubated together in buffer D. The holoenzyme forms immediately even when the subunits are mixed at 0°. The separated subunits can be assayed for cleavage activity as described above.

Assays of the Cleavage Activity of RNase P

Cleavage activity of the RNase P holoenzyme may be assayed by incubating either the native holoenzyme or the holoenzyme made by reconstitution from isolated purified subunits in buffer D in the presence of a radioactively labeled pre-tRNA substrate. When cleavage reactions are performed at 37°, a 5-min preincubation is carried out prior to addition of the substrate in order to allow time for conformational changes necessary for the formation of maximally active enzyme to occur.[26] Cleavage activity of the M1 RNA alone may be tested in a similar manner using buffer F.

[26] S. Altman and C. Guerrier-Takada, *Biochemistry* **25**, 1205 (1986).

After incubation, the reaction is stopped, and the products of the RNase P reaction are analyzed either on nondenaturing or denaturing polyacrylamide gels. For analysis on nondenaturing gels, the reaction mixture is added to one-fourth volume of "stop mix" that contains 0.4 M EDTA, 50 mM Tris-HCl (pH 8.0), 0.2% xylene cyanol (XC), and 0.2% bromphenol blue (BPB). For analysis on denaturing gels that contain 7 M urea, an aliquot of reaction mix is added to an equal volume of "stop mix" that contains 10 M urea, 0.05% XC, and 0.05% BPB.[7]

Large-Scale Reconstitution of M1 RNA and C5 Protein Subunits into RNase P Holoenzyme

Large-scale reconstitution of the holoenzyme is achieved by dialyzing a mixture of M1 RNA and C5 protein prepared in a buffer that contains 7 M urea (buffer C) against buffer D that contains 0.4 M NH$_4$Cl. Since C5 protein is usually included in molar excess of M1 RNA, any protein not bound to M1 RNA will precipitate out on dialysis as it is not soluble in the absence of urea.

Preparations of C5 protein vary in quality, and, in any given preparation, not all of the protein molecules are capable of binding to M1 RNA or of stimulating its catalytic activity. Therefore, C5 protein must be used in molar excess to maximize the yield of active holoenzyme on reconstitution. In order to determine the optimal molar ratio of C5 protein to M1 RNA for a given preparation, assays are performed using the small-scale, direct mixing method of reconstitution.

Prior to reconstitution, both the C5 protein and M1 RNA should be in buffer C. The two components are mixed in the molar ratio determined in the assay described above. For optimal results, dialysis should be carried out slowly. To achieve this, double dialysis is performed. The solution containing both components in buffer C is placed in an inner, smaller dialysis bag, and buffer C alone as well as the smaller dialysis tube are placed in a larger outer bag. This double bag is then placed in buffer D that contains 0.4 M NH$_4$Cl and dialyzed overnight. The buffer is changed 2 or 3 times during the course of the dialysis. Excess C5 will precipitate during the process; it can be removed by centrifugation of the suspension. The resulting suspension is transferred to a centrifuge tube, and the inner dialysis tube is rinsed twice with buffer D. Finally, the resulting solution of the holoenzyme may be concentrated. While concentration is not essential for assays of enzymatic activity, it is imperative that it be performed if the material is used for biophysical studies. There are several possible methods for concentrating the material: centrifugation in an Amicon Centriprep 10 concentrator, precipitation with ammonium sulfate followed by dialysis, and chromatography on DEAE-Sephadex.

Although most of the protein that is present in excess of M1 RNA can be removed by centrifugation, in order to remove any residual C5 protein that is not bound to M1 RNA, the sample (after concentration) is passed through a Sephacryl S-300 column equilibrated with buffer D that contains 0.4 M NH$_4$Cl. The complex is eluted with the same buffer. The eluate in the first large peak is collected and concentrated again. Cleavage assays are performed after every step to monitor the activity of the M1 RNA and/or the holoenzyme.

Concluding Remarks

Studies of $E.$ $coli$ RNase P have been hampered by difficulties in the purification of sufficient quantities of the enzyme and of radioactively labeled natural pre-tRNA substrates. The successful reconstitution of the separated subunits, M1 RNA and C5 protein, into the RNase P holoenzyme and the subsequent cloning and overproduction of both subunits, have played major roles in our attempts to elucidate the properties of this RNA processing enzyme. Furthermore, once large quantities of precursor tRNA substrates could be made in $vitro,$ as a result of the availability of synthetic DNA and the use of SP6 and T7 RNA polymerases, the study of RNase P advanced rapidly. (The preparation of such substrates is discussed in detail elsewhere in this volume.) Over the years it has also become apparent that analysis by polyacrylamide gel electrophoresis of the products of the reaction catalyzed by RNase P provides the only unambiguous method for distinguishing the true, specific endoribonucleolytic reaction of M1 RNA from nonspecific cleavages.

[47] Characterization of Self-Cleavage of Viroid and Virusoid Rnas

By ANTHONY C. FORSTER, CHRISTOPHER DAVIES, CHERYL J. HUTCHINS, and ROBERT H. SYMONS

Certain circular and linear low molecular weight, single-stranded plant pathogenic RNAs can self-cleave at specific sites *in vitro* in the presence of Mg^{2+}.[1-4] This property is considered to play an important role during their replication *in vivo* by a rolling circle mechanism. The reaction is transphosphorylation involving attack by the free $2'$-hydroxyl group on the $3',5'$-phosphodiester linkage with the generation of $5'$-hydroxyl and $2',3'$-cyclic phosphodiester termini at the cleavage site. It is clearly different from the self-splicing reactions involved in the removal of introns from mRNA and rRNA precursors that give excised fragments with $5'$-phosphate and $3'$-hydroxyl termini.[5,6]

Self-cleavage occurs in the plus and minus RNAs of the 247-nucleotide avocado sunblotch viroid (ASBV),[2] the 324-nucleotide virusoid (circular satellite RNA) of lucerne transient streak virus (vLTSV),[4] the 332-nucleotide virusoid of subterranean clover mottle virus,[7] and the 359-nucleotide linear satellite RNA of tobacco ringspot virus (sTRSV).[1,3] That such reactions are not limited to plant pathogenic RNAs is demonstrated by the self-cleavage *in vitro* of an RNA transcript of satellite 2 DNA from the newt[8] and of the circular RNA delta factor associated with hepatitis B virus.[9]

Approaches for Characterization of Self-Cleavage Reactions

This chapter describes methods that have been successfully used and the results obtained in the characterization of the self-cleavage of the

[1] J. M. Buzayan, W. L. Gerlach, and G. Bruening, *Nature (London)* **323**, 349 (1986).

[2] C. J. Hutchins, P. D. Rathjen, A. C. Forster, and R. H. Symons, *Nucleic Acids Res.* **14**, 3627 (1986).

[3] G. A. Prody, J. T. Bakos, J. M. Buzayan, I. R. Schneider, and G. Bruening, *Science* **231**, 1577 (1986).

[4] A. C. Forster and R. H. Symons, *Cell* **49**, 211 (1987).

[5] T. R. Cech and B. L. Bass, *Annu. Rev. Biochem.* **55**, 599 (1986).

[6] P. A. Sharp, *Science* **235**, 766 (1987).

[7] C. Davies, J. Haseloff, and R. H. Symons, unpublished.

[8] L. M. Epstein and J. G. Gall, *Cell* **48**, 535 (1987).

[9] L. Sharmeen, M. Y. P. Kuo, G. Dinter-Gottlieb, and J. Taylor, *J. Virol.* **62**, 2674 (1988).

RNAs of ASBV and vLTSV. Since the use of RNAs purified from infected plant tissue has severe limitations in that only circular plus RNAs can be readily obtained, with the minus species being present in much lower amounts, all approaches have made use of various cDNA clones of ASBV and vLTSV in transcription vectors. The flexibility offered by this approach is illustrated here.

Procedures

Reagents. SP6 RNA polymerase, T4 polynucleotide kinase, $[\alpha\text{-}^{32}\text{P}]\text{UTP}$ and $[\gamma\text{-}^{32}\text{P}]\text{ATP}$ are obtained from Bresatec Ltd. (Adelaide). Calf intestinal alkaline phosphatase is from Boehringer-Mannheim, T4 RNA ligase from Pharmacia, restriction enzymes from New England Biolabs (Beverly, MA) and Toyobo, RNase T1 from Sigma (St. Louis, MO), and RNase U2 from Calbiochem (San Diego, CA). RNase PhyM from *Physarum polycephalum* is purified as described by Donis-Keller,[10] and *Bacillus cereus* RNase according to Lockard *et al.*[11]

Preparation of cDNA Clones

Partial-Length cDNA Clones of vLTSV. A *Taq*I fragment derived from a cDNA clone of an Australian isolate of vLTSV[12] is cloned into the *Acc*I site of the phage DNA vector M13mp73. The insert plus short flanking vector sequences is excised with *Bam*HI and cloned in both orientations into *Bam*HI-cut vector pSP64.[2,13] Transcripts prepared with SP6 RNA polymerase contain 273 of the total of 324 nucleotides of vLTSV (residues 55 to 3).[4]

Dimeric cDNA Clones of ASBV. Tandem dimeric cDNA clones of ASBV are constructed from cloned monomeric DNA inserts in the phage DNA vector, M13mp93.[14] Full-length inserts are excised from replicative form DNA with *Sau*3A (residues 154 to 153)[15] and recloned in the same vector at varying ratios of insert to vector to ensure formation of dimeric inserts. The sequence and orientation of selected clones are confirmed by

[10] H. Donis-Keller, *Nucleic Acids Res.* **8**, 3133 (1980).
[11] R. E. Lockard, B. Alzner-Deweerd, J. E. Heckman, J. MacGee, M. W. Tabor, and U.L. RajBhandary, *Nucleic Acids Res.* **5**, 37 (1978).
[12] P. Keese, G. Bruening, and R. H. Symons, *FEBS Lett.* **159**, 185 (1983).
[13] P. Krieg and D. A. Melton, this series, Vol. 155, p. 397.
[14] J. M. Barker, J. L. McInnes, P. J. Murphy, and R. H. Symons, *J. Virol. Methods* **10**, 87 (1985).
[15] R. H. Symons, *Nucleic Acids Res.* **9**, 6527 (1981).

dideoxy sequence analysis.[16] The complete dimeric insert plus short flanking regions of the M13mp93 vector are excised with *Eco*RI and *Hin*dIII and recloned into the plasmid vectors pSP64 and pSP65[13] cut with the same enzymes so that, on transcription with SP6 RNA polymerase, both plus and minus ASBV transcripts can be produced.[2]

In Vitro Synthesis of Plus and Minus vLTSV and ASBV RNAs on Plasmid Templates

Nonradioactive transcription reactions containing 0.1 μg/μl template DNA, 0.5 unit/μl SP6 RNA polymerase, 40 mM tris(hydroxymethyl) aminomethane hydrochloride (Tris-HCl) (pH 7.5), 6 mM MgCl$_2$, 0.1 μg/μl bovine serum albumin (BSA), 10 mM dithiothreitol (DTT), and 0.5 mM of each of the four NTPs are incubated at 40° for 1–2 hrs. Radioactive transcription mixtures contain 25 μM UTP and up to 5 μCi/μl of [α-^{32}P]UTP. Reaction mixtures are fractionated as described below.

Gel Electrophoresis and Purification of RNAs

Transcription mixtures or self-cleavage reactions (see below) are added to an equal volume of *formamide loading solution* [95% deionized formamide, 10 mM ethylenediaminetetraacetic acid (EDTA), 0.02% xylene cyanol FF, 0.02% bromphenol blue], heated at 80° for 30 sec, and snap cooled on ice before loading on 35 × 20 × 0.05 cm, 7 M urea–polyacrylamide gels that are run in 90 mM Tris–borate (pH 8.3), 2 mM EDTA. Detection of labeled RNAs is by autoradiography at −70°, with or without an intensifying screen. Nonradioactive products are detected by staining with 0.05% toluidine blue and brief destaining in water. When required, RNAs are eluted from gel slices in 0.1% sodium dodecyl sulfate (SDS), 1 mM EDTA, 10 mM Tris-HCl (pH 7.5) for several hours at 37°, precipitated by the addition of 3 M sodium acetate (pH 5) to 0.3 M and 2.5 volumes of cold ethanol, stored at 0 or −20°, and later centrifuged.

Self-Cleavage of Purified Transcription Products

Purified transcripts are usually incubated in 50 mM Tris-HCl (pH 8.0), 5 mM MgCl$_2$, 0.5 mM EDTA, at 25 or 40°. In the case of vLTSV transcripts, prior heating in 1 mM EDTA at 80° for 1 min followed by snap cooling on ice is necessary to obtain self-cleavage. Reaction mixtures are examined by denaturing polyacrylamide gel electrophoresis.

---

[16] F. Sanger, A. R. Coulson, B. G. Barrell, A. J. H. Smith, and B. A. Roe, *J. Mol. Biol.* **143**, 161 (1980).

Removal of 2',3'-Cyclic Phosphodiester Groups from RNA Fragments

Purified fragments ($\sim$1 μg) are incubated in 10 μl of 10 mM HCl at 25° for 2 hr to decyclize the cyclic phosphate.[17,18] The 2'(3')-phosphate is then removed by incubation of the RNA with 0.5 μl of calf intestinal alkaline phosphatase [1 unit/μl in 50% glycerol, 125 mM NaCl, 1.25 mM triethanolamine, 40 nM MgCl$_2$, 4 μM ZnCl$_2$, 4.5 mM Tris-HCl (pH 7.5)] in 10 μl of 0.1 M Tris-HCl (pH 9.0) at 70° for 30 min. After extracting the solution by shaking with an equal volume of phenol–chloroform (1:1, v/v), the RNA is recovered by ethanol precipitation.

5'-Labeling of RNA Fragments

The usual procedure is to incubate the RNA fragments in 10 μl of 25 mM Tris-HCl (pH 9.0), 5 mM MgCl$_2$, 5 mM DTT, containing 0.5 units T4 polynucleotide kinase at 37° for 30 min. After addition of 5 μl of the gel loading solution and heating at 80° for 1 min followed by snap cooling on ice, the 5'-labeled RNA is purified on a 7 M urea–polyacrylamide gel, eluted, and used for sequence analysis.

3'-Labeling of RNA Fragments

Preparation of [5'-^{32}P]pCp. Twenty microliters (100 μCi) of [γ-^{32}P] ATP (3,000 Ci/mmol) is dried and dissolved in 20 μl of a reaction mixture containing 25 mM Tris-HCl (pH 9.0), 5 mM MgCl$_2$, 3 mM DTT, 5 μM 3'-CMP, 0.05 μg/μl BSA, and 6 units of T4 polynucleotide kinase. After incubation at 37° for 1 hr, the reaction is ended by heating at 65° for 3 min, and the reaction mixture is stored at $-20°$.

The efficiency of conversion of Cp to [5'-^{32}P]pCp is determined by thin-layer chromatography.[19] A small sample (0.2 μl) of the reaction mixture and a marker of [γ-^{32}P]ATP are spotted on a polyethyleneimine-cellulose thin-layer plate and developed in 0.8 M (NH$_4$)$_2$SO$_4$ until the solvent nears the top of the plate. After drying, the plate is autoradiographed. pCp has an R_f of approximately 0.6 and ATP of approximately 0.3. Essentially all of the [γ-^{32}P]ATP should be converted to product.

3'-Labeling of RNA with [5'-^{32}P]pCp. The reaction mixture of 20 μl contains 5 pmol [γ-^{32}P]pCp (2 μl of reaction mixture described above), approximately 5 pmol RNA, 50 mM N-2-hydroxyethylpiperazine-N'-2-ethanesulfonic acid (HEPES)–KOH (pH 7.5), 15 mM MgCl$_2$, 3.3 mM DTT, 15 μM ATP, 5 ng/μl BSA, 5% (v/v) redistilled dimethyl sulfoxide

[17] H. I. Abrash, C.-C. S. Chueng, and J. C. Davies, *Biochemistry* **6**, 1298 (1967).
[18] J. Hashimoto, K. Suzuki, and T. Uchida, *J. Gen. Virol.* **66**, 1545 (1985).
[19] T. E. England, A. B. Bruce, and O. C. Uhlenbeck, this series, Vol. 65, p. 65.

(DMSO), and 2 units of T4 RNA ligase. Incubation is at 0° overnight. The 3′-labeled RNA is then purified by gel electrophoresis.

Enzymatic Sequencing of 5′- and 3′-Labeled RNA Fragments

The purified, 5′- or 3′-labeled RNA fragment is dried and dissolved in 20 μl of water containing 50 μg *E. coli* tRNA as carrier. Six aliquots, each of 3 μl, are then dried *in vacuo* in Eppendorf tubes and resuspended as follows:

Tube N (no enzyme)
Tube T (RNase T1) } 9 μl of 20 mM sodium citrate (pH 5.0),
Tube P (RNase PhyM) } 1 mM EDTA, 7 M urea
Tube U (RNase U2) 9 μl of 20 mM sodium citrate (pH 3.5), 1 mM EDTA, 7 M urea
Tube B (*B. cereus* RNase) 5 μl of 20 mM sodium citrate (pH 5.0), 1 mM EDTA
Tube L (ladder); 6 μl of deionized formamide, 1 mM MgCl$_2$

Heat tubes N, T, P, U, and B at 80° for 1 min and snap cool on ice. Nucleases are then added as follows: tube N, nil; tube T, 1 μl 10 units/μl RNase T1; tube P, 1 μl RNase PhyM extract; tube U, 1 μl of 5 units/μl RNase U2; tube B, 1 μl *B. cereus* RNase extract. These tubes are then incubated at 50° for 20 min and then frozen at −70° or cooled on ice if to be used immediately for electrophoresis. Before loading onto a 7 M urea–polyacrylamide gel, formamide loading solution is added to 12 μl final volume in each tube, which is heated at 80° for 1 min and snap cooled on ice. Tube L is heated at 100° for 1.5 min to generate the ladder. The actual amount of nuclease required for each digestion may vary between batches of enzyme and should be determined experimentally.

After fractionation of the digests by gel electrophoresis on a 7 M urea–10 or 20% polyacrylamide gel, the labeled bands are detected by autoradiography. Bands terminate at G in the RNase T1 digest, at A in RNase U2, A plus U in RNase PhyM, and C plus U in *B. cereus* RNase.

5′-Terminal Deletion Analysis of 3′-End-Labeled RNA

The nonradioactive *Rsa*I complete transcript, derived from transcription of the partial-length vLTSV cDNA clone after linearization with *Rsa*I (Fig. 3), is purified by gel electrophoresis, 3′-end-labeled with [5′-³²P]pCp and repurified by gel electrophoresis from minor amounts of other RNAs. After elution and ethanol precipitation, the labeled RNA is dissolved in 50 mM NaHCO$_3$–Na$_2$CO$_3$ (pH 9.0), 1 mM EDTA and heated at 100° for 3 min. The resulting RNA fragments are ethanol precipitated and redissolved in 1 mM EDTA. One aliquot is incubated in 5 mM MgCl$_2$, 0.5 mM EDTA, 50 mM Tris-HCl (pH 7.5) for 1 min at 25°, conditions expected to promote self-cleavage, while another aliquot is incubated under the same conditions in the absence of Mg^{2+}. Reactions are terminated with formamide loading solution. Electrophoresis is on a 10% polyacrylamide–7 M urea–30% formamide gel in 90 mM Tris–borate, 2 mM EDTA (pH 8.3).

Results

Self-Cleavage of Plus and Minus RNA Transcripts of Partial Length vLTSV

Transcription Reactions. Plus and minus cDNA clones containing 273 residues of the 324-nucleotide virusoid of vLTSV were prepared in the plasmid vector pSP64; plus clones were designated as those which on transcription gave RNA with the same polarity as the infectious virusoid. When these circular plasmid clones were linearized with *Eco*RI and transcribed with SP6 RNA polymerase, full-length transcripts of 336 nucleotides were expected which contained 63 nucleotides of vector sequence in addition to the 273 nucleotides specific for vLTSV (Fig. 1A). However, in

FIG. 1. *In vitro* synthesis of RNAs containing plus and minus partial-length sequences of vLTSV (Australian isolate), and self-cleavage reactions of purified RNAs. (A) Diagram of plus and minus DNA templates and of the SP6 RNA polymerase products generated by transcription of the templates linearized with *Eco*RI. pSP64 plasmid vectors contained 273 nucleotides of a total of 324 nucleotides of plus or minus vLTSV (nucleotides 55 to 3). Vector sequences are striped, the SP6 promoter is black, and the vLTSV sequences are white. The transcription products are C$^+$ and C$^-$ (complete or full-length transcripts) and 5′F$^+$, 3′F$^+$, 5′F$^-$, and 3′F$^-$ (cleavage fragments from the 5′ and 3′ ends of C$^+$ and C$^-$). The site of cleavage is indicated by a thick arrow. Nucleotide numbers in vLTSV are given above the products, with their expected sizes underneath. Vector sequences at the 5′ and 3′ ends of C$^+$ are 39 and 24 nucleotides, respectively, and 40 and 23 nucleotides at the corresponding ends of C$^-$. (B) Analysis of SP6 RNA polymerase products by electrophoresis on a 5% polyacrylamide–7 M urea gel and autoradiography. The positions of the products are indicated at the right of each gel and correspond to those in (A). Lanes 2 and 8 are transcription reactions with linearized vector and lanes 3 and 9 with nonlinearized vector. Lanes 4 and 10 show self-cleavage of purified C$^+$ and C$^-$ after heating at 80° for 1 min in 1 mM EDTA, snap cooling on ice for 5 min, followed by incubation at 25° for 10 min in 5 mM MgCl$_2$, 0.5 mM EDTA, 50 mM Tris-HCl (pH 7.5). Lanes 5 and 11, same as lanes 4 and 10 but without heating and cooling. Lanes 6 and 12, same as lanes 4 and 10 but without MgCl$_2$ during incubation. Lanes 1 and 7, marker DNAs prepared by digesting pUC19 vector with *Hpa*II, end-labeling with [α-^{32}P]dCTP, and heat denaturing prior to electrophoresis; the number of nucleotides in each single-stranded fragment is given at the left of the gel. [Reproduced with permission from A.C. Forster and R. H. Symons, *Cell* **49**, 211 (1987).]

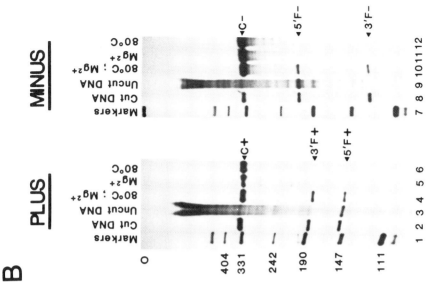

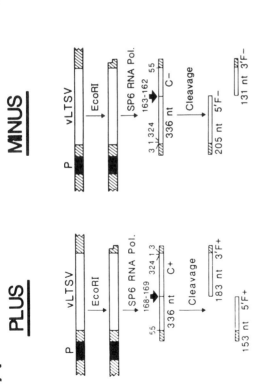

addition to these full-length transcripts C^+ and C^-, fragments 5'F and 3'F were found (Fig. 1A, Fig. 1B, lanes 2 and 8) that were present in approximately equimolar amounts as determined by scintillation counting. Furthermore, the sum of their sizes approximated to those of C^+ and C^-. Transcription of uncut vectors gave a smear of high molecular weight material and defined bands corresponding to 5'F$^+$ and 5'F$^-$ (Fig. 1B, lanes 3 and 9, respectively). These results indicated that 5'F and 3'F were the products of specific cleavage of the full-length transcript C, and that 3'F was not the product of premature termination of transcription.

Self-Cleavage Reactions. Although no plant protein was present in the transcription reactions of Fig. 1A,B, it was important to show that the specific cleavage observed could occur in the absence of protein. Hence, transcripts C^+ and C^- were purified from reaction mixtures by denaturing gel electrophoresis (Fig. 1B), eluted, and tested for specific self-cleavage under a range of conditions. When C^+ and C^- were incubated for 10 min at 25° at pH 7.5 in the presence of 5 mM MgCl$_2$ no self-cleavage occurred (Fig. 1B, lanes 5 and 11). However, efficient self-cleavage of C^+ and C^- occurred after heating at 80° in 1 mM EDTA, snap cooling on ice, addition of cold buffer containing 5 mM MgCl$_2$, and incubation at 25° for 10 min (lanes 4 and 10). No self-cleavage was seen when Mg^{2+} was omitted (lanes 6 and 12). Negligible self-cleavage was found when the snap cooling step was replaced with gradual cooling over 2 hr or when the snap-cooled transcripts were allowed to warm to room temperature before addition of the Mg^{2+} (data not shown).

Significant self-cleavage of C^+ and C^- occurred in the absence of added Mg^{2+} if the heating and snap cooling were carried out in water instead of 1 mM EDTA followed by incubation in 50 mM sodium phosphate (pH 7.0) at 20° for 5 min. However, these self-cleavage reactions were prevented when 1 mM EDTA replaced water, indicating that the EDTA removed trace amounts of divalent metal ions.

The self-cleavage reaction was also sensitive to a variety of other conditions.[20] For example, addition of increasing concentrations of NaCl during the prior heating and snap cooling step decreased subsequent self-cleavage, which was prevented at 50 mM NaCl for C^- and at 200 mM for C^+. The extent of self-cleavage showed little variation between pH 7.5 and 9.0 but decreased as the pH in reaction mixtures was lowered; it was eliminated in C^- at pH 4.5 and in C^+ at pH 4.0. Na$^+$, CO^{2+}, Ni^{2+}, and Zn^{2+} could not substitute for Mg^{2+} in self-cleavage reactions, where 1 mM Ca^{2+} and Mn^{2+} could. Spermine at concentrations from 0.1 to 10 mM did not cause self-cleavage, whereas spermidine gave significant self-cleavage of C^- but was inactive with C^+.

The efficiency of self-cleavage of C^+ and C^- usually varied between 10 and 35%. However, the remaining C^+ and C^- could then be eluted from gels and subjected to the usual heating–snap cooling steps followed by incubation with Mg^{2+} with further self-cleavage to a similar extent. This process can be repeated. To account for these and other self-cleavage data,[17] we have proposed[4] that heat denaturation of inactive C^+ and C^- followed by snap cooling generates a mixture of active and inactive structures, and self-cleavage of the active structures is essentially quantitative under all of the self-cleavage conditions. The reaction pathway can be represented by the following equation:

$$C_I \overset{\Delta}{\rightleftharpoons} C_A \xrightarrow{Mg^{2+}} 5'F + 3'F$$

where the subscript I represents the inactive conformation and the subscript A the active conformation. Evidence that different conformational forms of the 336-nucleotide C^+ and C^- transcripts exist was shown by gel electrophoresis at 4° under nondenaturing conditions. The patterns observed for purified, inactive C^+ and C^- resembled those found with the inactive, heated, and slowly cooled RNAs but differed significantly from the partially active, heated, and snap-cooled RNAs.[20]

Determination of Site of Self-Cleavage and Nature of 5' and 3' Termini. The site of self-cleavage was determined by direct RNA sequencing of the plus and minus 3'F and 5'F RNAs. $3'F^+$ and $3'F^-$, gel purified from self-cleavage reactions of nonradioactive C^+ and C^-, could be $5'-^{32}P$-labeled at pH 9 and 7.5 using T4 polynucleotide kinase and $[\gamma-^{32}P]ATP$ without prior treatment with calf intestinal alkaline phosphatase, indicating the presence of a free 5'-hydroxyl.[21] RNA sequence analysis[21–24] showed that the 5' end of $3'F^+$ was nucleotide 169 of plus vLTSV and that of $3'F^-$ was nucleotide 162 of minus vLTSV.[4] (For convenience, the same residue numbering is retained in the plus and minus RNAs.)

Gel-purified $5'F^+$ and $5'F^-$ from nonradioactive transcription or self-cleavage reactions could not be $3'-^{32}P$-labeled using T4 RNA ligase and $[5'-^{32}P]pCp$,[19] even after extensive treatment with alkaline phosphatase, indicating that the 3' termini were neither hydroxyl nor phosphomonoester groups. However, 3'-labeling was possible when $5'F^+$ and $5'F^-$ were preincubated in 10 mM HCl and then treated with phosphatase, indicating

[20] A. C. Forster, A. C. Jeffries, C. C. Sheldon, and R. H. Symons, *Cold Spring Harbor Symp. Quant. Biol.* **52**, 249 (1987).
[21] M. Silberklang, A. M. Gillum, and U. L. RajBhandary, this series, Vol. 59, p. 58.
[22] J. Haseloff and R. H. Symons, *Nucleic Acids Res.* **10**, 3681 (1982).
[23] H. Donis-Keller, A. M. Maxam, and W. Gilbert, *Nucleic Acids Res.* **4**, 2527 (1977).
[24] G. Krupp and H. J. Gross, *Nucleic Acids Res.* **6**, 3481 (1979).

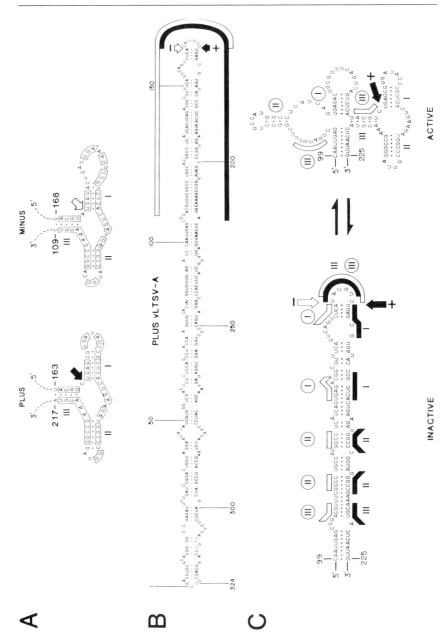

that they terminated with $2',3'$-cyclic phosphodiester groups.[17,18] RNA sequencing of these labeled fragments confirmed the plus and minus sites of self-cleavage and showed that there was no loss or gain of nucleotides during the reaction.

Proposed Secondary Structures around Self-Cleavage Sites. Hammerhead-shaped secondary structures can be drawn around the self-cleavage sites of the plus and minus RNAs (Fig. 2A). In spite of the plus and minus sequences being obtained from overlapping complementary strands (Fig. 2B), 43 residues are conserved (boxed in Fig. 2A) between the two structures out of a total of 55 (plus) and 60 (minus) in each. The self-cleavage domains within the proposed native secondary structure of vLTSV are indicated by black (plus) and white (minus) bars.[4]

Other important features are the central, interior loop formed by the junction of three base-paired stems I, II, and III in the hammerheads with the self-cleavage site in an identical position in both plus and minus structures. It is considered that, in the presence of Mg^{2+}, these hammerhead structures form active tertiary complexes which lower the activation energy sufficiently at the specific internucleotide bond of the self-cleavage site to allow nonhydrolytic cleavage, resulting in the characteristic $5'$-hydroxyl and $2',3'$-cyclic phosphate termini.

Proposed Structural Switches between Active and Inactive Conformations of Plus and Minus RNAs of vLTSV. The native structure of vLTSV (Fig. 2B) does not contain the hammerhead structure and would not be expected to self-cleave. The same holds for the minus vLTSV, which is almost a mirror image of the plus structure (not shown). Rearrangement of the inactive linear molecule to form the active plus hammerhead struc-

---

FIG. 2. Location of proposed self-cleavage domains of vLTSV and the proposed structural switching in these domains. (A) Proposed hammerhead-shaped secondary structures around the self-cleavage site of plus and minus vLTSV. Homologous nucleotides are boxed. The three base-paired stems are numbered, I, II, and III. (B) Self-cleavage domains within the proposed secondary structure of vLTSV-A. Black and white bars indicate the sequences that constitute these domains and that are required to form the plus hammerhead in (A) and the complement of the minus hammerhead in (A), respectively. Arrows show the plus and minus self-cleavage sites. (C) Proposed structural switching of the self-cleavage domains from an inactive to an active self-cleaving structure. In the left-hand structure, the black bars define the sequences that form the three stems of the plus hammerhead on structural switching to the right-hand structure. The white bars of the left-hand structure define the complementary sequences that form the three stems of the minus hammerhead in (A). Because of lack of G-U base pairs in the minus stems, nucleotides complementary to the minus stems are capable of complete base pairing in the plus RNA; only two of the minus stems (I and II) are formed in the right-hand structure since some of the nucleotides in minus stem III are used in the plus stem III. Only the site of self-cleavage of the plus RNA is given. [Reproduced with permission from A. C. Forster and R. H. Symons, *Cell* **49**, 211 (1987).]

ture is shown in Fig. 2C. The three stems of the plus structure and two of the minus structure are given; because of the overlap between the plus and minus stems III, only five of the six stems can be base-paired at the same time. The experimental evidence given above on the necessity for heating and snap cooling of purified C^+ and C^- to allow self-cleavage (Fig. 1B) can be explained on the basis of the conversion of the inactive linear structure in Fig. 2C to an active branched form containing the hammerhead structure.

3'-Terminal Deletion Analysis of Plus vLTSV RNA Transcripts. In order to examine which sequences on the 3' side of the self-cleavage site are required for self-cleavage, truncated transcripts were generated by transcription of the template after linearization with six restriction enzymes (Fig. 3A,B). These transcripts contained 183, 116, 74, 48, 33, and 31 nucleotides on the 3' side of the self-cleavage site. After purification by gel electrophoresis, the transcripts were subjected to self-cleavage conditions and analyzed by gel electrophoresis (Fig. 3C). The only transcripts that did not self-cleave were those derived from the templates linearized with *Hpa*II and *Sau*961 (Fig. 3A,C). The transcripts that self-cleaved gave the same 153-nucleotide 5'-terminal fragment and 3' fragments of varying sizes consistent with the sites of template restriction. Hence, as few as 48 nucleotides on the 3' side of the self-cleavage site are sufficient for self-cleavage, but the deletion of a further 15 nucleotides abolished activity. The smallest self-cleaving transcript terminated at residue 216, which is the conserved U in stem III of the hammerhead structure (Fig. 2A).

5'-Deletion Analysis of Plus vLTSV Transcripts. A novel and effective approach was used for 5'-terminal deletion analysis. The principle of the method (Fig. 4A) is to 3'-end-label a nonradioactive transcript to give a fixed reference point and then randomly cleave the transcript in the absence of Mg^{2+} to generate a series of radiolabeled fragments with 5'-terminal deletions. These fragments are then subjected to self-cleavage conditions without prior heating and snap cooling and then analyzed by gel electrophoresis. The 201-nucleotide *Rsa*I transcript was used for this work because it was the smallest transcript that self-cleaved in Fig. 3C. After 3'-end-labeling with [5'-^{32}P]pCp and RNA ligase, the 202-nucleotide product was purified by gel electrophoresis. No self-cleavage occurred during the end-labeling reactions, presumably because the RNA was in an inactive configuration. The RNA was then heated at 100° for 3 min in pH 9.0 buffer in the absence of Mg^{2+} to generate an OH^- ladder containing 5'-terminally deleted radioactive fragments with hydroxyl groups at their 5' termini (Fig. 4A). The mixture of fragments was ethanol precipitated and then incubated in the presence or absence of Mg^{2+} for 1 min at room temperature. Reactions were terminated by addition of EDTA and for-

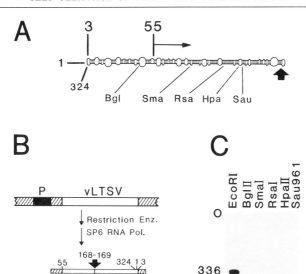

FIG. 3. Self-cleavage analysis of 3'-terminally deleted fragments of a vLTSV plus transcript. (A) Diagram of the vLTSV molecule indicating the sites of cloning (nucleotides 55 to
3) in the pSP64 vector of Fig. 1. The direction of RNA synthesis from the cloning site and the
self-cleavage site are indicated by arrows. The 3' termini of RNAs generated by transcription of the template cut with different restriction enzymes are indicated. (B) Diagram of
DNA template and full-length transcripts generated by SP6 RNA polymerase on the template linearized with different restriction enzymes. The site of self-cleavage is indicated by
an arrow. Nucleotide numbers are given above the products, and the expected size of the
full-length products and the restriction enzymes used are given below the products. Vector
sequences are striped, the SP6 promoter is black, and vLTSV sequences are white.
(C) Analysis of self-cleavage reactions of purified complete products by electrophoresis on a
5% polyacrylamide–7 M urea gel and autoradiography. Purified full-length products from the
template cut with different restriction enzymes were submitted to self-cleavage conditions
(Fig. 1) except that incubation was at 40° for 1 hr. The sizes of the EcoRI full-length RNA
and its 5' and 3' self-cleavage fragments are indicated next to lane 1. O, Origin of gel.
[Reproduced with permission from A. C. Forster and R. H. Symons, Cell **50**, 9 (1987).]

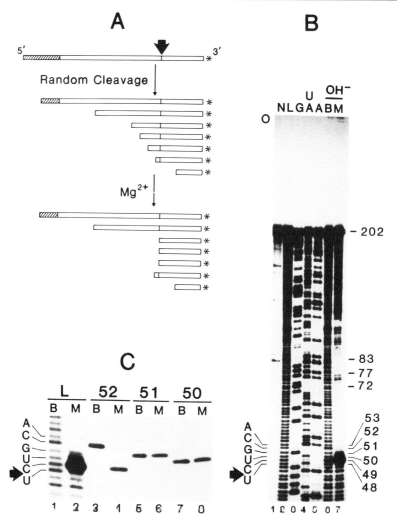

FIG. 4. Self-cleavage analysis of 5'-terminally deleted fragments of the *Rsa*I full-length RNA of Fig. 3. (A) General method used for 5'-terminal deletion analysis. RNA 3'-end-labeled is randomly cleaved to generate a series of 5'-terminally deleted radiolabeled fragments which are then subjected to self-cleavage conditions in the presence of Mg^{2+}. The self-cleavage site is indicated with an arrow and the labeled terminus marked with an asterisk. (B) 5'-Terminal deletion analysis of the 3'-end-labeled *Rsa*I full-length transcript by electrophoresis on a 10% polyacrylamide–7 M urea–30% formamide gel. The purified, nonradioactive, full-length transcript was 3'-end-labeled with [5'-³²P]pCP and further purified, and deletion fragments were prepared and submitted to self-cleavage as in Procedures. Lane 1, 3'-end-labeled *Rsa*I full-length transcript. Lanes 2 to 5, markers produced by enzymatic sequencing (Procedures) of end-labeled *Rsa*I transcript: lane 2, ladder; lane 3, RNase T₁; lane 4, *Physarum polycephalum* RNase; lane 5, RNase U2. Lane 6, ladder of 5'-terminally

mamide and analyzed by electrophoresis (Fig. 4B, lanes 6 and 7). Fragments were identified by comparison with markers produced by the direct enzymatic RNA sequence analysis of the 3'-end-labeled 202-nucleotide fragment (Fig. 4B, lanes 1–5).

As compared with the more or less uniform OH$^-$ ladder of fragments in lane 6, the same fragments after treatment with Mg^{2+} showed two windows in the ladder between nucleotides 52 and 72 and between 77 and 83. In addition, there was a major band at the position expected for the 3' self-cleavage fragment of 49 nucleotides. Hence, it appeared that self-cleavage occurred readily for most fragments between 52 and 83 nucleotides long, but there was little, if any, self-cleavage of longer fragments under the conditions used.

In order to determine the minimum number of nucleotides on the 5' side of the self-cleavage site necessary for self-cleavage, the three fragments above the 3' self-cleavage fragment of 49 nucleotides were excised from a ladder similar to the one in Fig. 4B, lane 6, and eluted. The individual fragments were then treated with pH 7.5 buffer or buffer containing Mg^{2+} as in lanes 6 and 7 (Fig. 4B). Gel analysis (Fig. 4C) showed that the 52-nucleotide fragment self-cleaved completely at the correct site within 1 min (lane 4), whereas the 51- and 50-nucleotide fragments were inactive (lanes 5 to 8). Thus, a minimum of three nucleotides on the 5' side of the self-cleavage site are needed for self-cleavage. The proposed hammerhead structure of the 52-nucleotide fragment is given in Fig. 5.

It is important to appreciate that the OH$^-$ ladder of 3' terminally labeled fragments also contains the nonradioactive 5' fragments released by the pH 9 treatment. Hence, it is possible that the labeled 52-nucleotide fragment in Fig. 4C, lane 3, may have been contaminated with nonradioactive RNAs of a similar size which acted in trans to catalyze the self-cleavage. Several aspects make this highly unlikely. The cleavage reactions were very rapid and efficient despite the very low concentration of all RNA species. For example, the concentration of the 52-nucleotide RNA in the reaction mixtures of lanes 3 and 4, Fig. 4C, was estimated

---

deleted fragments. Lane 7, fragments as in lane 6 but after treatment with Mg^{2+}. Sizes of some of the fragments in lane 6 are indicated. The position of 6 residues at the self-cleavage site (arrowed) is given on the left-hand side of the gel. (C) Treatment of purified 52-, 51-, and 50-nucleotide fragments with self-cleavage conditions. The fragments were excised from a ladder similar to that in lane 6 of (B), eluted, and precipitated with ethanol at $-20°$ for 12 hr, $-80°$ for 6 hr, then centrifuged. The fragments were treated with pH 7.5 buffer (lanes 3, 5, and 7) or with buffer containing Mg^{2+} (lanes 4, 6, and 8) for 1 min at room temperature prior to gel electrophoresis as in (B). [Reproduced with permission from A. C. Forster and R. H. Symons, *Cell* **50**, 9 (1987).]

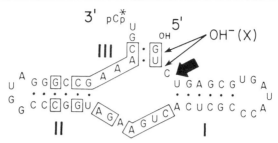

FIG. 5. Proposed hammerhead structure of the 52-nucleotide fragment of plus vLTSV which self-cleaved to completion in Fig. 4C. The sequence is derived from nucleotides 166 to 216 of vLTSV except for the labeled 3'-terminal nucleotide; the labeled phosphate is indicated by an asterisk. OH⁻(X) indicates that the deletion of one or two of the 5'-terminal nucleotides by the OH⁻ treatment abolished self-cleavage. Boxed nucleotides are those conserved in nine proposed hammerhead structures of plant viroid, virusoid, and satellite RNAs [A. C. Forster and R. H. Symons, *Cell* **49**, 211 (1987)]. [Reproduced with permission from A. C. Forster and R. H. Symons, *Cell* **50**, 9 (1987).]

from radioactivity to be only 10 pM. In addition, transcription of a synthetic DNA template with phage T7 RNA polymerase by the method of Milligan *et al.*[25] and Uhlenbeck[26] gave a 58-nucleotide transcript which cleaved completely during the transcription reaction.[19] The template was designed to direct the synthesis of an RNA which contained the same 52 3'-terminal nucleotides as in the *Rsa*I RNA transcript used in Fig. 4.[27] The six 5'-terminal nucleotides of the RNA were derived, by necessity, from the T7 RNA polymerase promoter sequence.

Self-Cleavage of Plus and Minus RNA Transcripts of Dimeric ASBV

Transcription Reactions. Plus and minus dimeric cDNA clones of the 247 nucleotide ASBV, prepared in the plasmid vectors pSP64 and pSP65, were used to prepare RNA transcripts for the characterization of self-cleavage (Fig. 6). On transcription with phage SP6 RNA polymerase, these templates were expected to give dimeric transcripts of plus and minus sense, respectively. In the case of the *Sma*I-linearized pSP64 plus clone, a complete transcript C of 528 nucleotides containing the plus dimer sequence (494 nucleotides) and flanking vector sequences was predicted (Fig. 6A). However, three major products, M, 5'E, and 3'E, in addition to lesser amounts of the putative full-length transcript C, were seen on analysis of reaction mixtures by gel electrophoresis (Fig. 7A, lane

[25] J. F. Milligan, D. R. Groebe, G. W. Witherell, and O. C. Uhlenbeck, *Nucleic Acids Res.* **15**, 8783 (1987).
[26] O. C. Uhlenbeck, *Nature (London)* **328**, 596 (1987).
[27] A. C. Forster and R. H. Symons, *Cell* **50**, 9 (1987).

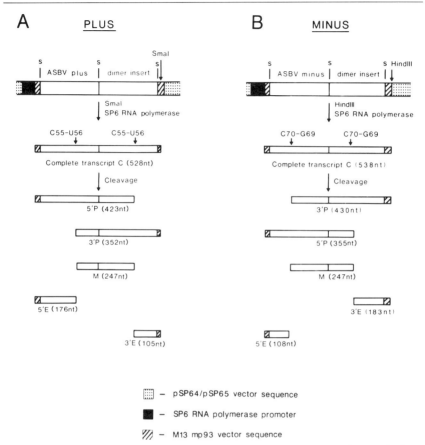

FIG. 6. Diagram of plus and minus dimeric cDNA clones of ASBV in pSP64 and pSP65 plasmid vectors and the SP6 RNA polymerase products generated by transcription of the vector linearized with SmaI (plus) and HindIII (minus). Products are depicted in order of decreasing size, with the length of each in parentheses. The positions of self-cleavage in the complete (full-length) transcripts C are indicated by short arrows together with residue numbers. Partially cleaved fragments are labeled P, monomeric fragments M, and end fragments E. [Reproduced with permission from C. J. Hutchins, P. D. Rathjen, A. C. Forster, and R. H. Symons, *Nucleic Acids Res.* **14**, 3627 (1986).]

2). Product M coelectrophoresed with a marker of purified ASBV linear monomer[15] (ML, not shown), indicating that a linear monomer was specifically cleaved from dimer transcript C.

Further identification of the transcription products was achieved using [γ-³²P]GTP as the sole labeled nucleotide; incorporation of label only into products 5'E and C was seen (Fig. 7A, lane 1), a result which also indicated that products M and 3'E (Fig. 7A, lane 2) were not the result of

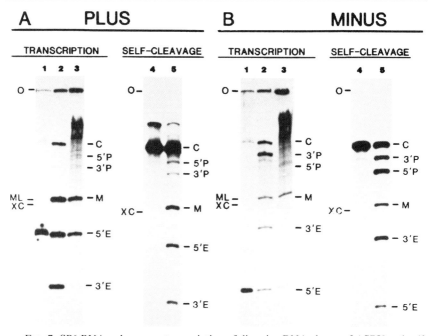

FIG. 7. SP6 RNA polymerase transcription of dimeric cDNA clones of ASBV and self-cleavage of purified products. Analysis of transcription and self-cleavage reactions was by electrophoresis on 5% polyacrylamide–7 M urea gels and autoradiography. The positions of products are indicated on the right-hand side of each gel and correspond to the diagram of Fig. 6. ML, Marker linear ASBV detected by staining with 0.02% toluidine blue; XC, xylene cyanol FF dye; O, origin of gel. In A (plus RNA) and B (minus RNA), lanes 1 and 2 are transcription reactions with linearized vector and lane 3 with nonlinearized vector. [γ-^{32}P]GTP was the only labeled nucleotide in reactions of lane 1 and [α-^{32}P]GTP the only labeled nucleotide in lanes 2 and 3. Lanes 4 contain purified C isolated from the corresponding lane 2; no self-cleavage was observed when C was incubated in 40 mM Tris-HCl (pH 8.0), 1 mM EDTA at 40° for 1 hr. Lanes 5 contain the products of the self-cleavage reaction in 40 mM Tris-HCl (pH 8.0), 6 mM MgCl$_2$, at 40° for 1 hr. [Reproduced with permission from C. J. Hutchins, P. D. Rathjen, A. C. Forster, and R. H. Symons, *Nucleic Acids Res.* **14**, 3627 (1986).]

premature termination of transcription. When nonlinearized vector was used as template using [α-^{32}P]GTP, only products M and 5'E and a smear of high molecular weight material were observed; bands corresponding to C and 3'E were absent (Fig. 7A, lane 3). It was concluded that product 5'E was the 5'-terminal fragment and product 3'E the 3'-terminal fragment resulting from excision of product M from complete transcript C (Fig. 6A). The minor amounts of products 5'P and 3'P in lane 2, Fig. 7A, were considered to arise by cleavage at only one of the two possible sites in C (Fig. 6A).

Similar results were obtained upon SP6 RNA polymerase transcription of the *Hind*III-linearized pSP65 ASBV minus dimer clone (Figs. 6B, 7B). In addition to the putative full-length 534-nucleotide transcript C (Fig. 6B; Fig. 7B, lane 2), four major products, 3'P, M, 3'E, and 5'E, were obtained. Further identification of products was shown with reaction mixtures containing [γ-^{32}P]GTP (Fig. 7B, lane 1) or a nonlinearized dimer minus clone (Fig. 7B, lane 3).

Self-Cleavage Reactions. The approaches used to characterize the transcription products of ASBV cDNA clones were similar to those applied to vLTSV transcripts. Hence, gel-purified complete plus and minus dimer transcripts C were tested for self-cleavage under a range of conditions. The only critical component required was a divalent cation, Mg^{2+}, Mn^{2+}, or Ca^{2+}. In contrast with vLTSV, no prior heating and snap cooling were required for self-cleavage. Thus, incubation of purified plus and minus C at 40° in 40 m*M* Tris-HCl, 6 m*M* MgCl$_2$ (pH 8.0) for 1 hr showed significant self-cleavage of both plus (Fig. 7A, lane 5) and minus (Fig. 7B, lane 5) transcripts. The presence of products 5'P and 3'P for both plus and minus transcripts indicated that these were formed by self-cleavage at only one of two possible sites (Fig. 6).

Incubation of purified plus and minus transcripts C for periods longer than 1 hr or at Mg^{2+} concentrations higher than 6 m*M* increased the extent of self-cleavage. For example, self-cleavage of the minus transcript C was essentially complete after 1 hr at 40° in 50 m*M* MgCl$_2$. Minus transcript C always self-cleaved more efficiently than plus transcript C (not shown).

Determination of Site of Self-Cleavage and Nature of 5' and 3' Termini. The sites of self-cleavage of the plus and minus transcripts were determined by direct enzymatic sequencing of 5'-^{32}P-labeled products, essentially as described for vLTSV. Thus, gel-purified, nonradioactive plus and minus products M and 3'E could be readily labeled using polynucleotide kinase and [γ-^{32}P]ATP without prior dephosphorylation. Sequence analysis showed that self-cleavage had occurred between nucleotides C55 and U56 of the plus species and between nucleotides C70 and G69 of the minus species.[2] (As in the case of vLTSV, the same residue numbers are retained in the plus and minus species.)

All evidence indicated that the plus and minus monomeric products M terminated in a 2',3'-cyclic phosphate. Ligation of [α-^{32}P]GTP-labeled plus and minus M with a partially purified wheat germ RNA ligase preparation gave circular monomeric ASBV in good yield;[2] this enzyme specifically requires a 2',3'-cyclic phosphate terminus for ligation with either a 5'-hydroxyl or a 5'-phosphorylated terminus.[28] In addition, product M

[28] P. Gegenheimer, J. H. Gabius, C. L. Peebles, and J. Abelson, *J. Biol. Chem.* **258**, 8365 (1983).

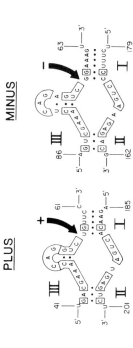

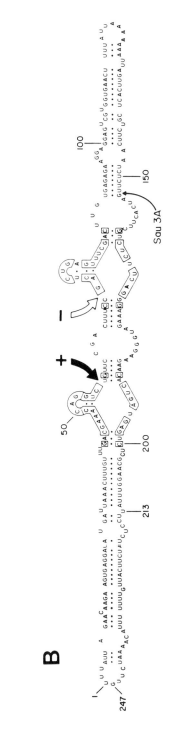

could not be $3'$-^{32}P-labeled with $[5'$-^{32}P]pCp using T4 RNA ligase, even after extensive treatment with alkaline phophatase, indicating that the $3'$-hydroxyl group was blocked and did not contain a singly esterified phosphate. Further, nuclease P1 digestion of $[\alpha$-^{32}P]CTP labeled plus product M gave a small amount of $[5'$-^{32}P] pC $2',3'$-cyclic phosphate, in addition to $[5'$-^{32}P]CMP.[2]

Proposed Secondary Structures around Self-Cleavage Sites. Single, hammerhead-shaped secondary structures around the plus and minus ASBV self-cleavage sites have been proposed (Fig. 8A).[2] Twenty-three nucleotides (boxed) are conserved between the plus and minus structures. Of these 23, 19 are conserved in identical positions in the plus and minus hammerheads of vLTSV (Figs. 2A, 5). The plus hammerhead may be formed by only a minor rearrangement of the secondary structure in the linear, rodlike molecule originally proposed for ASBV[15] (Fig. 8B). In a similar way, the minus hammerhead can be superimposed in the plus sequence (Fig. 8B). An important difference between the vLTSV and ASBV hammerheads is that each of the ASBV hammerheads is formed by two well-separated sections of the molecule whereas the vLTSV hammerheads are formed by contiguous sequences (Fig. 2B).

The stability of the hammerhead structures in Fig. 8 was questioned on the basis of thermodynamic considerations because of the short base-paired stem III with only a three-base loop.[2,27] Potentially much more stable active structures that contain two self-cleavage sites are given in Fig. 9;[29] in these, the stems III are theoretically stable.[30] Such structures, termed double-hammerheads, can be formed in trans by the association of two monomeric molecules of ASBV, or by the association in *cis* of appropriate sequences in a dimeric ASBV molecule (Fig. 6). The folded dimeric plus ASBV RNA containing the proposed double-hammerhead is schematically represented in Fig. 10B; for comparison, dimeric ASBV RNA

[29] A. C. Forster, C. Davies, C. C. Sheldon, A. C. Jeffries, and R. H. Symons, *Nature* (*London*) **334**, 265 (1988).

[30] S. M. Freier, R. Kierzek, J. A. Jaeger, N. Sugimoto, M. H. Caruthers, T. Neilson, and D. H. Turner, *Proc. Natl. Acad. Sci. U.S.A.* **83**, 9373 (1986).

FIG. 8. Hammerhead-shaped structures proposed around the plus and minus self-cleavage sites of ASBV. (A) Conserved residues between both structures are boxed and numbered as in (B). The site of self-cleavage is indicated by an arrow. (B) Proposed native secondary structure of ASBV [R. H Symons, *Nucleic Acids Res.* **9**, 6527 (1981)] modified to incorporate the plus self-cleavage structure and the complement of the minus self-cleavage structure. The *Sau*3A site used for the construction of the cDNA clones is indicated. [Reproduced with permission from C. J. Hutchins, P. D. Rathjen, A. C. Forster, and R. H. Symons, *Nucleic Acids Res.* **14**, 3627 (1986).]

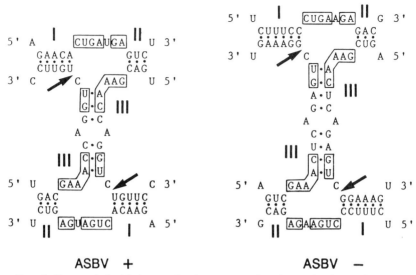

FIG. 9. Proposed double-hammerhead structures for plus and minus ASBV RNAs around the self-cleavage site. Residues conserved between ASBV and the self-cleaving plant satellite and newt RNAs are boxed [A. C. Forster, C. Davies, C. C. Sheldon, A. C. Jeffries, and R. H. Symons, *Nature (London)* **334**, 265 (1988)]. Stems I, II, and III are indicated, and the sites of self-cleavage shown by arrows.

folded to contain two single-hammerhead structures is given in Fig. 10C. The requirement for a double-hammerhead was demonstrated by the use of mutant dimeric plus ASBV RNAs (Figs. 10, 11).[29] In these experiments, cDNA dimeric clones were prepared in the pSP64 vector such that either or both of the conserved GAAAC sequences (labeled A and B, Fig. 10) in the hammerhead structure were changed to GAAC. To do this, a monomeric, full-length cDNA clone of ASBV in the phage M13 DNA vector (cloned via the single *Sau*3A site in ASBV)[14] was mutagenized by the oligonucleotide technique of Zoller and Smith,[31] the mutated monomer excised from the replicative form with *Sau*3A, and dimers with only one of the monomers mutated prepared by cloning in the presence of an equal amount of wild-type monomer into the pSP64 vector. Double-mutant dimers were prepared using only mutated monomers. Initial screening was done on the basis of transcription patterns, and then selected clones were subcloned in phage M13 DNA vector for sequencing.

The results obtained on transcription of the various linearized mutants are given in Fig. 11. When both GAAACs were mutated, cleavage at both

[31] M. J. Zoller and M. Smith, this series, Vol. 100, p. 468.

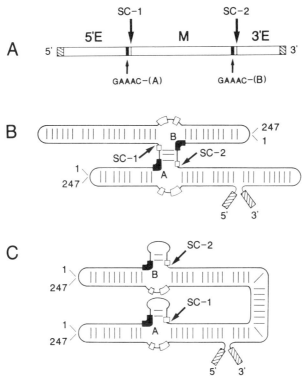

FIG. 10. Schematic representation of plus dimeric ASBV RNA transcript unfolded and folded in single- and double-hammerhead conformations. (A) Unfolded SP6 RNA polymerase transcript of plus dimeric ASBV clone (see Fig. 6A). Self-cleavage sites, labeled SC-1 and SC-2, are indicated by arrows; striped boxes designate vector sequences at 5' and 3' ends; closed boxes represent GAAAC sequences (see Figs. 8 and 9) labeled A and B (indicated by arrows). M, 247-nucleotide monomeric ASBV released on self-cleavage at both sites; 5'E, 5'-end fragment; 3'E, 3'-end fragment. (B) Schematic diagram of plus ASBV dimeric RNA drawn containing a double-hammerhead. Labeling is as in (A) except open boxes represent conserved nucleotides (see Fig. 8); base pairing is represented by lines between RNA strands. The sequence is numbered as in Fig. 8. (C) Schematic diagram of plus ASBV dimeric RNA transcript drawn to contain two single-hammerheads. Labeling is as in Fig. 10B. [Reproduced with permission from A. C. Forster, C. Davies, C. C. Sheldon, A. C. Jeffries, and R. H. Symons, *Nature (London)* **334,** 265 (1988).]

sites was abolished (Fig. 11B, lane 3), demonstrating the importance of this sequence for self-cleavage. When only the 5'-proximal GAAAC (A in Fig. 10) was mutated (i.e., nearest the 5'E, Fig. 10A), cleavage at the second self-cleavage site (SC-2) nearest to the 3'-end fragment of the transcript (3'E) was abolished whereas cleavage at the site SC-1 nearest the 5' end was unaffected (Fig. 11B, lane 4). In the single-hammerhead

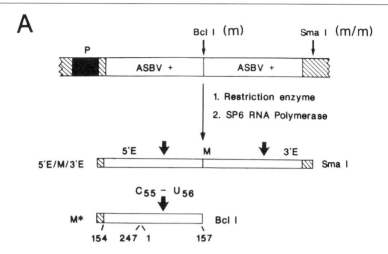

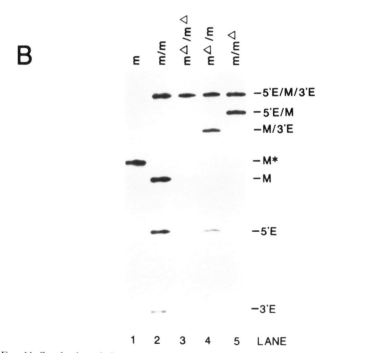

FIG. 11. Synthesis and cleavage of monomeric and dimeric plus ASBV RNAs; effect of a GAAAC to GAAC mutation on the ability of dimeric plus ASBV RNA to self-cleave. (A) Diagram of plus ASBV dimeric DNA template and full-length RNAs produced by SP6 RNA polymerase transcription of templates linearized with either *Sma*I (m/m) or *Bcl*I (m).

model, the reverse would be expected (see two structures in Fig. 10). Similarly, mutation of the 3'-proximal GAAAC (B in Fig. 10) abolished cleavage at the site SC-1 but did not affect cleavage at site SC-2 (Fig. 11B, lane 5). These results provide convincing evidence for the double-hammerhead model for plus ASBV (Fig. 10).

When the plus dimeric cDNA clone was linearized with *Bc*lI at the *Sau*3A junction between the two monomeric inserts (Fig. 11A), transcription with SP6 RNA polymerase gave a monomeric transcript that underwent negligible self-cleavage (Fig. 11B, lane 1), even though an intact self-cleavage sequence was present. This result supports the double-hammerhead model because the monomeric RNA molecules can form active self-cleavage structures (i.e., structures with stable stems III) only if two molecules become associated during the transcription reaction to form a double-hammerhead.

The double-hammerhead model is especially relevant to future efforts to elucidate the tertiary structure of such self-cleaving molecules. Attempts to utilize short RNAs containing plus ASBV-like sequences (i.e., those lacking stable stem IIIs) for X-ray crystallographic or NMR studies may be complicated by the requirement for intermolecular interaction and the presence of two self-cleavage sites per unit structure. Better substrates for such experiments may be structures, such as the vLTSV hammerhead, that contain stable stem IIIs and are active as single-hammerheads (Figs. 2A, 5).

Acknowledgments

This work was supported by the Australian Research Grants Scheme and by a Commonwealth Government Grant to the Adelaide University Centre for Gene Technology in the Department of Biochemistry.

Striped regions represent vector sequences; closed box, SP6 promoter; open boxes, ASBV sequence. Sites of self-cleavage are indicated by thick arrows, and the ASBV sequence is numbered as in Fig. 8. Self-cleavage at both sites of the full-length transcript of *Sma*I-linearized template (m/m) gives rise to a 5'-end fragment (5'E), a 247-nucleotide monomeric fragment (M), and a 3'-end fragment (3'E). (B) Autoradiogram of a polyacrylamide gel showing RNA transcripts of plus ASBV templates. Lane 1, transcript M* of wild-type dimeric ASBV template truncated with *Bc*lI (m). Lane 2, as for lane 1 but template linearized with *Sma*I (m/m). Lane 3, as for lane 2 but both template GAAAC sequences (A and B) (see Fig. 10A) mutated to GAAC (mΔ/mΔ). Lane 4, as for lane 3 but only GAAAC (A) mutated (mΔ/m). Lane 5, as for lane 3 but only GAAAC (B) mutated (m/mΔ). Labels down the margin represent RNA fragments, either full length (5'E/M/3'E) for *Sma*I-digested template, M* for *Bc*lI-digested template, or fragments resulting from self-cleavage. [Reproduced with permission from A. C. Forster, C. Davies, C. C. Sheldon, A. C. Jeffries, and R. H. Symons, *Nature (London)* **334**, 265 (1988).]

Author Index

Numbers in parentheses are footnote reference numbers and indicate that an author's work is referred to although the name is not cited in the text.

D

d'Aubenton-Carata, Y., 568
Dahlberg, A. E., 133, 357, 358(6), 366, 369(4)
Dahlberg, J. E., 37, 38(8), 40(8), 41(8), 42(8), 357, 358(6), 366, 369(4)
DaLio, A., 393
Daly, M., 154
Dani, C., 207
Daniels, C., 541, 542(16)
Daniels, L. P., 293, 303(3)
Danshin, A., 544, 545(29)
Darnell, J. E., 49, 169
Davanloo, P., 560
Davidson, B. L., 467
Davidson, N., 358
Davies, C., 583, 603, 604, 605, 607
Davies, J. C., 586, 593(17)
Davies, R. W., 521, 522(5)
Davis, R. W., 332, 333(8), 342(8), 343(8), 344(8), 345(8), 347(8), 349(8), 350(8)
de Haan, M., 540
de Haseth, P., 317
de la Salle, H., 542, 543(17), 546(17), 549, 556(47)
De Paolis, A., 510
de Robertis, E. M., 83
De Robertis, E., 219
De Segni, G., 440
Dean, M., 207
DeFranco, D., 391, 453
Delahodde, A., 544, 545(30, 31), 550(31)
Dembeck, P., 511
Dembek, P., 533
Dennis, E. S., 148
DePillis, G., 209
Derbyshire, K. M., 525
DeRobertis, E. M., 412
Desjarlais, L. J., 36, 37(7), 43(7)
Desrosiers, R. C., 216, 217(2), 285
Deutscher, M. P., 393, 394(50), 395, 421, 422, 423, 424(13, 15), 425(15), 426, 427, 429(6), 430(10), 432(23), 433, 434, 435(8), 436(7), 438(8), 439
Dhundale, A., 182
Diamond, D., 312
Dib-Hajj, S., 540, 541(7)
Dickson, G. R., 36

Dieckmann, C., 554, 557(75)
Dietrich, R., 546
Diez, J., 169
Digman, J. D., 7, 20, 21(1), 22(1), 28(1), 50, 76, 78(18), 183, 212, 239, 269, 270(15), 283, 288
Digweed, M., 216, 233, 285
Dihanich, M. E., 420
Dingermann, T., 391, 393, 453
Dingman, C. W., 133
Dinter-Gottlieb, G., 583
Dobersen, M. J., 97
Doly, J., 104
Domdey, H., 480
Domier, L. L., 209
Donis-Keller, H., 286, 286(46), 291(46), 309, 566, 569, 584, 591
Donovan, W. P., 422
Dorson, J., 162, 164(12), 165(12)
Dreyfus, M., 549, 556(47)
Dreyfuss, G., 4, 68, 294, 307, 317, 318, 319(2), 320, 322, 325(3), 326, 327, 328, 330, 331, 332, 336(6)
Drinkard, L., 380
Drutsa, V., 275
Dujardin, G., 549, 550, 551, 554(56), 556(47)
Dujon, B., 540, 544, 545
Dull, T. J., 356, 364
Dumas, L. B., 331
Dumont, J. N., 37
Dunn, E. J., 216
Dunn, J. J., 189, 190(3, 5), 192(10, 11), 194(10), 196, 197(3, 5, 10), 199(10), 560, 573

E

Ebel, J. P., 83
Ebert, P., 410, 416(30), 420(30)
Eckstein, F., 383
Edery, I., 210
Edmonds, M., 71, 74(70), 161, 162(1, 2), 164(2, 11), 165(1, 2), 166(1, 2), 167(1, 2), 169(2), 185, 188(29)
Edoh, D., 149
Edwards, S. W., 521, 522(5)
Efstratiadis, A., 44, 169, 182

Subject Index

A

Adenylyltransferase, HeLa, 493
Affinity chromatography. *See also* Immu-
      noaffinity chromatography
   with antibodies specific for m⁶A, for
         isolation of U2 and U4/U6 small
         nuclear ribonucleoproteins, 244–245
   anti-m₃, immunopurification of small
         nuclear ribonucleoproteins by, 241–
         244
   with biotinylated RNA, 97–121
      affinity chromatography materials,
            101–102
      binding and eluting RNA, 107–109,
            112–118
      buffers, 102–103
      detection of ribonucleoproteins bound
            to immobilized pre-mRNA, 109–
            110, 118–121
      materials, 101–103
      methods, 104–110
      principle of, 98–100
      reagents, 101
      solid-phase splicing, 109, 118
      solutions, 102–103
   of hnRNPs, 325–331
ATF
   assay for, 28–30
   purification of, 28–30
Avocado sun-blotch viroid
   dimeric cDNA clones of, 584–585, 598–
         607
   plus and minus RNAs
      5' and 3' termini, determination of
            nature of, 601–603
      gel electrophoresis and purification of,
            585
      in vitro synthesis on plasmid tem-
            plates, 585
      self-cleavage of, 598–607
      self-cleavage reactions, 601

self-cleavage site
   determination of, 601–603
   proposed secondary structures
         around, 602–607
   transcription reactions, 598–601
purified transcription products, self-
      cleavage of, 585
RNA, self-cleavage reactions, 583–584

B

Bacillus subtilis, ribonuclease M5, 366–374
   activity, 367
   assay for, 369–370
      procedure, 371
   purification of, 372–373
      and nonionic detergents, 370–371
Bacteriophage, for tRNA mutation of *E.
      coli*, 395–396
Bacteriophage T4, intron-containing *td*
      gene, transformation of *E. coli* with,
      522
Bacteriophage T7
   RNA polymerase
      in preparation of M1 RNA subunit of
            ribonuclease P from *E. coli*, 578–
            580
      RNA synthesis using, 560
   R1.1 RNA
      cleavage, with cloned and wild-type
            RNase III, 199–201
      mapping and synthesis of, 196–197
Biotin–avidin system, 97

C

Cap binding activity, assay for, 210–212
   preparation of probes, 211
   procedure, 211
   rationale of, 210
Capped RNA, preparation of, 211

plus and minus transcripts of, self-cleavage of, 588–598

plus and minus

active and inactive, proposed structural switches between, 592–594

in vitro synthesis on plasmid templates, 585

plus transcripts

5'-terminal deletion analysis of, 594–598

3'-terminal deletion analysis of, 594–595

purified transcription products, self-cleavage of, 585

self-cleavage reactions, 583–584, 588–591

self-cleavage site

determination of, 591–593

proposed secondary structure around, 592–593

transcription reactions, 588–590

Lymphoid cells, nuclear extract, active in polyadenylation, preparation of, 52

decay reactions, in cell-free system, 203, 205–207

half-life of, 202–203

in *E. coli*, 522, 534

precursors. *See* Pre-mRNA

splicing, role of cap structure in, 210

steady-state level of, 202–203

3'-end processing

factors involved in, chromatography for, 71–74

role of cap structure in, 210

turnover

in cell-free extracts, 202–209

advantages of, 203

in vitro system for study of, 203

assay of, 203, 207–208

Messenger RNA capping enzyme

cellular sources, 171

reactions catalyzed, 170–171

vaccinia virus, 170–180

for labeling RNA, 179–180

viral sources, 171

mRNA. *See* Messenger RNA

M

Maize. *See Zea mays*

Mammalian cells

cultured

isolation of 40 S hnRNP particles from, 294–297

nuclear isolation, 294–295

preparation of active nuclei from, 32–34

with detergent, 33–34

without detergent, 32–33

isolated nuclei, extraction of hnRNP from, 295–297

nuclear extracts

histone-specific RNA 3' processing in, 74–89

preparation of, 76–78

Mammalian components, active in pre-mRNA splicing, 3–19

Maturase. *See* Ribosomal RNA, terminal maturase; Yeast, maturases

Messenger RNA

cap structure, 170, 209–210

cap synthesis, 170–171

N

Neurospora crassa, RNA splicing in, 521

Nicotiana plumbaginifolia

leaf protoplasts

isolation of, 153

PEG-mediated transformation of, 154

pre-mRNA processing, 149

Nicotiana tabacum

leaf protoplasts, PEG-mediated transformation of, 154

pre-mRNA processing, 149

protoplasts, isolation from suspension cultures, 152

Northern analysis

of RNA, 291–292

of RNA splice products, 534

of yeast tRNA mutants, 388–389

Nuclear cap binding protein, from HeLa cells, 209–215

Nuclear extracts. *See also* HeLa cells, nuclear extracts; Mammalian cells, nuclear extracts

active fractions, assays for identification of, 70–71

W

Wheat germ RNA ligase, 480–499
activity, 481–482
assay, 481, 488–491
substrate preparations, 488–490
associated activities, 481–483
assay of, 491–492
discovery of, 480–481
ligation reaction, intermediates in, identification of, 497–499
purification of, 481–488
reaction product
analysis of nucleotide linkage of, 492–497
analysis using poly(A) sunstrates, 493–495
RNase T2-resistant, analysis of, 495–497
substrates, 481–482

X

Xenopus laevis
egg extracts
for analysis of U snRNPs, comparison of different extracts, 280–283
combination with U snRNAs synthesized *in vitro*, for analysis of U snRNPs, 273–274, 276, 283
preparation, 274–275
germinal vesicles
extracts, enzymatic activities in, 510
isolation of, 513–515
oocytes
active nuclei from
nonaqueous isolation of, 36–43
oil-isolated
characteristics of, 40
experimental manipulation of, 41–43
heteropairs, transport studies in, 42–43
RNA synthesized by, labeling and analysis of, 41
transcription of exogenous DNA by, 41–42
isolation of, 513
tRNA splicing endonuclease, 510–517
assays, 512–513

purification of
affinity column for, 511–512
buffers for, 511
materials for, 511–512
method, 515–517

Y

Yeast. *See also Schizosaccharomyces pombe*
conditional expression of suppressor tRNAs, screen for, 409, 411–412
conditional lethal
assay of *in vivo* incorporation of radioisotopes into precursor species, 401–402
cellular extracts, assay for tRNA processing activities, 404–405
RNA blot analysis, in screening for defects in pre-tRNA processing, 402–404
screening for defects in pre-tRNA processing, 401–405
consensus sequence mutants, and splicing efficiency, 133–137
extracts
fractionation of activities, 448–450
preparation of, 441–443
for tRNA gene transcription and processing, 439–450
growth of, for purification of tRNA splicing endonuclease, 473
HZ18-derived RNA species present in analysis of, 125–128
primer extension assay on, 131–133
intron degradation pathway, 181
intron-encoded proteins, genetic studies of, 543–546
maturases, 542–544
mitochondria, RNA splicing in, genetics of, 539–558
mitochondrial DNA, introns in, 540–542
mitochondrial mutants
backmutations, 547–548
nuclear suppressors of, 551
pseudorevertants
second-site suppressors in, 548–550
unstable, 550–551
revertants of, 546–550
second-site suppressors in, 548–550

.